Interdisciplinary Perspectives on Operations Management and Service Evaluation

Panagiotis Manolitzas
Ionian University, Greece

Constantin Zopounidis
Technical University of Crete, Greece

Michael Talias
Open University of Cyprus, Cyprus

Evangelos Grigoroudis
Technical University of Crete, Greece

Nikolaos Matsatsinis
Technical University of Crete, Greece

IGI Global
PUBLISHER of TIMELY KNOWLEDGE

A volume in the Advances in Logistics, Operations, and Management Science (ALOMS) Book Series

Published in the United States of America by
IGI Global
Business Science Reference (an imprint of IGI Global)
701 E. Chocolate Avenue
Hershey PA, USA 17033
Tel: 717-533-8845
Fax: 717-533-8661
E-mail: cust@igi-global.com
Web site: http://www.igi-global.com

Library of Congress Cataloging-in-Publication Data

Names: Manolitzas, Panagiotis, editor.
Title: Interdisciplinary perspectives on operations management and service evaluation / Panagiotis Manolitzas [and four others], editors.
Description: Hershey, PA : Business Science Reference, [2021] | Includes bibliographical references and index. | Summary: "This book focuses on operations management across several sectors and assessment strategies for the improvement of these industries"-- Provided by publisher.
Identifiers: LCCN 2020005740 (print) | LCCN 2020005741 (ebook) | ISBN 9781799854425 (hardcover) | ISBN 9781799854432 (paperback) | ISBN 9781799854449 (ebook)
Subjects: LCSH: Management. | Industrial management.
Classification: LCC HD31.2 .I57 2021 (print) | LCC HD31.2 (ebook) | DDC 658--dc23
LC record available at https://lccn.loc.gov/2020005740
LC ebook record available at https://lccn.loc.gov/2020005741

This book is published in the IGI Global book series Advances in Logistics, Operations, and Management Science (ALOMS) (ISSN: 2327-350X; eISSN: 2327-3518)

British Cataloguing in Publication Data
A Cataloguing in Publication record for this book is available from the British Library.

Advances in Logistics, Operations, and Management Science (ALOMS) Book Series

John Wang
Montclair State University, USA

ISSN:2327-350X
EISSN:2327-3518

Mission

Operations research and management science continue to influence business processes, administration, and management information systems, particularly in covering the application methods for decision-making processes. New case studies and applications on management science, operations management, social sciences, and other behavioral sciences have been incorporated into business and organizations real-world objectives.

The **Advances in Logistics, Operations, and Management Science** (ALOMS) Book Series provides a collection of reference publications on the current trends, applications, theories, and practices in the management science field. Providing relevant and current research, this series and its individual publications would be useful for academics, researchers, scholars, and practitioners interested in improving decision making models and business functions.

Coverage

- Risk Management
- Operations Management
- Finance
- Marketing engineering
- Decision analysis and decision support
- Networks
- Services management
- Political Science
- Organizational Behavior
- Computing and information technologies

IGI Global is currently accepting manuscripts for publication within this series. To submit a proposal for a volume in this series, please contact our Acquisition Editors at Acquisitions@igi-global.com or visit: http://www.igi-global.com/publish/.

The Advances in Logistics, Operations, and Management Science (ALOMS) Book Series (ISSN 2327-350X) is published by IGI Global, 701 E. Chocolate Avenue, Hershey, PA 17033-1240, USA, www.igi-global.com. This series is composed of titles available for purchase individually; each title is edited to be contextually exclusive from any other title within the series. For pricing and ordering information please visit http://www.igi-global.com/book-series/advances-logistics-operations-management-science/37170. Postmaster: Send all address changes to above address.

Titles in this Series

For a list of additional titles in this series, please visit: http://www.igi-global.com/book-series/advances-logistics-operations-management-science/37170

Agile Leadership Methods for Industry 4.0
Bülent Akkaya (Manisa Celal Bayar University, Turkey)
Business Science Reference • © 2021 • 300pp • H/C (ISBN: 9781799842941) • US $195.00

Relating Information Culture to Information Policies and Management Strategies
Denise A.D. Bedford (Georgetown University, USA) and Wioleta Kucharska (Gdańsk University of Technology, Poland)
Business Science Reference • © 2021 • 314pp • H/C (ISBN: 9781799843153) • US $175.00

AI and Future Management Tools for Business Transformation Emerging Research and Opportunities
Rami Shaheen (IAIDL, United Arab Emirates)
Business Science Reference • © 2021 • 220pp • H/C (ISBN: 9781799844174) • US $185.00

Encyclopedia of Organizational Knowledge, Administration, and Technology
Mehdi Khosrow-Pour D.B.A. (Information Resources Management Association, USA)
Business Science Reference • © 2021 • 4000pp • H/C (ISBN: 9781799834731) • US $3,950.00

Handbook of Research on Sustainable Supply Chain Management for the Global Economy
Ulas Akkucuk (Boğaziçi University, Turkey)
Business Science Reference • © 2020 • 409pp • H/C (ISBN: 9781799846017) • US $265.00

Leadership Strategies for Global Supply Chain Management in Emerging Markets
Ashish Dwivedi (The University of Hull, UK) and Mohammed Saad Alshamrani (King Fahd Naval College, Saudi Arabia)
Business Science Reference • © 2020 • 333pp • H/C (ISBN: 9781799828679) • US $215.00

Managerial Issues in Digital Transformation of Global Modern Corporations
Thangasamy Esakki (Poompuhar College (Autonomous), India)
Business Science Reference • © 2020 • 270pp • H/C (ISBN: 9781799824022) • US $195.00

Contemporary Approaches to Military Operations Research
Mumtaz Karatas (National Defense University, Turkey) and Hakan Tozan (Istanbul Medipol University, Turkey)
Information Science Reference • © 2020 • 300pp • H/C (ISBN: 9781799844051) • US $195.00

IGI Global
PUBLISHER of TIMELY KNOWLEDGE

701 East Chocolate Avenue, Hershey, PA 17033, USA
Tel: 717-533-8845 x100 • Fax: 717-533-8661
E-Mail: cust@igi-global.com • www.igi-global.com

Table of Contents

Preface xv

Acknowledgment xviii

Chapter 1
Day Ahead Electricity Price Forecasting in Coupled Markets: An Application in the Italian Market... 1
Konstantinia Daskalou, Hellenic Open University, Greece & Watt and Volt SA, Greece
Christina Diakaki, Hellenic Open University, Greece & Technical University of Crete, Greece

Chapter 2
Multi-Attribute Decision Making Using Interval Type-2 Fuzzy AHP and Its Application on Borsa Istanbul (BIST) 30
Beyza Ahlatcioglu Ozkok, Yildiz Technical University, Turkey
Hale Gonce Kocken, Yildiz Technical University, Turkey

Chapter 3
Customer and Order Prioritization With Multi-Criteria Decision Making: An Application in Metal Industry 57
Emre Aslan, Tokat Gaziosmanpasa University, Turkey

Chapter 4
Clustering Techniques for Categorical Data: Correspondence Analysis 76
Stelios Zimeras, University of the Aegean, Greece
Manolis Kalligeris, University of the Aegean, Greece

Chapter 5
Metaheuristics Approaches for the Travelling Salesman Problem on a Spherical Surface 94
Yusuf Sahin, Department of Business Administration, Burdur Mehmet Akif Ersoy University, Turkey
Erdal Aydemir, Department of Industrial Engineering, Suleyman Demirel University, Turkey
Kenan Karagul, Department of Logistics, Pamukkale University, Turkey
Sezai Tokat, Department of Computer Engineering, Pamukkale University, Turkey
Burhan Oran, Department of Computer Engineering, Pamukkale University, Turkey

Chapter 6
Routing Searching Algorithm for Combined Freight Transport Trips in South East Europe Region 114
Panagiotis Otapasidis, University of the Aegean, Greece
Christina Arampantzi, University of the Aegean, Greece
Vasileios Zeimpekis, University of the Aegean, Greece

Chapter 7
Trigonometric Grey Prediction Method for Turkey's Electricity Consumption Prediction 136
Adem Tuzemen, Tokat Gaziosmanpaşa Üniversitesi, Turkey

Chapter 8
Analyzing the Efficiency of Renewable Energy Sources in Terms of Electricity Generation: The Case of Turkey 155
Talip Arsu, University of Aksaray, Turkey

Chapter 9
"My Regular Pharmacy": A Study on the Evaluative Criteria Consumers Use in Selecting an Independent Community Pharmacy as Their Regular One 177
Niki Glaveli, University of the Aegean, Greece

Chapter 10
Quality Management in Primary Dental Care: Methods and Challenges............... 192
A. I. Syngelakis, Faculty of Dentistry, University of Athens, Greece & Department of Dentistry, European University of Cyprus, Cyprus
Maria Kamariotou, University of Macedonia, Greece
Fotis C. Kitsios, University of Macedonia, Greece
Chrystala Charalambous, European University of Cyprus, Cyprus
Argy Polychronopoulou, National and Kapodistrian University of Athens, Greece

Chapter 11
SERVQUAL-Based Evaluation of Service Quality in Turkish Health Industry With Fuzzy Logic .. 213
Hasan Dinçer, Istanbul Medipol University, Turkey
Serhat Yüksel, Istanbul Medipol University, Turkey
Yaşar Gökalp, Istanbul Medipol University, Turkey
Serkan Eti, Istanbul Medipol University, Turkey

Chapter 12
Economic Sustainable Health Information Systems............... 234
Angelos I. Stoumpos, Open University of Cyprus, Cyprus
Michael A. Talias, Open University of Cyprus, Cyprus

Chapter 13
Chronic Diseases and Socioeconomic Factors: Findings From the SHARE 50+ Database 252
Athanassios Vozikis, Laboratory of Health Economics and Management (LabHEM), University of Piraeus, Greece
Nikoletta Panou, University of Piraeus, Greece
Iris-Panagiota Efthymiou, Laboratory of Health Economics and Management (LabHEM), University of Piraeus, Greece

Chapter 14
No One Left Behind? Migrant and Refugee Health in the COVID-19 Crisis in Greece.................... 270
Athanassios Vozikis, Laboratory of Health Economics and Management (LabHEM), University of Piraeus, Greece
Theodoros Fouskas, University of West Attica, Greece
Symeon Sidiropoulos, Laboratory of Health Economics and Management (LabHEM), University of Piraeus, Greece

Chapter 15
Agro-Tourism Customer Satisfaction Analysis Based on the Theory of Attractive Quality 290
Maria Lachnidaki, Technical University of Crete, Greece
Evangelos Grigoroudis, Technical University of Crete, Greece
Constantin Zopounidis, Technical University of Crete, Greece & Audencia Business School, Greece

Chapter 16
Assessing the Impact of Cigarette Taxation on the Supply Chain Stakeholders' Revenue Shares in Greece 313
Athanassios Vozikis, Laboratory of Health Economics and Management (LabHEM), University of Piraeus, Greece
Yannis A. Pollalis, University of Piraeus, Greece
Archontoula Armoutaki, University of Piraeus, Greece

Chapter 17
An Overview of the Cruise Industry and Its Economic Impacts: Lessons From Europe 329
Georgia Papadopoulou, Al Ghurair University, UAE

Compilation of References 343

About the Contributors 391

Index 399

Detailed Table of Contents

Preface.. xv

Acknowledgment... xviii

Chapter 1
Day Ahead Electricity Price Forecasting in Coupled Markets: An Application in the Italian Market... 1
Konstantinia Daskalou, Hellenic Open University, Greece & Watt and Volt SA, Greece
Christina Diakaki, Hellenic Open University, Greece & Technical University of Crete, Greece

Day ahead electricity price forecasting is an extensively studied problem, and several statistical, intelligence-based, and other techniques have been proposed in literature to address it. However, the liberalization of the electricity market taking place during the last decades and the market coupling pursued within the European Union reshape the problem and create the need to confirm the effectiveness and/or revise existing methods and solution techniques, and/or invent new approaches. Given that complete integration has not achieved yet, both relevant data and studies of forecasting considering integration are still rather sparse. It has thus been the aim of this chapter to contribute to filling this gap by focusing on and studying the market integration effects in day ahead electricity price forecasting. To this end, an artificial neural network has been developed and used under several, with respect to inputs, forecasting scenarios considering the Italian electricity market.

Chapter 2
Multi-Attribute Decision Making Using Interval Type-2 Fuzzy AHP and Its Application on Borsa Istanbul (BIST) 30
Beyza Ahlatcioglu Ozkok, Yildiz Technical University, Turkey
Hale Gonce Kocken, Yildiz Technical University, Turkey

Analytic hierarchy process (AHP) is a widely used multi-attribute decision-making (MADM) approach. Due to the complexity and uncertainty involved in real world problems, decision makers might be prefer to make fuzzy judgments instead of crisp ones. Furthermore, even when people use the same words, individual judgments of events are invariably subjective, and the interpretations that they attach to the same words may differ. This is why fuzzy numbers has been introduced to characterize linguistic variables. Fuzzy AHP methods have recently been extended by using type-2 fuzzy sets. Type-2 fuzzy set theory incorporates the uncertainty of membership functions into the fuzzy set theory. In this chapter, the authors firstly provide a short review on applications of interval type-2 fuzzy AHP on MADM problems. Then, they present a very efficient MADM technique, interval type-2 fuzzy AHP, to solve the portfolio selection problem that is to decide which stocks are to be chosen for investment and in what proportions they will be bought. And finally, they provided a case study on BIST.

Chapter 3
Customer and Order Prioritization With Multi-Criteria Decision Making: An Application in Metal Industry 57
Emre Aslan, Tokat Gaziosmanpasa University, Turkey

Customer and order prioritization is prioritizing the orders according to features of customers and orders. In make-to-order production systems, especially when demand exceeds the capacity, a need arises for prioritizing the orders. As a customer may have more than one order and as one type of order may belong to different customers, every job order must be prioritized according to both customer and order features. Prioritization values may be used for accepting and scheduling orders or selecting customers as a reflection of supplier selection. In this chapter, the production manager and customer relationship manager of a metal processing firm were interviewed for determining the criteria for prioritizing customers and orders. After determining the criteria, the managers were asked to pairwise compare the criteria for weighting with analytical hierarchy process (AHP). Then by considering scores of customers and orders for each criterion, orders were prioritized with technique for order preference by similarity to ideal solution (TOPSIS).

Chapter 4
Clustering Techniques for Categorical Data: Correspondence Analysis 76
Stelios Zimeras, University of the Aegean, Greece
Manolis Kalligeris, University of the Aegean, Greece

Categorical data are generally thought to consist of contingency tables, which are data tables created whenever categorical data are cross-classified. Correspondence analysis is a statistical visualization method for picturing the associations between the levels of a two-way contingency table. In this work, the correspondence analysis for categorical data is analyzed using statistical and mathematical techniques when big data must be organized and analyzed due to the huge amount of data, and a MATLAB package (simple correspondence analysis) for this technique is presented.

Chapter 5
Metaheuristics Approaches for the Travelling Salesman Problem on a Spherical Surface 94
Yusuf Sahin, Department of Business Administration, Burdur Mehmet Akif Ersoy University, Turkey
Erdal Aydemir, Department of Industrial Engineering, Suleyman Demirel University, Turkey
Kenan Karagul, Department of Logistics, Pamukkale University, Turkey
Sezai Tokat, Department of Computer Engineering, Pamukkale University, Turkey
Burhan Oran, Department of Computer Engineering, Pamukkale University, Turkey

Traveling salesman problem in which all the vertices are assumed to be on a spherical surface is a special case of the conventional travelling salesman problem. There are exact and approximate algorithms for the travelling salesman problem. As the solution time is a performance parameter in most real-time applications, approximate algorithms always have an important area of research for both researchers and engineers. In this chapter, approximate algorithms based on heuristic methods are considered for the travelling salesman problem on the sphere. Firstly, 28 test instances were newly generated on the unit sphere. Then, using various heuristic methods such as genetic algorithms, ant colony optimization, and fluid genetic algorithms, the initial solutions for solving test instances of the traveling salesman problem are obtained in Matlab®. Then, the initial heuristic solutions are used as input for the 2-opt algorithm. The performances and time complexities of the applied methods are analyzed as a conclusion.

Chapter 6
Routing Searching Algorithm for Combined Freight Transport Trips in South East Europe Region 114
Panagiotis Otapasidis, University of the Aegean, Greece
Christina Arampantzi, University of the Aegean, Greece
Vasileios Zeimpekis, University of the Aegean, Greece

In this chapter, the authors propose a mixed integer linear programming model (MILP) to address the shortest path problem when trans-shipments exist on combined freight transport operations. The objective is to minimize trip duration, cost, and CO2 emissions. For solving this problem, we propose an extension of Dijkstra's algorithm that identifies the optimal way of cargo movement with trans-shipment. We applied the proposed algorithm in real-life data by embracing three different types of transportation modes, namely road, rail, and inland waterways (IWW), and by testing it in three different cases where (1) goods should cross 2 or 3 South East Europe (SEE) countries, (2) goods should cover a distance more or less than 500 km, and (3) goods with origin the Port of Piraeus and destination four capital cities of SEE. The results show that the use of road transport is the most favorable solution if trip duration is the main criterion, whereas combined transport is selected when minimization of cost and CO2 emissions is the objective.

Chapter 7
Trigonometric Grey Prediction Method for Turkey's Electricity Consumption Prediction 136
Adem Tuzemen, Tokat Gaziosmanpaşa Üniversitesi, Turkey

Industry and technology continue to develop rapidly in today's world. The indisputable most important source of this development, energy is among the indispensables of daily life. Since it is one of the determining factors for the country's economy, the future forecast of electricity demand means calculating the future steps. Based on this, to forecast Turkey's electricity demand, it was benefited from grey model (GM) and trigonometric GM (TGM) techniques. The data set includes annual electricity consumption for the period 1970-2018. The performances of the methods determined were compared based on the forecast evaluation criteria (MSE, MAD, MAPE, and RMSE). Short-term forecasting analysis was carried out by determining the method that gives these values to a minimum. In the future forecast, it has been determined that electricity consumption will increase continuously.

Chapter 8
Analyzing the Efficiency of Renewable Energy Sources in Terms of Electricity Generation: The Case of Turkey 155
Talip Arsu, University of Aksaray, Turkey

Electricity generation, one of the renewable energy sources (RES), delivers a solution for various problems such as energy efficiency, energy supply security, reducing foreign dependency, and especially, environmental concerns. However, the solutions provided for these problems bring along the question of which RESs are produced more effectively. Therefore, in this research, RESs used for electricity generation in Turkey were analyzed by using generation data to show which one is more effective. Bi-objective multiple-criteria data envelopment analysis (BiO-MCDEA) method, a goal programming-based efficiency determination method, was used for the efficiency analysis conducted for five years between the years of 2014 and 2018. As a result of the analysis, geothermal energy came into prominence as the most effective RES for all of the years included in the solution. Geothermal energy was followed by biomass energy, wind energy, hydroelectric, and solar energy, respectively.

Chapter 9
"My Regular Pharmacy": A Study on the Evaluative Criteria Consumers Use in Selecting an Independent Community Pharmacy as Their Regular One 177
Niki Glaveli, University of the Aegean, Greece

The objectives of this chapter are (1) to develop an instrument that incorporates the core attributes of evaluative criteria that are used by patients/customers in selecting and repeatedly visiting a community pharmacy, (2) to analyze the proposed attributes in order to put forward valid dimensions of evaluative criteria, and (3) to assess the reliability of this instrument. The selection criteria incorporated in the final list were selected based on an extensive literature review and on experts' and customers' opinions. To collect the data, a survey was conducted in the area of Macedonia, Greece. In total, 223 questionnaires were collected. Confirmatory factor analyses revealed a four-factor structure of the proposed instrument namely: price, availability of products, staff competence, and pharmaceutical care services. Moreover, reliability was assessed through Cronbach's alpha coefficient and average variance extracted. The current study outcomes can guide pharmacists' strategic actions in boosting customer patronage behavior in the highly competitive pharmaceutical sector.

Chapter 10
Quality Management in Primary Dental Care: Methods and Challenges.. 192
A. I. Syngelakis, Faculty of Dentistry, University of Athens, Greece & Department of Dentistry, European University of Cyprus, Cyprus
Maria Kamariotou, University of Macedonia, Greece
Fotis C. Kitsios, University of Macedonia, Greece
Chrystala Charalambous, European University of Cyprus, Cyprus
Argy Polychronopoulou, National and Kapodistrian University of Athens, Greece

In dental care services, quality is an important factor that affects decision making, the planning of health strategies and policies, the cost of health services, and the evaluation of them. The evaluation of quality in dental services using the assessment methods that are used in other services of primary healthcare is difficult due to the special characteristics of dentistry. However, the improvement and the evaluation of primary oral healthcare services is a complicated issue because it involves many factors that affect it. Therefore, the purpose of this chapter is to provide a complete overview of the literature using Webster and Watson's methodology. Fifty peer-reviewed papers were analyzed and the results of this review revealed that the number of publications in this domain has increased in the last decade, and there is a need to foster research (especially empirical) in this field.

Chapter 11
SERVQUAL-Based Evaluation of Service Quality in Turkish Health Industry With Fuzzy Logic .. 213
Hasan Dinçer, Istanbul Medipol University, Turkey
Serhat Yüksel, Istanbul Medipol University, Turkey
Yaşar Gökalp, Istanbul Medipol University, Turkey
Serkan Eti, Istanbul Medipol University, Turkey

The aim of this study is to analyze the service quality in the Turkish health sector. In this context, a literature review is conducted, and 16 different criteria based on SERVQUAL are determined. Fuzzy DEMATEL method is used to determine which of these criteria are more important. The findings indicate that operating hours and quality of equipment have the highest weights. In addition to these items, experience of personnel and individual attention to patient are other important factors with respect to the service quality in Turkish health industry. Hence, it is strongly recommended that hospitals in Turkey should operate in an extended time because Turkish people prefer to go to the hospitals after working hours. It is also obvious that Turkish hospitals should make investment to quality equipment. Finally, it is a good idea for new hospitals to have qualified employee. This situation has a positive contribution to the service quality of these hospitals.

Chapter 12
Economic Sustainable Health Information Systems 234
Angelos I. Stoumpos, Open University of Cyprus, Cyprus
Michael A. Talias, Open University of Cyprus, Cyprus

Public health systems have adopted computer health and information technology as a dynamic transformational tool both to improve real-time surveillance systems and to communicate and exchange information between different organizations. Health information systems (HIS) incorporate data collection, processing, report creation, and the use of information necessary to improve the efficiency and effectiveness of health services through better management of these services at every level. The sustainable development of these systems depends on their interoperability, the combination of the "three pillars of sustainability" (economic, social, environmental). At the global health level, there are variations between countries in terms of the application and implementation of sustainable systems. However, taking advantage of the available knowledge and technology, and with proper management of the economy, the viability of health information systems can be sustainable.

Chapter 13
Chronic Diseases and Socioeconomic Factors: Findings From the SHARE 50+ Database 252
Athanassios Vozikis, Laboratory of Health Economics and Management (LabHEM), University of Piraeus, Greece
Nikoletta Panou, University of Piraeus, Greece
Iris-Panagiota Efthymiou, Laboratory of Health Economics and Management (LabHEM), University of Piraeus, Greece

The socioeconomic and behavioral risk factors, as well as advances in technology and medicine, have led to an unprecedented increase in the prevalence of chronic diseases, which have become the leading causes of morbidity and mortality. In this chapter, the authors aim to unveil the relation between socioeconomic and behavioral risk factors and chronic diseases among the Greek population. For this purpose, data from the SHARE 50+ database (specifically from the easySHARE v.6.1.1) were analyzed using the IBM SPSS

26 software. The research sample comprises of 3,584 Greek citizens of age between 50-75 years. The research findings indicate that 60.1% of the research sample have at least one chronic condition, with low education level to associate with a greater prevalence of chronic diseases. Chronic diseases also are correlated to poor self-perceived health and quality of life, especially to those with a low socioeconomic status. The research results are in accordance with the findings in the international literature.

Chapter 14
No One Left Behind? Migrant and Refugee Health in the COVID-19 Crisis in Greece.................... 270
Athanassios Vozikis, Laboratory of Health Economics and Management (LabHEM), University of Piraeus, Greece
Theodoros Fouskas, University of West Attica, Greece
Symeon Sidiropoulos, Laboratory of Health Economics and Management (LabHEM), University of Piraeus, Greece

Asylum seekers, refugees, and migrants, who are living in RICs, are faced with multiple challenges and vulnerabilities that must be taken into consideration when responding to the COVID-19 pandemic. The chapter focuses on the COVID-19 pandemic concerns over increasing cases recorded in the RICs in Greece. The impact of migration on public health is of particular concern to Greek migration policy, as migrants in the country have a completely different epidemiological profile and higher risks to public health due to the poor living conditions in their countries of origin and during their stay. They live mostly in overcrowded reception and identification centers and accommodation centers under deplorable conditions, lack of proper shelter, extremely unhygienic living conditions. The urgent decongestion of the overcrowded RICs and accommodation centers is required to avoid the risk of rapid spread of the infection.

Chapter 15
Agro-Tourism Customer Satisfaction Analysis Based on the Theory of Attractive Quality 290
Maria Lachnidaki, Technical University of Crete, Greece
Evangelos Grigoroudis, Technical University of Crete, Greece
Constantin Zopounidis, Technical University of Crete, Greece & Audencia Business School, Greece

Customer satisfaction measurement is an important and critical issue of all businesses in the tourism sector, since it gives a better understanding of the needs and requirements of tourists. The Kano model is one of the most widely used approaches in analyzing customer satisfaction, aiming to categorize the quality attributes of products/services in different quality dimensions. The main aim of this chapter is to analyze customer satisfaction from an agro-tourism farm and prioritize potential improvement actions. The preliminary analysis has identified a large set of customer satisfaction criteria that refer to the location, the personnel, the activities, the cost, and the "green" aspects of the agro-tourism farm. The MUSA method has been applied in the context of Kano's theory of attractive quality. The main results show which satisfaction criteria mostly affect satisfaction and/or dissatisfaction, identifying the strong and the weak points of the farm. Finally, combining the previous results, it is possible to prioritize improvement actions.

Chapter 16
Assessing the Impact of Cigarette Taxation on the Supply Chain Stakeholders' Revenue Shares in Greece 313
Athanassios Vozikis, Laboratory of Health Economics and Management (LabHEM), University of Piraeus, Greece
Yannis A. Pollalis, University of Piraeus, Greece
Archontoula Armoutaki, University of Piraeus, Greece

The chapter aims to analyze the impact of cigarette taxes evolution over the period 1992-2017 on the revenue share of the main tobacco supply chain stakeholders in Greece. This empirical analysis uses a pooled time series from 1992 through 2017 including a data set of retail prices, three tax groups levied on cigarettes (specific tax, ad valorem tax, and V.A.T.), and revenue shares for three categories of stakeholders. The results indicate that the revenue share of the stakeholders is decreasing over the whole period, and specifically, their shares drop by half in the last 15 years. The regression results show that the revenue shares of tobacco companies are most affected by both excise taxes in a negative way, and similarly, both price and the excises affect significantly the revenue share of retailers, whereas the findings regarding distributors' revenues are insignificant. This knowledge is likely to be useful for policymakers in the development of effective tobacco control policies.

Chapter 17
An Overview of the Cruise Industry and Its Economic Impacts: Lessons From Europe 329
Georgia Papadopoulou, Al Ghurair University, UAE

The cruise industry is one of the dynamic markets with high investment in port facilities worldwide and in building luxury cruise ships as well. It incorporates elements of the product market, educational activities, and has developed into a spectacular destination. Cruising includes a great part in the tourism chain, including transport, catering, tourism, leisure, and travel. The aim of the chapter is to examine the economic and statistical data of the cruise industry and cruise tourism through a European approach, based on secondary data and research. It further explores the economic impact of the European cruise industry on the European economies. Moreover, the chapter tries to compare the economic impact of the European cruise among the previous years and to ascertain whether these impacts benefit the European economies or not. The evolution of the direct economic impacts are captured and a discussion about the total economic impacts of the European cruise sector is given. This study would offer facilitation in directing future research relevant to cruise tourism.

Compilation of References 343

About the Contributors 391

Index 399

Preface

The economic crisis forced the researchers across the globe to develop and implement mathematical models in order to optimize the services in several sectors. The scope of this book is to present recent methods and case studies in the area of Operations Management and Service Evaluation. In addition, the book provides a helpful tool for policy makers, researchers and the academic community in general interested in the area of Management Science.

The current book contains 17 chapters. More analytically in chapter 1, Daskalou and Diakaki analyze the market integration effects in day ahead electricity price forecasting. To this end, an Artificial Neural Network has been developed and used under several, with respect to inputs, forecasting scenarios considering the Italian electricity market.

In the second chapter, Ozkok and Kocken present an efficient MADM technique: interval type-2 fuzzy AHP to solve the portfolio selection problem that is to decide which stocks must be selected for investment and in what proportions they will be bought.

In the third chapter, the author (Emre Aslan) used the Analytical Hierarchy process in order to determine the criteria for prioritizing customers and orders. After determining the criteria, managers were asked to pairwise compare the criteria for weighting with Analytical Hierarchy Process (AHP). Then by considering scores of customers and orders for each criterion, orders were prioritized with Technique for Order Preference by Similarity to Ideal Solution (TOPSIS).

Zimeras and Kalligeris, in the fourth chapter present the correspondence analysis, which is a statistical visualization method for picturing the associations between the levels of a two way contingency table. More specifically in this chapter, the correspondence analysis for categorical data will be analyzed using statistical and mathematical techniques when big data must be organized, and analyzed due to the huge amount of data and a MATLAB package (Simple CORrespondence ANalysis) for this technique would be presented.

In chapter 5, approximate algorithms based on heuristic methods are considered for the Travelling Salesman Problem on the sphere. Firstly, 28 test instances were newly generated on the unit sphere. Then, using various heuristic methods such as Genetic Algorithms, Ant Colony Optimization and Fluid Genetic Algorithms, the initial solutions for solving test instances of the Traveling Salesman Problem are obtained in Matlab. Then, the initial heuristic solutions are used as input for the 2-opt algorithm. The performances and time complexities of the applied methods are analyzed as a conclusion.

In the next chapter Otapasidis et al., propose a mixed integer linear programming model (MILP) to address the shortest path problem when trans-shipments exist on combined freight transport operations. The objective is to minimize trip duration, cost, and CO2 emissions. For solving this problem, they propose an extension of Dijkstra's algorithm that identifies the optimal way of cargo movement with trans-shipment. The results show that the use of road transport is the most favorable solution if trip duration is the main criterion, whereas combined transport is selected when minimization of cost and CO2 emissions is the objective.

In Chapter 7, Tuzemen A., uses a trigonometric Grey Prediction Model in order to forecast the electricity consumption in Turkey. On the other hand, in chapter 8, the authors evaluate the Renewable Energy Sources in Turkey in order to reveal which one is more effective. Bi-Objective Multiple Criteria Data Envelopment Analysis (BiO-MCDEA) method, a goal programming based efficiency determination method, was used for the efficiency analysis conducted for five years between the years of 2014 and 2018. As a result of the analysis, geothermal energy came into prominence as the most effective RES for all of the years included in the solution. Geothermal energy was followed by biomass energy, wind energy, hydroelectric and solar energy respectively.

In Chapter 9, Glaveli N., developed an instrument that incorporates the core attributes of evaluative criteria that are used by patients/customers in selecting and repeatedly visiting a community pharmacy. Confirmatory factor analysis revealed a four-factor structure of the proposed instrument namely: price, availability of products, staff competence and pharmaceutical care services. Moreover, reliability was assessed through Cronbach's alpha coefficient and average variance extracted. The current study outcomes can guide pharmacists' strategic actions in boosting customer patronage behavior in the highly competitive pharmaceutical sector.

In chapter 10, the authors provide a complete overview of the literature using Webster's and Watson's (2002) methodology. Fifty peer-reviewed papers were analyzed and the results of this review revealed that the number of publications in this domain has increased in the last decade, and there is a need to foster research (especially empirical) in this field.

In the next chapter (chapter 11), the authors analyze the service quality in the Turkish health sector. In this context, a literature review conducted, and 16 different criteria based on SERVQUAL are determined. Fuzzy DEMATEL method is used to determine which of these criteria are more important. The findings indicate that operating hours and quality of equipment have the highest weights.

In Chapter 12 the authors through the analysis reveal that Sustainable technologies can effectively contribute to a country's development goals. In chapter 13, Vozikis et al. aim to unveil the relation between socioeconomic and behavioral risk factors and chronic diseases among the Greek population. For this purpose, data from the SHARE 50+ database (specifically from the easySHARE v.6.1.1), were analyzed using the IBM SPSS 26 software. The research sample comprises of 3,584 Greek citizens of age between 50-75 years. The research findings indicate that 60,1% of the research sample have at least one chronic condition, with low education level to associate with a greater prevalence of chronic diseases. Chapter 14 focuses on the COVID-19 pandemic concerns over increasing cases recorded in the RICs in Greece.

In Chapter 15, the main cope of the authors is to analyze the customer satisfaction from an agro-tourism farm and prioritize potential improvement actions. The preliminary analysis has identified a large set of customer satisfaction criteria that refer to the location, the personnel, the activities, the cost, and the "green" aspects of the agro-tourism farm. The MUSA method has been applied in the context of Kano's theory of attractive quality. The main results show which satisfaction criteria mostly affect satisfaction and/or dissatisfaction, identifying the strong and the weak points of the farm. Finally, combining the previous results, it is possible to prioritize improvement actions.

The next chapter (Chapter 16) analyzes the impact of cigarette taxes evolution over the period 1992-2017 on the revenue share of the main tobacco supply chain stakeholders in Greece. This empirical analysis uses a pooled time series from 1992 through 2017 including a data set of retail prices, three tax groups levied on cigarettes (specific tax, ad valorem tax and V.A.T.) and revenue shares for three categories of stakeholders. Our results indicate that the revenue share of the stakeholders is decreasing over the whole period and specifically, their shares drop by half in the last fifteen years. The regression results show that the revenue shares of tobacco companies are most affected by both excise taxes in a negative way and similarly, both price and the excises affect significantly the revenue share of retailers, whereas the findings regarding distributors' revenues are insignificant. This knowledge is likely to be useful for policy-makers in the development of effective tobacco control policies.

The last chapter of the book (Chapter 17) examines the economic and statistical data of the cruise industry and cruise tourism through a European approach, based on secondary data and research. It further explores the economic impact of the European Cruise industry on the European economies. In addition, the chapter tries to compare the economic impact of the European cruise among the previous years and to ascertain whether these impacts benefit the European economies or not. The evolution of the direct economic impacts are captured and a discussion about the total economic impacts of the European Cruise Sector is given. This study would offer facilitation in directing future research relevant to cruise tourism.

The edited volume may be used as a textbook for courses covering topics like management and strategy, decision making, health management, tourism management and operational research in several research sectors. In addition, policy makers, researchers, and academics will find this text useful in order to enrich their knowledge in the scientific area of Management Science. We hope that this book will provide a useful resource of ideas, techniques, and methods for additional research on this topic.

The editors of the book would like to express their sincere thanks to all the authors who have devoted considerable time and effort for their comprehensive contributions. We thank also IGI Global and the publishing team for their encouragement and support during the preparation of the book.

Panagiotis Manolitzas
Ionian University, Greece

Constantin Zopounidis
Technical University of Crete, Greece

Michael Talias
Open University of Cyprus, Cyprus

Evangelos Grigoroudis
Technical University of Crete, Greece

Nikolaos Matsatsinis
Technical University of Crete, Greece

Acknowledgment

The editors of the book would like to express their sincere thanks to all the authors who have devoted considerable time and effort for their comprehensive contributions. We also thank IGI GLOBAL and the publishing team for their encouragement and support during the preparation of the book.

Panagiotis Manolitzas
Ionian University, Greece

Constantin Zopounidis
Technical University of Crete, Greece

Michael A. Talias
Open University of Cyprus, Cyprus

Evangelos Grigoroudis
Technical University of Crete, Greece

Nikolaos Matsatsinis
Technical University of Crete, Greece

Chapter 1
Day Ahead Electricity Price Forecasting in Coupled Markets:
An Application in the Italian Market

Konstantinia Daskalou
Hellenic Open University, Greece & Watt and Volt SA, Greece

Christina Diakaki
https://orcid.org/0000-0003-1724-0193
Hellenic Open University, Greece & Technical University of Crete, Greece

ABSTRACT

Day ahead electricity price forecasting is an extensively studied problem, and several statistical, intelligence-based, and other techniques have been proposed in literature to address it. However, the liberalization of the electricity market taking place during the last decades and the market coupling pursued within the European Union reshape the problem and create the need to confirm the effectiveness and/or revise existing methods and solution techniques, and/or invent new approaches. Given that complete integration has not achieved yet, both relevant data and studies of forecasting considering integration are still rather sparse. It has thus been the aim of this chapter to contribute to filling this gap by focusing on and studying the market integration effects in day ahead electricity price forecasting. To this end, an artificial neural network has been developed and used under several, with respect to inputs, forecasting scenarios considering the Italian electricity market.

INTRODUCTION

In the last two decades, the liberalization and deregulation of the electricity markets have completely reshaped the landscape of the traditionally monopolistic and government-controlled power sectors. European Union (EU) has passed several laws trying to achieve a single internal electricity market through the implementation of a target model for the operation of the EU electricity markets (Daskalou, 2019). At the same time, electricity displays a set of characteristics that are uncommon to other markets; it is

DOI: 10.4018/978-1-7998-5442-5.ch001

non-storable and requires a constant balance between consumption and production, in order to ensure the stability of the power system (Carrion et al., 2007). Moreover, the electricity consumption is influenced by the weather conditions and the intensity of business of everyday activities, which, on their turn, result in prices that exhibit seasonality at a daily, weekly and annual basis (Weron, 2014). As a result, accurate electricity price forecasting, especially for the day ahead, is one of the biggest challenges for all market participants (generators, power suppliers, investors, traders) in order to maximize their profitability.

The day ahead electricity price forecasting problem is not a new one, and several statistical, intelligence-based and other techniques have been proposed in literature to address it, at all required levels that is daily, weekly and annual (see Weron (2014) and Nowotarski and Weron (2018) for relevant comprehensive reviews). What is, however, still largely lacking, are methods that take into account the forthcoming market integration. Given that complete integration has not achieved yet, both relevant data and studies of forecasting considering integration are still rather sparse. Thus, most relevant studies focus mainly in methodological issues in an effort to develop and/or refine forecasting models and techniques, while little effort (Daskalou, 2019) has been placed in investigating the effects that integration may have, and whether it is indeed useful and/or necessary to include in the forecasting model of a country inputs from the electricity markets of the countries with which it is already or will eventually be integrated with.

It is the aim of this chapter to focus on and study the integration effects in day ahead electricity price forecasting in an effort to contribute in filling the aforementioned gap. To this end, an Artificial Neural Network (ANN) is developed, and several, with respect to inputs, forecasting scenarios are used considering the Italian electricity market. Italy has been chosen for this study because its electricity market is already coupled (Gestore Mercati Energetici, n.d.) with some of its neighboring countries, thus data are available. This was not, however, the only reason. Another equally important reason is that the electricity system and therefore the market of Italy is divided into zones (Gestore Mercati Energetici, n.d.), the behavior of which resembles the behavior of coupled markets. More specifically, the zones correspond to geographical areas, which, for reasons of system security, present limited electricity exchange between them. Based on its own supply and demand, an electricity price is determined daily for each of these zones, and a weighted, by zonal-purchased electricity quantity, average is then used to determine the national single price (PUN-Prezzo Unico Nazionale). This means that more insight can be gained on the integration effects, by studying not only the behavior of the PUN price, but the behavior of the zonal prices as well.

BACKGROUND

The EU Electricity Market

The process of liberalization of the European electricity market started more than 20 years ago aiming at enhancing the competition in the provision of electricity, thus reducing prices (Pepermans, 2019). Before the liberalization process, most EU member states had one vertically integrated firm, running mainly under government control, which owned the production units (generation), as well as the transmission and distribution systems; the same firm was also the single electricity provider. It was the goal of liberalization to allow competition in generation and retail, while the transmission and distribution activities have been kept regulated since economies of scale are typically important for them (Pepermans, 2019).

Electricity market liberalization has been an incremental process with policies spread over three directives promulgated by the European Commission (EC) with general objective the development of the EU internal energy market (Daskalou, 2019):

- The first directive was Directive 96/92/EC. Known as *First Energy Package*, it was promulgated in 1996, and comprised the milestone for the creation of a single European market, since it introduced common rules for the generation, transmission and distribution of electricity. More specifically, this directive introduced the roles of the Transmission System Operators (TSOs) and Distribution System Operators (DSOs), emphasizing in the non-discriminatory treatment of system users for the TSOs and customers for the DSOs. Additionally, it obliged DSOs to give priority to the energy produced by Renewable Energy Sources (RES).
- The aforementioned directive was succeeded by Directive 2003/54/EC promulgated in 2003. This directive, known as *Second Energy Package*, included provisions to overcome shortcomings that emerged after the implementation of Directive 96/92/EC, mainly as regards the market dominance, the non-discriminatory network access and the protection of vulnerable customers. Furthermore, this directive obligated member states to designate an independent regulatory authority, responsible for monitoring the market TSOs and DSOs, and submitting reports on market dominance, predatory and anti-competitive behavior to the EC.
- Directive 2003/54/EC was then replaced in 2009 by Directive 2009/72/EC, the *Third Energy Package*, which aimed at further liberalizing and integrating Europe's energy markets. In particular, it contained provisions for the establishment of the Agency for the Cooperation of Energy Regulators (ACER), the independence of regulators, the strengthening of cross-border cooperation between TSOs and the creation of European networks for gas and electricity (ENTSO-G and ENTSO-E, respectively), the contract switching and customer protection.

The aforementioned directives for the liberalization of the electricity markets and the need for "the completion of a fully-functioning, interconnected and integrated internal energy market which is central to Europe's competitiveness and must not be fragmented" (Barroso, 2013) have driven EC to promote a *target model* for electricity markets. The target model has been based on guidelines and codes developed and/or promoted by ACER, ENTSO-E, and EC aiming at the harmonization of cross-border exchanges and the operation of wholesale electricity markets. It is based on two broad principles (Keay, 2013):

- Energy only regional markets, organized in zones in which the price of each marginal unit of energy supplied, determines the generators' revenues.
- Market coupling, which is a way of linking zonal day ahead spot markets into a virtual market, so that the lowest priced bids are accepted up to the point where congestion constraints limit further trade.

Market coupling is a rather complex process which requires a step by step approach to achieve the ultimate goal of a single European market (Daskalou, 2019). The first step took place in 2006 with the integration of French, Belgium and the Netherlands day ahead markets; the so-called *Tri-Lateral Market Coupling*. The next step was achieved in 2010 with the market coupling of Netherlands, Belgium and France with Germany, known as Central West European (CWE) Market Coupling. In parallel, CWE market was coupled with the Nordic-Baltic region.

However, the most important step towards the integration of the European electricity markets took place in 2014, with the launch of Price Coupling of Regions (PCR), a project undertaken by the European Power Exchanges (EPE) in order to develop a single price coupling algorithm that can be used to calculate electricity prices across Europe. PCR was launched in North Western Europe (NWE), which covered the region of CWE, Great Britain, the Nordics and the Baltics. Later in the same year, however, Spain and Portugal were coupled with NWE, while in 2015 Italy was coupled with Austria and Slovenia.

Nowadays, PCR is used to couple the countries shown in Table 1 (Energy Exchange Group, n.d.). In addition, under the provisions of Act 4425/2016 for the Implementation of the Target Model, the Greek energy market was envisaged to be coupled in 2018 first with Italy and then with Bulgaria. Apparently, 2018 was a very optimistic scenario. According to stakeholders of the Hellenic electricity market, the implementation of the target model in Greece is expected in the mid-2020, while after this launch, market coupling with Italy and Bulgaria will take place (Daskalou, 2019).

Table 1. PCR coupled markets

No	Country	No	Country	No	Country	No	Country	No	Country
1	Austria	6	Estonia	11	Italy	16	The Netherlands	21	Slovakia
2	Belgium	7	France	12	Ireland	17	Norway	22	Slovenia
3	Croatia	8	Finland	13	Latvia	18	Poland	23	Spain
4	Czech Republic	9	Germany	14	Lithuania	19	Portugal	24	Sweden
5	Denmark	10	Hungary	15	Luxembourg	20	Romania	25	UK

Electricity Price Forecasting Approaches

Overview

Since the introduction of the deregulated wholesale electricity markets, electricity price forecasting has become a crucial issue for all market participants (generators, power suppliers, investors, traders) because they require accuracy in order to maximize their profitability. In power exchanges (PX), power producers submit generation bids and their corresponding biddings prices, and consumption companies (retailers and traders) do the same with consumption bids. The clearing of the market is implemented by the market operator with the use of a market clearing tool that determines the market clearing price for each hour of the day ahead. Electricity prices series in the spot market have complex features such as non-stationarity, nonlinearity, and high volatility due to the basic characteristics of electricity that is its non-storable nature, the requirement of maintaining a continuous balance between supply and demand and rigid nature of demand over a short time period. Therefore, the developed models should be able to capture this complexity (Carrion et al., 2007).

Current research on electricity price forecasting can be classified in terms of the planning horizon's duration and in terms of the adopted forecasting technique. From time horizon point of view, it is common to define short-term (STPF), medium-term (MTPF) and long-term (LTPF) price forecasting (Cerjan et al., 2013):

- In LTPFs, times are typically measured in years. LTPF is applied in planning and investment profitability analyses, such as decision making for future investments of power plants.
- In MTPFs, times are typically measured in months. The most common usages of MTPFs is on balance sheet calculations, risk management, and derivatives pricing. MTPF focuses on the estimation of the distributions of future prices over certain time periods rather than actual point forecasts.
- In STPFs, lead times are measured from a few hours to a few days. STPF is of high importance for market participants and it focuses on providing accurate forecasts of the actual point prices.

Obviously, predicting the day ahead price requires STPF methods.

In terms of the adopted techniques, price forecasting literature is typically divided into five areas (Weron, 2014):

1. Multi-agent (multi-agent simulation, equilibrium, game theoretic) models, which simulate the operation of a system of various entities (generating units, companies) interacting with each other, and calculate the price using a procedure of matching the demand and supply in the market.
2. Reduced form (quantitative, stochastic) models, whose main goal is not to provide accurate price forecasts but rather to characterize the statistical properties of electricity prices over time. Such models are used for derivatives pricing risk management.
3. Structural methods, which describe the price fluctuations by modeling the impacts of important technical and economic parameters on the price of electricity.
4. Statistical (econometric, technical analysis) approaches, which use a mathematical combination of previous prices and other exogenous factors typically consumption and production figures or weather variables.
5. Computational intelligence (artificial intelligence-based, non-parametric, non-linear statistical) techniques, which combine elements of learning, evolution and fuzziness to create approaches that are capable of adapting to complex dynamic systems and may be regarded as intelligent in this sense.

Day Ahead Electricity Price Forecasting

As far as day ahead price forecasting is concerned, the best forecasting results, according to Weron's (2014) comprehensive review, have been reported for statistical, artificial intelligence and hybrid approaches; hybrid approaches combine more than one techniques.

In statistical approaches, statistical models are developed by combining historical price values and/or values of exogenous factors such as consumption and production and/or weather variables. These models are then used for the prediction of the current price. The models, used typically in this respect, are autoregressive (AR) and autoregressive with exogenous terms (ARX), autoregressive moving average models (ARMA) and autoregressive integrated moving average models (ARIMA), dynamic regression, transfer functions, semi/non-parametric models, threshold autoregressive exogenous (TARX) models, generalized autoregressive conditional heteroscedasticity-based models (GARCH) and double seasonal Holtz-Winter model. In some cases, hybrid statistical models are also developed, such as wavelet transform (for more information, see Weron (2014), and Nowotarski and Weron (2018)).

Table 2 summarizes statistical models, developed and applied in day ahead electricity price forecasting, which have contributed significantly to the relevant literature and constitute until today the reference point for the evaluation of other forecasting techniques (Daskalou, 2019). The table provides also information on the market and forecasting period for each model, as well as the resulting forecasting accuracy in terms of the mean absolute percentage error (MAPE), the most commonly used performance metric, computed as the mean of the absolute percentage errors that is the differences between actual and predicted values over the forecasting horizon.

Since statistical models are usually linear forecasters, they perform adequately in normal, non-spiky periods, due to the seasonality that electricity prices show in similar days of the week with similar consumption demands. However, the overall nonlinear behavior of electricity prices, as a result of the complicated bidding strategies used by participants especially in periods of high demand, decreases their efficiency in day ahead electricity price forecasting (Amjady and Hemmati, 2006). To address poor performance in such cases, methods adopting artificial intelligence (AI) techniques have been developed (Weron, 2014; Lago et al., 2018a).

Table 2. Overview of representative statistical model-based forecasting studies

Model	Market	Forecasting period	MAPE (forecasting period duration)
Dynamic regression & transfer function (Nogales et al., 2002)	Spain	21/8/2000-27/8/2000, 13/11/2000-19/11/2000	5.0% (2 weeks)
	California	3/4/2000-9/4/2000	3.0% (1 week)
ARIMA (Contreras et al., 2003)	Spain	25/5/2000-31/5/2000, 25/8/2000-31/8/2000, 13/11/2000-19/11/2000	10.0% (3 weeks)
	California	3/4/2000-9/4/2000	5.00% (1 week)
GARCH (Garcia et al., 2003)	Spain	25/5/2000-31/5/2000, 25/8/2000-31/8/2000, 9/11/2000-15/11/2000	9.00% (3 weeks)
	California	3/4/2000-9/4/2000	9.00% (1week)
Wavelet transform and ARIMA (Conejo et al., 2005)	Spain	18/2/2002-24/2/2002, 20/5/2002-26/5/2002, 19/8/2002-25/8/2002, 18/11/2002-24/11/2002	8.00% (4 weeks)
Threshold AR/ARX (Misiorek et al., 2006)	California	3/4/2000-3/12/2000	15.80% (1 week)
Extended ARMA models with Gaussian-mixture and switching-regime approaches (Swider and Weber, 2007)	Germany	1/6/2002-31/5/2004	13.30% (1 week)

AI is a very diverse group of nature-inspired computational techniques that combines elements of learning, evolution and fuzziness to create flexible models that can handle complexity and non-linearity. That is why they are promising for STPF and numerous authors have reported their excellent performance (Weron, 2014).

The most commonly used AI techniques are artificial neural networks (ANNs), radial basis function networks (RBFNs), fuzzy networks and support vector regressors (Lago et al., 2018a, 2018b). Deep neural networks (DNNs) seem also very promising (Lago et al., 2018a, 2018b). Table 3 summarizes the characteristics, techniques, and performance of representative AI-based forecasting studies (Daskalou, 2019).

In recent years, an increasing number of authors combine statistical and AI-based techniques and develop hybrid models in an effort to exploit the features, characteristics and thus strengths of each approach in detecting different patterns in the data and conjointly yield a more accurate result (Cerjan et al., 2019). Table 4 summarizes the characteristics, combined techniques, and performance of representative such examples (Daskalou, 2019).

All the aforementioned forecasting techniques have been adopted in the day ahead electricity forecasting of single markets. However, due to the pursued EU electricity market integration, it is very likely that the prices of a country will be affected at a higher or lower degree by the prices of other countries with which it is already coupled or just interconnected in the sense of electricity imports and/or exports. Since complete integration has not been achieved yet, both data and corresponding studies are still rather sparse. A review of the relevant literature (Daskalou, 2019) identified three only studies considering market integration (Ziel et al., 2015; Panapakidis and Dagoumas, 2016; Lago et al., 2018b), all of which conclude that forecasting accuracy improves when considering prices from coupled markets. Still however, the main focus of these studies has been on the methodological part, rather than the integration aspect, which is the focus of the present study, as discussed in the following section.

Table 3. Overview of representative AI-based forecasting studies

Technique	Market	Forecasting period	MAPE (duration of forecasting period)
Fuzzy ANN (Amjady, 2006)	Spain	18/2/2002-24/2/2002, 20/5/2002-26/5/2002, 19/8/2002-25/8/2002, 18/11/2002-24/11/2002	7.50% (4 weeks)
ANN (Catalao et al., 2007)	Spain	18/2/2002-24/2/2002, 20/5/2002-26/5/2002, 19/8/2002-25/8/2002, 18/11/2002-24/11/2002	8.90% (4 weeks)
	California	3/4/2000-9/4/2000	3.10% (1 week)
RBFN (Lin, et al., 2010)	Pennsylvania-New Jersey-Maryland (PJM)	22/7/2002, 21/7/2002, 18/3/2002-24/3/2002, 24/6/2002-30/6/2002, 23/9/2002-29/9/2002, 23/12/2002-29/12/2002	6.30% (5 weeks)
Singular spectrum analysis and wavelet ANN (Xiao et al., 2017)	New South Wales, Australia	10/4/2009-16/4/2009, 15/7/2009-21/7/2009, 20/10/2009-26/10/2009, 1/1/2009-7/1/2009	6.76% (4 weeks)

FORECASTING THE ITALIAN DAY AHEAD ELECTRICITY PRICE

The Italian Electricity Market

The electricity market of Italy is divided into zones. A zone represents a part of the electricity network where the transfers of electricity among geographical areas are physically limited for reasons of system security. There are three categories of zones:

- **Geographical zones:** They are part of the national network.

- **National virtual zones:** They represent a set of generating units, whose installed capacity is higher than their operational capacity and is restricted by the insufficient transmission capacity of the network.
- **Foreign virtual zones:** They represent the points of interconnection with neighboring foreign countries.

Currently, the Italian electricity market comprises six active geographical zones (see Figure 1), and thirteen virtual, one national and twelve foreign. Table 5 provides the details of this division (Gestore Mercati Energetici, n.d.).

The Italian day ahead market (Mercato del Giorno Prima - MGP) is part of the spot electricity market (along with intraday, daily products and ancillary services markets), which hosts most of the electricity sale and purchase transactions. In the Italian MGP, hourly energy blocks are traded for the next day. Its sitting opens at 08:00 of the ninth day before the day of delivery and closes at 24:00 of the day exactly before the day of delivery. Within this period, the market participants submit bids/asks whereby they declare the quantity, as well as the minimum/maximum price at which they are willing to sell/purchase. Based on the economic merit-order criterion and considering transmission capacity limits between zones, the submitted bids/asks are accepted after the closure of the market sitting.

Table 4. Overview of representative hybrid model-based forecasting studies

Combined models	Market	Forecasting period	MAPE (duration of forecasting period)
Wavelet transform, ARIMA and Radial Basis Function ANN (Shafiekhah et al., 2010)	Spain	18/2/2002-24/2/2002, 20/5/2002-26/5/2002, 19/8/2002-25/8/2002, 18/11/2002-24/11/2002	5.74% (4 weeks)
Wavelet transform, kernel extreme learning machine based on particle swarm optimization and ARMA (Yang et al., 2017)	Pennsylvania-New Jersey-Maryland (PJM)	23/2/2004-29/2/2004, 17/5/2004-23/5/2004, 23/8/2004-29/8/2004, 22/11/2004-28/11/2004, 1/2/2006-7/2/2006 22/2/2006-28/2/2006	8.27% (2 weeks in 2006); 5.23% (4 weeks in 2004)
	Australia	1/1/2014-31/12/2014	3.74% (1 year)
	Spain	26/1/2009-1/2/2009, 27/4/2009-3/5/2009, 27/7/2009-2/8/2009, 26/10/2009-1/11/2009, 1/1/2015-31/12/2015	4.25% (5 weeks)
Statistical method of similar days and ANN (Cerjan et al., 2019)	Hungarian Power Exchange (HUPX)	1/5/2017-30/4/2018	2.10% to 5.48% (1 year)
Improved empirical mode decomposition, ARMA with exogenous terms, exponential GARCH and adaptive network-based fuzzy inference system (Zhang et al., 2019)	Spain	26/1/2009-1/2/2009, 27/4/2009-3/5/2009, 27/7/2009-2/8/2009, 26/10/2009-1/11/2009, 1/1/2015-31/12/2015, weeks in 2016	3.30% (4 weeks in 2009, 1 year and weeks in 2016)
	Australia	28/1/2013-3/2/2013, 29/4/2013-5/5/2013, 29/7/2013-4/8/2013, 29/7/2013-4/8/2013, 28/10/2013-3/11/2013	3.14% (5 weeks)

Figure 1. The active geographical zones of the Italian electricity market

The marginal price of each zone is determined by the intersection of the demand and supply curves. The zonal prices are different from zone to zone only when transmission constraints are enabled. The supply offers, the demand bids of pumping units and the demand bids of consuming units, which belong to foreign virtual zones are valued at the zonal clearing price to which they belong. On the other hand, the demand bids of consuming units that belong to Italian geographical zones are valued at the national single price PUN, which is computed as the weighted average of the prices of the geographical zones; weighting is based on the quantities purchased in these zones.

Day Ahead PUN Price Forecasting Methods, Techniques and Input Scenarios

Given the market operation described in the previous section, and in order to predict the day ahead PUN prices two alternative methods are examined and compared (Daskalou, 2019):

- **Historical data-based PUN price forecasting:** This method follows the typical forecasting rationale. In other words, under this method, PUN prices are predicted based on past, historical values of the same variable, plus inputs on other variables such e.g. loads and natural gas prices, which are considered to have a significant impact in the PUN price determination.

Table 5. The active zones of the Italian electricity market

Name	Acronym	Details
Geographical zones		
Central-Northern Italy	CNOR	Marche, Toscana, Umbria
Central-Southern Italy	CSUD	Lazio, Abruzzo, Campania
Northern Italy	NORD	Val D'Aosta, Piemonte, Liguria, Lombardia, Trentino, Veneto, Friuli Venezia Giulia, Emilia Romagna
Sardinia	SARD	Sardinia
Sicily	SICI	Sicily
Southern Italy	SUD	Molise, Puglia, Basilicata, Calabria
National virtual zones		
Rossano	ROSN	
Foreign virtual zones		
Austria	AUST	
Austria Coupling	XAUS	Represents the interconnection dedicated to Italy-Austria market coupling
Corsica	CORS	
Corsica AC	COAC	
France	FRAN	
France Coupling	XFRA	Represents the interconnection dedicated to Italy-France market coupling
Greece	GREC	
Slovenia	SLOV	
Slovenia Coupling	BSP	Represents the interconnection dedicated to Italy-Slovenia market coupling
Switzerland	SVIZ	
Switzerland Coupling	XSVI	Represents the interconnection dedicated to Italy-Switzerland market coupling
Malta	MALT	

- **Zone-based PUN price forecasting:** This method adopts the rationale that MGP uses to compute the PUN price. More specifically, under this method, price prediction is first performed individually for each of the six active geographical zones in Italy (see Table 5 or Figure 1), and then, the PUN price prediction is computed as the weighted average of the prices predicted for the six geographical zones; weighting is based on the load requirements of the zones. For each zone, price is predicted based on past, historical values plus inputs of other important variables, as in the aforementioned historical data-based PUN price forecasting method.

Both methods adopt as forecasting technique an ANN, which has been developed for this purpose, and are applied on the basis of four alternative input scenarios, which have been specifically developed to account for potential effects in forecasting accuracy, due to market coupling and/or just interconnection (Daskalou, 2019). These scenarios carry the following characteristics:

- **Scenario 1:** Taking into account the periodicity of the electricity prices (Weron, 2014), as well as the effects that natural gas prices (Panapakidis, 2016) and RES may have on it, this first, reference scenario considers as inputs, historical day ahead electricity prices, loads, predicted RES generation, and natural gas prices.
- **Scenario 2:** To account for potential effects due to market coupling, this scenario considers, in addition to the inputs of scenario 1, the day ahead system marginal prices (SMPs) of the electricity markets already coupled with the Italian one, that is of Austria, France, Slovenia and Switzerland (see Table 5).
- **Scenario 3:** To account for potential effects due to interconnection, this scenario considers, in addition to the inputs of scenario 1, the day ahead prices of Greece, the electricity system of which is interconnected to the Italian, but the two markets are not coupled as yet.
- **Scenario 4:** This scenario merges the inputs of scenarios 2 and 3, in order to account for potential effects of both market coupling and interconnection.

In each of the aforementioned scenarios, data of the period 01.01.2017-31.12.2018 are used. More specifically, for the necessary natural gas prices, the 2017-2018 hourly day ahead prices of the Italian spot gas market (MGP-GAS) are used in both methods, while for the historical day ahead electricity prices, loads and predicted RES generation, data utilization depends on the applied method (Daskalou, 2019):

- In the historical data-based PUN forecasting method, the 2017-2018 hourly day ahead PUN prices are used, while the hourly load and predicted RES generation for 2017-2018 are computed as the summation of the corresponding hourly loads and RES generation of the six geographical zones.
- In the zone-based PUN price forecasting method, the 2017-2018 hourly, zone-specific day ahead electricity prices, loads and predicted RES generation are used to predict the day ahead electricity prices of each zone, which are then used to compute the hourly day ahead PUN prices.

The necessary load and price data, as well as the predicted RES generation data have been downloaded from the transparency platform of ENTSO-E, while natural gas prices have been downloaded from the website of the Italian market operator (Daskalou, 2019).

The Forecasting Technique

Basics of Artificial Neural Networks

ANNs may be defined as structures comprised of densely interconnected simple processing elements, the artificial neurons or nodes, which are capable of performing massively parallel computations and knowledge representation (Hecht-Nielsen, 1990; Schalkoff, 1997). Their attractiveness emanates from their ability to emulate the remarkable information processing characteristics of biological systems such as nonlinearity, high parallelism, robustness, fault and failure tolerance, learning, as well as their ability to handle imprecise and fuzzy information and to generalize (Jain et al., 1996).

The fundamental building block of an ANN is the *simple neuron*, which comprises two processes, a *summer* and a *transformation*, as displayed in Figure 2. The summer produces an output $n = \mathbf{WP+b}$, with **P** an inputs vector, **W** a weights matrix and b a bias, which is then fed to a transfer function f that produces the final neuron output $a = f(\mathbf{WP}+b)$. The weights matrix **W**, and bias b as well, are adjustable parameters; they are adjusted by some learning rule, via a procedure called *training*, so that the neuron input/output relationship meets some specific goal. The transfer function f, on the other hand, is chosen by the user so as to satisfy problem-specific requirements, and can take several forms such e.g. linear, sigmoid, etc. (Hagan et al., 2014).

Figure 2. A simple neuron

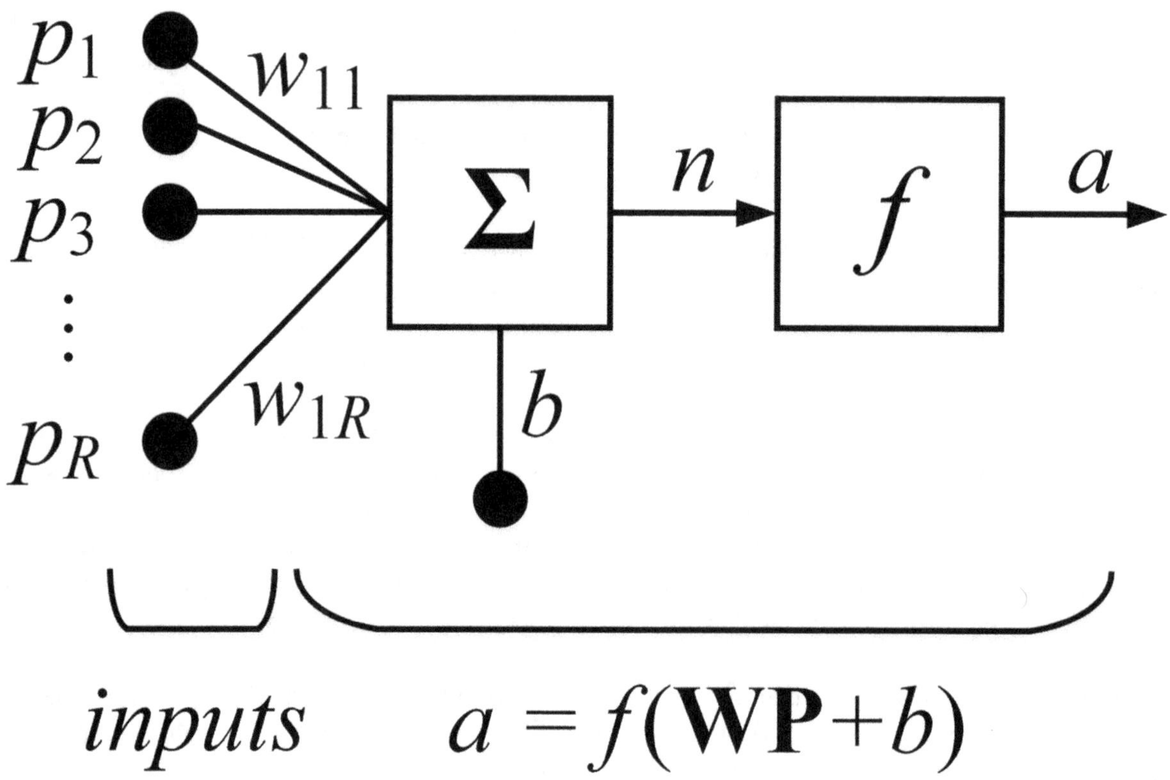

To develop ANNs, simple neurons are first organized in parallel, forming layers. All neurons of a layer receive the same inputs **P**, but each produces its own output. Then, two or more layers are connected in series. The last in the series layer is called *output layer*, while all others are called *hidden*. The first in the series layer receives the inputs **P**, the output layer produces the ANNs outputs **a**, and each intermediary hidden layer receives as input the output of its preceding layer, and its output is used as input by the layer that follows.

In terms of architecture, ANNs may be classified into two main categories, *feedforward* and *recurrent* neural networks (Weron, 2014). In feedforward networks, the outputs are computed directly from the inputs in one pass; no feedback is involved. In recurrent networks, on the other hand, feedback is involved, as some of the outputs are connected to the inputs. The feedforward ANNs are generally preferred for forecasting, whereas recurrent ANNs excel in optimization, pattern classification and categorization (Jain et al., 1996; Rutkowski, 2008).

Determining the number of neurons and the number of layers is a key issue in designing a neural network. *Multilayer* networks are more powerful than *single layer* networks. The increase, however, in the number of layers increases the complexity of the network. For this reason, most practical problems consider two to three layers only. The increase in the number of neurons also increases the network performance but at the same time the complexity and the need to adjust an increased number of parameters (weights and biases). In addition, when large numbers of neurons are used, the network may overfit data, that is, perform well with the training data set, but poorly with new, and as a consequence, its generalization may be poor. Given these, the number of neurons, as well as layers should be adjusted so as to fit the complexity of the given problem and data (Hagan et al., 2014).

The process of repeatedly modifying the weights and biases until an ANN fits to the given data and learns to perform a certain task is called, as mentioned earlier, training. ANN training may be performed in three modes (Hagan et al., 2014):

- **Supervised Learning:** In supervised learning the training algorithm is provided with a set of inputs and corresponding outputs, and adjusts the weights and biases of the network so as to move the ANN's outputs as close as possible to the given ones.
- **Reinforcement learning:** In reinforcement learning, rather than feeding the training algorithm with sets of inputs and corresponding outputs, a performance index is computed, which measures the performance of the network. The training algorithm adjusts then the weights and biases so as to improve the ANN's performance as expressed by this performance index.
- **Unsupervised Learning:** In unsupervised learning the training algorithm adjusts the weights and biases in response to network inputs only. This type of learning is used for clustering operations.

The Day Ahead Electricity Price Forecasting ANN

For the needs of the present study, a feedforward ANN (see Figure 3), which as mentioned earlier is preferred for forecasting, with one hidden and one output layer has been developed in MATLAB (Daskalou, 2019).

Taking into account that the typical number of neurons proposed by MATLAB is 10 and that a small number of neurons may reduce the performance of the ANN, the number of neurons was set double the typical value for the hidden layer. As regards to the output layer, one only neuron was needed, since the output is one single matrix with the 24 day ahead prices. In addition, backpropagation with Bayesian regularization was chosen as training algorithm.

Backpropagation is a supervised learning training algorithm, and the most popular training algorithm for feedforward ANNs (Hagan and Menhaj, 1994). Nevertheless, the algorithm may converge slowly and cause overfitting; the ANN may memorize the data in the training set and then fail to generalize to new situations (Hagan et al., 2014). To overcome such problems, regularization techniques are also employed (Saini, 2008; MacKay, 1992) with most commonly used techniques the Levenberg-Marquardt regularization (LM) and the Bayesian regularization (BR). LM was especially developed for faster convergence, while BR aims mainly at avoiding overfitting and achieving generalizable models, thus chosen for this particular application (Burden and Winkler, 2008). As a result of BR utilization, which requires a differentiable transfer function, the log-sigmoid transfer function has been used for all neuros.

Figure 3. The feedforward ANN

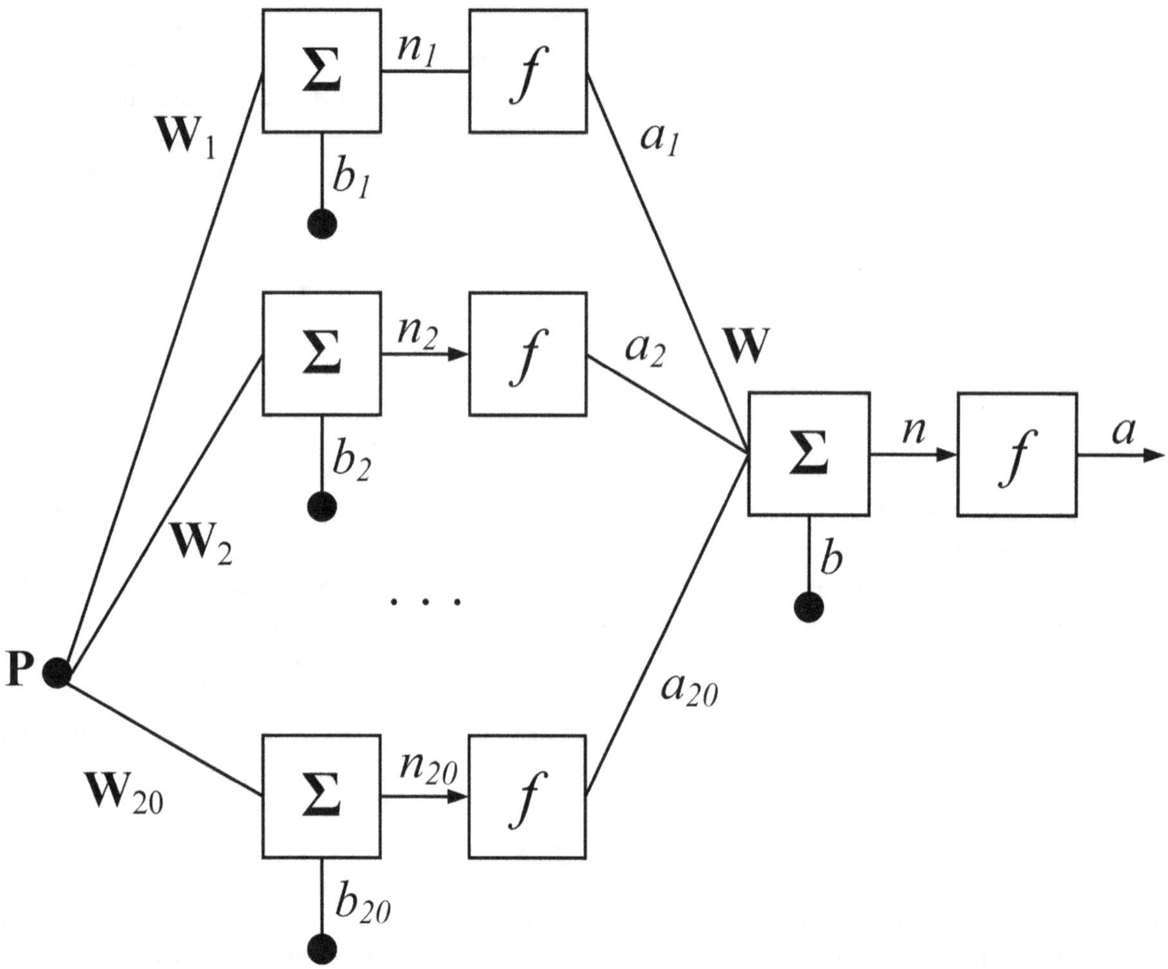

In order to train the network, the data have been randomly divided in order for 70% of them to be used for training, 15% for validation and 15% for testing, according to the typically used division (Hagan et al., 2014). The target of the training process is to move the ANN outputs to the given ones. This is performed via the minimization of their difference, which is typically measured with the Mean Square Error (MSE) function, also used in this application (Hagan et al., 2014).

The training error almost never becomes absolutely equal to zero. For this reason, other stopping criteria are also defined. The MATLAB toolbox provides the following stopping criteria when ANN training involves the BR technique:

- The training error, as measured by a performance index like MSE, reaches a target value.
- The maximum number of epochs (iterations) is reached.

- The maximum amount of time is surpassed.
- The performance gradient falls below a specific value.

For these criteria, MATLAB provides default values. Since, however, the default value of training error is zero, goal that as mentioned earlier is practically never achieved, the training error for this application was set to 10^{-25}. Moreover, the default value of the maximum number of epochs was set to double its default value that is 1000, since a small number of epochs may reduce the performance of the network. For the rest of the stopping criteria, the default MATLAB values have been adopted.

In addition, since the ANN output comprises electricity prices, while for inputs, variables of different scaling (prices lie in the range 0-200, load requirements lie in the range 0-20,000, etc.) are used, in order to improve the ANN's understanding of their correlations, inputs and target outputs were first normalized, and then used for the training of the ANN.

The trained network was finally used for forecasting. More specifically, the developed ANN was used to predict the day ahead hourly prices of PUN or a specific zone, depending on the method, for a whole month, based on all past available input data starting from January 1, 2017. For example, the data of period 01.01.2017-31.12.2017 were used to predict the hourly prices for each day of January 2018, the data of period 01.01.2017-31.01.2018 were used to predict the hourly prices for each day of February 2018, and so on so forth. Figure 4, displays the flow chart of the ANN code developed in MATLAB, while Table 6 summarizes the inputs fed to the developed ANN for each of the considered input data scenario. It should be noted here that the sine and cosine functions of weekdays and hours have been used rather than their absolute values, in order to help the ANN understand the weekly and daily periodicity of loads and prices.

Forecasting Quality Evaluation

To evaluate the quality of the forecasting methods described previously, the mean absolute percentage error (MAPE), which is the most widely used performance measure (Weron, 2014), has been chosen. MAPE is computed for each month m for which the hourly day ahead prices are predicted according to the formula:

$$\text{MAPE}_m = \frac{1}{H_m}\sum_{i=1}^{H_m}\frac{\left|P_i^a - P_i^f\right|}{P_i^a}$$

where P_i^a and P_i^f stand for the actual and predicted prices of hour i, and H_m for the total hours of month m. It should be noted here that MAPE is calculated based only on the hours of a month where actual prices are different that zero.

Figure 4. Flow chart of the ANN MATLAB code

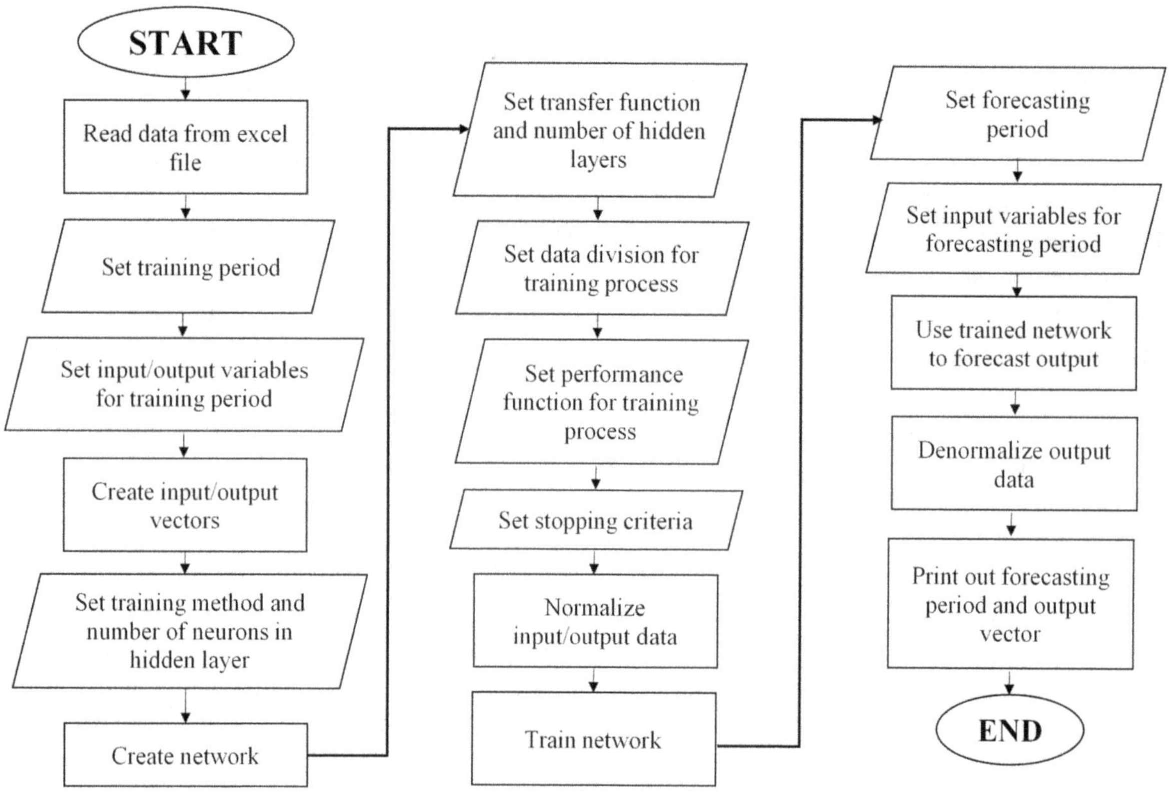

Despite the fact that MAPE is the most popular measure of accuracy, it may become misleading in case of close to zero or spike electricity prices (Weron, 2014). In the former case, it may become very high, and in the latter very small, regardless the actual absolute error $\left|P_i^a - P_i^f\right|$ value. To trace such situations, the absolute error is also computed according to the following formula:

$$\mathrm{AE}_m = \frac{1}{H_m}\sum_{i=1}^{H_m}\left|P_i^a - P_i^f\right|$$

Forecasting Results and Analysis

Historical Data-Based PUN Price Forecasting

In this first method, PUN prices are predicted based on own past values plus other inputs according to the considered input data scenario (see Table 6). Table 7 summarizes the corresponding results (Daskalou, 2019) in terms of the MAPE and AE performance measures.

Table 6. ANN's inputs per input data scenario

Scenario ID	Inputs	Description
1	$\sin(d)$	Sine of weekday d, computed as $\sin(2\bullet\pi\bullet d/7)$, where $d = 1, 2, \ldots, 7$, for Monday, Tuesday, …, Sunday, respectively
	$\cos(d)$	Cosine of weekday d of week, computed as $\cos(2\bullet\pi\bullet d/7)$
	$\sin(h)$	Sine of hour h of day, computed as $\sin(2\bullet\pi\bullet h/24)$, where $h = 1, 2,, \ldots, 24$ for hour 01:00, 02:00, …, 24:00, respectively
	$\cos(h)$	Cosine of the hour h of day, computed as $\cos(2\bullet\pi\bullet h/24)$
	$H(d)$	Binary indicator taking the value 1 if weekday d is holiday and 0 otherwise
	$L(d, h)$	Predicted load of hour h of target weekday d (in MWh)
	$L(d-1, h)$	Load of hour h of weekday d–1, i.e. the weekday before the target one (in MWh)
	$L(d-7, h)$	Load of hour h of weekday d–7, i.e. the same weekday with the target one, but of the previous week (in MWh)
	$\overline{L}(d-1)$	Average load of weekday d–1 (in MWh)
	$P(d-1, h)$	Price of hour h of weekday d–1 (in €/MWh)
	$P(d-7, h)$	Price of hour h of weekday d–7 (in €/MWh)
	$\overline{P}(d-1)$	Average price of weekday d–1 (in €/MWh)
	$RES(d)$	Predicted RES generation of target weekday d (in MWh)
	$P_{NG}(d, h)$	Natural gas price of hour h of target weekday d (in €/MWh)
	$P_{NG}(d-1, h)$	Natural gas price of hour h of weekday d–1 (in €/MWh)
2	Scenario 1 inputs	All inputs of scenario 1
	$SMP_{AU}(d, h)$	Austria SMP of hour h of target weekday d (in €/MWh)
	$SMP_{FR}(d, h)$	France SMP of hour h of target weekday d (in €/MWh)
	$SMP_{SL}(d, h)$	Slovenia SMP of hour h of target weekday d (in €/MWh)
	$SMP_{SW}(d, h)$	Switzerland SMP of hour h of target weekday d (in €/MWh)
3	Scenario 1 inputs	All inputs of scenario 1
	$SMP_{GR}(d, h)$	Greece SMP of the hour h of target weekday d (in €/MWh)
4	Scenario 1 inputs	All inputs of scenario 1
	Scenario 2 inputs	All inputs of scenario 2, used in addition to those of scenario 1
	Scenario 3 inputs	All inputs of scenario 3, used in addition to those of scenario 1

Applying the first input data scenario, a yearly average MAPE of 10.57% is reached, with the first 5 months to lie in the range 11.45-17.55% and the following 7 months to be reduced by almost half, thus lying in the range of 6.14-9.85%. During the first 5 months, the actual hourly prices are occasionally presenting extreme, low or high values as the example of Figure 5 displays. Such values, however, affect as mentioned earlier the MAPE values, and may explain the weak performance suggested by the corresponding MAPE values, despite the fact that the same performance is not present in the corresponding AE values. According to AE, performance does not deviate much around the average value of each scenario (see average and standard deviation lines in Table 7).

Table 7. MAPE and AE results of historical data-based PUN price forecasting

Month	MAPE (%)				AE (€/MWh)			
	Scenario 1	Scenario 2	Scenario 3	Scenario 4	Scenario 1	Scenario 2	Scenario 3	Scenario 4
Jan-18	13.27	8.61	13.18	9.42	5.62	3.65	5.69	3.78
Feb-18	11.45	14.74	11.38	10.41	6.47	9.14	7.21	6.38
Mar-18	15.37	10.40	13.96	9.79	9.01	5.84	8.95	5.59
Apr-18	17.55	15.42	18.25	15.85	7.40	6.33	7.95	6.64
May-18	14.70	14.08	13.58	12.24	6.99	6.84	6.57	5.96
Jun-18	9.85	7.75	10.00	7.74	5.34	3.98	5.49	4.06
Jul-18	6.73	6.89	7.24	7.12	4.19	4.26	4.40	4.40
Aug-18	6.14	5.87	6.50	5.09	4.20	3.99	4.45	3.47
Sep-18	6.96	5.99	6.48	6.51	5.49	4.75	5.15	5.17
Oct-18	7.93	5.65	6.74	5.12	5.79	4.19	4.99	3.79
Nov-18	7.24	6.42	6.99	7.27	5.04	4.40	4.80	5.02
Dec-18	9.71	8.01	10.26	7.37	6.04	4.95	6.35	4.53
Average	10.58	9.15	10.38	8.66	5.97	5.19	6.00	4.90
Standard deviation	3.86	3.64	3.80	3.11	1.37	1.60	1.44	1.07

Figure 5. Actual hourly PUN price trajectory of April 29, 2018

Table 8. MAPE and AE results of CNOR price projection

Month	MAPE (%)				AE (€/MWh)			
	Scenario 1	Scenario 2	Scenario 3	Scenario 4	Scenario 1	Scenario 2	Scenario 3	Scenario 4
Jan-18	13.57	11.39	14.16	11.19	6.20	4.50	5.75	4.39
Feb-18	14.08	12.58	14.52	12.30	8.92	8.18	9.13	7.78
Mar-18	13.33	11.79	15.04	13.71	8.04	6.78	8.67	7.86
Apr-18	28.50	25.19	23.17	24.09	9.24	8.31	7.19	7.88
May-18	16.27	14.53	15.83	15.62	7.82	6.80	7.34	7.16
Jun-18	13.49	9.95	12.21	11.16	7.10	5.05	6.38	5.65
Jul-18	8.28	7.70	7.78	8.04	4.83	4.48	4.64	4.73
Aug-18	6.03	4.82	6.20	4.88	3.98	3.15	4.10	3.22
Sep-18	8.18	7.96	8.68	7.02	6.60	6.36	7.01	5.61
Oct-18	7.30	6.55	8.07	6.97	5.69	4.99	6.19	5.30
Nov-18	10.69	8.57	8.99	8.97	7.23	5.51	6.00	5.93
Dec-18	9.47	8.21	9.96	8.20	5.83	4.70	5.86	5.01
Average	12.38	10.73	12.01	10.97	6.76	5.71	6.50	5.86

Applying the second input data scenario, which considers the coupling of the Italian market with the markets of other countries (see Table 6), the forecasting accuracy seems to improve for practically all months. However, similarly with the first scenario, the forecasting of the first 5 months, seems to be weaker than the forecasting of the last 7 (see Table 7). Moreover, a closer look at the monthly MAPE values reveals that there are months where the consideration of the markets coupled with the Italian one has a greater impact in PUN price prediction than in other months. For example, MAPE value of January decreased from 13.27% to 8.61%, while MAPE value of May and July decreased and increased respectively from 14.70% to 14.08%, and 6.73% to 6.89%, only. In fact, market coupling does not affect the actual PUN prices at all times with the same strength, and this may cause confusion to the ANN, as the contradictory results of months January, May and July suggest.

Leading to the neurons of the ANN the inputs of the third scenario, which considers the interconnection of Italy with Greece, MAPE is improved compared to scenario 1, but deteriorates, even slightly compared to scenario 2 (see Table 7). In addition, a pattern with worse performance in the first months of the year is observed similarly to the previous scenarios, while considering the MAPE values; the same pattern does not appear as far as AE is concerned. Moreover, similarly with the second scenario, there are months where the consideration of the interconnection of Italy with Greece has a greater impact in improving PUN price prediction than in other months; see e.g. the MAPE values of March vs December 2018.

Finally, applying the fourth input data scenario, which considers both coupling and interconnection, the forecasting accuracy improves compared to all previous scenarios (see Table 7). MAPE yearly average is the lowest observed among the different scenarios and so does standard deviation, which means that performance has consistently improved for all the months of the year.

Table 9. MAPE and AE results of CSUD price projection

Month	MAPE (%)				AE (€/MWh)			
	Scenario 1	Scenario 2	Scenario 3	Scenario 4	Scenario 1	Scenario 2	Scenario 3	Scenario 4
Jan-18	13.44	13.40	12.78	10.73	5.56	5.74	4.92	4.31
Feb-18	13.26	13.50	19.27	13.91	8.40	7.89	13.23	8.12
Mar-18	20.17	14.66	17.85	15.98	10.90	6.93	9.08	8.32
Apr-18	25.42	28.09	25.96	22.00	7.34	8.70	8.30	8.12
May-18	12.27	14.46	13.42	14.37	6.22	7.46	6.79	7.31
Jun-18	13.60	12.39	12.69	12.09	7.47	7.05	6.99	6.80
Jul-18	8.09	8.40	7.75	7.53	4.95	5.14	4.68	4.65
Aug-18	5.67	5.55	5.16	5.45	3.83	3.78	3.47	3.67
Sep-18	8.60	8.39	8.40	8.81	6.73	6.60	6.62	6.90
Oct-18	7.55	7.22	7.95	7.82	5.58	5.22	5.83	5.67
Nov-18	11.81	9.86	10.74	9.78	7.84	6.48	7.13	6.38
Dec-18	10.21	8.52	10.14	9.82	6.32	5.20	6.01	6.01
Average	12.47	11.98	12.59	11.48	6.74	6.32	6.86	6.33

Zone-Based PUN Price Forecasting

In the second PUN price forecasting method, the day ahead zonal prices are first predicted, and their weighted, by corresponding load requirements, average is then used to compute the PUN price. Tables 8-13 provide the corresponding results (Daskalou, 2019), which reveal the following:

Figure 6. Actual hourly SICI price trajectory for September 2018

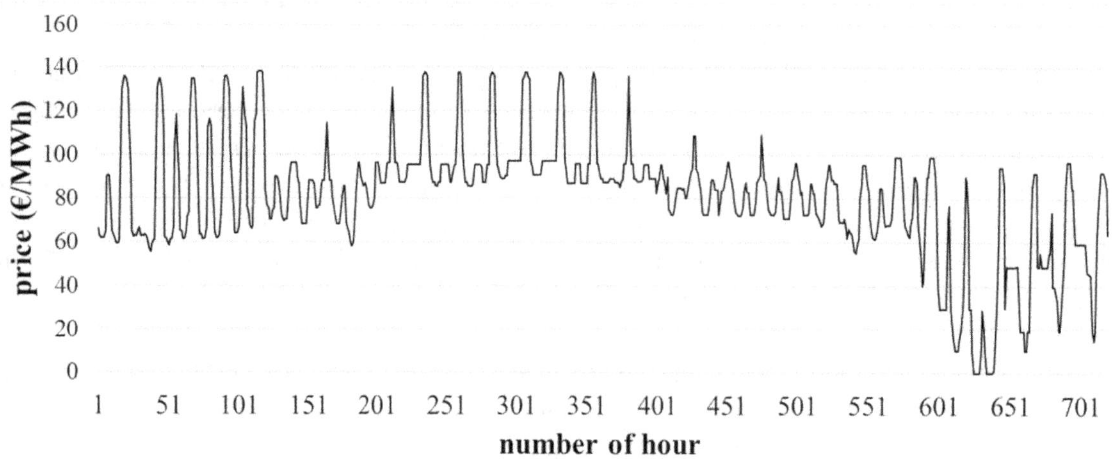

Table 10. MAPE and AE results of NORD price projection

Month	MAPE (%)				AE (€/MWh)			
	Scenario 1	Scenario 2	Scenario 3	Scenario 4	Scenario 1	Scenario 2	Scenario 3	Scenario 4
Jan-18	13.78	10.34	14.13	14.17	5.61	4.15	6.23	5.58
Feb-18	11.44	11.87	12.46	10.83	7.33	7.75	7.55	7.02
Mar-18	18.49	11.23	14.99	11.99	12.25	6.67	9.32	7.63
Apr-18	20.85	17.36	20.59	18.62	9.21	7.32	9.29	8.09
May-18	14.77	12.92	15.66	13.69	7.09	5.97	7.47	6.22
Jun-18	11.57	8.87	11.55	9.58	5.93	4.26	6.01	4.65
Jul-18	9.95	7.71	9.65	8.12	5.92	4.47	5.59	4.73
Aug-18	7.89	4.64	6.02	4.55	5.30	3.01	3.94	2.95
Sep-18	7.10	7.16	6.94	6.80	5.74	5.72	5.62	5.47
Oct-18	9.19	5.52	8.06	5.73	7.04	4.13	6.28	4.39
Nov-18	8.87	7.32	9.31	6.68	6.28	5.11	6.49	4.57
Dec-18	10.95	8.36	11.46	8.20	6.73	4.78	6.84	4.73
Average	12.08	9.41	11.73	9.90	7.04	5.25	6.71	5.49

Table 11. MAPE and AE results of SARD price projection

Month	MAPE (%)				AE (€/MWh)			
	Scenario 1	Scenario 2	Scenario 3	Scenario 4	Scenario 1	Scenario 2	Scenario 3	Scenario 4
Jan-18	21.70	20.26	22.81	18.50	6.27	5.78	5.87	5.22
Feb-18	12.75	14.37	15.58	12.63	7.44	9.63	8.79	7.42
Mar-18	17.68	16.10	18.91	17.40	9.05	7.89	10.20	8.66
Apr-18	23.43	23.08	22.54	26.75	9.23	8.21	8.46	8.44
May-18	16.23	16.23	17.19	17.15	7.63	7.81	7.77	8.19
Jun-18	17.16	15.83	17.48	15.81	8.62	7.77	8.60	7.82
Jul-18	8.38	7.72	7.75	7.65	5.04	4.74	4.75	4.70
Aug-18	5.79	5.56	6.34	5.92	3.89	3.81	4.20	3.99
Sep-18	8.76	8.49	8.01	8.30	6.80	6.65	6.32	6.53
Oct-18	12.49	12.92	11.87	12.49	6.23	6.31	5.92	6.22
Nov-18	10.46	10.27	10.80	10.70	6.73	6.69	6.98	6.92
Dec-18	11.49	9.65	10.20	9.56	6.84	5.62	6.02	5.72
Average	13.86	13.35	14.10	13.56	6.97	6.71	6.97	6.64

Table 12. MAPE and AE results of SICI price projection

Month	MAPE (%)				AE (€/MWh)			
	Scenario 1	Scenario 2	Scenario 3	Scenario 4	Scenario 1	Scenario 2	Scenario 3	Scenario 4
Jan-18	25.07	29.44	24.84	27.08	13.08	15.23	13.19	13.61
Feb-18	37.99	33.81	40.12	33.93	12.94	10.93	12.14	12.19
Mar-18	42.31	36.10	37.74	39.34	11.60	9.32	9.29	10.64
Apr-18	24.97	26.51	24.72	28.19	8.35	9.59	8.99	10.12
May-18	21.22	20.45	20.32	23.03	13.01	12.61	11.89	13.81
Jun-18	13.77	13.92	13.00	13.45	8.17	8.61	7.77	8.34
Jul-18	16.41	17.68	16.01	17.78	12.66	13.55	12.44	13.43
Aug-18	17.24	18.46	17.30	18.11	14.94	16.21	16.02	15.79
Sep-18	410.68	421.78	445.32	1111.64	14.40	14.14	14.71	14.14
Oct-18	24.40	28.73	27.50	27.73	8.72	9.45	9.08	9.28
Nov-18	11.49	11.76	11.93	11.77	7.68	7.81	7.84	7.71
Dec-18	21.33	20.06	23.03	21.71	13.47	13.37	14.95	14.40
Average	54.93	55.93	57.77	112.84	11.60	11.76	11.54	11.97

- The pattern with the worse performance in the first months of the year observed in the MAPE values computed in the previous PUN price forecasting method appears also to the MAPE values of all scenarios for all but the SICI and SUD price predictions. Similarly also to the previous method, the same behavior is not present when considering AE rather than MAPE.

- In SICI, in particular, an overall weak performance of the ANN is observed in all considered input scenarios (see Table 12). The actual prices in this zone present high volatility among very low and very high prices, as the example of Figure 6 presents. This volatility leads to poor predictions as suggested by both MAPE and AE, which get values that are significantly higher than the corresponding values of the other zones. As suggested by the investigations performed herein, the best prediction for the SICI zone comes under scenario 1 that is without consideration of coupling and/or interconnection. However, predicting SICI price, requires undoubtedly further investigation.

- In SUD on the other hand (see Table 13), an overall weak performance is also observed, which is however improved under the third scenario that is while considering the interconnection with Greece, which is close to this zone. In contrast to this, performance deteriorates when considering the coupled markets that are farther away from the SUD zone (i.e. in scenarios 2 and 4).
- The behavior observed in the case of SUD reverses in the cases of the CNOR and NORD zones (see Tables 8 and 10, respectively), the predictions of which improve significantly when considering the coupled markets (i.e. in scenario 2), which are geographically close to them, and deteriorate when considering Greece (i.e. in scenarios 3 and 4), which is farther away. The same behavior is observed for SARD (see Table 11), which is also father away from Greece.

- Finally, in the case of the CSUD zone (see Table 9), which is located somehow in the middle of the other zones, thus preserving approximately similar distances to both Greece and the coupled markets, the predictions improve when considering all these markets, that is under input scenario 4.

Table 13. MAPE and AE results of SUD price projection

Month	MAPE (%)				AE (€/MWh)			
	Scenario 1	Scenario 2	Scenario 3	Scenario 4	Scenario 1	Scenario 2	Scenario 3	Scenario 4
Jan-18	12.21	15.73	13.74	12.36	5.32	6.58	5.62	5.45
Feb-18	16.23	12.79	10.96	15.49	9.47	8.29	6.10	9.67
Mar-18	19.02	24.84	19.28	17.67	7.99	11.92	8.51	7.74
Apr-18	23.89	25.79	27.91	20.14	7.90	7.87	7.69	7.00
May-18	14.93	16.45	14.52	15.14	7.53	8.24	7.40	7.84
Jun-18	11.22	13.42	10.38	13.37	6.43	7.74	5.86	7.59
Jul-18	9.42	9.06	9.41	10.12	5.51	5.40	5.58	5.88
Aug-18	6.29	6.76	6.61	7.12	4.13	4.45	4.36	4.71
Sep-18	16.79	21.73	17.38	23.42	7.60	8.31	8.24	9.22
Oct-18	29.43	26.57	27.35	25.32	9.47	9.72	9.36	9.37
Nov-18	8.88	9.59	8.80	8.67	5.40	5.99	5.35	5.28
Dec-18	13.93	12.12	12.90	13.25	7.59	6.60	6.89	7.23
Average	15.18	16.25	14.95	15.15	7.01	7.59	6.75	7.23

The above observations suggest that the geographical proximity of a zone to the markets coupled or interconnected to the Italian one affects the prediction quality; the closer the better.

Table 14 summarizes the input scenarios that yielded the best forecasting accuracy in terms of yearly MAPE values in predicting zonal prices. Using the results of these scenarios for PUN price prediction, the results of Table 15 are obtained (Daskalou, 2019).

Table 14. Input scenarios with best forecasting performance per zone

Zone	Scenario ID
Central-Northern Italy (CNOR)	2
Central-Southern Italy (CSUD)	4
Northern Italy (NORD)	2
Sardinia (SARD)	2
Sicilia (SICI)	1
Southern Italy (SUD)	3

The zone-based PUN price forecasting resulted in a yearly MAPE value of 8.5% that is a slight improvement compared to the best MAPE results obtained under scenario 4 in historical data-based PUN price forecasting method (average 8.64%; see scenario 4 in Table 7). Given the fact that the actual PUN price is determined by the Italian MGP based on the zone prices, and that the price predictions of the zones seem to be affected by the geographical proximity of the zones to the coupled and interconnected markets, the zone-based PUN price forecasting method achieved better results by allowing the consideration of this proximity.

Table 15. MAPE and AE results of zone-based PUN price forecasting

Month	MAPE (%)	AE (€/MWh)
Jan-18	9.88	3.93
Feb-18	9.84	6.28
Mar-18	9.70	5.67
Apr-18	16.17	6.63
May-18	11.42	5.52
Jun-18	7.69	3.95
Jul-18	6.50	4.02
Aug-18	4.47	3.04
Sep-18	6.33	5.03
Oct-18	5.55	4.05
Nov-18	6.82	4.62
Dec-18	7.84	4.57
Average	8.50	4.76

SOLUTIONS, RECOMMENDATIONS AND FUTURE RESEARCH DIRECTIONS

The investigations performed herein, confirm like other past studies (see e.g. Panapakidis and Dagoumas 2016; Lago et al., 2018b) that consideration of market coupling and/or interconnection may indeed improve the forecasting performance. They also suggest however that in case of countries, which like Italy have a zone-based electricity system and market, coupling should be considered with care, since geographical proximity may affect the forecasting accuracy of the individual zones and consequently the forecasting accuracy at country level. This behavior suggests also that inter-zone effects may also exist, which were not considered herein, but should be examined in the future.

In addition, since the focus of the present study was on the effects of market coupling and interconnection, little effort was placed in the forecasting technique, which was based on a relatively simple ANN. Forecasting accuracy could thus be improved via an improved network configuration obtained through experimentation and investigation of other training algorithms, different number of neurons in the hidden layer, etc. Moreover, forecasting accuracy could be improved via employment of other more advanced and at the same time complex techniques like deep ANNs and hybrid approaches (see e.g. Lago et al., 2018b).

Regardless the forecasting technique, forecasting of extreme (high or low) price values needs also further investigation, since their existence, as reported in literature (Weron, 2014; Nowotarski and Weron, 2018) and also confirmed by the investigations presented herein, may lead to poor forecasting results.

CONCLUSION

The liberalization of the electricity market and the pursued coupling at EU level have completely reshaped the landscape of the traditionally monopolistic and government-controlled power sectors and calls for revised and/or new approaches as far as the day ahead electricity price forecasting problem is concerned. This forecasting problem is not a new one, and several statistical, intelligence-based and other techniques have been proposed in literature to address it, at all required levels that is daily, weekly and annual. What is, however, still largely lacking, are methods that take into account the pursued market integration. Given that complete integration has not achieved yet, both relevant data and studies of forecasting considering integration are still rather sparse. Thus, most relevant studies focus mainly in methodological issues in an effort to develop and/or refine forecasting models and techniques, while little effort has been placed in investigating the effects that integration may have, and whether it is indeed useful and/or necessary to include in the forecasting model of a country inputs from the countries with which it is already or will eventually be integrated with.

It has been the aim of this chapter to focus on and study the integration effects in day ahead electricity price forecasting in an effort to contribute in filling the aforementioned gap. To this end, an Artificial Neural Network (ANN) has been developed, and several, with respect to inputs, forecasting scenarios have been used considering the Italian electricity market. The Italian electricity market is divided in zones that correspond to geographical areas, which, for reasons of system security, present limited electricity exchange between them. Based on its own supply and demand, an electricity price is determined daily for each of these zones, and a weighted, by zonal-purchased electricity quantity, average is then used to determine the national single price PUN, which is used to value the electricity demand bids of consuming units in Italy. At the same time, the Italian electricity market is coupled with the markets of Austria, France, Slovenia, and Switzerland, while coupling with Greece is still expected, although the two markets are already interconnected in the sense of electricity imports and/or exports. Due to these features, the Italian electricity market lends itself to investigations of market coupling and/or interconnection effects to the day ahead electricity price forecasting problem.

The performed investigations confirmed like other past studies that the consideration of market coupling and/or interconnection may indeed improve the day ahead electricity price forecasting accuracy. They also, however, suggested that the geographical proximity of considered markets may affect the forecasting results; the closer the markets, the stronger the effects. Thus market coupling effects need further investigation.

REFERENCES

Amjady, N. (2006). Day-ahead price forecasting of electricity markets by a new fuzzy neural. *IEEE Transactions on Power Systems*, *21*(2), 887–896. doi:10.1109/TPWRS.2006.873409

Amjady, N., & Hemmati, M. (2006). Energy price forecasting - problems and proposals for such predictions. *IEEE Power & Energy Magazine*, *4*(2), 20–29. doi:10.1109/MPAE.2006.1597990

Barroso, J. M. (2013). *Letter of President Barroso to the Members of the European Council*. MEMO/13/416. Retrieved January 3, 2020, https://europa.eu/rapid/press-release_MEMO-13-416_en.htm

Burden, F., & Winkler, D. (2008). Bayesian regularization of neural networks. *Methods in Molecular Biology (Clifton, N.J.)*, *458*, 25–44. doi:10.1007/978-1-60327-101-1_3 PMID:19065804

Carrion, M., Conejo, A., & Arroyo, J. (2007). Forward contracting and selling price determination for retailer. *IEEE Transactions on Power Systems*, *22*(4), 2105–2114. doi:10.1109/TPWRS.2007.907397

Catalao, J., Mariano, S., Mendes, V., & Ferreira, L. (2007). Short-term electricity prices forecasting in a competitive market: A neural network approach. *Electric Power Systems Research*, *77*(10), 1297–1304. doi:10.1016/j.epsr.2006.09.022

Cerjan, M., Krželj, I., Vidak, M., & Delimar, M. (2013). A literature review with statistical analysis of electricity price forecasting methods. In I. Kuzle, T. Capuder, & H. Pandžić (Eds.), *Eurocon 2013* (pp. 756–763). IEEE. doi:10.1109/EUROCON.2013.6625068

Cerjan, M., Petricic, A., & Delimar, M. (2019). HIRA model for short-term electricity. *Energies*, *12*(3), 568–600. doi:10.3390/en12030568

Conejo, A., Plazas, M., Espínola, R., & Molina, A. (2005). Day-ahead electricity price forecasting using the wavelet transform and ARIMA models. *IEEE Transactions on Power Systems*, *20*(2), 1035–1042. doi:10.1109/TPWRS.2005.846054

Contreras, J., Espínola, R., Nogales, F., & Conejo, A. (2003). ARIMA models to predict next-day electricity prices. *IEEE Transactions on Power Systems*, *18*(3), 1014–1020. doi:10.1109/TPWRS.2002.804943

Daskalou, K. (2019). *Forecasting of day ahead electricity prices in coupled markets. Implementation in the Italian day ahead market* (MBA Dissertation). Hellenic Open University, Patra, Greece.

Energy Exchange Group. (n.d.). *Price coupling of regions (PCR)*. Retrieved January 3, 2020, from http://www.enexgroup.gr/agores/agores-monteloy-stochoy/syzeyxi-perifereiakon-agoron/

Garcia, R., Contreras, J., Akkeren, M., & Garcia, J. (2003). A GARCH forecasting model to predict day-ahead electricity prices. *IEEE Transactions on Power Systems*, *2*, 867–874.

Gestore Mercati Energetici. (n.d.). *Zones*. Retrieved January 3, 2020, from https://www.mercatoelettrico.org/En/Mercati/MercatoElettrico/Zone.aspx

Hagan, M., & Menhaj, M. (1994). Training feedforward networks with the Marquardt algorithm. *IEEE Transactions on Neural Networks*, *5*(6), 989–993. doi:10.1109/72.329697 PMID:18267874

Hagan, M. T., Demuth, H. B., Beale, M. H., & De Jesús, O. (n.d.). *Neural network design* (2nd ed.). Retrieved January 3, 2020, from http://hagan.okstate.edu/NNDesign.pdf

Hecht-Nielsen, R. (1990). *Neurocomputing*. Addison-Wesley.

Jain, A., Mao, J., & Mohiuddin, K. (1996). Artificial neural networks: A tutorial. *IEEE Computer*, *29*(3), 31–44. doi:10.1109/2.485891

Keay, M. (2013). *The EU "Target Model" for electricity markets: fit for purpose?* Oxford Energy Comment, The Oxford Institute for Energy Studies. Retrieved from https://www.oxfordenergy.org/publications/the-eu-target-model-for-electricity-markets-fit-for-purpose/?v=f214a7d42e0d

Lago, J., Ridder, F., & Schutter, B. (2018a). Forecasting spot electricity prices: Deep learning approaches and empirical comparison of traditional algorithms. *Applied Energy*, *221*, 386–405. doi:10.1016/j.apenergy.2018.02.069

Lago, J., Ridder, F., Vrancx, P., & Schutter, B. (2018b). Forecasting day-ahead electricity prices in Europe: The importance of considering market integration. *Applied Energy*, *211*, 890–903. doi:10.1016/j.apenergy.2017.11.098

Lin, W.-M., Gow, H.-J., & Tsai, M.-T. (2010). An enhanced radial basis function network for short-term electricity price forecasting. *Applied Energy*, *87*(10), 3226–3234. doi:10.1016/j.apenergy.2010.04.006

MacKay, D. (1992). Bayesian Interpolation. *Neural Computation*, *4*(3), 415–447. doi:10.1162/neco.1992.4.3.415

Misiorek, A., Trueck, S., & Weron, R. (2006). Point and interval forecasting of spot electricity prices: Linear vs. non-linear time series models. *Studies in Nonlinear Dynamics and Econometrics*, *10*, 1–34.

Nogales, F., Contreras, J., Conejo, A., & Espínola, R. (2002). Forecasting next-day electricity prices by time series models. *IEEE Transactions on Power Systems*, *17*(2), 342–348. doi:10.1109/TPWRS.2002.1007902

Nowotarski, J., & Weron, R. (2018). Recent advances in electricity price forecasting: A review of probabilistic forecasting. *Renewable & Sustainable Energy Reviews*, *81*, 1548–1568. doi:10.1016/j.rser.2017.05.234

Panapakidis, I., & Dagoumas, A. (2016). Day-ahead electricity price forecasting via the application of artificial neural network based models. *Applied Energy*, *172*, 132–151. doi:10.1016/j.apenergy.2016.03.089

Pepermans, G. (2019). European energy market liberalization: Experiences and challenges. *International Journal of Economic Policy Studies*, *13*(1), 3–26. doi:10.100742495-018-0009-0

Rutkowski, L. (2008). *Computational intelligence: Methods and techniques*. Springer-Verlag. doi:10.1007/978-3-540-76288-1

Saini, L. (2008). Peak load forecasting using Bayesian regularization: Resilient and adaptive backpropagation learning based artificial neural networks. *Electric Power Systems Research*, *78*(7), 1302–1310. doi:10.1016/j.epsr.2007.11.003

Schalkoff, R. J. (1997). *Artificial neural networks*. McGraw-Hill.

Shafie-khah, M., Parsa Moghaddam, M., & Sheikh-El-Eslami, M. K. (2010). Price forecasting of day-ahead electricity markets using a hybrid forecast method. *Energy Conversion and Management*, *52*(5), 2165–2169. doi:10.1016/j.enconman.2010.10.047

Swider, D., & Weber, C. (2007). Extended ARMA models for estimating price developments on day-ahead electricity markets. *Electric Power Systems Research*, *77*(5-6), 583–593. doi:10.1016/j.epsr.2006.05.013

Weron, R. (2014). Electricity price forecasting: A review of the state-of-the-art with a look into the future. *International Journal of Forecasting*, *30*(4), 1030–1081. doi:10.1016/j.ijforecast.2014.08.008

Xiao, L., Shao, W., Yu, M., Ma, J., & Jin, C. (2017). Research and application of a hybrid wavelet neural network model with the improved cuckoo search algorithm for electrical power system forecasting. *Applied Energy*, *198*, 203–222. doi:10.1016/j.apenergy.2017.04.039

Yang, Z., Ce, L., & Lian, L. (2017). Electricity price forecasting by a hybrid model, combining wavelet transform, ARMA and kernel-based extreme learning machine methods. *Applied Energy*, *190*, 291–305. doi:10.1016/j.apenergy.2016.12.130

Zhang, J.-L., Zhang, Y.-J., Li, D.-Z., Tan, Z.-F., & Ji, J.-F. (2019). Forecasting day-ahead electricity prices using a new integrated model. *International Journal of Electrical Power & Energy Systems*, *105*, 541–548. doi:10.1016/j.ijepes.2018.08.025

Ziel, F., Steinert, R., & Husmann, S. (2015). Forecasting day ahead electricity spot prices: The impact of the EXAA to other European electricity markets. *Energy Economics*, *51*, 430–444. doi:10.1016/j.eneco.2015.08.005

ADDITIONAL READING

Brockwell, P. J., & Davis, R. A. (2016). *Introduction to time series and forecasting* (3rd ed.). Springer. doi:10.1007/978-3-319-29854-2

Haykin, S. (2009). *Neural networks and learning machines* (3rd ed.). Prentice-Hall.

Joskow, P. L. (2006). Introduction to electricity sector liberalization: Lessons learned from cross-country studies. In F. P. Sioshansi, & W. Pfaffenberger, (Eds.), Elsevier Global Energy Policy and Economics Series, Electricity Market Reform (pp. 1-32). Elsevier. Retrieved January 3, 2020, from doi:10.1016/B978-008045030-8/50002-3

Newbery, D., Strbac, G., & Viehoff, I. (2016). The benefits of integrating European electricity markets. *Energy Policy*, *94*, 253–263. doi:10.1016/j.enpol.2016.03.047

Pollitt, M. G. (2019). The European single market in electricity: An economic assessment. *Review of Industrial Organization*, *55*(1), 63–87. Retrieved January 3, 2020, from. doi:10.100711151-019-09682-w

The European Parliament and the Council of the European Union. (1997). Directive 96/92/EC concerning common rules for the internal market in electricity. *Official Journal L 027*, 20-29. Retrieved January 3, 2020, from https://eur-lex.europa.eu/legal-content/EN/TXT/?uri=CELEX:31996L0092

The European Parliament and the Council of the European Union. (2003). Directive 2003/54/EC concerning common rules for the internal market in electricity and repealing Directive 96/92/EC. *Official Journal L 176*, 37-56. Retrieved January 3, 2020, from https://eur-lex.europa.eu/legal-content/EN/TXT/?uri=CELEX%3A32003L0054

The European Parliament and the Council of the European Union. (2009). Directive 2009/72/EC concerning common rules for the internal market in electricity and repealing Directive 2003/54/EC. *Official Journal L 211*, 55-93. Retrieved January 3, 2020, from https://eur-lex.europa.eu/legal-content/EN/ALL/?uri=CELEX%3A32009L0072

KEY TERMS AND DEFINITIONS

Artificial Intelligence: A diverse group of nature-inspired computational techniques that combine elements of learning, evolution, and fuzziness to create flexible models that can handle complexity and non-linearity of the studied phenomena.

Artificial Neural Network: Structures comprised of densely interconnected simple processing elements, the artificial neurons, or nodes, which are capable of performing massively parallel computations and knowledge representation.

Day Ahead Electricity Price: The spot electricity prices of the day ahead market.

Electricity Market Coupling: A way of linking day ahead spot markets into a virtual market, so that the lowest priced bids are accepted up to the point where congestion constraints limit further trade.

EU Electricity Target Model: An EU operation framework aimed at leading to a single internal electricity market, which is competitive, secure and sustainable.

Forecasting: The process of making predictions of the future by analyzing the trends, which are present in time series of given historical data.

PUN Price: The national single price (Prezzo Unico Nazionale) used to value the electricity demand bids of consuming units in Italy.

Chapter 2
Multi-Attribute Decision Making Using Interval Type-2 Fuzzy AHP and Its Application on Borsa Istanbul (BIST)

Beyza Ahlatcioglu Ozkok
Yildiz Technical University, Turkey

Hale Gonce Kocken
https://orcid.org/0000-0003-1121-7099
Yildiz Technical University, Turkey

ABSTRACT

Analytic hierarchy process (AHP) is a widely used multi-attribute decision-making (MADM) approach. Due to the complexity and uncertainty involved in real world problems, decision makers might be prefer to make fuzzy judgments instead of crisp ones. Furthermore, even when people use the same words, individual judgments of events are invariably subjective, and the interpretations that they attach to the same words may differ. This is why fuzzy numbers has been introduced to characterize linguistic variables. Fuzzy AHP methods have recently been extended by using type-2 fuzzy sets. Type-2 fuzzy set theory incorporates the uncertainty of membership functions into the fuzzy set theory. In this chapter, the authors firstly provide a short review on applications of interval type-2 fuzzy AHP on MADM problems. Then, they present a very efficient MADM technique, interval type-2 fuzzy AHP, to solve the portfolio selection problem that is to decide which stocks are to be chosen for investment and in what proportions they will be bought. And finally, they provided a case study on BIST.

DOI: 10.4018/978-1-7998-5442-5.ch002

INTRODUCTION

Decision-making is a very important part of our lives, which often involves many and contradictory criteria that we face at any time in our daily lives. Multi-criteria decision making (MCDM) problems can be examined in two categories: multi-attribute decision making (MADM) and multi-objective decision making (MODM).

AHP is a MCDM method that has a wide range of applications and can be easily applied to real-life problems that allow weighting alternatives when choosing from existing alternatives. The AHP, which many researchers often use to address selection problems, is criticized for whether it reflects the uncertainty of the problems addressed in the solution process sufficiently because decision makers express their evaluations in exact or crisp judgments (numbers). However, decision-makers may not be willing to make crisp and precise evaluations when making their decisions due to uncertainty in real-life problems. Moreover, even if decision makers use the same words for their evaluation, the meanings they add to these words may be different from each other. Furthermore, even if an evaluation is well defined, the limit criterion that determines whether this evaluation belongs to the set is often not clear or ambiguous. Therefore, fuzzy numbers and fuzzy sets have been proposed to define linguistic variables which cannot be expressed by numbers. A linguistic variable is a variable expressed in words or sentences. The evaluations in AHP can be more accurately reflected in the process with fuzzy decision-making approaches because they are based on human judgment. (Chen et al. (2008), Tiryaki and Ahlatcioglu (2009), Ozkok and Pappalardo (2013)).

Portfolio selection problem, which has been handled by many researchers since 1950s, is one of the most important multi attribute decision making problems of modern finance. In the literature, there are several approaches to select the best portfolio for the investors. The first of the solution methods developed for the portfolio selection problem was the mean variance model proposed by Markowitz (1952). Markowitz's mean variance model has pioneered many other portfolio selection models in the literature, most notably the Capital Asset Pricing Model (CAPM), the Sharpe-Lintner model, the black model, and the two-factor model (Alexander et al. (2003), Tiryaki and Ahlatcioglu (2009)). In this part of our study, we summarize some important mathematical models proposed in the literature for the portfolio selection problem. Tanaka and Guo (1999), using two possibility distributions, the upper and lower possibility distributions, reflected expert evaluations for the portfolio selection problem. To address the portfolio selection problem, Xia et al. (2000) proposed a methodology based on genetic algorithms. In Tanaka et al. (2000), two portfolio selection models based on fuzzy probabilities and probability distributions have been proposed. (2000). Inuiguchi and Tanino (2000) developed a probabilistic programming method based on the worst regret in portfolio selection. Parra et al. (2001) developed a fuzzy goal programming approach and considered return, risk and liquidity as the criteria to be reflected in the model. Ong et al. (2005) proposed a portfolio selection model based on gray and probabilistic regression models. Tiryaki and Ahlatcıoğlu (2005) implemented a new ranking and weighting method to determine the most rational portfolio for investors from ISE stocks. Lacagnina and Pecorella (2006) have developed a method for portfolio selection problem that uses soft constraints based on stochastic fuzzy programming. Huang et al. (2006), to determine the optimal portfolio selection under an uncertain situation, they revisited the conventional mean-variance method. Terol et al. (2006) presented a portfolio selection model based on Sharpe's single index model and using Fuzzy Compromise Programming. Giove et al. (2006), in their studies have considered the prices of securities as interval variables. Zhang et al. (2007) proposed two types of portfolio selection models based on lower and upper possibilistic means and possibilistic variance.

In Tiryaki and Ahlatcioglu (2009), they combined the fuzzy AHP with the portfolio selection problem. In order to measure the asymmetry of fuzzy portfolio return, a concept of skewness is defined as the third central moment in Li et al. (2010), and its mathematical properties are studied. In Bhattacharyya et al. (2011), they utilized the concept of interval numbers in fuzzy set theory to extend the classical mean-variance portfolio selection model into a mean-variance–skewness model with consideration of transaction cost. Based on the membership function, Li et al. (2015) redefined the concepts of mean and variance for fuzzy numbers, also proposed the concept of distortion, and formulated a fuzzy mean-variance-skewness portfolio selection model. They studied the stock portfolio selection problem based on varying conservative-neutral-aggressive attitudes in Zhou et al. (2018). In their paper, the return rates of stocks are characterized by fuzzy variables. The Pareto-optimal solutions are obtained by maximizing the return and minimizing the risk subject to constraints of transaction cost and value at risk.

As can be seen in the development of literature for portfolio selection problem, as in many decision problems, fuzzy decision making is a very useful tool. Type-1 fuzzy sets were developed in 1965 by Zadeh to model the uncertainties and imprecise data. Type-1 methods have been applied in many different areas such as economics, engineering, business, etc. In 1975 Zadeh extended type-1 fuzzy sets to type-2 fuzzy sets to handle uncertainty in a better way.

In this chapter, after the introduction section, we present a general look to MADM using type-1 fuzzy AHP and provide a short review on applications of type-1 fuzzy AHP. After presenting the steps of type-1 fuzzy AHP, interval type-2 fuzzy AHP on MADM problems and its applications handle. We also present a very efficient multi-attribute technique: interval type-2 fuzzy AHP to solve the portfolio selection problem that is to decide which stocks are to be chosen for investment and in what proportions they will be bought and we provide a case study on Borsa Istanbul (BIST).

MADM USING TYPE-1 FUZZY AHP

A Short Review on Type-1 Fuzzy AHP

There is a vast literature for fuzzy AHP in various application areas. In this subsection, we aim to present a short review of type-1 fuzzy AHP (Ozkok and Pappalardo (2013)). Laarhoven and Pedrycz (1983) presented is the pioneer study in this research field. The fuzzy weights and fuzzy performance scores are obtained by using Lootsma's logarithmic least square method and in this paper, all the evaluations in pairwise comparison matrices are taken as triangular fuzzy numbers. And the authors state that the proposed method can also be used to aggregate multiple decision makers' judgments. After this study, in Buckley (1985), all the evaluations in pairwise comparison matrices are made by trapezoidal fuzzy numbers and Laarhoven and Pedrycz's method is discussed in two aspects: (1) the proposed method can be used with only triangular fuzzy numbers. (2) The system of linear equations in the proposed method may have no solution. Laarhoven and Pedrycz's method also criticized in Boender et al. (1989), and the authors proposed a modification for this method. Ruoning and Xiaoyan (1992) gave a fuzzy extension of Saaty (1980)'s AHP. In this study, step-form fuzzy numbers are used and they showed the fuzzy AHP pairwise decision matrices are also perfectly consistent as Saaty's original AHP. After this study, Buckley et al. (2001) discussed the method in Ruoning and Xiaoyan (1992) from the perspective of not being the matrices perfectly consistent but only reasonably consistent. The authors stated that for crisp data, different weights will be obtained compared to Saaty's original method. By using fuzzy relational equations, the fuzzy AHP is modelled in Mohanty and Singh (1994). A new and general methodology for fuzzy AHP is proposed by Mon et al. (1994), which is based on entropy weight. This paper also provided a real-life application on the weapon systems. Cheng and Mon (1994) developed a new algorithm to evaluate weapon systems. Their algorithm is based on fuzzy AHP whose pairwise comparison matrices comprise of triangular fuzzy numbers. Wang et al. (1995) gave a new definition of consistency of the pairwise comparison matrices in the fuzzy AHP and all fuzzy weights are obtained under the condition of consistency. Ruoning and Xiaoyan (1996) focused on the main ranking method in AHP, namely the logarithmic least squares method. They discussed in detail to obtain the fuzzy weights from the fuzzy pairwise comparison matrices. In Stam et al. (1996), two artificial neural network formulations are developed to evaluate the preference ratings from the judgment matrices. The authors also emphasized the importance and power of applying the feed-forward neural network formulations to analyze the multi-criteria decision problems with imprecise or fuzzy preference judgments for selecting discrete alternatives. Chang (1996) focused on the fuzzy AHP with triangular fuzzy numbers to make the evaluations of criteria and alternatives and developed a new method by introducing the synthetic extent analysis of the fuzzy pairwise comparisons. Chen (1996) provided a case study for the performance evaluation of weapon systems by using tactical missile systems selection. The AHP method presented in Chen (1996) is based on simplified fuzzy arithmetic operations and the author claimed that the execution of its method is faster than the method in Mon et al. (1994) that uses the complicated entropy weight calculations. Cheng (1997) also focused on a similar case study. The fuzzy AHP in Cheng (1997) is based on the grade value of membership function and used for evaluating naval tactical missile systems. A common feature of Cheng (1997), Cheng and Mon (1994) and Mon et al. (1994) is changing a positive fuzzy pairwise comparison matrix into a crisp matrix using alpha cut concept and convex combinations, and then computing the eigenvector by using a matrix whose all entries are crisp numbers. In other words, these methods do not result in a fuzzy vector. Buckley et al. (2001) presented a criticism about these papers since the result of the method generates a crisp solution vector, not a fuzzy one. Gogus and Boucher (1997) gave an analysis of using fuzzy pairwise comparisons in multi-criteria decision making. For their analysis, they focused on the problem in Van Laarhoven and Pedrycz (1983) and Boender et al. (1989). They also state that in these earlier articles, normalized weights obtained from evaluations based on ratio scales can exhibit an irrational ordering. Weck et al. (1997) combined the traditional AHP method with fuzzy mathematics to utilize different production cycle alternatives. Any production cycle utilized by this method results in a fuzzy set. Defuzzification of the final outcome can be made by determining the center of gravity of any fuzzy set, and the alternative production cycles can be ordered according to the main objective set. In Yang and Lee (1997), an AHP decision model for facility location selection is developed for the organizations searching for a site for a new facility, or relocation of an existing one. The main principle of the developed AHP model is providing flexibility to match the preferences with the site characteristics, to combine the decision makers' managerial experience and judgment. For fuzzy judgment matrices in AHP, Gogus and Boucher (1998) redefined the strong transitivity and the weak monotonicity, which are two important axioms in the theory of value and utility measurement. Using the strong transitivity, they developed a measuring system to check the consistency when preferences are given by fuzzy numbers. Kwiesielewicz (1998) proved that the fuzzy pairwise-comparison problem by Laarhoven and Pedrycz (1983) may be decomposed into a modal values problem and an interval analysis, which may be solved separately. And based on this decomposition, they developed a pseudo-inverse approach which gives a possibility to calculate a solution in the general case and provides the condition to check the existence of the fuzzy solution. Lee et al. (1999) discussed the main principles of AHP and its various generalizations. They introduced the concept of comparison interval to achieve consistency and to accommodate the fuzzy nature of the comparison process by using a stochastic method. And they proposed an iterative approximation method to deal with the interdependence in the super-matrix technique. Cheng et al. (1999) presented an application on evaluating weapon systems by using a fuzzy AHP methodology that uses linguistic variables. Considering the main idea of fuzzy AHP is ranking the fuzzy numbers, Zhu et al. (1999) focused on the theory of triangular fuzzy numbers and present an improvement for the comparison of them. They implemented this improvement to the extent analysis method and mentioned about the applications of AHP. Xu (2000) focused on finding the weights of alternatives from a fuzzy pairwise comparison matrix by the least square method. Csutora and Buckley (2001) dealt with the λm_{ax} method and presented a direct fuzzification of this method to obtain the final fuzzy weights in fuzzy AHP. Revisiting the form of the Csutora and Buckley (2001), Buckley et al. (2001) provided a test to their new method by using crisp weights. Integrating a fuzzy rating expression with an absolute term linearization technique, Yu (2002) presented a Goal Programming-AHP model for a group decision making problem by treating an AHP problem as a goal programming problem. Mikhailov (2003) focused on the alpha-cut decomposition of a fuzzy number and presented a series of interval comparisons. Using these comparisons, a new approach is proposed to obtain the final weights from fuzzy evaluations matrices in fuzzy AHP. Kahraman et al. (2003) and Bozdag et al. (2003) presented four different fuzzy multi-attribute techniques for a group decision making problem. These methods are based on the fuzzy model of group decision proposed by Blin, the fuzzy synthetic evaluation, the Yager's weighted goals method, and the fuzzy AHP. Wang and Lin (2003) gave a multi-criteria group decision-making model for the selection of configuration items in software development. This model constructs imprecise preference relations among candidate items based on the concept of fuzzy majority and determines ranking orders among them. Mikhailov (2004b) presented a new group decision making methodology based on fuzzy preference programming to obtain group priorities from crisp pairwise comparison matrices. Buyukozkan et al. (2004) focused on the difficulties that a decision-maker faces in the uncertainty and vagueness from subjective perception and experience in the decision process and proposed a new procedure to use a selection framework based on fuzzy multi-criteria evaluations by considering both economic and quality factors, and finally provided an application on software development strategy selection. Mikhailov (2004a) proposed a new fuzzy programming approach to obtain priorities from interval pairwise comparison matrices and conclude that the proposed method has substantial potential for solving complex decision making problems under uncertainty. Mikhailov and Tsvetinov (2004) provided an application on the service evaluation process under an uncertain environment. Kahraman et al. (2004) presented a fuzzy optimization model which is based on the fuzzy quality function deployment method to determine the technical product requirements of a new product. In the optimization model, fuzzy analytic network process is used to determine the coefficients of the objective function. Enea and Piazza (2004) dealt with the problem of selection of a project among a set of possible alternatives by presenting a fuzzy extension of AHP. Their approach aimed to consider all the available information and it is showed that considering all the information deriving from the constraints results better in terms of certainty and reliability. Kulak and Kahraman (2005) provided an integrated method of AHP and multi-attribute axiomatic design in a fuzzy environment for the problem of multi-attribute transportation company selection. The corelations between competitive advantages, competitive priorities, and competencies of a firm in the context of technology management are investigated by fuzzy AHP in Erensal et al. (2006). Golec and Taskin (2007) compared some complex fuzzy methodologies with fuzzy AHP for evaluating the performance of a manufacturing system and concluded that the best result is obtained by fuzzy AHP. In Tiryaki and Ahlatcioglu (2009), the methods of Enea and Piazza (2004) are modified with the aim of ranking a set of stocks in a fuzzy environment and demonstrated the usefulness of fuzzy methodology in financial problems. In Dağdeviren et al. (2009), a real world application is conducted for the weapon selection problem by using fuzzy AHP integrated with fuzzy TOPSIS. Güngör et al. (2009) and Kilincci et al. (2011) presented real life applications of fuzzy AHP on personnel and supplier selection, respectively. To help the industrial practitioners for the performance evaluation in a fuzzy environment, Sun (2010) developed an integrated technique of fuzzy AHP and fuzzy TOPSIS where linguistic variables are taken as triangular fuzzy numbers. Under fuzzy environment, Kutlu and Ekmekçioğlu (2012) provided a hypothetical case study on failure mode and effects analysis which is for designing, identifying and eliminating known and/or potential failures, by using fuzzy TOPSIS method integrated with fuzzy AHP. Shaw et al. (2014) developed an integrated approach for the problem of supplier selection addressing the carbon emission issue, using fuzzy-AHP and fuzzy multi-objective linear programming. After applying fuzzy AHP for analyzing the weights of the multiple factors, these weights are used in fuzzy multi-objective linear programming for supplier selection and quota allocation. Junior et al. (2014) presented a comparative analysis of Fuzzy TOPSIS and Fuzzy AHP methods concerning the problem of supplier selection based on a set of seven criteria. The authors concluded that Fuzzy TOPSIS performed better in regard to changes of alternatives and criteria, agility and number of criteria and alternatives. Chou et al (2019) investigated the human resource in science and technology performance of Southeast Asian countries by using AHP and TOPSIS under fuzzy environment. Focusing on the hospital management, Singh and Prasher (2019) presented an application of fuzzy AHP with the view of service quality to find out the priorities of healthcare service quality attributes.

Type-1 Fuzzy Sets

In this section, brief information about the trapezoidal fuzzy numbers and some basic definitions are presented (Zimmermann (2001)).

Definition 1: If X is a collection of objects denoted by x then a fuzzy set $\tilde{A}$ in X is a set of ordered pairs: $\tilde{A} = \{(x, \mu_{\tilde{A}}(x)) | x \in X\}$ where $\mu_{\tilde{A}}(x)$ is called the membership function or grade of membership of x in $\tilde{A}$ which maps X to the membership space M (M contains [0,1]).

Definition 2: A fuzzy number $\tilde{A} = (a,b,c,d)$ is said to be an arbitrary trapezoidal fuzzy number if its membership function is given by

$$\mu_{\tilde{A}}(x) = \begin{cases} \dfrac{x-a}{b-a} & , a \le x < b \\ 1 & , b \le x \le c \\ \dfrac{d-x}{d-c} & , c < x \le d \\ 0 & , \text{Otherwise} \end{cases}$$

where $a, b, c, d \in R$ and $a \leq b \leq c \leq d$.

Definition 3: An arbitrary trapezoidal fuzzy number $\tilde{A} = (a,b,c,d)$ is called positive (negative), denoted by $\tilde{A} > 0$ $(\tilde{A} < 0)$ if its membership function $\mu_{\tilde{A}}(x) = 0, \forall x < 0$ $(\forall x > 0)$.

Definition 4: Two arbitrary trapezoidal fuzzy numbers $\tilde{A} = (a,b,c,d)$ and $\tilde{B} = (e,f,g,h)$ are said to be equal if and only if $a=e$, $b=f$, $c=g$ and $d=h$.

Definition 5: A ranking function *R* is a mapping from a fuzzy set *F*(*R*) to real numbers where a natural order exists. Let $\tilde{A} = (a,b,c,d)$ be a trapezoidal fuzzy number then $R(\tilde{A}) = \dfrac{a+b+c+d}{4}$.

Definition 6: Let $\tilde{A} = (a,b,c,d)$ and $\tilde{B} = (e,f,g,h)$ be arbitrary trapezoidal fuzzy numbers then

(i) $\tilde{A} \prec \tilde{B}$ iff $R(\tilde{A}) < R(\tilde{B})$

(ii) $\tilde{A} \succ \tilde{B}$ iff $R(\tilde{A}) > R(\tilde{B})$

(iii) $\tilde{A} \cong \tilde{B} \Rightarrow R(\tilde{A}) = R(\tilde{B})$

Arithmetic Operations

Some algebraic operations on trapezoidal fuzzy numbers are defined as follows:

Let $\tilde{A} = (a,b,c,d)$ and $\tilde{B} = (e,f,g,h)$ be arbitrary two triangular fuzzy numbers,

(i) $\tilde{A} + \tilde{B} = (a+e,\ b+f,\ c+g, d+h)$

(ii) $-\tilde{A} = (-d,-c,-b,-a)$

(iii) $\tilde{A} - \tilde{B} = (a-h,\ b-g,\ c-f, d-e)$.

(iv) The fuzzy multiplication based on the extension principle is performed via the following equation:

$$\tilde{A} * \tilde{B} = (l, m^1, m^2, u)$$

where l=min{ae, ah, de, dh}, m^1=min{bf, bg, cf, cg}, m^2=max{bf, bg, cf, cg} and u=max{ae, ah, de, dh}.

Definition 7: Suppose $\tilde{A}, \tilde{B} \in F(R)$ then $R(\tilde{A} + \tilde{B}) = R(\tilde{A}) + R(\tilde{B})$.

This property can be generalized for *n* fuzzy numbers $\tilde{A}_1, \tilde{A}_2, \ldots, \tilde{A}_n$ i.e.

$$R\left(\sum_{j=1}^{n} \tilde{A}_j\right) = \sum_{j=1}^{n} R(\tilde{A}_j).$$

Buckley's Fuzzy AHP

Firstly, the 1-9 scale in AHP proposed by Saaty was fuzzified by Van Laarhoven and Pedrycz (1983) by expressing it with fuzzy triangular numbers. In their study, evaluations of decision makers for pairwise comparison matrices are expressed with fuzzy triangular numbers. The comparison and solution steps of the fuzzy AHP proposed by Laarhoven and Pedrycz are the same as Saaty's AHP. In doing so, they used Lootsma's logarithmic least square method to obtain fuzzy weights and fuzzy performance scores. However, Buckley (1985) criticized the method of Laarhoven and Pedrycz with the view of the linear equation systems that obtained during the solution procedure are not always has a solution and that their decision-making evaluations can be incorporated into the solution procedure with only triangular fuzzy numbers. Boender et al. (1989) studied how Laarhoven and Pedrycz's fuzzy AHP can be corrected by identifying errors in the solution approach. Buckley (1985) proposed a fuzzy AHP method using trapezoidal fuzzy numbers, which is the more general form of triangular fuzzy numbers for fuzzy AHP. In this method, the geometric mean was used to obtain fuzzy weights and performance scores.

In this section, Buckley (1985) the three solution steps are explained (Chen et al. (1992)):

Step 1: Consult the decision maker and obtain the comparison matrix A whose elements $\overline{a}_{ij} = \left(a_{ij}, b_{ij}, b_{ij}, d_{ij}\right)\ \forall\ i, j$ are trapezoidal fuzzy numbers.

Step 2: The fuzzy weights w_i can be calculated as follows. The geometric mean for each row is determined as:

$$z_i = \left[a_{i1} \odot \ldots \odot a_{in}\right]^{1/n}\ \forall i \tag{1}$$

where the sign "⊙" represents fuzzy multiplication. The fuzzy weight w_i is given as

$$w_i = z_i \odot \left(z_1 \oplus \ldots \oplus z_n\right)^{-1} \tag{2}$$

where the sign "⊕"represents fuzzy addition. In the following discussion will detail the derivation of fuzzy weight w_i. Let the left leg and right leg of $\overline{a}_{ij}$ defined as:

$$g_i(\alpha) = \left[\prod_{j=1}^{n}\left((c_{ij} - d_{ij})\alpha + b_{ij}\right)\right]^{1/n},\ \alpha \in [0,1] \tag{3}$$

$$f_i(\alpha) = \left[\prod_{j=1}^{n}\left((b_{ij} - a_{ij})\alpha + a_{ij}\right)\right]^{1/n},\ \alpha \in [0,1] \tag{4}$$

respectively. Furthermore, let

$$a_i = \left[\prod_{j=1}^{n} a_{ij} \right]^{1/n} \tag{5}$$

and

$$a = \sum_{i=1}^{m} a_i \tag{6}$$

Similarly, we can define b_i and b, c_i and c, and d_i and d. The fuzzy weight w_i is determined as:

$$w_i = \left(\frac{a_i}{d}, \frac{b_i}{c}, \frac{c_i}{b}, \frac{d_i}{a} \right), \forall i \tag{7}$$

where the membership function $\mu_{w_i}(x)$ is defined as follows let x be a real number on the horizontal axis. The $\mu_{w_i}(x)$ can be summarized as

$$\mu_{W_i}(x) = \begin{cases} 0 & , x \le a_i / d \\ 1 & , b_i / c < x \le c_i / b \\ \alpha \in [0,1] & , a_i / d < x \le b_i / c \\ \alpha \in [0,1] & , c_i / b < x \le d_i / a \\ 0 & , x > d_i / a \end{cases}$$

When $x \in [a_i/d, b_i/c]$ the x is calculated as:

$$x = f_i(\alpha) / g(\alpha) \tag{8}$$

When $x \in [c_i/b, d_i/a]$ the x is determined as:

$$x = g_i(\alpha) / f(\alpha) \tag{9}$$

where

$$f(\alpha) = \sum_{i=1}^{m} f_i(\alpha) \tag{10}$$

$$g(\alpha) = \sum_{i=1}^{m} g_i(\alpha) \tag{11}$$

Step 2 is repeated until the fuzzy performance scores r_{ij}, $\forall i,j$, are obtained in a similar manner.

Step 3: The fuzzy weights and fuzzy performance scores are aggregated as a fuzzy MADM problem. The fuzzy utilities U_i, $\forall_i$ are obtained based on

$$U_i = \sum_{j=1}^{n} w_j r_{ij}, \forall i \tag{12}$$

MADM USING TYPE-2 FUZZY AHP

A Short Review on Type-2 Fuzzy AHP

In Type 1 fuzzy sets, elements have memberships within the interval [0,1] assigned by membership functions. Zadeh (1965) introduced Type-2 Fuzzy sets as an extension of Type-1 fuzzy sets. The most striking difference of this proposed expansion is that the membership functions of type-1 fuzzy sets are two-dimensional, whereas the membership functions of type-2 fuzzy sets are three-dimensional. The third dimension introduced here is the additional degrees of freedom, which allows for direct handling of uncertainty. (Kahraman et al. (2014)). Using interval Type-2 fuzzy sets, Type-2 fuzzy AHP was first introduced to the literature by Kahraman et al. (2014). Taking advantage of the flexibility of Type-2 fuzzy sets to express uncertainty, the linguistic scale used in pairwise comparison matrices is expressed with Type-2 fuzzy numbers. They also proposed a new defuzzification method that can be used in triangular and trapezoidal Type-2 fuzzy sets. Çelik, Gümüş and Alegoz (2014) identified and evaluated the success factors critical to the effective management of humanitarian relief logistics operations. With the help of Type-2 fuzzy AHP, which uses Type-2 fuzzy sets that reflect the uncertainty arising from this problem to the model with the additional degree of freedom, they have developed a powerful solution method in modeling effective and the fuzziness of real-life. Abdullah and Najib (2014) have proposed interval type-2 fuzzy sets to facilitate and enable decision-making in decision-making environments involving fuzziness. This model includes linguistic variables in interval type-2 fuzzy sets and a rank value method for normalizing upper and lower memberships of interval type-2 fuzzy sets. To illustrate the applicability of their methods, they presented a numerical example for work safety evaluation. Cevik Onar, Öztayşi and Kahraman (2014) developed an MCDM method to address the uncertainty and complexity inherent in the nature of strategic decisions. In the first step, Type-2 fuzzy AHP was used to determine the weights of the factors and they tried to select the best strategy with the help of Hesitant Fuzzy TOPSIS using the weights obtained Öztaysi (2015) studied the problem of EIS selection which is a multi-criteria decision-making problem. In this study, they discussed how fuzzy AHP can handle group decision situations using interval type-2 sets and applied the method they developed for selecting an ERP with six criteria and four alternatives. The originality of this study is the use of interval type-2 fuzzy sets for the decision problem discussed. In group decision-making, type-2 fuzzy sets provides more efficient results than type-1 fuzzy sets, mainly because type-2 fuzzy sets address uncertainty better than type 1 fuzzy sets. Abdullah and Zulkifli (2015) proposed an integrated model that includes linguistic variables in interval type-2 fuzzy sets and expected value for normalizing upper and lower memberships for fuzzy AHP and interval type-2 fuzzy DEMATEL. The proposed method is implemented by a case study in human resources management. Gul et al. (2016) proposed an integration of computer simulation, MCDM methods of interval type-2 fuzzy AHP, and ELECTRE for an Emergency department system in a university hospital. Due to the high uncertainty in Emergency department operations, interval type-2 fuzzy sets are used in the MCDM process. Celik and Akyuz (2018) presented a comprehensive MCDM method by integrating AHP and TOPSIS under the interval type-2 fuzzy sets environment. Moreover, it provides an application to make an investment decision for an appropriate ship loader type in a dry bulk cargo shipment.

Type-2 Fuzzy Sets

In this section, brief information about interval type-2 fuzzy sets is presented (Kahraman et al. (2014)).

Definition 8: A type-2 fuzzy set $\tilde{\tilde{A}}$ in the universe of discourse X can be represented by a type-2 membership function $\mu_{\tilde{\tilde{A}}}(x)$, shown as follows:

$$\tilde{\tilde{A}} = \left\{ \left((x,u), \mu_{\tilde{\tilde{A}}}(x,u) \right) \middle| \forall x \in X, \middle| \forall u \in J_x \subseteq [0,1], 0 \le \mu_{\tilde{\tilde{A}}}(x,u) \le 1 \right\}$$

where $J_x \in [0,1]$ denotes an interval [0, 1]. The type-2 fuzzy set $\tilde{\tilde{A}}$ also can be represented as follows:

$$\tilde{\tilde{A}} = \int_{x \in X} \int_{u \in J_x} \mu_{\tilde{\tilde{A}}}(x,u) / (x,u)$$

where $J_x \subseteq [0,1]$ and $\iint$ denotes union overall admissible x and u.

Definition 9: The upper and lower membership functions of an interval type-2 fuzzy set are type-1 membership functions. A trapezoidal interval type-2 fuzzy set is illustrated as

$$\tilde{\tilde{A}}_i = \left(\tilde{A}_i^U ; \tilde{A}_i^L \right) = \left(\left(a_{i1}^U, a_{i2}^U, a_{i3}^U, a_{i4}^U ; H_1\left(\tilde{A}_i^U\right), H_2\left(\tilde{A}_i^U\right) \right), \left(a_{i1}^L, a_{i2}^L, a_{i3}^L, a_{i4}^L ; H_1\left(\tilde{A}_i^L\right), H_2\left(\tilde{A}_i^L\right) \right) \right)$$

where $\tilde{A}_i^U$ and $\tilde{A}_i^L$ are type-1 fuzzy sets, $a_{i1}^U, a_{i2}^U, a_{i3}^U, a_{i4}^U, a_{i1}^L, a_{i2}^L, a_{i3}^L$ and a_{i4}^L are the reference points of the interval type-2 fuzzy set $\tilde{\tilde{A}}_i$, $H_j\left(\tilde{A}_i^U\right)$; denotes the membership value of the element $a_{i(j+1)}^U$ in the upper trapezoidal membership function $\tilde{A}_i^U$, $1 \le j \le 2$, $H_j\left(\tilde{A}_i^L\right)$; $1 \le j \le 2$ denotes the membership value of the element $a_{i(j+1)}^{L}$ in the lower trapezoidal membership function $\tilde{A}_i^L$, $1 \le j \le 2$, $H_1\left(\tilde{A}_i^U\right) \in [0,1]$, $H_2\left(\tilde{A}_i^U\right) \in [0,1]$, $H_1\left(\tilde{A}_i^L\right) \in [0,1]$, $H_2\left(\tilde{A}_i^L\right) \in [0,1]$ and $1 \le i \le n$ (See Figure 1).

Figure 1. The membership function of trapezoidal interval type-2 fuzzy set (Kahraman et al. (2014)).

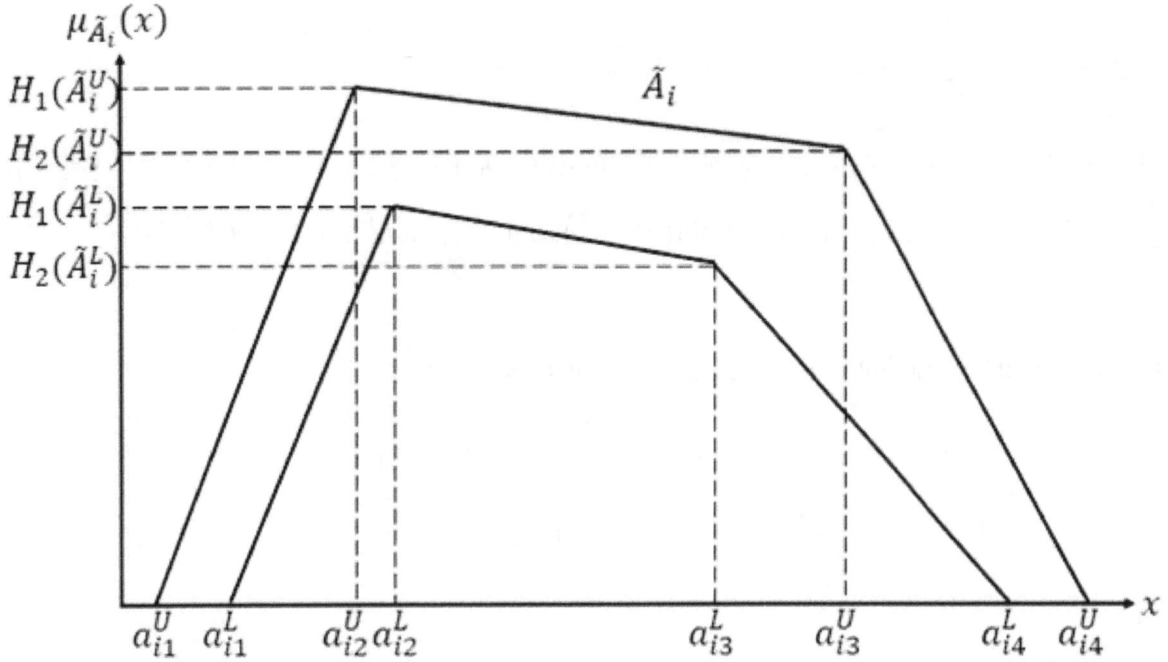

Let

$$\tilde{\tilde{A}}_1 = \left(\tilde{A}_1^U ; \tilde{A}_1^L\right) = \left(\left(a_{11}^U, a_{12}^U, a_{13}^U, a_{14}^U ; H_1\left(\tilde{A}_1^U\right), H_2\left(\tilde{A}_1^U\right)\right), \left(a_{11}^L, a_{12}^L, a_{13}^L, a_{14}^L ; H_1\left(\tilde{A}_1^L\right), H_2\left(\tilde{A}_1^L\right)\right)\right)$$

and

$$\tilde{\tilde{A}}_2 = \left(\tilde{A}_2^U ; \tilde{A}_2^L\right) = \left(\left(a_{21}^U, a_{22}^U, a_{23}^U, a_{24}^U ; H_1\left(\tilde{A}_2^U\right), H_2\left(\tilde{A}_2^U\right)\right), \left(a_{21}^L, a_{22}^L, a_{23}^L, a_{24}^L ; H_1\left(\tilde{A}_2^L\right), H_2\left(\tilde{A}_2^L\right)\right)\right)$$

be two arbitrary trapezoidal interval type-2 fuzzy sets.

Definition 10: The addition of $\tilde{\tilde{A}}_1$ and $\tilde{\tilde{A}}_2$ is defined as follows:

$$\tilde{\tilde{A}}_1 \oplus \tilde{\tilde{A}}_2 = \left(a_{11}^U + a_{21}^U, a_{12}^U + a_{22}^U, a_{13}^U + a_{23}^U, a_{14}^U + a_{24}^U; \min\left(H_1\left(\tilde{A}_1^U\right), H_1\left(\tilde{A}_2^U\right)\right), \min\left(H_1\left(\tilde{A}_1^U\right), H_1\left(\tilde{A}_2^U\right)\right)\right),$$
$$\left(a_{11}^L + a_{21}^L, a_{12}^L + a_{22}^L, a_{13}^L + a_{23}^L, a_{14}^L + a_{24}^L; \min\left(H_1\left(\tilde{A}_1^L\right), H_1\left(\tilde{A}_2^L\right)\right), \min\left(H_1\left(\tilde{A}_1^L\right), H_1\left(\tilde{A}_2^L\right)\right)\right).$$

Definition 11: The difference of $\tilde{\tilde{A}}_1$ and $\tilde{\tilde{A}}_2$ is defined as follows:

$$\tilde{\tilde{A}}_1 \Theta \tilde{\tilde{A}}_2 = \left(a_{11}^U - a_{24}^U, a_{12}^U - a_{23}^U, a_{13}^U - a_{22}^U, a_{14}^U - a_{21}^U; \min\left(H_1\left(\tilde{A}_1^U\right), H_1\left(\tilde{A}_2^U\right)\right), \min\left(H_1\left(\tilde{A}_1^U\right), H_1\left(\tilde{A}_2^U\right)\right)\right),$$
$$\left(a_{11}^L - a_{24}^L, a_{12}^L - a_{23}^L, a_{13}^L - a_{22}^L, a_{14}^L - a_{21}^L; \min\left(H_1\left(\tilde{A}_1^L\right), H_1\left(\tilde{A}_2^L\right)\right), \min\left(H_1\left(\tilde{A}_1^L\right), H_1\left(\tilde{A}_2^L\right)\right)\right).$$

Definition 12: The multiplication of $\tilde{\tilde{A}}_1$ and $\tilde{\tilde{A}}_2$ is defined as follows:

$$\tilde{\tilde{A}}_1 \otimes \tilde{\tilde{A}}_2 \cong \left(a_{11}^U \otimes a_{21}^U, a_{12}^U \otimes a_{22}^U, a_{13}^U \otimes a_{23}^U, a_{14}^U \otimes a_{24}^U; \min\left(H_1\left(\tilde{A}_1^U\right), H_1\left(\tilde{A}_2^U\right)\right), \min\left(H_1\left(\tilde{A}_1^U\right), H_1\left(\tilde{A}_2^U\right)\right)\right)$$
$$\left(a_{11}^L \otimes a_{21}^L, a_{12}^L \otimes a_{22}^L, a_{13}^L \otimes a_{23}^L, a_{14}^L \otimes a_{24}^L; \min\left(H_1\left(\tilde{A}_1^L\right), H_1\left(\tilde{A}_2^L\right)\right), \min\left(H_1\left(\tilde{A}_1^L\right), H_1\left(\tilde{A}_2^L\right)\right)\right).$$

Definition 13: The scalar multiplication and division of $\tilde{\tilde{A}}_1$ with a scalar k>0 can be given as:

$$k\tilde{\tilde{A}}_1 = \left(\left(ka_{11}^U, ka_{12}^U, ka_{13}^U, ka_{14}^U; H_1\left(\tilde{A}_1^U\right), H_2\left(\tilde{A}_1^U\right)\right), \left(ka_{11}^L, ka_{12}^L, ka_{13}^L, ka_{14}^L; H_1\left(\tilde{A}_1^L\right), H_2\left(\tilde{A}_1^L\right)\right)\right)$$

$$\frac{\tilde{\tilde{A}}_1}{k} = \left(\left(\frac{a_{11}^U}{k}, \frac{a_{12}^U}{k}, \frac{a_{13}^U}{k}, \frac{a_{14}^U}{k}; H_1\left(\tilde{A}_1^U\right), H_2\left(\tilde{A}_1^U\right)\right), \left(\frac{a_{11}^L}{k}, \frac{a_{12}^L}{k}, \frac{a_{13}^L}{k}, \frac{a_{14}^L}{k}; H_1\left(\tilde{A}_1^L\right), H_2\left(\tilde{A}_1^L\right)\right)\right).$$

Definition 14: The reciprocal of $\tilde{\tilde{A}}_1$ is defined as follows:

$$\frac{1}{\tilde{\tilde{A}}_1} = \left(\tilde{\tilde{A}}_1\right)^{-1} = \left(\left(\frac{1}{a_{14}^U}, \frac{1}{a_{13}^U}, \frac{1}{a_{12}^U}, \frac{1}{a_{11}^U}; H_1\left(\tilde{A}_1^U\right), H_2\left(\tilde{A}_1^U\right)\right), \left(\frac{1}{a_{14}^L}, \frac{1}{a_{13}^L}, \frac{1}{a_{12}^L}, \frac{1}{a_{11}^L}; H_1\left(\tilde{A}_1^L\right), H_2\left(\tilde{A}_1^L\right)\right)\right).$$

Definition 15: In Kahraman et al. (2014) the defuzzified value $DTraT\left(\tilde{\tilde{A}}_1\right)$ of a trapezoidal type-2 fuzzy set $\tilde{\tilde{A}}_1$ is given as

$$DTraT\left(\tilde{\tilde{A}}_1\right)=\frac{\frac{\alpha}{4}+a_{11}^U+\frac{\beta}{4}+a_{11}^L}{2} \tag{13}$$

where

$$\alpha=\left(a_{14}^U-a_{11}^U\right)+\left(H_1\left(\tilde{A}_1^U\right)a_{12}^U-a_{11}^U\right)+\left(H_2\left(\tilde{A}_1^U\right)a_{13}^U-a_{11}^U\right)$$

and

$$\beta=\left(a_{14}^L-a_{11}^L\right)+\left(H_1\left(\tilde{A}_1^L\right)a_{12}^L-a_{11}^L\right)+\left(H_2\left(\tilde{A}_1^L\right)a_{13}^L-a_{11}^L\right).$$

By using the ranking function *DTraT* in Eq. 13, trapezoidal interval type-2 fuzzy sets can be ordered easily. Here we would like to point out that many ranking functions are defined for type-2 fuzzy sets in the literature. The function given by Eq. 13 used in this chapter is one of the most common used.

Type-2 Fuzzy AHP

Type-2 fuzzy AHP method is firstly proposed in Kahraman et al. (2014). The linguistic variables and their trapezoidal interval type-2 fuzzy scales are given in Table 1.

Table 1. Interval type 2 fuzzy scales of the linguistic variables (Kahraman et al. (2014)).

Linguistic variables	Trapezoidal interval type-2 fuzzy scales
Absolutely Strong (AS)	(7,8,9,9;1,1) (7.2,8.2,8.8,9;0.8,0.8)
Very Strong (VS)	(5,6,8,9;1,1) (5.2,6.2,7.8,8.8;0.8,0.8)
Fairly Strong (FS)	(3,4,6,7;1,1) (3.2,4.2,5.8,6.8;0.8,0.8)
Slightly Strong (SS)	(1,2,4,5;1,1) (1.2,2.2,3.8,4.8;0.8,0.8)
Exactly Equal (E)	(1,1,1,1;1,1) (1,1,1,1;1,1)
If factor has one of the above linguistic variables assigned to it when compared with factor *j*, then *j* has the reciprocal value when compared with *i*	Reciprocals of above

By using the scale in Table 1, the main steps of this method can be given as follows:

Step 1. Construct all fuzzy pairwise comparison matrices among all alternatives and criteria. Let a fuzzy binary comparison matrix is given as follows:

$$\tilde{\tilde{A}} = \begin{bmatrix} 1 & \tilde{\tilde{a}}_{12} & \cdots & \tilde{\tilde{a}}_{1n} \\ \tilde{\tilde{a}}_{21} & 1 & \cdots & \tilde{\tilde{a}}_{2n} \\ \vdots & \vdots & \ddots & \vdots \\ \tilde{\tilde{a}}_{n1} & \tilde{\tilde{a}}_{n2} & \cdots & 1 \end{bmatrix} = \begin{bmatrix} 1 & \tilde{\tilde{a}}_{12} & \cdots & \tilde{\tilde{a}}_{1n} \\ 1/\tilde{\tilde{a}}_{12} & 1 & \cdots & \tilde{\tilde{a}}_{2n} \\ \vdots & \vdots & \ddots & \vdots \\ 1/\tilde{\tilde{a}}_{1n} & 1/\tilde{\tilde{a}}_{2n} & \cdots & 1 \end{bmatrix}$$

Step 2. Check the consistency of each type-2 fuzzy pairwise comparison matrix by converting each type-2 fuzzy entry to a crisp one using *DTraT* in Eq. 13. Apply the traditional consistency calculations of AHP. If the defuzzified matrix is consistent, then the corresponding type-2 fuzzy pairwise comparison matrix is consistent. If all type-2 fuzzy pairwise comparison matrices are consistent, go to the next step. Otherwise, reconstruct the matrices not consistent.

Step 3. For each type-2 fuzzy pairwise comparison matrix, calculate the geometric mean of each row as

$$\tilde{\tilde{r}}_i = \left[\tilde{\tilde{a}}_{i1} \otimes \tilde{\tilde{a}}_{i2} \otimes \cdots \otimes \tilde{\tilde{a}}_{in} \right]^{1/n} \tag{14}$$

where

$$\sqrt[n]{\tilde{\tilde{a}}_{ij}} = \left(\left(\sqrt[n]{a_{ij1}^U}, \sqrt[n]{a_{ij2}^U}, \sqrt[n]{a_{ij3}^U}, \sqrt[n]{a_{ij4}^U}; H_1^U\left(a_{ij}\right), H_2^U\left(a_{ij}\right) \right), \left(\sqrt[n]{a_{ij1}^L}, \sqrt[n]{a_{ij2}^L}, \sqrt[n]{a_{ij3}^L}, \sqrt[n]{a_{ij4}^L}; H_1^L\left(a_{ij}\right), H_2^L\left(a_{ij}\right) \right) \right).$$

Normalize the geometric means and obtain the fuzzy weight of the *i*th criterion by

$$\tilde{\tilde{w}}_i = \tilde{\tilde{r}}_i \otimes \left[\tilde{\tilde{r}}_1 \oplus \ldots \oplus \tilde{\tilde{r}}_i \oplus \cdots \oplus \tilde{\tilde{r}}_n \right]^{-1} \tag{15}$$

where the power (-1) represents the reciprocal of the corresponding trapezoidal interval type-2 fuzzy set.

Step 4. Combine the fuzzy weights and fuzzy performance scores by using the formula in Eq. 16

$$\tilde{\tilde{U}}_i = \sum_{j=1}^{n} \left(\tilde{\tilde{w}}_j \otimes \tilde{\tilde{r}}_{ij} \right) \tag{16}$$

where $\tilde{\tilde{U}}_i$ is the fuzzy utility of alternative *i*; $\tilde{\tilde{w}}_j$ is the weight of the criterion *j*, and $\tilde{\tilde{r}}_{ij}$ is the performance score of alternative *i* with respect to criterion *j*.

Step 5. Defuzzify the fuzzy utility of alternatives by using *DTraT* in Eq. 13 and normalize these values to determine the best alternative.

AN APPLICATION ON BIST USING TYPE-2 FUZZY AHP

One of the most important techniques of a company's financial analysis is financial ratios. Financial ratios allow quantification of a company's profitability, liquidity, debt situation and growth with quantitative data. From financial statements, information is calculated as financial ratios, which are quantitative measures that allow a company to be analyzed in terms of profitability, liquidity, debt and growth. These numerical data provide the opportunity to compare different companies operating in the same sector with each other and guide the investor in choosing the best investment alternative (Goel (2015)).

The Portfolio Selection problem is an MCDM problem and is inherently ambiguous. In this section, we will first present the financial ratios for the evaluation of alternative stocks. In order to demonstrate the applicability of our methodology in this part of our study, five stocks are selected from Borsa İstanbul by using the past price movements (Ozkok and Kocken (2019)). In order to evaluate our alternative stocks, we selected five important financial ratios as follows:

C1: Market to book value ratio: The book-to-market ratio is used to find the value of a company by comparing its book value with its market value. The book value of a company is calculated by looking at the historical cost or accounting value of the company. (Retrieved from https://www.investopedia.com/terms/b/booktomarketratio.asp).

C2: Net profit margin: The net profit margin is equal to how much net income or profit is generated as a percentage of revenue. Net profit margin is the ratio of the net profit of a company or business segment to income. The net profit margin indicates how much of each dollar in revenue that a company has turned into profit. (Retrieved from https://www.investopedia.com/terms/n/net_margin.asp).

C3: Return on capital employed (ROCE): Return on capital employed (ROCE) is a financial ratio that measures a company's profitability and the efficiency with which its capital is used. In other words, the ratio measures how well a company is generating profits from its capital (Retrieved from https://www.investopedia.com/terms/r/roce.asp).

C4: Return on assets (ROA): Return on assets is a profitability ratio that provides how much profit a company is able to generate from its assets. In other words, return on assets (ROA) measures how efficient a company's management is in generating earnings from its economic resources or assets on its balance sheet (Retrieved from https://www.investopedia.com/ask/answers/031215/what-formula-calculating-return-assets-roa.asp).

C5: Price-to-Earnings Ratio (P/E Ratio): The price-to-earnings ratio (P/E ratio) is the ratio for valuing a company that measures its current share price relative to its per-share earnings (EPS) (Retrieved from https://www.investopedia.com/terms/p/price-earningsratio.asp).

We obtained C1: Market to book value ratio, C2: Net profit margin, C3: Return on capital employed (ROCE), C4: Return on assets (ROA), C5: Price-to-Earnings Ratio (P/E Ratio) financial ratio values from the website of finnet.com.tr on the date of 01.02.2019 for each selected alternative stocks. Our decision maker, who provided these values filled the evaluation matrices. After this stage, we apply the steps of the proposed interval type-2 fuzzy AHP method in the following:

Step 1. Table 2 and Table 3 - Table 7 present pairwise comparison matrices for criteria and alternatives using linguistic terms, respectively. Then, fuzzy pairwise comparison matrices are constructed by using the interval type-2 fuzzy scales of the linguistic variables in Table 1.

Table 2. Pairwise comparison matrix for criteria.

	C1	C2	C3	C4	C5
C1	E	SS	FS	VS	AS
C2	1/SS	E	SS	SS	VS
C3	1/FS	1/SS	E	SS	SS
C4	1/VS	1/SS	1/SS	E	SS
C5	1/AS	1/VS	1/SS	1/SS	E

Table 3. Pairwise comparison matrix for alternatives w.r.t. C1.

	A1	A2	A3	A4	A5
A1	E	FS	AS	SS	VS
A2	1/FS	E	FS	1/SS	SS
A3	1/AS	1/FS	E	1/VS	1/SS
A4	1/SS	SS	VS	E	FS
A5	1/VS	1/SS	SS	1/FS	E

Table 4. Pairwise comparison matrix for alternatives w.r.t. C2.

	A1	A2	A3	A4	A5
A1	E	SS	VS	AS	FS
A2	1/SS	E	FS	VS	SS
A3	1/VS	1/FS	E	SS	1/SS
A4	1/AS	1/VS	1/SS	E	1/FS

Table 5, Pairwise comparison matrix for alternatives w.r.t. C3.

	A1	A2	A3	A4	A5
A1	E	SS	FS	AS	VS
A2	1/SS	E	SS	VS	VS
A3	1/FS	1/SS	E	FS	SS
A4	1/AS	1/VS	1/FS	E	1/SS
A5	1/VS	1/VS	1/SS	SS	E

Step 2. Using the *DTraT* method, the defuzzified matrices are obtained. All defuzzified matrices have been checked for its consistency that is the consistency ratio of each matrix has been found under 0.1.

Table 6. Pairwise comparison matrix for alternatives w.r.t. C4.

	A1	A2	A3	A4	A5
A1	E	SS	FS	AS	SS
A2	1/SS	E	VS	VS	SS
A3	1/FS	1/VS	E	SS	1/SS
A4	1/AS	1/VS	1/SS	E	1/FS
A5	1/SS	1/SS	SS	FS	E

Table 7. Pairwise comparison matrix for alternatives w.r.t. C5.

	A1	A2	A3	A4	A5
A1	E	1/SS	1/FS	FS	SS
A2	SS	E	1/SS	VS	FS
A3	FS	SS	E	AS	VS
A4	1/FS	1/VS	1/AS	E	1/SS
A5	1/SS	1/FS	1/VS	SS	E

Step 3. The geometric mean of each row of comparison matrices is calculated by using Eq. 14. For the representative calculations, Table 3 is used, which includes the type-2 fuzzy sets of pairwise comparisons of alternatives with respect to C1. For example, the geometric mean of the first row is calculated as:

$$\tilde{\tilde{r}}_1 = \left[\tilde{\tilde{a}}_{11} \otimes \tilde{\tilde{a}}_{12} \otimes \tilde{\tilde{a}}_{13} \otimes \tilde{\tilde{a}}_{14} \otimes \tilde{\tilde{a}}_{15}\right]^{1/5} = \left[(1,1,1,1;1,1)(1,1,1,1;1,1)\right.$$
$$\otimes(3,4,6,7;1,1)(3.2,4.2,5.8,6.8;0.8,0.8)\otimes(7,8,9,9;1,1)(7.2,8.2,8.8,9;0.8,0.8)$$
$$\left.\otimes(1,2,4,5;1,1)(1.2,2.2,3.8,4.8;0.8,0.8)\otimes(5,6,8,9;1,1)(5.2,6.2,7.8,8.8;0.8,0.8)\right]^{1/5}$$
$$=(2.54,3.29,4.44,4.90;1,1)(2.70,3.42,4.33,4.81;0.8,0.8)$$

The geometric means of the remaining rows of Table 3 are given in Table 8.

Table 8. Geometric means $\tilde{\tilde{r}}_i$ of pairwise comparison matrix for alternatives w.r.t. C1.

A1	(2.54,3.29,4.44,4.90;1,1)(2.70,3.42,4.33,4.81;0.8,0.8)
A2	(0.61,0.80,1.25,1.64;1,1)(0.65,0.84,1.19,1.53;0.8,0.8)
A3	(0.20,0.23,0.30, 0.39;1,1)(0.21,0.23,0.29,0.37;0.8,0.8)
A4	(1.25,1.64,2.49,3.16;1,1)(1.33,1.72,2.39,2.99;0.8,0.8)
A5	(0.32,0.40,0.61,0.80;1,1)(0.33,0.42,0.58,0.75;0.8,0.8)

The priority weights of criteria and alternatives are determined by using Eq. 15. For example the priority weights of alternatives w.r.t. C1 can be calculated as follows:

$$\tilde{\tilde{w}}_1 = \tilde{\tilde{r}}_1 \otimes \left[\tilde{\tilde{r}}_1 \oplus \tilde{\tilde{r}}_2 \oplus \cdots \oplus \tilde{\tilde{r}}_5 \right]^{-1} = (2.54,3.29,4.44,4.90;1,1)(2.70,3.42,4.33,4.81;0.8,0.8)$$
$$\otimes\big[(2.54,3.29,4.44,4.90;1,1)(2.70,3.42,4.33,4.81;0.8,0.8)\oplus(0.61,0.80,1.25,1.64;1,1)$$
$$(0.65,0.84,1.19,1.53;0.8,0.8)\oplus(0.20,0.225,0.30,\ 0.39;1,1)(0.21,0.23,0.29,0.37;0.8,0.8)$$
$$\oplus(1.25,1.64,2.49,3.159;1,1)(1.33,1.72,2.39,2.99;0.8,0.8)\oplus(0.316,0.40,0.608,0.80;1,1)$$
$$(0.33,0.42,0.58,0.75;0.8,0.8)\big]^{-1} = (0.23,0.36,0.70,1;1,1)(0.26,0.39,0.65,0.92;0.8,0.8).$$

Normalized geometric means of the pairwise comparison matrices are given in Table 9.

Step 4. The fuzzy weights and fuzzy performance scores are aggregated by Eq. 16, and the results are given in Table 10.

Step 5. By using *DTraT*, the fuzzy utility of alternatives are defuzzified and then normalized in Table 10.

Considering the normalized values, the results show that 48%, 28%, 14%, 7%, 4% of the principal should be invested in A3, A2, A1, A5, and A4, respectively.

CONCLUSION

The main purpose of this chapter is to present general and satisfactory information on type-1 fuzzy AHP and type-2 fuzzy AHP. In doing so, brief literature reviews of the most well-known studies in the literature about type-1 fuzzy AHP and type-2 fuzzy AHP are presented. Since type-2 fuzzy AHP studies proposed in the literature are based on Buckley's fuzzy AHP method, the steps of this method are also presented.

Moreover, in this chapter, the portfolio selection problem, which is included in the scope of finance theory and examined very widely, is discussed. There are some methods that can be used effectively in the current financial literature for portfolio selection. However, as is known, the methods that reflect the uncertainty arising from the future in the best way to the model are evaluated and preferred effectively. The movement of stocks in stock exchanges cannot be known in advance. However, past price movements give insight into shares. From this point of view, it is clear that the approaches that make an evaluation in terms of the portfolio in the fuzzy environment according to financial ratios and which are preferred among alternatives. Within the scope of our study, the main goal was to sort a set of stocks in a fuzzy environment and thus to illustrate the usefulness of fuzzy methodology in financial problems. The ranking or preference numbers obtained by interval type-2 fuzzy AHP, proposed by Kahraman et al. (2014), show how to allocate the available monetary resources among the given stocks.

Table 9. Type 2 fuzzy weights of the criteria and alternatives(Priority weights).

With respect to the Goal	
C1	(0.23,0.37,0.73,1.07;1.00,1.00)(0.26,0.40,0.68,0.98;0.80,0.80)
C2	(0.09,0.16,0.38,0.64;1.00,1.00)(0.11,0.18,0.34,0.57;0.80,0.80)
C3	(0.05,0.08,0.19,0.33;1.00,1.00)(0.05,0.09,0.17,0.29;0.80,0.80)
C4	(0.03,0.05,0.11,0.22;1.00,1.00)(0.03,0.05,0.10,0.19;0.80,0.80)
C5	(0.02,0.03,0.06,0.11;1.00,1.00)(0.02,0.03,0.05,0.09;0.80,0.80)
With respect to C1	
A1	(0.23,0.36,0.70,1.00;1.00,1.00)(0.26,0.39,0.65,0.92;0.80,0.80)
A2	(0.06,0.09,0.20,0.33;1.00,1.00)(0.06,0.10,0.18,0.29;0.80,0.80)
A3	(0.02,0.02,0.05,0.08;1.00,1.00)(0.02,0.03,0.04,0.07;0.80,0.80)
A4	(0.11,0.18,0.39,0.64;1.00,1.00)(0.13,0.20,0.36,0.57;0.80,0.80)
A5	(0.03,0.04,0.10,0.16;1.00,1.00)(0.03,0.05,0.09,0.14;0.80,0.80)
With respect to C2	
A1	(0.23,0.36,0.70,1.00;1.00,1.00)(0.26,0.39,0.65,0.92;0.80,0.80)
A2	(0.11,0.18,0.39,0.64;1.00,1.00)(0.13,0.20,0.36,0.57;0.80,0.80)
A3	(0.03,0.04,0.10,0.16;1.00,1.00)(0.03,0.05,0.09,0.14;0.80,0.80)
A4	(0.02,0.02,0.05,0.08;1.00,1.00)(0.02,0.03,0.04,0.07;0.80,0.80)
A5	(0.06,0.09,0.20,0.33;1.00,1.00)(0.06,0.10,0.18,0.29;0.80,0.80)
With respect to C3	
A1	(0.23,0.36,0.69,0.97;1.00,1.00)(0.26,0.39,0.64,0.90;0.80,0.80)
A2	(0.13,0.19,0.41,0.66;1.00,1.00)(0.14,0.21,0.38,0.59;0.80,0.80)
A3	(0.06,0.09,0.19,0.32;1.00,1.00)(0.06,0.09,0.18,0.29;0.80,0.80)
A4	(0.02,0.02,0.05,0.08;1.00,1.00)(0.02,0.03,0.04,0.07;0.80,0.80)
A5	(0.03,0.04,0.09,0.14;1.00,1.00)(0.03,0.04,0.08,0.13;0.80,0.80)
With respect to C4	
A1	(0.17,0.30,0.65,0.99;1.00,1.00)(0.19,0.33,0.60,0.90;0.80,0.80)
A2	(0.13,0.20,0.45,0.76;1.00,1.00)(0.14,0.22,0.41,0.67;0.80,0.80)
A3	(0.03,0.05,0.10,0.18;1.00,1.00)(0.03,0.05,0.09,0.16;0.80,0.80)
A4	(0.02,0.03,0.05,0.09;1.00,1.00)(0.02,0.03,0.05,0.08;0.80,0.80)
A5	(0.06,0.10,0.24,0.46;1.00,1.00)(0.07,0.11,0.22,0.40;0.80,0.80)
With respect to C5	
A1	(0.06,0.09,0.20,0.33;1.00,1.00)(0.06,0.10,0.18,0.29;0.80,0.80)
A2	(0.11,0.18,0.39,0.64;1.00,1.00)(0.13,0.20,0.36,0.57;0.80,0.80)
A3	(0.23,0.36,0.70,1.00;1.00,1.00)(0.26,0.39,0.65,0.92;0.80,0.80)
A4	(0.02,0.02,0.05,0.08;1.00,1.00)(0.02,0.03,0.04,0.07;0.80,0.80)
A5	(0.03,0.04,0.10,0.16;1.00,1.00)(0.03,0.05,0.09,0.14;0.80,0.80)

Table 10. U Vector.

	$\tilde{\tilde{U}}_i$	Defuzzified values	Normalized Values
A1	(0.13,0.20,0.44,0.80;1.00,1.00)(0.15,0.21,0.40,0.69;0.80,0.80)	0.36	0.14
A2	(0.27,0.40,0.87,1.55;1.00,1.00)(0.30,0.43,0.79,1.35;0.80,0.80)	0.72	0.28
A3	(0.55,0.80,1.55,2.41;1.00,1.00)(0.60,0.86,1.44,2.17;0.80,0.80)	1.24	0.48
A4	(0.04,0.05,0.11,0.19;1.00,1.00)(0.05,0.06,0.10,0.17;0.80,0.80)	0.09	0.04
A5	(0.07,0.10,0.21,0.39;1.00,1.00)(0.07,0.11,0.19,0.34;0.80,0.80)	0.18	0.07

ACKNOWLEDGMENT

This research has been supported by Yıldız Technical University Scientific Research Projects Coordination Department. Project Number: FBA-2017-3073.

REFERENCES

Abdullah, L., & Najib, L. (2014). A new type-2 fuzzy set of linguistic variables for the fuzzy analytic hierarchy process. *Expert Systems with Applications*, *41*(7), 3297–3305. doi:10.1016/j.eswa.2013.11.028

Abdullah, L., & Zulkifli, N. (2015). Integration of fuzzy AHP and interval type-2 fuzzy DEMATEL: An application to human resource management. *Expert Systems with Applications*, *42*(9), 4397–4409. doi:10.1016/j.eswa.2015.01.021

Alexander, G. J., Sharpe, W. F., & Bailey, J. V. (2003). *Fundamentals of investments/fundamentos de inversiones*. Pearson Educación.

Bhattacharyya, R., Kar, S., & Majumder, D. D. (2011). Fuzzy mean–variance–skewness portfolio selection models by interval analysis. *Computers & Mathematics with Applications (Oxford, England)*, *61*(1), 126–137. doi:10.1016/j.camwa.2010.10.039

Bilbao-Terol, A., Pérez-Gladish, B., Arenas-Parra, M., & Rodríguez-Uría, M. V. (2006). Fuzzy compromise programming for portfolio selection. *Applied Mathematics and Computation*, *173*(1), 251–264. doi:10.1016/j.amc.2005.04.003

Boender, C. G. E., De Graan, J. G., & Lootsma, F. A. (1989). Multi-criteria decision analysis with fuzzy pairwise comparisons. *Fuzzy Sets and Systems*, *29*(2), 133–143. doi:10.1016/0165-0114(89)90187-5

Bozdağ, C. E., Kahraman, C., & Ruan, D. (2003). Fuzzy group decision making for selection among computer integrated manufacturing systems. *Computers in Industry*, *51*(1), 13–29. doi:10.1016/S0166-3615(03)00029-0

Buckley, J. J. (1985). Fuzzy hierarchical analysis. *Fuzzy Sets and Systems*, *17*(3), 233–247. doi:10.1016/0165-0114(85)90090-9

Buckley, J. J., Feuring, T., & Hayashi, Y. (2001). Fuzzy hierarchical analysis revisited. *European Journal of Operational Research*, *129*(1), 48–64. doi:10.1016/S0377-2217(99)00405-1

Buyukozkan, G., Kahraman, C., & Ruan, D. (2004). A fuzzy multi-criteria decision approach for software development strategy selection. *International Journal of General Systems*, *33*(2-3), 259–280. doi:10.1080/03081070310001633581

Celik, E., & Akyuz, E. (2018). An interval type-2 fuzzy AHP and TOPSIS methods for decision-making problems in maritime transportation engineering: The case of ship loader. *Ocean Engineering*, *155*, 371–381. doi:10.1016/j.oceaneng.2018.01.039

Celik, E., Gumus, A. T., & Alegoz, M. (2014). A trapezoidal type-2 fuzzy MCDM method to identify and evaluate critical success factors for humanitarian relief logistics management. *Journal of Intelligent & Fuzzy Systems*, *27*(6), 2847–2855. doi:10.3233/IFS-141246

Cevik Onar, S., Oztaysi, B., & Kahraman, C. (2014). Strategic decision selection using hesitant fuzzy TOPSIS and interval type-2 fuzzy AHP: A case study. *International Journal of Computational Intelligence Systems*, *7*(5), 1002–1021. doi:10.1080/18756891.2014.964011

Chang, D. Y. (1996). Applications of the extent analysis method on fuzzy AHP. *European Journal of Operational Research*, *95*(3), 649–655. doi:10.1016/0377-2217(95)00300-2

Chen, M. F., Tzeng, G. H., & Ding, C. G. (2008). Combining fuzzy AHP with MDS in identifying the preference similarity of alternatives. *Applied Soft Computing*, *8*(1), 110–117. doi:10.1016/j.asoc.2006.11.007

Chen, S. J., & Hwang, C. L. (1992). Fuzzy multiple attribute decision making methods. In *Fuzzy multiple attribute decision making* (pp. 289–486). Springer. doi:10.1007/978-3-642-46768-4_5

Chen, S. M. (1996). Evaluating weapon systems using fuzzy arithmetic operations. *Fuzzy Sets and Systems*, *77*(3), 265–276. doi:10.1016/0165-0114(95)00096-8

Cheng, C. H. (1997). Evaluating naval tactical missile systems by fuzzy AHP based on the grade value of membership function. *European Journal of Operational Research*, *96*(2), 343–350. doi:10.1016/S0377-2217(96)00026-4

Cheng, C. H., & Mon, D. L. (1994). Evaluating weapon system by analytical hierarchy process based on fuzzy scales. *Fuzzy Sets and Systems*, *63*(1), 1–10. doi:10.1016/0165-0114(94)90140-6

Cheng, C. H., Yang, K. L., & Hwang, C. L. (1999). Evaluating attack helicopters by AHP based on linguistic variable weight. *European Journal of Operational Research*, *116*(2), 423–435. doi:10.1016/S0377-2217(98)00156-8

Chou, Y. C., Yen, H. Y., Dang, V. T., & Sun, C. C. (2019). Assessing the Human Resource in Science and Technology for Asian Countries: Application of Fuzzy AHP and Fuzzy TOPSIS. *Symmetry*, *11*(2), 251. doi:10.3390ym11020251

Csutora, R., & Buckley, J. J. (2001). Fuzzy hierarchical analysis: The Lambda-Max method. *Fuzzy Sets and Systems*, *120*(2), 181–195. doi:10.1016/S0165-0114(99)00155-4

Dağdeviren, M., Yavuz, S., & Kılınç, N. (2009). Weapon selection using the AHP and TOPSIS methods under fuzzy environment. *Expert Systems with Applications*, *36*(4), 8143–8151. doi:10.1016/j.eswa.2008.10.016

Enea, M., & Piazza, T. (2004). Project selection by constrained fuzzy AHP. *Fuzzy Optimization and Decision Making*, *3*(1), 39–62. doi:10.1023/B:FODM.0000013071.63614.3d

Erensal, Y. C., Öncan, T., & Demircan, M. L. (2006). Determining key capabilities in technology management using fuzzy analytic hierarchy process: A case study of Turkey. *Information Sciences*, *176*(18), 2755–2770. doi:10.1016/j.ins.2005.11.004

Giove, S., Funari, S., & Nardelli, C. (2006). An interval portfolio selection problem based on regret function. *European Journal of Operational Research*, *170*(1), 253–264. doi:10.1016/j.ejor.2004.05.030

Goel, S. (2015). *Financial ratios*. Business Expert Press.

Gogus, O., & Boucher, T. O. (1997). A consistency test for rational weights in multi-criterion decision analysis with fuzzy pairwise comparisons. *Fuzzy Sets and Systems*, *86*(2), 129–138. doi:10.1016/0165-0114(95)00410-6

Gogus, O., & Boucher, T. O. (1998). Strong transitivity, rationality and weak monotonicity in fuzzy pairwise comparisons. *Fuzzy Sets and Systems*, *94*(1), 133–144. doi:10.1016/S0165-0114(96)00184-4

Göleç, A., & Taşkın, H. (2007). Novel methodologies and a comparative study for manufacturing systems performance evaluations. *Information Sciences*, *177*(23), 5253–5274. doi:10.1016/j.ins.2007.06.024

Gul, M., Celik, E., Gumus, A. T., & Guneri, A. F. (2016). Emergency department performance evaluation by an integrated simulation and interval type-2 fuzzy MCDM-based scenario analysis. *European Journal of Industrial Engineering*, *10*(2), 196–223. doi:10.1504/EJIE.2016.075846

Güngör, Z., Serhadlıoğlu, G., & Kesen, S. E. (2009). A fuzzy AHP approach to personnel selection problem. *Applied Soft Computing*, *9*(2), 641–646. doi:10.1016/j.asoc.2008.09.003

Huang, J. J., Tzeng, G. H., & Ong, C. S. (2006). A novel algorithm for uncertain portfolio selection. *Applied Mathematics and Computation*, *173*(1), 350–359. doi:10.1016/j.amc.2005.04.074

Inuiguchi, M., & Tanino, T. (2000). Portfolio selection under independent possibilistic information. *Fuzzy Sets and Systems*, *115*(1), 83–92. doi:10.1016/S0165-0114(99)00026-3

Junior, F. R. L., Osiro, L., & Carpinetti, L. C. R. (2014). A comparison between Fuzzy AHP and Fuzzy TOPSIS methods to supplier selection. *Applied Soft Computing*, *21*, 194–209. doi:10.1016/j.asoc.2014.03.014

Kahraman, C., Cebeci, U., & Ruan, D. (2004). Multi-attribute comparison of catering service companies using fuzzy AHP: The case of Turkey. *International Journal of Production Economics*, *87*(2), 171–184. doi:10.1016/S0925-5273(03)00099-9

Kahraman, C., Öztayşi, B., Sarı, İ. U., & Turanoğlu, E. (2014). Fuzzy analytic hierarchy process with interval type-2 fuzzy sets. *Knowledge-Based Systems*, *59*, 48–57. doi:10.1016/j.knosys.2014.02.001

Kahraman, C., Ruan, D., & Doğan, I. (2003). Fuzzy group decision-making for facility location selection. *Information Sciences*, *157*, 135–153. doi:10.1016/S0020-0255(03)00183-X

Kilincci, O., & Onal, S. A. (2011). Fuzzy AHP approach for supplier selection in a washing machine company. *Expert Systems with Applications*, *38*(8), 9656–9664. doi:10.1016/j.eswa.2011.01.159

Kulak, O., & Kahraman, C. (2005). Fuzzy multi-attribute selection among transportation companies using axiomatic design and analytic hierarchy process. *Information Sciences*, *170*(2-4), 191–210. doi:10.1016/j.ins.2004.02.021

Kutlu, A. C., & Ekmekçioğlu, M. (2012). Fuzzy failure modes and effects analysis by using fuzzy TOPSIS-based fuzzy AHP. *Expert Systems with Applications*, *39*(1), 61–67. doi:10.1016/j.eswa.2011.06.044

Kwiesielewicz, M. (1998). A note on the fuzzy extension of Saaty's priority theory. *Fuzzy Sets and Systems*, *95*(2), 161–172. doi:10.1016/S0165-0114(96)00329-6

Lacagnina, V., & Pecorella, A. (2006). A stochastic soft constraints fuzzy model for a portfolio selection problem. *Fuzzy Sets and Systems*, *157*(10), 1317–1327. doi:10.1016/j.fss.2005.10.002

Lee, M., Pham, H., & Zhang, X. (1999). A methodology for priority setting with application to software development process. *European Journal of Operational Research*, *118*(2), 375–389. doi:10.1016/S0377-2217(98)00320-8

Li, X., Guo, S., & Yu, L. (2015). Skewness of fuzzy numbers and its applications in portfolio selection. *IEEE Transactions on Fuzzy Systems*, *23*(6), 2135–2143. doi:10.1109/TFUZZ.2015.2404340

Li, X., Qin, Z., & Kar, S. (2010). Mean-variance-skewness model for portfolio selection with fuzzy returns. *European Journal of Operational Research*, *202*(1), 239–247. doi:10.1016/j.ejor.2009.05.003

Markowitz, H. M. (1952). Portfolio Selection/Harry Markowitz. *The Journal of Finance*, *7*(1), 77–91.

Mikhailov, L. (2003). Deriving priorities from fuzzy pairwise comparison judgements. *Fuzzy Sets and Systems*, *134*(3), 365–385. doi:10.1016/S0165-0114(02)00383-4

Mikhailov, L. (2004a). A fuzzy approach to deriving priorities from interval pairwise comparison judgements. *European Journal of Operational Research*, *159*(3), 687–704. doi:10.1016/S0377-2217(03)00432-6

Mikhailov, L. (2004b). Group prioritization in the AHP by fuzzy preference programming method. *Computers & Operations Research*, *31*(2), 293–301. doi:10.1016/S0305-0548(03)00012-1

Mikhailov, L., & Tsvetinov, P. (2004). Evaluation of services using a fuzzy analytic hierarchy process. *Applied Soft Computing*, *5*(1), 23–33. doi:10.1016/j.asoc.2004.04.001

Mohanty, B. K., & Singh, N. (1994). Fuzzy relational equations in analytical hierarchy process. *Fuzzy Sets and Systems*, *63*(1), 11–19. doi:10.1016/0165-0114(94)90141-4

Mon, D. L., Cheng, C. H., & Lin, J. C. (1994). Evaluating weapon system using fuzzy analytic hierarchy process based on entropy weight. *Fuzzy Sets and Systems*, *62*(2), 127–134. doi:10.1016/0165-0114(94)90052-3

Ong, C. S., Huang, J. J., & Tzeng, G. H. (2005). A novel hybrid model for portfolio selection. *Applied Mathematics and Computation, 169*(2), 1195–1210. doi:10.1016/j.amc.2004.10.080

Ozkok, B. A., & Kocken, H. G. (2019). An Investment Decision Making Methodology Based on Type-2 Fuzzy AHP: A Case Study on BORSA ISTANBUL (BIST). *Proceedings of the International Conference on Applied Research in Management, Business and Economics*, 50-61.

Ozkok, B. A., & Pappalardo, E. (2013). Fuzzy Analytic Hierarchy Process and Its Application to E-Marketplace Selection. In Enterprise Business Modeling, Optimization Techniques, and Flexible Information Systems (pp. 26-44). IGI Global.

Oztaysi, B. (2015). A Group Decision Making Approach Using Interval Type-2 Fuzzy AHP for Enterprise Information Systems Project Selection. *Journal of Multiple-Valued Logic & Soft Computing, 24*(5).

Parra, M. A., Terol, A. B., & Urıa, M. R. (2001). A fuzzy goal programming approach to portfolio selection. *European Journal of Operational Research, 133*(2), 287–297. doi:10.1016/S0377-2217(00)00298-8

Ruoning, X., & Xiaoyan, Z. (1992). Extensions of the analytic hierarchy process in fuzzy environment. *Fuzzy Sets and Systems, 52*(3), 251–257. doi:10.1016/0165-0114(92)90236-W

Ruoning, X., & Xiaoyan, Z. (1996). Fuzzy logarithmic least squares ranking method in analytic hierarchy process. *Fuzzy Sets and Systems, 77*(2), 175–190. doi:10.1016/0165-0114(95)00073-9

Saaty, T. L. (1980). The Analytic Hierarchy Process McGraw-Hill New York. *Agricultural Economics Review*, 70.

Shaw, K., Shankar, R., Yadav, S. S., & Thakur, L. S. (2012). Supplier selection using fuzzy AHP and fuzzy multi-objective linear programming for developing low carbon supply chain. *Expert Systems with Applications, 39*(9), 8182–8192. doi:10.1016/j.eswa.2012.01.149

Singh, A., & Prasher, A. (2019). Measuring healthcare service quality from patients' perspective: Using Fuzzy AHP application. *Total Quality Management & Business Excellence, 30*(3-4), 284–300. doi:10.1080/14783363.2017.1302794

Stam, A., Sun, M., & Haines, M. (1996). Artificial neural network representations for hierarchical preference structures. *Computers & Operations Research, 23*(12), 1191–1201. doi:10.1016/S0305-0548(96)00021-4

Sun, C. C. (2010). A performance evaluation model by integrating fuzzy AHP and fuzzy TOPSIS methods. *Expert Systems with Applications, 37*(12), 7745–7754. doi:10.1016/j.eswa.2010.04.066

Tanaka, H., & Guo, P. (1999). Portfolio selection based on upper and lower exponential possibility distributions. *European Journal of Operational Research, 114*(1), 115–126. doi:10.1016/S0377-2217(98)00033-2

Tanaka, H., Guo, P., & Türksen, I. B. (2000). Portfolio selection based on fuzzy probabilities and possibility distributions. *Fuzzy Sets and Systems, 111*(3), 387–397. doi:10.1016/S0165-0114(98)00041-4

Tiryaki, F., & Ahlatcioglu, B. (2009). Fuzzy portfolio selection using fuzzy analytic hierarchy process. *Information Sciences, 179*(1-2), 53–69. doi:10.1016/j.ins.2008.07.023

Tiryaki, F., & Ahlatcioglu, M. (2005). Fuzzy stock selection using a new fuzzy ranking and weighting algorithm. *Applied Mathematics and Computation*, *170*(1), 144–157. doi:10.1016/j.amc.2004.10.092

Van Laarhoven, P. J., & Pedrycz, W. (1983). A fuzzy extension of Saaty's priority theory. *Fuzzy Sets and Systems*, *11*(1-3), 229–241. doi:10.1016/S0165-0114(83)80082-7

Wang, J., & Lin, Y. I. (2003). A fuzzy multicriteria group decision making approach to select configuration items for software development. *Fuzzy Sets and Systems*, *134*(3), 343–363. doi:10.1016/S0165-0114(02)00283-X

Wang, X., Kerre, E. E., & Ruan, D. (1995). Consistency of judgement matrix and fuzzy weights in fuzzy analytic hierarchy process. *International Journal of Uncertainty, Fuzziness and Knowledge-based Systems*, *3*(01), 35–46. doi:10.1142/S0218488595000050

Weck, M., Klocke, F., Schell, H., & Rüenauver, E. (1997). Evaluating alternative production cycles using the extended fuzzy AHP method. *European Journal of Operational Research*, *100*(2), 351–366. doi:10.1016/S0377-2217(96)00295-0

Xia, Y., Liu, B., Wang, S., & Lai, K. K. (2000). A model for portfolio selection with order of expected returns. *Computers & Operations Research*, *27*(5), 409–422. doi:10.1016/S0305-0548(99)00059-3

Xu, R. (2000). Fuzzy least-squares priority method in the analytic hierarchy process. *Fuzzy Sets and Systems*, *112*(3), 395–404. doi:10.1016/S0165-0114(97)00376-X

Yang, J., & Lee, H. (1997). An AHP decision model for facility location selection. *Facilities*, *15*(9/10), 241–254. doi:10.1108/02632779710178785

Yu, C. S. (2002). A GP-AHP method for solving group decision-making fuzzy AHP problems. *Computers & Operations Research*, *29*(14), 1969–2001. doi:10.1016/S0305-0548(01)00068-5

Zadeh, L. A. (1965). Fuzzy sets. *Information and Control*, *8*(3), 338–353. doi:10.1016/S0019-9958(65)90241-X

Zhang, W. G., Wang, Y. L., Chen, Z. P., & Nie, Z. K. (2007). Possibilistic mean–variance models and efficient frontiers for portfolio selection problem. *Information Sciences*, *177*(13), 2787–2801. doi:10.1016/j.ins.2007.01.030

Zhou, X., Wang, J., Yang, X., Lev, B., Tu, Y., & Wang, S. (2018). Portfolio selection under different attitudes in fuzzy environment. *Information Sciences*, *462*, 278–289. doi:10.1016/j.ins.2018.06.013

Zhu, K. J., Jing, Y., & Chang, D. Y. (1999). A discussion on extent analysis method and applications of fuzzy AHP. *European Journal of Operational Research*, *116*(2), 450–456. doi:10.1016/S0377-2217(98)00331-2

Zimmermann, H. J. (2011). *Fuzzy set theory—and its applications*. Springer Science & Business Media.

ADDITIONAL READINGS

Castillo, O., Melin, P., Kacprzyk, J., & Pedrycz, W. (2007, November). Type-2 fuzzy logic: theory and applications. *In 2007 IEEE International Conference on Granular Computing (GRC 2007)* (pp. 145-145). IEEE. 10.1109/GrC.2007.118

Karnik, N. N., & Mendel, J. M. (2001). Operations on type-2 fuzzy sets. *Fuzzy Sets and Systems*, *122*(2), 327–348. doi:10.1016/S0165-0114(00)00079-8

Karnik, N. N., Mendel, J. M., & Liang, Q. (1999). Type-2 fuzzy logic systems. *IEEE Transactions on Fuzzy Systems*, *7*(6), 643–658. doi:10.1109/91.811231

Kilic, M., & Kaya, İ. (2015). Investment project evaluation by a decision making methodology based on type-2 fuzzy sets. *Applied Soft Computing*, *27*, 399–410. doi:10.1016/j.asoc.2014.11.028

Liang, Q., & Mendel, J. M. (2000). Interval type-2 fuzzy logic systems: Theory and design. *IEEE Transactions on Fuzzy Systems*, *8*(5), 535–550. doi:10.1109/91.873577

KEY TERMS AND DEFINITIONS

Analytical Hierarcy Process: A multi-criteria decision-making technique that results with the relative evaluation of alternatives based on the specified criteria.

Fuzzy Decision Making: Decision making process involves taking decisions under uncertain environments where information can be handled by fuzzy sets and systems.

Portfolio Selection Problem: The problem that is to decide which stocks are to be chosen for investment and in what proportions they will be bought considering the specific criteria.

Type-1 Fuzzy Set: A set addressing uncertainty by assigning a membership function value among [0,1] to each element.

Type-2 Fuzzy Set: A set which is the extension of type-1 fuzzy sets has three-dimensional membership functions that provide additional degrees of freedom to make it possible to directly model uncertainties.

Chapter 3
Customer and Order Prioritization With Multi-Criteria Decision Making:
An Application in Metal Industry

Emre Aslan
https://orcid.org/0000-0002-1744-8077
Tokat Gaziosmanpasa University, Turkey

ABSTRACT

Customer and order prioritization is prioritizing the orders according to features of customers and orders. In make-to-order production systems, especially when demand exceeds the capacity, a need arises for prioritizing the orders. As a customer may have more than one order and as one type of order may belong to different customers, every job order must be prioritized according to both customer and order features. Prioritization values may be used for accepting and scheduling orders or selecting customers as a reflection of supplier selection. In this chapter, the production manager and customer relationship manager of a metal processing firm were interviewed for determining the criteria for prioritizing customers and orders. After determining the criteria, the managers were asked to pairwise compare the criteria for weighting with analytical hierarchy process (AHP). Then by considering scores of customers and orders for each criterion, orders were prioritized with technique for order preference by similarity to ideal solution (TOPSIS).

INTRODUCTION

In make-to-order (MTO) production systems, customers and orders are needed to be prioritized especially when demand is more than capacity. Prioritizing only orders may not be enough, when the same type of orders are waiting for processing from different customers. On the other hand, a customer may have more than one order of different types, in this case the orders will have different priorities according to the features of different orders. So, both orders and customers need to be prioritized. Like supplier selec-

DOI: 10.4018/978-1-7998-5442-5.ch003

tion or evaluation, customers also need to be evaluated depending on their features and profiles such as loyalty, profitability, payment performance, good relation, being solution oriented etc.

In this chapter a model will be proposed for customer and order prioritization. Although there are too many studies about supplier selection/evaluation, there are very few studies about the downstream of the supply chain, customer selection/evaluation. This topic is the reflection of supplier selection/evaluation on the downstream side of supply chain as customer selection/evaluation or in other words customer and order prioritization. For this purpose, the customers and orders of a common use facility working in metal industry with an MTO manufacturing environment where demand exceeds capacity, customers and orders are prioritized according to determined criteria. The target of the facility is sustainability rather than profitability, so while prioritizing customers and orders criteria about customer potential or long-term relations are used rather than revenue, profitability.

The rest of the chapter is as follows: Literature about customer and order prioritization is reviewed, classified and summarized. Afterwards prioritization and related subjects like order acceptance, demand, revenue, capacity management mentioned. Then application in a metal facility is carried with AHP and TOPSIS. Lastly conclusions are presented.

LITERATURE

Studies on customer and order prioritization are rare. When related literature is reviewed, order prioritization is found under topics of order acceptance or job selection and due date determination. The literature is concentrated on order acceptance/rejection with the subjects of revenue, order and demand management. Slotnick (2011) reviewed and classified the literature about order acceptance and scheduling. Some accessible studies in the related literature are given below. Following studies are related studies in which "order prioritization" phrase used. Guerrero & Kern (1988) introduced and discussed the essential elements for establishing a demand management system and major decisions about order accumulation, order prioritization and capacity allocation involved in demand management. Through careful implementation of demand management procedures, the answers about accepting or rejecting an order, resources allocated to orders and future orders can be clarified. Wang, Yang & Lee (1994) defined a multi criteria decision method for order acceptance and an artificial neural network-based preference model for order prioritization in an environment that demand exceeds capacity. Homburg, Droll & Totzek (2008), studied whether it is worth prioritizing customers. They state that prioritizing customers lead to higher average customer profitability and higher sales return according to their research on two independent samples of 310 firms. The reason for it is indicated as while customer prioritization positively affects relations with upper level customers, there is no effect on lower level customers, furthermore marketing and sales costs decrease. Chamodrakas, Alexopoulou & Martakos (2009) developed a method based on fuzzy sets and TOPSIS for customer evaluation, order acceptance and prioritization. Kalantari, Rabbani & Ebadian (2011) proposed a 5-step decision support system for order acceptance or rejection in a hybrid (MTS/MTO) production environment. At first step customers are prioritized with fuzzy TOPSIS method. In the second step capacity and inventory status are calculated and some undesired orders are rejected. In the next step price and due dates of accepted orders are determined with mathematical model. In the fourth step suggestions guiding negotiations with customers about price and due date are given. In the last step customer approved orders are added to production schedule, unapproved orders are rejected. Huang (2012) proposed a heuristic model for prioritizing customer according to their forecasted lifelong value. It is stated that forecasting lifelong value of a customer correctly is not easy and giving first class service to forecasted upper level customers and economic service to rest of the customers is risky. Manavizadeh, Goodarzi, Rabbani & Jolai (2013) developed a decision support system for order acceptance or rejection in an MTO environment. If the system has not adequate capacity these steps are applied: Firstly, customers are prioritized according to their current and future profitability. Undesired orders are rejected in this step. In the next step price and due dates are determined with a mathematical model for the orders that were not rejected. If the customer satisfaction level is low about price and due date, negotiations are carried on. If negotiation ends with deal the order is accepted and added to the schedule. Schedule of accepted orders are done by minimizing the excess workload. Akyıldız, Kadaifci & Topcu (2015) determined the criteria for prioritizing customer orders with Analytical Network Process in a structural steel manufacturing firm. They determined the criteria as profit rate per unit of time, compatibility of order with available capacity, potential future order with higher profit, customer credit of future business opportunity and negotiability of schedule for the order.

Whybark (1994) examined the influence of marketing on manufacturing practices of companies in four different region of the world; North America, Western Europe, South Korea, China and found that on forecasting stage marketing has a great impact while the impact is weak on the shop floor. Korpela, Kylaheiko, Lehmusvaara & Tuominen (2002) examined customer and supplier relationship in a supply chain in terms of production capacity allocation. Their aim is to allocate limited production capacity to the customers based on strategic importance and the risk involved. They used AHP to evaluate strategic importance and risk of customers and mixed integer programming to allocate capacity to customers.

Studies about job selection or order acceptance are encountered in the literature. Balackandran & Schaefer (1981) developed a non-linear programming model for accepting optimal mix of jobs considering processing times of the products and profit per unit of each product. Matsui (1982) compared two policies (Simple Selection Policy and Sequential Selection Policy) of order selection for maximizing the long-run average return. In another study, Matsui (1985) compared six policies (Single Sales Order Policy, Static Selection Policy, Fixed Selection Policy, Sequential Selection Policy, Periodic Selection Policy, Dynamic Selection Policy) for order selection. In terms of reward rate Dynamic Selection Policy found to be superior to others. Philipoom & Fry (1992) tested three methods whether to accept or reject an order in a congested shop. Pourbabai (1992) proposed a mathematical model for optimal selection of customer orders to maximize net operational profit in group technology and just-in-time production environment. Slotnick & Morton (1996) tried to find an answer to the question "if there are more jobs available to be processed than can be done without incurring time-related penalties such as lateness, how do we decide which ones to accept?" They developed an optimal method and two heuristic approaches maximizing total net profit, defined as revenue of each job minus weighted lateness. In addition to this study Ghosh (1997) showed that the problem is NP-hard and provided two dynamic programming algorithms. Further a fully polynomial time approximation scheme that generates a solution within a specified percentage of the optimal was provided. Lewis & Slotnick (2002) developed a dynamic programming and heuristic algorithms for job selection with the objective of maximizing profit. Calosso, Cantamessa, Vu & Villa (2003) developed a mixed integer programming model for order acceptance and negotiation between firms in e-commerce. Barut & Sridharan (2005) developed a heuristic which attempts to maximize revenue by selectively accepting or rejecting customer orders in an order driven (MTO) production system. The heuristic generates a higher profit when compared to first-come-first-served (FCFS) policy. Ebben, Hans & Weghuis (2005) compared four approaches for workload-based order acceptance in MTO job shop environment. Lin & Chang (2008) developed a method based on mixed integer programming and

TOPSIS for order selection and pricing. Arredondo & Martinez (2010) developed a method considering criteria like product mix, price, size and due date for dynamic order acceptance or rejection in MTO environment. Martinez & Arredondo (2010) used an intelligent decision rule for order acceptance in MTO production system for revenue management. Mestry, Damodaran & Chen (2011) developed a mixed integer linear programming model for order acceptance in an MTO system which has fixed due dates for maximizing overall profit. To solve the MILP model a branch-and-price model is devised. de Man, Andersen & Strandhagen (2017) developed a conceptual performance measurement method for order acceptance. They used utilization of resources, output, flexibility for measuring performance and found a close relationship between order acceptance and sales and operations planning. Aouam, Gerly, Kumar & Brahimi (2018) examined demand flexibility in order acceptance problem by developing a combination of mixed integer linear programming and heuristic models.

Studies about job selection or order acceptance with scheduling are encountered in the literature. Wester, Wijngaard & Zijm (1992) compared three approaches for order acceptance strategies and discussed interdependence between order acceptance, production planning and scheduling. De, Ghosh & Wells (1993) used dynamic programming for job selection and scheduling in a single machine environment with random processing times and deadline. The objective is to select a subset of jobs in order to maximize profit. ten Kate (1994) compared two coordinating mechanisms focusing on interaction of order acceptance and production scheduling. Yang & Geunes (2007) developed heuristic method to maximize profit for job selecting and scheduling in a single resource environment. Oğuz, Salman & Yalçın et al. (2010) developed a mixed integer linear programming model and heuristic method to maximize total revenue for order acceptance and scheduling in a single machine environment with sequence dependent setup times. Zhang, Lu & Yuan (2010) developed dynamic programming algorithm for single machine scheduling problem under job rejection constraint and analyzed computational complexities of the problems for distinct objective functions. Cesaret, Oğuz & Salman (2012) developed a tabu search algorithm for order acceptance and scheduling problem on a single machine with release dates and sequence dependent setup times. Zhong, Ou & Wang (2014) developed a model considering machine availability constraints for order acceptance and scheduling problem. Rahman, Sarker & Essam (2015) developed a simple heuristic and genetic algorithm approach for dynamic order acceptance and scheduling problem in permutation flow shop environment. Esmaeilbeigi, Charkhgard & Charkhgard (2016) developed a mixed integer programming model for order acceptance and scheduling in two machine flow shop environment. Wang, Qi, Cui & Zhang (2019) developed a mathematical model and hybrid algorithm for order acceptance and scheduling problem for a job shop MTS / MTO production environment. A variable dimension particle swarm optimization algorithm selects orders and then Giffler-Thompson algorithm is used for scheduling the selected orders.

Studies about job selection or order acceptance with due date determination or tardiness objective are encountered in the literature. Lawrence (1994) presented a methodology for negotiating due dates with customers and producers in complex manufacturing environments. Slotnick & Morton (2007) developed optimal and heuristic methods solving order acceptance and sequencing problems in a facility with limited capacity, late delivery discount and no penalty or reward in early delivery. Gharehgozli, Rabbani, Zaerpour & Razmi (2008) developed a two-step decision making structure to manage incoming orders. At the first step orders are evaluated according to their due dates and material arrival times and at the second step orders are accepted or rejected according to manufacturing system capacity with AHP and TOPSIS. Rom and Slotnick (2009) used genetic algorithm to maximize profit for order acceptance with tardiness penalties. Xiao, Zhang, Zhao & Kaku (2012) developed a simulated annealing-based solution procedure for permutation flow shop scheduling problem with order acceptance and weighted tardiness.

DEMAND, ORDER, REVENUE AND CAPACITY MANAGEMENT

There are different production policies like Make-To-Stock (MTS), Make-To-Order (MTO), Assemble-To-Order (ATO) and Engineer-To-Order (ETO), each of which have some positive and negative aspects. MTS satisfies customer demand with finished good inventory, short delivery time, high holding cost and low flexibility. MTO, on the other hand, makes production according to customer order with long delivery time, low holding cost and high flexibility. In MTS production environment push system is used, production planning is done according to forecasted demand and order acceptance is out of concern. However, in MTO production environment which uses pull system, order acceptance is the starting point of production planning (Kalanatari et al., 2011: 1363-4). If the product diversity is large it's better to produce in an MTO manner rather than MTS. MTO systems works with pull system in which production starts with customer order (Gharehgozli et al., 2008: 1016).

Divesification of customer needs and wants, mass customization, pull systems, lean manufacturing are the drivers of MTO systems (Barut & Sridharan, 2005: 287). Small batches, large variety of product and routing are the characteristics of MTO systems. In some cases, order features are not completely known during order acceptance, so evaluating the effect of accepting order to the shop floor is difficult (Ebben et al., 2005: 108). MTO manufacturing firms often face demand exceeding capacity. Capacity allocation between competing customers and orders, achieving on-time delivery is the challenging part of the situation (Sridharan, 1998: 601). While working with low finished product inventory, MTO systems has an important disadvantage because of limited capacity. To stay competitive firms must reduce lead time of orders and this is up to usage of a regular production plan determining which orders will be accepted and scheduled (Esmaeilbeigi et al. 2016: 419). Businesses working with MTO has three goals: Smoothing the workload at each station, providing a constant utilization rate for all parts and minimizing setup costs (Manavizadeh et al., 2013: 2532).

Marketing and manufacturing have a special relation; while they are working closely, they may also conflict as most business functions (Clare & Sanford, 1984; Crittenden, 1992; Konijnendijk, 1994). While concern of marketing is about external environment like satisfying customers, concern of manufacturing is about internal environment like cost control, product quality, on time delivery (Sridharan, 1998: 601). Order prioritization, order acceptance, capacity management, scheduling and due date determination processes are interdependent and related with demand, order, revenue and capacity management (Figure 1). Some of these concepts are in the field of marketing, some of them are in the field of manufacturing, and some of them are in both.

Figure 1. Order prioritization and acceptance in marketing and manufacturing interface

Demand management is the critical link between marketing and manufacturing. It requires material coordination, matching sales level with available capacity (Sridharan, 1998: 602). While demand management and order management used synonymously, demand management is broader and include order management. Order management consists of arrival, accumulation, prioritization and acceptance of orders considering capacity allocation. Order management tries to balance shop floor capacity and workload to realize satisfactory due dates. Order management problems can be described with four dimensions (Wang et al., 1994: 1-4):

- The characteristics of the order stream (MTO or MTS)
- The structure of products to be manufactured (MRP or non-MRP)
- The level of planning decision (Aggregate vs. individual level)
- The flexibility of order acceptance (Negotiable vs. non-negotiable)

The success of MTO system largely depends on the selectivity of the order acceptance policy, which will maximize the average revenue relative to the capacity needed. An arriving order may be accepted, negotiated or rejected depending on the revenue contribution and schedule availability. Revenue Management evaluates the opportunity costs of accepting and rejecting an arriving order (Martinez & Arredondo, 2010: 1189-90).

Revenue management besides a tool for service operations like airline, railway, hotel, healthcare, electric utility, car rental, advertising; has potential for MTO manufacturing environments. It tries to maximize revenue from existing capacity by using price discrimination, capacity reallocation and overbooking for order acceptance or rejection. (Harris & Pinder,1995: 299). While accepting an order; revenue, resource usage and delay in other jobs must be compared. Sales department bring order and manufacturing deals with limited capacity. Late deliveries result in lost orders in the future besides penalties (Slotnick & Morton, 2007: 3029). According to customer relationship management, lifetime value of customers are heterogenous and this value can and must be measured. It is generally accepted to pamper upper class customers and to ignore less profitable customers (Huang, 2012: 497-8).

Order acceptance is a critical decision-making problem between customer relationship management and production planning for MTO firms. To cope with the challenges of MTO, the order management system should use decision support tools to combine selective order negotiation with planning and scheduling to react quickly. Successful MTO firms should carefully choose which orders are worth accepting or negotiating, or not, from an opportunity cost and revenue management perspective. Success of an MTO firm is up to a successful order acceptance policy (Arredondo & Martinez, 2010: 70).

New order acceptance decisions cannot be made without considering the congestion of shop floor (ten Kate, 1994: 141). Orders arrive irregularly and have features such as processing time, due date. Selection (accept or reject) of orders is an important problem and generally solved to maximize total profit rather than production rate (Matsui, 1982: 201). Order acceptance is also related with revenue management (Mestry et al., 2011: 482). As a tactical managerial decision, order acceptance also deal with due date and price determination (Ebben et al., 2005: 107).

While order acceptance and scheduling decisions must be made simultaneously, two problems arise: One is selection of a subset among orders arrived and the other is how to sequence and schedule selected orders with an objective (Xiao et al., 2012: 7913). If on-time delivery is a significant part of competitive strategy of a company, rejecting some orders may be crucial to avoid delays. On the other hand, long term effect of rejection orders on firm prestige must be considered (Lewis & Slotnick, 2002: 1082). In practice order acceptance and production planning are separated. Generally, sales department are responsible for order acceptance and they have a tendency for accepting all orders without considering the capacity. On the other side production department plans the production by trying to maximize capacity utilization and on-time delivery (Ebben et al., 2005: 108). Accepting all orders will lead to an over-load in shop floor and unmet deadlines. Overtime and outsourcing may be a solution in short term which increase costs; tardy deliveries may also occur penalty costs and unsatisfied customers (Mestry et al., 2011: 480-481). The optimal order acceptance and rejection decisions between arriving orders have crucial impact on profit, market share and production planning of a company. Choosing best set of orders among arriving orders constitute the main point of a good order acceptance policy (Kalanatari et al., 2011: 1364). Factors that affect order acceptance decisions in an MTO environment are decision period, order size, due date requirement, shop capacity constraint and order preference (Wang et al., 1994: 5).

Order acceptance and scheduling decisions are done at the same time in MTO systems because of limited capacity and on-time delivery. Orders are accepted with considering their effect on capacity. Otherwise accepting orders more than capacity may cause delays or due date violations, resulting in penalties, unsatisfied customers and lost revenues in the long run (Cesaret et al., 2012: 1197). If an MTO manufacturing system integrates order acceptance and capacity planning, it has an opportunity to maximize profits (Mestry et al., 2011: 495). In the condition of demand is higher than capacity the most reasonable option seems like accepting most profitable customers / orders and rejecting or ignoring less profitable ones. However, there are other criteria than profit maximization. There may be a minimum service level for all products to stay in business (Sridharan, 1998: 604). Order acceptance or prioritization based on profitability while on the one hand increases profits, on the other hand may cause penalties of missing due dates (Slotnick & Morton, 1996: 132).

Capacity is allocated between customers and orders considering delivery performance when demand is higher than capacity. Increasing capacity is a long-term decision, and cost of such decision is high. So, in short term the objective is allocating the capacity to maximize profit and on-time delivery (Sridharan, 1998: 602). If the capacity utilization is high, accepting new orders may cause the overload and congestion in shop floor, thus it is useful to consider rejecting new orders. While rejection may cause dissatisfaction, accepting and failing to deliver on time of an order probably causes more dissatisfaction. Accepting an order and setting a delivery time that exceeds the requirements of customer may be an alternative to rejection (Philipoom & Fry, 1992: 2562).

In service operations lallocating capacity between competing customers (demand) is especially done with price adjustments (Sridharan, 1998: 603). A reserve capacity may be hold for emerging orders from customers willing to pay extra. In this manner reserve capacity represents a revenue opportunity rather than a capacity loss. Revenue management models of capacity allocation considers relative contribution margin of customer segments. Revenue management for manufacturing operations makes a tradeoff between maximizing scheduled capacity utilization and maximizing revenue (Harris & Pinder, 1995: 300).

Capacity rationing, discriminating to maximize a specified objective may be used in case of excess demand. MTO firms dynamically allocate the available capacity to different priority customers and orders. This result in rejecting some lower priority customers and orders. However rationing increases profit by sacrificing service level for lower priority (Sridharan, 1998: 604-6). Prioritization based not only on profit maximization but criteria considering long term customer relations and sustainability may be more useful. Customer prioritization leads to higher average customer profitability and higher sales return. While customer prioritization positively affects high level customers, there is almost no effect on low level customers. On the other hand, there are some counter argument about customer prioritization. Low level customers may complain, and this may reduce sales and profits in long-term. Focusing on limited number of customers may lead to ignore economies of scale (Homburg et al. 2008: 110). Customer prioritization is especially used in call centers and banking sector. When a call center is called from personal phone number the system automatically recognizes the caller and may prioritize according to the profile of the caller. Bank offices meets the customers with q-matics, customers take a number with their debit cards or customer number and the number is given according to the selected operation and profile of the customer.

APPLICATION IN METAL INDUSTRY

The application of the study is done in a metal shop of a Common Use Facility (CUF) that serves to small industrialists. It works in an MTO manufacturing environment and has one 5-axis CNC machine used for surface treatment, grooving etc. As it is the single machine in the region with advanced technology, by the help of promotion activities and participation in related fairs, number of orders increased within time and exceeding the capacity of the machine. When needed, 2 operators work at 2 shifts for delivering orders on time. Nevertheless, orders may be accepted and postponed to a forward date. In this case prioritizing customers and orders waiting for processing is considered.

For this reason, the criteria for prioritization are determined, Analytical Hierarchy Process (AHP) is used for weighting the criteria and Technique for Order Preference by Similarity to Ideal Solution (TOPSIS) is used for final prioritization of orders considering customers. AHP was devised by Thomas L. Saaty, as a general theory of measurement used to derive ratio scales from both discrete and continuous paired comparisons which found its widest applications in multicriteria decision making (Saaty, 1987: 161). AHP is used in many diverse areas such as supplier selection (Chan, 2003), convention site selection (Chen, 2006), electric power plant selection (Akash et al., 1999), site location selection (Mohajeri & Amin, 2010; Roig-Tierno et al., 2013), project selection (Palcic & Lalic, 2009), design concept selection (Hambali et al., 2009), construction systems ranking (Lam et al., 2007), selection of collection center location for reverse logistics (Kannan et al., 2008). A study reviewing the applications of AHP categorized the studies under the topics; selection, evaluation, benefit-cost, allocation, planning and development, priority and ranking, decision making, forecasting, medicine and QFD (Vaidya & Kumar, 2006). TOPSIS was proposed by Hwang and Yoon based on the concept of chosen alternative to have shortest distance from ideal solution and farthest from the negative ideal solution (Hwang & Yoon, 1981: 128). TOPSIS is used in many diverse areas such as information system selection (Huang, 2008), evaluation of weapon systems (Hui & Bifeng, 2009), optimal configuration selection for manufacturing system (Goyal et al, 2012), mutual fund performance evaluation (Chang et al., 2010), construction process safety solutions evaluation (Liaudanskiene et al., 2015), material selection for microelectromechanical systems (Yazdani & Payam, 2015), sustainable development analysis of countries (Balcerzak & Pietrzak, 2016). A study reviewing the applications of TOPSIS categorized the studies under the topics; supply chain management and logistics, design, engineering and manufacturing systems, business and marketing management, health, safety and environment management, human resource management, energy management, chemical engineering, water resource management (Behzadian et al., 2012).

In the determination of the criteria, the opinions of the production manager and customer relationship manager are taken, and criteria were established accordingly. In this step, 5 criteria were determined as in Table 1.

Table 1. Criteria for customer and order prioritization

Abbreviation	Criteria
PAP	Payment Performance
DRP	Delivery and Retrieval Performance
COO	Continuity of Order
ORF	Order Features
PFP	Potential for Problems

Past experiences show that some customers may delay the payment after delivery. And it may cause a financial problem for the firm. So, Payment Performance (PAP) is determined as a criterion, it is scored out of 10 and a new customer starts from 5. With timely payments the score is increased, delayed payments decrease the score. The firm serves to the industrialists in the region, so works on materials or parts that belong to the customer. After the order is accepted the customer is informed about the time interval of processing and expected to deliver and retrieve on-time since the stocking area of the facility is limited. So, Delivery and Retrieval Performance (DRP) is determined as a criterion, it is scored out of 10 and a new customer starts from 5. With on time delivery and retrieval the score is increased, late delivery and retrieval decreases the score. Continuity of Order (COO) is determined as a criterion for both rewarding the loyalty of customers and ongoing or potential continuity of order. A new customer and a new order is scored 1, if the customer and order comes again they are scored separately and the average is calculated with maximum score of 10. Order Features (ORF) consists of some features of order such as processing time, setup time, difficulty of the process, due date of the order etc. ORF is scored from 1 to 5, as the process and setup time and difficulty increase, ORF score increases. Potential for Problems (PFP) is determined as a criterion because of past experiences of problems with some customers like customer want to change the approved design after the process, consuming too much time of the designer or operator. A new customer starts with 1 and scored up to 5.

Table 2. The fundamental scale of AHP

Value	Definition
1	Equal importance
3	Moderate importance
5	Essential or strong importance
7	Very strong importance
9	Extreme importance
2,4,6,8	Intermediate values

Table 3. Pairwise comparison matrix of production manager

	PAP	DRP	COO	ORF	PFP
PAP	1	2	1	2	4
DRP	0.5	1	0.333	2	4
COO	1	3	1	2	3
ORF	0.5	0.5	0.5	1	4
PFP	0.25	0.25	0.333	0.25	1

Table 4. Pairwise comparison matrix of customer relationship manager

	PAP	DRP	COO	ORF	PFP
PAP	1	2	1	0.5	3
DRP	0.5	1	0.333	0.5	2
COO	1	3	1	0.25	3
ORF	2	2	4	1	3
PFP	0.333	0.5	0.333	0.333	1

Two managers made pairwise comparisons between these 5 criteria. For a set of 5 criteria each manager made $\binom{5}{2} = 10$ comparisons. In pairwise comparison the scale in Table 2 is used to indicate preference of one criterion over another (Saaty, 1987: 163).

Table 5. Combined matrix of two managers

	PAP	DRP	COO	ORF	PFP
PAP	1	2	1	1	3.464
DRP	0.5	1	0.333	1	2.828
COO	1	3	1	0.707	3
ORF	1	1	1.414	1	3.464
PFP	0.289	0.354	0.333	0.289	1

Pairwise comparison made by production manager and customer relationship manager are transformed to pairwise comparison matrices respectively in Table 3 and 4 below. The values in the matrices indicates how the criteria in the row is important compared to the criteria in the column. As every criterion is equally important when compared to itself, the diagonal values in the matrix are 1. The matrices are reciprocal ($a_{ij} = 1/a_{ji}$)

For reciprocal property to be protected, geometric means of pairwise comparisons of two managers are calculated and a single dominance matrix is obtained as in Table 5.

All the calculations for both methods were done in MS Excel. Pairwise comparisons are found to be consistent. Consistency Ratio of production manager's pairwise comparisons is calculated as 0,069; consistency ratio of customer relationship manager is calculated as 0,058 and consistency ratio of geometric means of these two is calculated as 0,040. Consistency ratio lower than 0.10 is assumed to be consistent.

Criteria weights are calculated as in Table 6 by synthesizing process of AHP. Synthesizing is simply normalizing the matrix by dividing every element of the matrix to the sum of that column. Average row values of normalized matrix give the weights of each criteria belonging that row. Continuity of Order (COO), Payment Performance (PAP) and Order Features (ORF) has similar weights.

Table 6. Criteria weights for customer and order prioritization

Criteria	PAP	DRP	COO	ORF	PFP
Weight	0.257	0.161	0.262	0.250	0.070

Order data for a month period is taken and 10 sample orders are selected, and priorities of orders are calculated with TOPSIS. Table 7 shows the scores of orders. There are 5 type of orders (A, B, C, D, E) coming from 7 different customers (C1, ..., C7) and total 10 separate orders. Same customer may have different orders and same type of order may be given by different customers. This table constitutes the decision matrix for TOPSIS. In first three criteria (PAP, DRP, COO) the score is desired to be higher for prioritizing customer and order, however in last two criteria (ORF, PFP) the score is desired to be lower.

Table 7. Order information and decision matrix

Order	Order Type	Customer	PAP	DRP	COO	ORF	PFP
O1	A	C1	7	8	5.5	2	2
O2	B	C2	8	8	7.5	4	1
O3	C	C3	4	5	4.5	3	3
O4	A	C4	6	3	3.5	2	4
O5	D	C1	7	8	6.5	1	2
O6	E	C5	5	5	2.5	5	1
O7	C	C6	2	6	4.5	3	5
O8	D	C3	4	5	4	1	3
O9	B	C7	9	7	7.5	4	1
O10	E	C2	8	8	6	5	1

As in the steps of TOPSIS (Hwang & Yoon, 1981: 128-140), decision matrix is normalized. Normalized decision matrix is calculated by dividing every value to square root of sum of squares of all values in that column. In the next step every column of normalized decision matrix is weighted with the weights determined with AHP (Table 6), so normalized decision matrix is calculated. Ideal and negative ideal solutions are determined for every criterion, for the first three criteria ideal solutions are maximum values of each column and negative ideal solutions are minimum values of each column, and vice versa for the las two criteria. Distance to ideal and negative ideal solutions of every order is calculated by square root of sum of square of differences between order values and ideal or negative ideal values. Lastly relative closeness to ideal solution of every order is calculated. Relative closeness of every order is calculated by distance to negative ideal solution divided by sum of distance to ideal and negative ideal solution. Orders are ranked in a descending order of relative closeness; Table 8 shows the final priorities of orders.

Table 8. Final priorities of orders

Order	O5	O1	O9	O2	O8	O10	O4	O3	O7	O6
Relative Closeness	0.817	0.711	0.641	0.627	0.541	0.519	0.502	0.407	0.343	0.283

The relative closeness values can be considered as prioritizing rule for scheduling orders and due dates can be determined by this way. An order with a priority value less than a critical value may be rejected, so order prioritization may be used as an order acceptance or rejection rule.

CONCLUSION

Customer and order prioritization is prioritizing the orders according to features of customers and orders. In make-to-order production systems in condition of demand exceeding capacity, arriving orders must be prioritized and selected according to different criteria. Accepting orders more than capacity will lead to overload, congestion, delays on delivery time and unsatisfied customers. Profit maximization is a generally accepted criteria for this situation especially in service sector. Profit maximization may be useful in short-term but, in the long-term using other criteria for sustainability reasons would be more helpful.

Both customer and order features must be used together for prioritization, since a customer may have more than one order and as one type of order may be given by different customers. Prioritization values may be used for accepting and scheduling orders or selecting customers as a reflection of supplier selection. Customers need to be evaluated depending on their features and profiles such as their loyalty, continuity, profitability, payment performance; similar to supplier evaluation / selection.

In this chapter customers and orders of a metal processing facility are prioritized according to determined criteria. The target of the facility is sustainability rather than profitability, so while prioritizing customers and orders criteria about customer potential or long-term relations are used rather than revenue, profitability. Production manager and customer relationship manager of the facility were interviewed for determining the criteria for prioritizing customers and orders. After determining the criteria, managers were asked to pairwise compare the criteria for weighting with Analytical Hierarchy Process (AHP). Then by considering scores of customers and orders for each criterion, orders were prioritized with Technique for Order Preference by Similarity to Ideal Solution (TOPSIS).

The prioritization of customers and orders may be used for order selection or acceptance by selecting the orders that have a priority more than a predetermined value. For sequencing and scheduling jobs it can be used as a priority dispatching rule. It may also be used for evaluating customers as a tool of customer relationship management that indicates which customers to focus on. Lastly not only in condition of demand exceeding capacity, but also even in condition of demand lower than capacity, prioritization can be used for serving desired customers first in order to maintain long-term and sustainable relations.

REFERENCES

Akash, B. A., Mamlook, R., & Mohsen, M. S. (1999). Multi-criteria selection of electric power plants using analytical hierarchy process. *Electric Power Systems Research*, *52*(1), 29–35.

Akyildiz, B., Kadaifci, C., & Topcu, I. (2015). A decision framework proposal for customer order prioritization: A case study for a structural steel company. *International Journal of Production Economics*, *169*, 21–30. doi:10.1016/j.ijpe.2015.07.004

Aouam, T., Geryl, K., Kumar, K., & Brahimi, N. (2018). Production planning with order acceptance and demand uncertainty. *Computers & Operations Research*, *91*, 145–159. doi:10.1016/j.cor.2017.11.013

Arredondo, F., & Martinez, E. (2010). Learning and adaptation of a policy for dynamic order acceptance in make-to-order manufacturing. *Computers & Industrial Engineering*, *58*(1), 70–83. doi:10.1016/j.cie.2009.08.005

Balackandran, K., & Schaefer, M. E. (1981). Optimal acceptance of job orders. *International Journal of Production Research, 19*(2), 195–200. doi:10.1080/00207548108956641

Balcerzak, A. P., & Pietrzak, M. B. (2016). Application of TOPSIS method for analysis of sustainable development in European Union countries. *Chapters, 1*, 82–92.

Barut, M., & Sridharan, V. (2005). Revenue management in order-driven production systems. *Decision Sciences, 36*(2), 287–316. doi:10.1111/j.1540-5414.2005.00074.x

Behzadian, M., Otaghsara, S. K., Yazdani, M., & Ignatius, J. (2012). A state-of the-art survey of TOPSIS applications. *Expert Systems with Applications, 39*(17), 13051–13069. doi:10.1016/j.eswa.2012.05.056

Calosso, T., Cantamessa, M., Vu, D., & Villa, A. (2003). Production planning and order acceptance in business to business electronic commerce. *International Journal of Production Economics, 85*(2), 233–249. doi:10.1016/S0925-5273(03)00112-9

Cesaret, B., Oğuz, C., & Salman, F. S. (2012). A tabu search algorithm for order acceptance and scheduling. *Computers & Operations Research, 39*(6), 1197–1205. doi:10.1016/j.cor.2010.09.018

Chamodrakas, I., Alexopoulou, N., & Martakos, D. (2009). Customer evaluation for order acceptance using a novel class of fuzzy methods based on TOPSIS. *Expert Systems with Applications, 36*(4), 7409–7415. doi:10.1016/j.eswa.2008.09.050

Chan, F. S. (2003). Interactive selection model for supplier selection process: An analytical hierarchy process approach. *International Journal of Production Research, 41*(15), 3549–3579.

Chan, F. S. (2003). Interactive selection model for supplier selection process: An analytical hierarchy process approach. *International Journal of Production Research, 41*(15), 3549–3579. doi:10.1080/0020754031000138358

Chang, C. H., Lin, J. J., Lin, J. H., & Chiang, M. C. (2010). Domestic open-end equity mutual fund performance evaluation using extended TOPSIS method with different distance approaches. *Expert Systems with Applications, 37*(6), 4642–4649. doi:10.1016/j.eswa.2009.12.044

Chen, C. F. (2006). Applying the analytical hierarchy process (AHP) approach to convention site selection. *Journal of Travel Research, 45*(2), 167–174. doi:10.1177/0047287506291593

Clare, D. A., & Sanford, D. G. (1984). Cooperation and conflict between industrial sales and production. *Industrial Marketing Management, 13*(3), 163–169. doi:10.1016/0019-8501(84)90029-4

Crittenden, V. L. (1992). Close the marketing/manufacturing gap. *MIT Sloan Management Review, 33*(3), 41.

De, P., Ghosh, J. B., & Wells, C. E. (1993). Job selection and sequencing on a single machine in a random environment. *European Journal of Operational Research, 70*(3), 425–431. doi:10.1016/0377-2217(93)90252-I

De Man, J. C., Andersen, B. S., & Strandhagen, J. O. (2017). Addressing the performance of order acceptance. Procedia CIRP, 63, 301–306. doi:10.1016/j.procir.2017.03.101

Ebben, M. J., Hans, E. W., & Weghuis, F. O. (2005). Workload based order acceptance in job shop environments. *OR-Spektrum*, *27*(1), 107–122. doi:10.100700291-004-0171-9

Esmaeilbeigi, R., Charkhgard, P., & Charkhgard, H. (2016). Order acceptance and scheduling problems in two-machine flow shops: New mixed integer programming formulations. *European Journal of Operational Research*, *251*(2), 419–431. doi:10.1016/j.ejor.2015.11.036

Gharehgozli, A. H., Rabbani, M., Zaerpour, N., & Razmi, J. (2008). A comprehensive decision-making structure for acceptance/rejection of incoming orders in make-to-order environments. *International Journal of Advanced Manufacturing Technology*, *39*(9-10), 1016–1032. doi:10.100700170-007-1275-6

Ghosh, J. B. (1997). Job selection in a heavily loaded shop. *Computers & Operations Research*, *24*(2), 141–145. doi:10.1016/S0305-0548(96)00045-7

Goyal, K. K., Jain, P. K., & Jain, M. (2012). Optimal configuration selection for reconfigurable manufacturing system using NSGA II and TOPSIS. *International Journal of Production Research*, *50*(15), 4175–4191. doi:10.1080/00207543.2011.599345

Guerrero, H. H., & Kern, G. M. (1988). How to more effectively accept and refuse orders. *Production and Inventory Management Journal*, *29*(4), 59–62.

Hambali, A., Sapuan, S. M., Ismail, N., & Nukman, Y. (2009). Application of analytical hierarchy process in the design concept selection of automotive composite bumper beam during the conceptual design stage. *Scientific Research and Essays*, *4*(4), 198–211.

Harris, F. H. D., & Pinder, J. P. (1995). A revenue management approach to demand management and order booking in assemble-to-order manufacturing. *Journal of Operations Management*, *13*(4), 299–309. doi:10.1016/0272-6963(95)00029-1

Homburg, C., Droll, M., & Totzek, D. (2008). Customer prioritization: Does it pay off, and how should it be implemented? *Journal of Marketing*, *72*(5), 110–130. doi:10.1509/jmkg.72.5.110

Huang, C. Y. (2012). To model, or not to model: Forecasting for customer prioritization. *International Journal of Forecasting*, *28*(2), 497–506. doi:10.1016/j.ijforecast.2011.04.004

Huang, J. (2008, September). Combining entropy weight and TOPSIS method for information system selection. In *2008 IEEE Conference on Cybernetics and Intelligent Systems* (pp. 1281-1284). IEEE. doi:10.1109/ICCIS.2008.4670971

Hui, G., & Bifeng, S. (2009). Study on effectiveness evaluation of weapon systems based on grey relational analysis and TOPSIS. *Journal of Systems Engineering and Electronics*, *20*(1), 106–111.

Hwang, C. L., & Yoon, K. (1981). Multiple attribute decision making: Methods and Applications (*Vol. 104*). Springer-Verlag. doi:10.1007/978-3-642-48318-9

Kalantari, M., Rabbani, M., & Ebadian, M. (2011). A decision support system for order acceptance/rejection in hybrid MTS/MTO production systems. *Applied Mathematical Modelling*, *35*(3), 1363–1377. doi:10.1016/j.apm.2010.09.015

Kannan, G., Haq, A. N., & Sasikumar, P. (2008). An application of the analytical hierarchy process and fuzzy analytical hierarchy process in the selection of collecting centre location for the reverse logistics multicriteria decision-making supply chain model. *International Journal of Management and Decision Making*, *9*(4), 350–365. doi:10.1504/IJMDM.2008.019360

Konijnendijk, P. A. (1994). Coordinating marketing and manufacturing in ETO companies. *International Journal of Production Economics*, *37*(1), 19–26. doi:10.1016/0925-5273(94)90004-3

Korpela, J., Kyläheiko, K., Lehmusvaara, A., & Tuominen, M. (2002). An analytic approach to production capacity allocation and supply chain design. *International Journal of Production Economics*, *78*(2), 187–195. doi:10.1016/S0925-5273(01)00101-3

Lam, P. T., Chan, A. P., Wong, F. K., & Wong, F. W. (2007). Constructability rankings of construction systems based on the analytical hierarchy process. *Journal of Architectural Engineering*, *13*(1), 36–43. doi:10.1061/(ASCE)1076-0431(2007)13:1(36)

Lawrence, S. R. (1994). Negotiating due-dates between customers and producers. *International Journal of Production Economics*, *37*(1), 127–138. doi:10.1016/0925-5273(94)90013-2

Lewis, H. F., & Slotnick, S. A. (2002). Multi-period job selection: Planning work loads to maximize profit. *Computers & Operations Research*, *29*(8), 1081–1098. doi:10.1016/S0305-0548(00)00105-2

Liaudanskienė, R., Ustinovičius, L., & Bogdanovičius, A. (2015). Evaluation of construction process safety solutions using the TOPSIS method. *The Engineering Economist*, *64*(4).

Lin, H. T., & Chang, W. L. (2008). Order selection and pricing methods using flexible quantity and fuzzy approach for buyer evaluation. *European Journal of Operational Research*, *187*(2), 415–428. doi:10.1016/j.ejor.2007.03.003

Manavizadeh, N., Goodarzi, A. H., Rabbani, M., & Jolai, F. (2013). Order acceptance/rejection policies in determining the sequence in mixed model assembly lines. *Applied Mathematical Modelling*, *37*(4), 2531–2551. doi:10.1016/j.apm.2012.06.012

Martínez, E., & Arredondo, F. (2010). Order acceptance for revenue management and capacity allocation in make-to-order batch plants. *Computer-Aided Chemical Engineering*, *28*, 1189–1194. doi:10.1016/S1570-7946(10)28199-3

Matsui, M. (1982). Job-shop model: A M/(G, G)/1 (N) production system with order selection. *International Journal of Production Research*, *20*(2), 201–210. doi:10.1080/00207548208947761

Matsui, M. (1985). Optimal order-selection policies for a job shop production system. *International Journal of Production Research*, *23*(1), 21–31. doi:10.1080/00207548508904688

Mestry, S., Damodaran, P., & Chen, C. S. (2011). A branch and price solution approach for order acceptance and capacity planning in make-to-order operations. *European Journal of Operational Research*, *211*(3), 480–495. doi:10.1016/j.ejor.2011.01.002

Mohajeri, N., & Amin, G. R. (2010). Railway station site selection using analytical hierarchy process and data envelopment analysis. *Computers & Industrial Engineering*, *59*(1), 107–114. doi:10.1016/j.cie.2010.03.006

Oğuz, C., Salman, F. S., & Yalçın, Z. B. (2010). Order acceptance and scheduling decisions in make-to-order systems. *International Journal of Production Economics*, *125*(1), 200–211. doi:10.1016/j.ijpe.2010.02.002

Palcic, I., & Lalic, B. (2009). Analytical Hierarchy Process as a tool for selecting and evaluating projects. *International Journal of Simulation Modelling*, *8*(1), 16–26. doi:10.2507/IJSIMM08(1)2.112

Philipoom, P. R., & Fry, T. D. (1992). Capacity-based order review/release strategies to improve manufacturing performance. *International Journal of Production Research*, *30*(11), 2559–2572. doi:10.1080/00207549208948176

Pourbabai, B. (1992). Optimal selection of orders in a just-in-time manufacturing environment: A loading model for a computer integrated manufacturing system. *International Journal of Computer Integrated Manufacturing*, *5*(1), 38–44. doi:10.1080/09511199208547564

Rahman, H. F., Sarker, R., & Essam, D. (2015). A real-time order acceptance and scheduling approach for permutation flow shop problems. *European Journal of Operational Research*, *247*(2), 488–503. doi:10.1016/j.ejor.2015.06.018

Roig-Tierno, N., Baviera-Puig, A., Buitrago-Vera, J., & Mas-Verdu, F. (2013). The retail site location decision process using GIS and the analytical hierarchy process. *Applied Geography (Sevenoaks, England)*, *40*, 191–198. doi:10.1016/j.apgeog.2013.03.005

Rom, W. O., & Slotnick, S. A. (2009). Order acceptance using genetic algorithms. *Computers & Operations Research*, *36*(6), 1758–1767. doi:10.1016/j.cor.2008.04.010

Saaty, R. W. (1987). The analytic hierarchy process—What it is and how it is used. *Mathematical Modelling*, *9*(3-5), 161–176. doi:10.1016/0270-0255(87)90473-8

Slotnick, S. A. (2011). Order acceptance and scheduling: A taxonomy and review. *European Journal of Operational Research*, *212*(1), 1–11. doi:10.1016/j.ejor.2010.09.042

Slotnick, S. A., & Morton, T. E. (1996). Selecting jobs for a heavily loaded shop with lateness penalties. *Computers & Operations Research*, *23*(2), 131–140. doi:10.1016/0305-0548(95)00015-E

Slotnick, S. A., & Morton, T. E. (2007). Order acceptance with weighted tardiness. *Computers & Operations Research*, *34*(10), 3029–3042. doi:10.1016/j.cor.2005.11.012

Sridharan, S. V. (1998). Managing capacity in tightly constrained systems. *International Journal of Production Economics*, *56*, 601–610. doi:10.1016/S0925-5273(98)00045-0

ten Kate, H. A. (1994). Towards a better understanding of order acceptance. *International Journal of Production Economics*, *37*(1), 139–152. doi:10.1016/0925-5273(94)90014-0

Vaidya, O. S., & Kumar, S. (2006). Analytic hierarchy process: An overview of applications. *European Journal of Operational Research*, *169*(1), 1–29. doi:10.1016/j.ejor.2004.04.028

Wang, J., Yang, J. Q., & Lee, H. (1994). Multicriteria order acceptance decision support in over-demanded job shops: A neural network approach. *Mathematical and Computer Modelling*, *19*(5), 1–19. doi:10.1016/0895-7177(94)90086-8

Wang, Z., Qi, Y., Cui, H., & Zhang, J. (2019). A hybrid algorithm for order acceptance and scheduling problem in make-to-stock/make-to-order industries. *Computers & Industrial Engineering*, *127*, 841–852. doi:10.1016/j.cie.2018.11.021

Wester, F. A. W., Wijngaard, J., & Zijm, W. H. M. (1992). Order acceptance strategies in a production-to-order environment with setup times and due-dates. *International Journal of Production Research*, *30*(6), 1313–1326. doi:10.1080/00207549208942959

Whybark, D. C. (1994). Marketing's influence on manufacturing practices. *International Journal of Production Economics*, *37*(1), 41–50. doi:10.1016/0925-5273(94)90006-X

Xiao, Y. Y., Zhang, R. Q., Zhao, Q. H., & Kaku, I. (2012). Permutation flow shop scheduling with order acceptance and weighted tardiness. *Applied Mathematics and Computation*, *218*(15), 7911–7926. doi:10.1016/j.amc.2012.01.073

Yang, B., & Geunes, J. (2007). A single resource scheduling problem with job-selection flexibility, tardiness costs and controllable processing times. *Computers & Industrial Engineering*, *53*(3), 420–432. doi:10.1016/j.cie.2007.02.005

Yazdani, M., & Payam, A. F. (2015). A comparative study on material selection of microelectromechanical systems electrostatic actuators using Ashby, VIKOR and TOPSIS. *Materials & Design*, *65*, 328–334.

Zhang, L., Lu, L., & Yuan, J. (2010). Single-machine scheduling under the job rejection constraint. *Theoretical Computer Science*, *411*(16-18), 1877–1882. doi:10.1016/j.tcs.2010.02.006

Zhong, X., Ou, J., & Wang, G. (2014). Order acceptance and scheduling with machine availability constraints. *European Journal of Operational Research*, *232*(3), 435–441. doi:10.1016/j.ejor.2013.07.032

ADDITIONAL READING

Adenso-Díaz, B., González-Torre, P., & Garcia, V. (2002). A capacity management model in service industries. *International Journal of Service Industry Management*, *13*(3), 286–302. doi:10.1108/09564230210431983

Belton, V., & Stewart, T. (2002). *Multiple criteria decision analysis: an integrated approach*. Springer Science & Business Media. doi:10.1007/978-1-4615-1495-4

Chen, I. J., Calantone, R. J., & Chung, C. H. (1992). The marketing-manufacturing interface and manufacturing flexibility. *Omega*, *20*(4), 431–443. doi:10.1016/0305-0483(92)90018-3

Croxton, K. L., Lambert, D. M., García-Dastugue, S. J., & Rogers, D. S. (2002). The demand management process. *International Journal of Logistics Management*, *13*(2), 51–66. doi:10.1108/09574090210806423

Fiigueira, J., Greco, S., & Ehrgott, M. (2005). *Multi Criteria Decision Analysis: State of the Art Surveys*. Springer. doi:10.1007/b100605

McGill, J. I., & Van Ryzin, G. J. (1999). Revenue management: Research overview and prospects. *Transportation Science*, *33*(2), 233–256. doi:10.1287/trsc.33.2.233

Satty, T. L., & Vargas, L. G. (2012). Models, methods, concepts and applications of the analytic hierarchy process. *International Series in Operations Research & Management Science*, 175.

Yu-Lee, R. T. (2002). *Essentials of capacity management* (Vol. 17). John Wiley & Sons.

KEY TERMS AND DEFINITIONS

Analytical Hierarchy Process: A method for weighting criteria or alternatives according to pairwise comparison and synthesizing process.

Customer Prioritization: Prioritizing customers according to the features of customers such as profitability, lifelong value, long-term relations.

Demand Management: Manipulating the demand with price under the conditions of positive or negative difference between demand and capacity.

Make to Order: Manufacturing system working according to pull system or orders received.

Make to Stock: Manufacturing system working according to push system or forecasted demand.

Multi-Criteria Decision Making: Selection, classification or ranking of alternatives according to more than one criterion.

Order Management: The whole process of order negotiation, prioritization, acceptance or rejection.

Order Prioritization: Prioritizing orders according to the features of orders such as profitability, continuity, lead-time.

Revenu Management: Maximizing revenue by selecting most profitable orders when demand exceeds capacity.

TOPSIS: A method ranking alternatives according to closeness to ideal and distant to negative ideal solution.

Chapter 4
Clustering Techniques for Categorical Data: Correspondence Analysis

Stelios Zimeras
University of the Aegean, Greece

Manolis Kalligeris
University of the Aegean, Greece

ABSTRACT

Categorical data are generally thought to consist of contingency tables, which are data tables created whenever categorical data are cross-classified. Correspondence analysis is a statistical visualization method for picturing the associations between the levels of a two-way contingency table. In this work, the correspondence analysis for categorical data is analyzed using statistical and mathematical techniques when big data must be organized and analyzed due to the huge amount of data, and a MATLAB package (simple correspondence analysis) for this technique is presented.

INTRODUCTION

This occurs frequently in the collection and presentation of social survey data based on interviews and questionnaires (Moser and Kalton 1971, Dixon and Leach 1978).

Correspondence analysis (CA) is a multivariate statistical technique proposed by Hirschfeld (1935) and later developed by Jean-Paul Benzécri (1973). It is conceptually similar to principal component analysis, but applies to categorical rather than continuous data.

In a two-way contingency table, the observed association of two traits is summarized by the cell frequencies, and a typical inferential aspect is the study of whether certain levels of one charactertistic are associated with some levels of another. Correspondence analysis is a geometric technique for displaying the rows and columns of a two-way contingency table as points in a low-dimensional space, such that

DOI: 10.4018/978-1-7998-5442-5.ch004

the positions of the row and column points are consistent with their associations in the table. The goal is to have a global view of the data that is useful for interpretation (Greenacre, 2007).

In this work, the correspondence analysis for categorical data would be analyzed using statistical and mathematical techniques (like SVD process) when big data must be organised, and analysed due to the huge amount of data and a MATLAB package (SCORAN – Simple **COR**respondence **AN**alysis) for this technique would be presented..

BACKGROUND

Categorical data analysis is a methodology where data, especially big data, can be summarized based on contingency tables. Due to this process, analysis could be more effective and organization of the final results could be introduced in an easier manner. The contingency tables introduce frequencies in a IxJ table (matrix) where n_{ij} is the number of individuals with the particular characteristic, defined as cell frequency.

These frequencies are based on particular distributions where data could be part of a sample. The most important distributions for categorical data analysis are:

Poisson Distribution (Pois(λ_i))

The Poisson distribution is a discrete probability distribution for the counts of events that occur randomly in a given interval of time (or space). The Poisson model assumes that the number of characteristics $X_i = n_i$ ($n_i \cong n_{ij}$) into the cells in a contingency table are Poisson independent random variables (i.i.d.) with

$$(X_i = n_i) \sim \text{Poisson}(\lambda_i),\ \lambda_i > 0,\ i=1,\ldots,N=IxJ$$

where λ_i is the mean number of events per cell. We say X follows a Poisson distribution with parameter λ. The probability density function (The probability of observing n_i characteristic in a given cell i(=i,j) is given by

$$P(X_i = n_i) = \frac{e^{-\lambda_i} \cdot \lambda_i^{n_i}}{n_i!},\ i=1,\ldots,N$$

with $E(X_i = n_i) = \lambda_i$ and $Var(X_i = n_i) = \lambda_i$ (Figure 1)

Figure 1. Graphical presentation of Poisson distribution for different parameters N and λ. a. N=15, λ= 5; b. N=80, λ= 3; c. N=200, λ= 50.

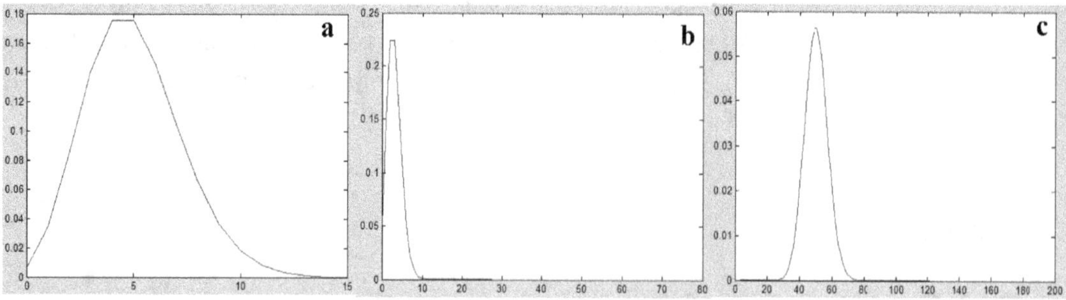

Binomial Distribution (Bin(n,p))

In this case, the final result has two outcomes: success (with probability p), and fail (with probability q=1-p). If X is the number of success in n-trials and p the probability of success then the random variable

$X_i \sim \text{Binomial}(n,p)$

considering the following conditions:

1: The number of observations n is fixed.
2: Each observation is independent.
3: Each observation represents one of two outcomes ("success" or "failure").
4: The probability of "success" p is the same for each outcome.

The probability density function is given by

$$P(X_i = x_i) = \binom{n}{x_i} p^{x_i} (1-p)^{n-x_i}$$

with $\binom{n}{x_i} = \frac{n!}{x_i!(1-x_i)!}$, $E(X_i) = n \bullet \pi$ and $Var(X_i) = n \bullet \pi \bullet (1-\pi)$ (Figure 2)

Figure 2. Graphical presentation of binomial distribution for different parameters n and p. a. n=20, p= 0.25; b. n=80, p= 0.5; c. n=200, p= 0.8.

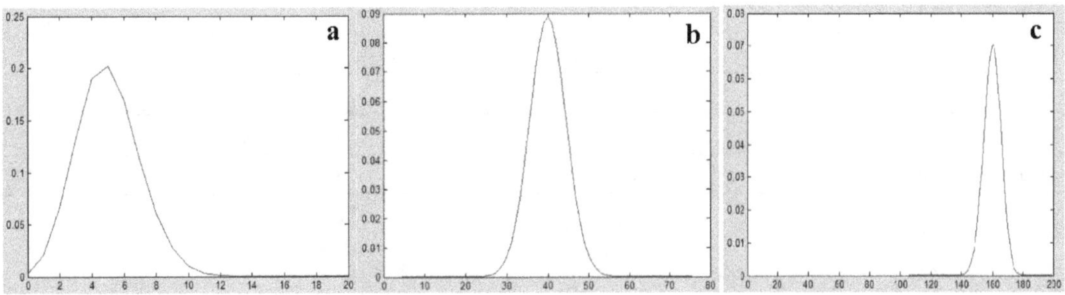

Multinomial Distribution (Multi(n, π_i))

A multinomial process is a sequence of independent, identically distributed random variables $X=(X_1,X_2,\ldots)$ each taking k-possible values. The probability density function of the trial variables is given by $\pi_i = \Pr(X_j = i)$, i=1,2,,...,k. with $\sum_{j=1}^{k} \pi_j = 1$.

If X_i is the number of sucess that the j characteristic is appeared under n-trial then the random variable

$(X_j= n_j) \sim \text{Multinomial}(n, \pi_j)$, j=1,...,k

The probability density function is given by

$$P(X_j= n_j) = \left(\frac{n!}{n_1!\cdot n_2!\cdots n_k!}\right)\pi_1^{n_1}\pi_2^{n_2}\cdots\pi_k^{n_k}$$

with $E(X_j= n_j) = n\bullet\pi_j$, $Var(X_j= n_j) = n\bullet\pi_j\bullet(1-\pi_j)$ and $Cov(n_i, n_j)=n\bullet p_i\bullet p_j$

MAIN FOCUS OF THE CHAPTER

Correspondence Analysis is a statistical technique that belongs to the family of multivariate statistical methods and relates to categorical data analysis. Its purpose is to study the data of a table (usually a frequency table) by plotting it on a graph called a biplot.

Through correspondence analysis the researcher has the ability to study dataset with effective visualization of big data and therefore to retrieve information that was not initially visible. The basic keys of the techniques are:

Table 1. Contingency table

	columns 1	columns 2	...	columns c	Total columns
row 1	n_{11}	n_{12}	...	n_{1c}	$n_{1.}$
row 2	n_{21}	n_{22}	...	n_{2c}	$n_{2.}$
⋮	⋮	⋮	⋮	⋮	⋮
row r	n_{r1}	n_{r2}	...	n_{rc}	$n_{r.}$
Total rows	$n_{.1}$	$n_{.2}$	...	$n_{.c}$	total N

Table 2. Contingency table example

Age class	**Status**			
	No diagnosis	*Severe headache*	*migraine*	**total**
0-19	$n_{11}= 77$	$n_{12}= 52$	$n_{13}= 20$	$n_{1.}= 149$
20-29	$n_{21}= 66$	$n_{22}= 82$	$n_{23}= 68$	$n_{2.}= 216$
30-39	$n_{31}= 31$	$n_{32}= 49$	$n_{33}= 37$	$n_{3.}= 117$
40-49	$n_{41}= 20$	$n_{42}= 44$	$n_{43}= 31$	$n_{4.}= 95$
50-59	$n_{51}= 27$	$n_{52}= 52$	$n_{53}= 25$	$n_{5.}= 104$
total	$n_{.1}= 221$	$n_{.2}= 279$	$n_{.3}= 181$	$N= 681$

Table 3. Contingency table with relative frequencies

Age class	**Status**			
	No diagnosis	*Severe headache*	*migraine*	**total**
0-19	$p_{11}= 0.113$	$p_{12}= 0.076$	$p_{13}= 0.029$	$rm_{1.}= 0.219$
20-29	$p_{21}= 0.097$	$p_{22}= 0.120$	$p_{23}= 0.01$	$rm_{2.}= 0.317$
30-39	$p_{31}= 0.046$	$p_{32}= 0.072$	$p_{33}= 0.054$	$rm_{3.}= 0.172$
40-49	$p_{41}= 0.029$	$p_{42}= 0.065$	$P3_{43}= 0.046$	$rm_{4.}= 0.139$
50-59	$p_{51}= 0.04$	$p_{52}= 0.076$	$p_{5}= 0.038$	$rm_{5.}= 0.152$
total	$cm_{.1}= 0.325$	$cm_{.2}= 0.41$	$cm_{.3}= 0.266$	1

- The rows and columns of the table are a point in the multidimensional space.
- Human perception fails to represent multidimensional spaces, and so we try to reduce the magnitude of the problem to a smaller dimension

First, we expect any relationships between rows and columns to emerge. In the graphical representation of the analysis, points that are very close together indicate a correlation between the corresponding rows and columns

Table 4. Line masses

$rm_{1.} = 0.219$
$rm_{2.} = 0.317$
$rm_{3.} = 0.172$
$rm_{4.} = 0.139$
$rm_{5.} = 0.152$

- For the calculation of column masses the process is the following:

Contingency Table

In correspondence analysis, the relevance table is called the contingency table and is the source of the analysis. Example of one of this table is given in Table 1.

The example to which we apply the correspondence analysis relates to 681 people who suffer from a headache and are diagnosed in one of the following 3 groups: no diagnosis, severe headaches, migraine (Table 2).

From the table the relative frequencies are calculated by dividing the frequency of each cell by the sample size N given by (Table 3)

$$p_{ij} = \frac{n_{ij}}{N},\ rm_{i.} = \frac{n_i}{N},\ cm_{.j} = \frac{n_{.j}}{N},\ i=1,..,r,\ j=1,\dots,c$$

Mazes

The marginal rows and columns of a relative frequency table are called line masses and column masses, respectively.

- For the calculation of line masses the process is the following:

Divide the margins of each line by the total sample size N. The mass of each line will be denoted by $rm_{i.}$

$$rm_{i.} = \frac{n_{i.}}{N},\ i=1,\dots,r.$$

Table 5. Column masses

column masses	$cm_{.1} = 0.325$	$cm_{.2} = 0.41$	$cm_{.3} = 0.266$

Table 6. Line profiles

Age class	Status			
	No diagnosis	*Severe headache*	*migraine*	**total**
0-19	$z_{11}= 0.517$	$z_{12}= 0.35$	$z_{13}= 0.134$	1
20-29	$z_{21}= 0.306$	$z_{22}= 0.38$	$z_{23}= 0.315$	1
30-39	$z_{31}= 0.265$	$z_{32}= 0.419$	$z_{33}= 0.316$	1
40-49	$z_{41}= 0.211$	$z_{42}= 0.463$	$z_{43}= 0.326$	1
50-59	$z_{51}= 0.26$	$z_{52}= 0.5$	$z_{53}= 0.24$	1
total	$cm_{.1}= 0.325$	$cm_{.2}= 0.41$	$cm_{.3}= 0.266$	

For our example the results are given in Table 4.

Divide the margins of each column by the total sample size N. The mass of each column will be denoted by $cm_{.j.}$

$$cm_{.j} = \frac{n_{.j}}{N}, \text{ j=1,...,c.}$$

For our example the results are given in Table 5.

Masses are a very important part of correspondence analysis as they will be used as weighting factors for distance calculation.

Profiles

There are two types of profiles: the line profiles and column profiles. Methodology calculation of the profiles is given bellow.

- Lines profiles

Table 7. Column profiles

Age class	Status			
	No diagnosis	*Severe headache*	*migraine*	**total**
0-19	$k_{11}= 0.348$	$k_{12}= 0.186$	$k_{13}= 0.111$	$rm_{1.}= 0.219$
20-29	$k_{21}= 0.299$	$k_{22}= 0.294$	$k_{23}= 0.376$	$rm_{2.}= 0.317$
30-39	$k_{31}= 0.14$	$k_{32}= 0.176$	$k_{33}= 0.204$	$rm_{3.}= 0.172$
40-49	$k_{41}= 0.091$	$k_{42}= 0.158$	$k_{43}= 0.171$	$rm_{4.}= 0.139$
50-59	$k_{51}= 0.122$	$k_{52}= 0.186$	$k_{53}= 0.138$	$rm_{5.}= 0.152$
total	1	1	1	

We convert the relative frequencies p_{ij} in $z_{ij}=\frac{p_{ij}}{rm_{i.}}$. That is to find the line profiles we have to divide the relative frequencies p_{ij} by the corresponding rm_i line masses. The results are illustrated is Table 6.

- Column profiles

We convert the relative frequencies p_{ij} in $k_{ij}=\frac{p_{ij}}{cm_{.j}}$. That is to find the column profiles we have to divide the relative frequencies p_{ij} by the corresponding $cm_{.j}$ column masses. The results are illustrated is Table 7.

The average in each column of the row profile column (or columns) is the corresponding column mass. These values are called centroid or 'average line profile' (respectively 'average column profile')

Distances

To calculate the difference between two row or column profiles we need to define a distance measure. A well-known measure of distance is the Euclidean with form

$$d_{ij}=\sqrt{\sum_{k=1}^{c}\left(z_{ik}-z_{jk}\right)^2}$$

where z is the row elements of the table, i=1,2,…, j=1,2,… $i^{1}j$ and k the column.

However, this has a serious disadvantage. The number of observations is not taken into account in measuring distance. The resulting is that the differences in small frequency in cells are having the same weight as the differences in large frequency in cells. For that reason a new measure is proposed, calling the mass. In particular the division $\sum_{k=1}^{c}\left(z_{ik}-z_{jk}\right)^2$ with the column mass $cm_{.j}$ and taking the square root of $\sqrt{\frac{\sum_{k=1}^{c}\left(z_{ik}-z_{jk}\right)^2}{cm_{.j}}}$, this measure is better than Euclidean. Based on this measure, the disadvantage is eliminated by weighting the distance with the combination of the masses. The form

$$d_{ij}=\sqrt{\frac{\sum_{k=1}^{c}\left(z_{ik}-z_{jk}\right)^2}{cm_{.j}}} \quad i=1,\dots,r,\ j=1,\dots,c,$$

is called the x^2 distance.

Table 8. Row distances

Age class	Distance dr_i	Distance dr_i^2
0-19	0.434	0.188
20-29	0.111	0.012
30-39	0.144	0.021
40-49	0.247	0.061
50-59	0.188	0.035

Mahalanobis Distance

The Mahalanobis distance of the observations $\underline{x}=(x_1,x_2,\ldots,x_N)^T$ with mean $\underline{\mu}=(\mu_1,\mu_2,\ldots,\mu_N)^T$ and variance – covariance matrix S given by

$$D_M(\underline{x})=\sqrt{(\underline{x}-\underline{\mu})^T S^{-1}(\underline{x}-\underline{\mu})}$$

If the covariance matrix is diagonal then Mahalanobis distance is called normalized Euclidean given by

$$d(\underline{x},\underline{y})=\sqrt{\sum_{i=1}^{N}\frac{(x_i-y_i)^2}{s_i^2}}$$

For the correspondence analysis, the selection of x^2 distance is taking place by calculating the profiles row distance from the center (centroid) using the following form:

$$dr_i=\sqrt{\frac{\sum_{k=1}^{c}\left(z_{ik}-cm_{.j}\right)^2}{cm_{.j}}},\ i=1,\ldots,I,\ j=1,\ldots,J$$

The corresponding calculation for the profiles column distance from the center (centroid) is given by

$$dc_j=\sqrt{\frac{\sum_{k=1}^{c}\left(k_{ik}-rm_{i.}\right)^2}{rm_{i.}}},\ i=1,\ldots,I,\ j=1,\ldots,J$$

Based on the appropriate forms the calculating distances dr_i are given in Table 8.

Inertia

Inertia can measure the variability of the points displayed in the graphical representation of the table profiles. It is defined as

$$I= \sum_{i=1}^{r} rm_{i.} dr_i^2$$

where $rm_{i.}$ is the masses of the line profiles and dr_i is the distance of each line profile from its centerline.

The closer the points are to each other, the smaller the inertia is. If I = 0 means that the profiles are the same for all lines. Inactivity is related to Pearson's x^2 independence test through the relationship

$$I= \frac{X^2}{N}$$

where the Pearson's x^2 independence test is given by

$$X^2= \sum_{i=1}^{I}\sum_{j=1}^{J} \frac{\left(n_{ij} - e_{ij}\right)^2}{e_{ij}} \sim x^2_{(I-1)(J-1)}$$

with $e_{ij}= \frac{n_{i.} \cdot n_{.j}}{n}$. Based on the example the inertia is given by Table 9

Table 9. Inertia table

Age class	distance dr_i^2	Masses row $rm_{i.}$	$dr_i^2 \cdot rm_{i.}$
0-19	0.188	0.219	0.041
20-29	0.012	0.317	0.004
30-39	0.021	0.12	0.003
40-49	0.061	0.14	0.009
50-59	0.035	0.153	0.005
Inertia (I) = 0.041 + 0.004 + 0.003 + 0.009 + 0.005= 0.062			

The Pearson's x^2 independence test is given by X^2= 42.556. Based on the value from x^2 – distribution το $x^2_{(5-1)(3-1),1-0,05} = x^2_{8,0.95}$. So X^2= 42.556 > 2.733= $x^2_{8,0.95}$ then we reject the hypothesis H_0 for independence then the dimensions can be reduced.

Inertia (I)= $\frac{X^2}{N} = \frac{42.556}{681} = 0.062$

Normalized Residuals

Residuals are defined u_i, $i=1,\ldots,k$ as the difference between the observed values Y_i and the estimated value $\widehat{Y}_i$.

$$u_i = Y_i - \widehat{Y}_i$$

Also the normalized residuals could be calculated based on the form

$$u_{si} = \frac{u_i}{\sqrt{MSR}}$$

Finally the calculation of the normalized residuals matrix A_{rxc} can be calculated using the formula

$$A = D_{rm}^{-\frac{1}{2}} \bullet Z \bullet D_{cm}^{-\frac{1}{2}}$$

where D_{rm} a rxr diagonial matrix with elements the row masses

$$D_{rm} = \begin{pmatrix} rm_1 & 0 & \cdots & 0 \\ 0 & rm_2 & \cdots & 0 \\ \vdots & \vdots & \ddots & \vdots \\ 0 & 0 & \cdots & rm_r \end{pmatrix},$$

the D_{cm} a rxr diagonial matrix with elements the column masses

$$D_{cm} = \begin{pmatrix} cm_1 & 0 & \cdots & 0 \\ 0 & cm_2 & \cdots & 0 \\ \vdots & \vdots & \ddots & \vdots \\ 0 & 0 & \cdots & cm_r \end{pmatrix},$$

and Z_{rxc} the matrix of row profiles with elements z_{ij}, $i=1,\ldots,r$, $j=1,\ldots,c$

Based on our example, the normalized residuals are given by

$$D_{rm} = \begin{pmatrix} 0.219 & 0 & 0 & 0 & 0 \\ 0 & 0.317 & 0 & 0 & 0 \\ 0 & 0 & 0.172 & 0 & 0 \\ 0 & 0 & 0 & 0.14 & 0 \\ 0 & 0 & 0 & 0 & 0.153 \end{pmatrix}$$

$$D_{cm} = \begin{pmatrix} 0.325 & 0 & 0 \\ 0 & 0.41 & 0 \\ 0 & 0 & 0.266 \end{pmatrix}$$

$$\mathbf{Z}= \begin{pmatrix} 0.517 & 0.349 & 0.134 \\ 0.306 & 0.38 & 0.315 \\ 0.265 & 0.419 & 0.316 \\ 0.211 & 0.463 & 0.326 \\ 0.26 & 0.5 & 0.24 \end{pmatrix}$$

$$\mathbf{A}= \begin{pmatrix} 0.158 & -0.044 & -0.119 \\ -0.019 & -0.027 & 0.054 \\ -0.043 & 0.006 & 0.041 \\ -0.075 & 0.031 & 0.044 \\ -0.045 & 0.055 & -0.019 \end{pmatrix}$$

$$\mathbf{A^2}= \begin{pmatrix} 0.025 & 0.002 & 0.014 \\ 0 & 0.001 & 0.003 \\ 0.002 & 0 & 0.002 \\ 0.006 & 0.001 & 0.002 \\ 0.002 & 0.003 & 0 \end{pmatrix}$$

Single Value Decomposition (SVD)

Using the normalized residuals, the process for calculating the reduction of matrix A is given by:

Step 1: Calculation of matrix $(A^TA)_{cxc}$ with $A^TA = V \begin{bmatrix} ó_{11}^2 & 0 & \cdots & 0 \\ 0 & ó_{22}^2 & \cdots & 0 \\ \vdots & \vdots & \ddots & \vdots \\ 0 & 0 & \cdots & ó_{cc}^2 \end{bmatrix} V^T$, where $ó_{ij}^2$ are the eigenvalues. So $V = [v_1| v_2| \cdots |v_c]$ where V_{cxc} the eigenvectors of matrix A^TA.

Step 2: Calculation of matrix $(AA^T)_{rxr}$ with $AA^T = U \begin{bmatrix} ó_{11}^2 & 0 & \cdots & 0 \\ 0 & ó_{22}^2 & \cdots & 0 \\ \vdots & \vdots & \ddots & \vdots \\ 0 & 0 & \cdots & ó_{rr}^2 \end{bmatrix} U^T$, where $ó_{ij}^2$ are the eigenvalues. So $U= [u_1| u_2| \cdots |u_r]$ where U_{rxr} the eigenvectors of matrix A^TA.

Step 3: Based on A^TA and AA^T the calculation of matrix variance – covariance Σ_{rxc} can be defined by

$$\Sigma= \begin{bmatrix} ó_{11} & 0 & \cdots & 0 \\ 0 & ó_{22} & \cdots & 0 \\ \vdots & \vdots & \ddots & \vdots \\ 0 & 0 & \cdots & ó_{rc} \end{bmatrix}$$

Normalization Methods

Normalization procedures in co analysis are used to determine when and how the similarities of the variable rows and columns, as well as the relationships between them, can be interpreted in terms of coordinates and columns and the principle of the axes of a graph. The normalization methods help to create the principal components row and column based on what we want to study.

The principal components of the row is given by

$$T= D_{rm}^{-\frac{1}{2}} \bullet RU \bullet R\Sigma^{\alpha}$$

where $(D_{rm})_{rxr}$ the diagonal matrix with the elements $rm_{i.}$ the rows masses

$$D_{rm} = \begin{pmatrix} rm_{1.} & 0 & \cdots & 0 \\ 0 & rm_{2.} & \cdots & 0 \\ \vdots & \vdots & \ddots & \vdots \\ 0 & 0 & \cdots & rm_{r.} \end{pmatrix}$$

the RU_{rxk} the matrix from SVD method, $R\Sigma_{kxk}$ the matrix of radial values, and a the appropriate value of normalization considering: $\alpha=\frac{1}{2}$, $\beta=\frac{1}{2}$ for Symmetrical normalization, α=1, β=1 for Principal normalization, **:** α=1, β=0 for Row normalization and α=0, β=1 for Column normalization.

The principal components of the column is given by

$$Q= D_{cm}^{-\frac{1}{2}} \bullet RV \bullet R\Sigma^{\beta}$$

where $(D_{cm})_{cxc}$ the diagonal matrix with the elements $rm_{i.}$ the column masses

$$D_{cm} = \begin{pmatrix} cm_{.1} & 0 & \cdots & 0 \\ 0 & cm_{.2} & \cdots & 0 \\ \vdots & \vdots & \ddots & \vdots \\ 0 & 0 & \cdots & cm_{.c} \end{pmatrix}$$

the RV_{cxk} the matrix from SVD method, $R\Sigma_{kxk}$ the matrix of radial values, and β the appropriate value of normalization considering: $\alpha=\frac{1}{2}$, $\beta=\frac{1}{2}$ for Symmetrical normalization, α=1, β=1 for Principal normalization, **:** α=1, β=0 for Row normalization and α=0, β=1 for Column normalization.

Considering the example, the values for U, V and Σ are given by

$$U= \begin{pmatrix} -0.859 & -0.06 & -0.466 & 0.04 & -0.202 \\ 0.166 & 0.595 & -0.604 & -0.099 & 0.493 \\ 0.248 & 0.133 & -0.241 & -0.644 & -0.669 \\ 0.386 & -0.161 & -0.47 & 0.681 & -0.375 \\ 0.156 & -0.774 & -0.373 & -0.332 & 0.358 \end{pmatrix}$$

$$\Sigma= \begin{pmatrix} 0.236 & 0 & 0 \\ 0 & 0.082 & 0 \\ 0 & 0 & 0 \\ 0 & 0 & 0 \\ 0 & 0 & 0 \end{pmatrix}$$

$$V= \begin{pmatrix} -0.785 & 0.245 & 0.57 \\ 0.237 & -0.731 & 0.64 \\ 0.573 & 0.637 & 0.516 \end{pmatrix}$$

Table 10. Coordinated rows elements

Age class	Coordinate 1	Coordinate 2
0-19	-0. 8920	-0. 0366
20-29	0. 1431	0. 3028
30-39	0. 2912	0. 0918
40-49	0. 5022	-0. 1233
50-59	0. 1945	-0. 5671

Also, the calculation based on the symmetric normalization is

$$T= D_{rm}^{-\frac{1}{2}} \bullet RU \bullet R\Sigma^{\frac{1}{2}}$$

$$Q= D_{cm}^{-\frac{1}{2}} \bullet RV \bullet R\Sigma^{\frac{1}{2}}$$

$$RU=\begin{pmatrix} -0.859 & -0.06 \\ 0.166 & 0.595 \\ 0.248 & 0.133 \\ 0.386 & -0.161 \\ 0.156 & -0.774 \end{pmatrix}$$

$$R\Sigma=\begin{pmatrix} 0.236 & 0 \\ 0 & 0.082 \\ 0 & 0 \\ 0 & 0 \\ 0 & 0 \end{pmatrix}$$

Table 11. Coordinated columns elements

Age class	Coordinate 1	Coordinate 2
NO DIAGNOSIS	-0. 6692	0. 1232
INTELLIGENT HEAD	0. 1795	-0. 3272
MIGRAINE	0. 5404	0. 3538

Figure 3. Biplot graph

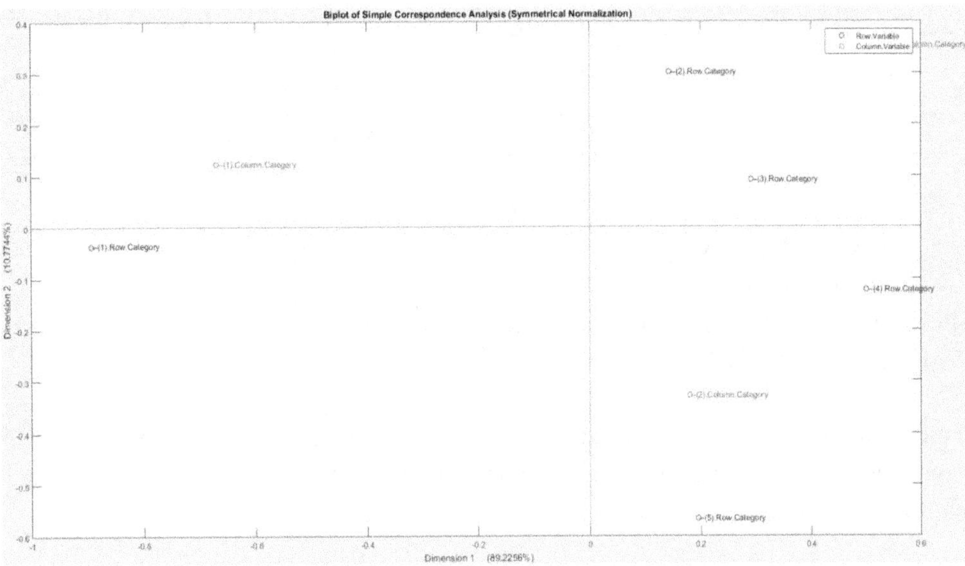

$$RV=\begin{pmatrix} -0.785 & 0.245 \\ 0.237 & -0.731 \\ 0.573 & 0.637 \end{pmatrix}$$

The coordinated rows and columns elements are given in Tables 10, and 11.

SOLUTIONS AND RECOMMENDATIONS

So having the coordinates of the two main axes we can represent the data on a map. The graph to use for this representation is called a biplot. (Figure 3). The biplot is an illustrative graph that illustrates in a few dimensions the points of the rows and columns of a rectangular table at the same time.

Considering our example, the symmetric type bitplot, the results are:

Table 12. Contribution of the i-row profiles in inertia of j-coordinate

Age class	Coordinate 1	Coordinate 2
0-19	73.7329	0.3574
20-29	2.7505	35.4435
30-39	6.1694	1.7644
40-49	14.9012	2.5858
50-59	2.4460	59.8489

Table 13. Contribution of the j-columns profiles in inertia of i-coordinate

Age class	Coordinate 1	Coordinate 2
NO DIAGNOSIS	61.5433	6.0044
INTELLIGENT HEAD	5.5912	53.4397
MIGRAINE	32.8655	40.5559

- 0-19 not show headache symptoms
- 20-29 appear to have migraines
- 30-49 are divided between severe headaches and migraines
- 50-59 have severe headaches

Based on the above, we can conclude that the degree of headache is related to the age as it appears not to in younger but in older age groups. The contribution of the inertia based on rows and columns into the bitplot is calculated based on:

- Contribution of the i-row profiles in inertia of j-coordinate (%)

$$\mathrm{IR}_{rxk} = \frac{rm_i.-\left(t_{ij}\right)^2}{\left[\left(r\sigma_{ij}\right)^2\right]^a} \bullet 100,\ i=1,\ldots,r,\ j=1,..,k$$

- Contribution of the j-columns profiles in inertia of i-coordinate (%)

Figure 4. Correspondence analysis program

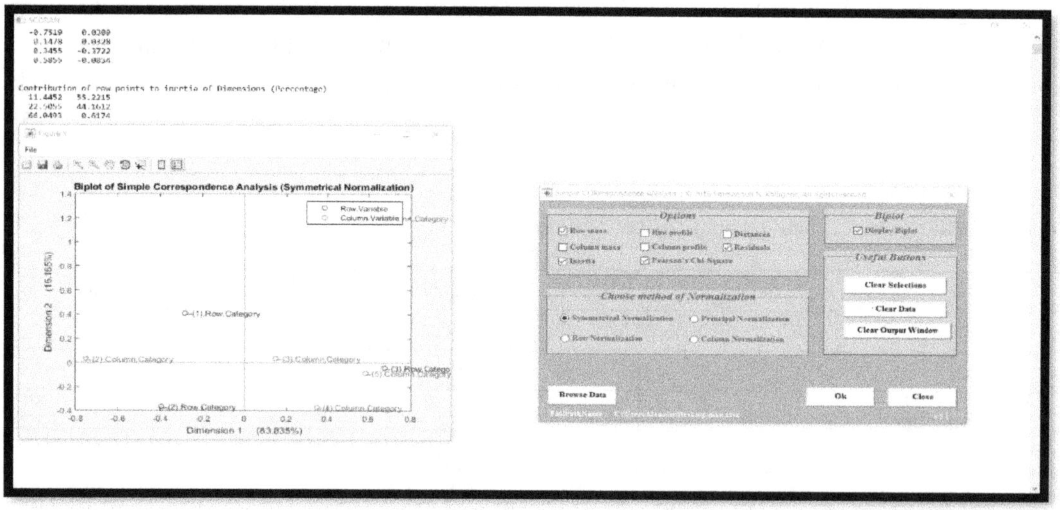

$$IC_{rxk} = \frac{cm_{.j} - \left(q_{ij}\right)^2}{\left[\left(r\sigma_{ij}\right)^2\right]^{\beta}} \bullet 100,\ i=1,\dots,r,\ j=1,..,k$$

where α and β are the appropriate value of normalization.

For our example, the tables contributions are given in Table 12 and 13.

In the table above we can see that the first axis is mainly based on the age group 0-19 while the second axis is based on the age group 50-59. In addition, we observe that when the contribution of one point is high on one axis, it has a smaller contribution on the other axis.

Similar comments to the above can be made in this table where the points NO DIAGNOSIS and INTELLIGENT HEAD have a higher contribution to the inertia of axes 1 and 2 respectively

Kalligeris (2016), created a program called SCORAN (Simple CORrespondence Analysis) based on MatLab language, with which one can quickly and easily implement simple correspondence analysis. The user will first need to install it on their computer and then be able to use it. Opening it can select exactly what it takes to apply simple match analysis. At the bottom of the program is the path from which the user entered the data. Finally, there are three key buttons that help the user either clear their workspace, clear what they chose to display, or clear the results that appeared (Figure 4).

CONCLUSION

Its purpose is to analyze the correspondence analysis method using appropriate theoretical background and examples. The work thoroughly analyzes the categorical data processing method using the correspondence analysis method. Masses, Distances and Biplot are some of the basic tools used in understanding and processing the data. In conclusion, we introduce the program " SCORAN " which is based solely on MatLab coding and is designed to implement and process data using the correspondence method.

REFERENCES

Benzécri, J.-P. (1973). L'Analyse des Données.: Vol. II. *L'Analyse des Correspondances*. Dunod.

Dixon, C.I., & Leach, B. (1978). Sampling methods for geographical research. *CATMOG 17*.

Greenacre, M. (2007). *Correspondence Analysis in Practice* (2nd ed.). Chapman & Hall/CRC. doi:10.1201/9781420011234

Hirschfeld, H. O. (1935). A connection between correlation and contingency. *Proceedings of the Cambridge Philosophical Society*, *31*(4), 520–524. doi:10.1017/S0305004100013517

Moser, C. A., & Kalton, G. (1971). *Survey methods in Social Investigation*. Gower.

Chapter 5
Metaheuristics Approaches for the Travelling Salesman Problem on a Spherical Surface

Yusuf Sahin
Department of Business Administration, Burdur Mehmet Akif Ersoy University, Turkey

Erdal Aydemir
https://orcid.org/0000-0003-4834-725X
Department of Industrial Engineering, Suleyman Demirel University, Turkey

Kenan Karagul
Department of Logistics, Pamukkale University, Turkey

Sezai Tokat
Department of Computer Engineering, Pamukkale University, Turkey

Burhan Oran
Department of Computer Engineering, Pamukkale University, Turkey

ABSTRACT

Traveling salesman problem in which all the vertices are assumed to be on a spherical surface is a special case of the conventional travelling salesman problem. There are exact and approximate algorithms for the travelling salesman problem. As the solution time is a performance parameter in most real-time applications, approximate algorithms always have an important area of research for both researchers and engineers. In this chapter, approximate algorithms based on heuristic methods are considered for the travelling salesman problem on the sphere. Firstly, 28 test instances were newly generated on the unit sphere. Then, using various heuristic methods such as genetic algorithms, ant colony optimization, and fluid genetic algorithms, the initial solutions for solving test instances of the traveling salesman problem are obtained in Matlab®. Then, the initial heuristic solutions are used as input for the 2-opt algorithm. The performances and time complexities of the applied methods are analyzed as a conclusion.

DOI: 10.4018/978-1-7998-5442-5.ch005

INTRODUCTION

Describe the general perspective of the chapter. End by specifically stating the objectives of the chapter.

Although the origins of the Travelling Salesman Problem (TSP) are not exactly known, the early mathematical problems related to TSP have been put forward by W.R. Hamilton and T.P. Kirkman in the 1800s. In 1932, Menger (1932) made the mathematical definition of TSP and published it as the "messenger problem" (das boten problem) of finding an optimal delivery route. This problem was later called the TSP by Robinson (1949), and has become widespread in the literature with this term. Then, Dantzig et al. (1954) solved the TSP of 49 cities using linear programming methods. Also, Karp (1972) showed that problems involving the Hamilton tour were NP-class problems.

In TSP, one salesman needs to visit each of a list of *n* cities exactly once and then returns to the starting city. The importance of the TSP is that it represents a larger class of problems known as combinatorial optimization problems that has grown into a very mature field with strong links to various other disciplines like discrete mathematics (graph theory, combinatorics, etc.), computer science (complexity theory, data structures,etc.), probability theory, continuous optimization, and many other application areas. TSP also belongs to the class of such problems known as NP-complete. Thus, due to its nature, TSP is still attracting the attention of many researchers from various fields through clever combinations of ideas for finding an efficient algorithm in polynomial-time. It is known that the solution time in TSP extends exponentially as the problem size grows. Today, with the inventions and developments in the commercial and open-source software packages and hardware, combinatorial optimization problems are exactly solved. However, instead of all these improvements, it is still not possible to solve all large-scale combinatorial optimization problems and or to solve in reasonable time. Thus, the meta-heuristics are generally used to achieve reasonable solution times instead of exact solutions (Croes 1958; Holland 1975; Laporte 1992; Sureja 2012).

The Hamiltonian cycle problem is to decide whether a given graph has a Hamiltonian cycle in which each node is visited exactly once. In TSP, the problem is to find a minimum weight Hamiltonian Cycle. For a problem of *n* points, there are $(n\text{-}1)!/2$ Hamiltonian Tours if the distance matrix is symmetric, and $n!/2$, otherwise. The mathematical model of the TSP can be described as follows (Croes 1958). The objective function can be modeled as

$$\min Z = \sum_{i=1}^{n}\sum_{j=1}^{n} X_{ij} * d_{ij} \tag{1}$$

with the constraints

$$u_i - u_j + nX_{ij} \le n-1 \qquad 1 \le i \ne j \le n \tag{2}$$

$$\sum_{i=1\ i\neq i}^{n} X_{ij} = 1, j = 1, 2, \ldots\ldots, n \tag{3}$$

$$\sum_{i=1}^{n} {}_{i \neq i} X_{ij} = 1, i = 1, 2, \ldots\ldots, n \tag{4}$$

where d_{ij} is the distance from nodes of i to j, n is the total number of nodes, u is the artificial parameter for a guaranteed tour and the X_{ij} is the decision variable given as:

$$X_{ij} = \begin{cases} 1, & \text{if visited from} i \text{ to } j \\ 0, & \text{otherwise} \end{cases} \tag{5}$$

BACKGROUND

In the scientific literature, there are a lot of artificial intelligence-based methods such as genetic algorithms (GA) (Freisleben and Merz 1996; Larranaga et al. 1999; Tsai et al. 2004; Chowdury et al. 2013), simulated annealing (SA) (Wang et al. 2013), tabu search (TS) (Fiechter 1994), ant colony optimization (ACO) (Dorigo and Gambardella 1997; Stüzle and Hoos 1997; Montgomery and Randall, 2003; Chu et al. 2004; Mavrovouniotis and Yang 2011; Gulcu et al. 2018), particle swarm optimization (PSO) (Wang et al. 2003; Golbarg et al. 2008; Strak et al. 2018), discrete firefly algorithm (DFA) (Jati 2011), discrete cuckoo search algorithm (DCSA) (Ouaarab et al. 2014), hybrid immune system algorithm (Pan et al. 2016) and 2-opt based harmony search (HS) (Karagul et al. 2016) that were used to solve TSPs for minimizing the tour lengths.

It is known that the surface of the earth is very close to a sphere. In reality, the world is not exactly an ellipsoid or sphere. The Earth is an irregular flattened sphere from the poles literally known as a geoid. The isometry of the sphere maintains the distance between points in existing state and it is a mapping of the sphere to itself (Ugur et al. 2009). Therefore, in modeling many real-world problems, sphere can be a good approximation of a system. The aim of this paper is to examine a spherical TSP with GA, ACO, and Fluid GA (FGA) on the newly generated datasets. Then 2-opt post improvement is applied for each method. About spherical TSP, a GA approach (Ugur et al. 2009), a novel discrete cuckoo search algorithm (Ouyang et al. 2017), and the combining glowworm swarm optimization with complete 2-opt algorithm as a hybrid algorithm (Chen et al. 2017) were developed for spherical TSP using Euclidian distances on a sphere surface.

The rest of this paper is organized as follows: The detail of the spherical TSP is given in Section 2. The solution approaches, GA, ACO and FGA are given including their parameter tuning operations in Section 3. Then, the solution models and test results are given with 2-opt post improvement methods on the initial heuristic solutions in Section 4 with detailed modeling, and Section 5 is also the conclusion.

MAIN FOCUS OF THE CHAPTER

Spherical Travelling Salesman Problem (3D-TSP)

In the spherical TSP, the solution is obtained by calculating the geodesics (shortest distances) between all pairs of points (cities/customers) on the sphere surface. By taking r as the radius of the sphere, a point $P(x, y, z)$ on the sphere surface has the following property:

$$x^2 + y^2 + z^2 = r^2 \tag{6}$$

All points on a sphere's surface can be characterized as a parametric form as:

$$P(u,v) = (x(u,v), y(u,v), z(u,v)) \tag{7}$$

which is the cartesian vector point function (Ugur et al. 2009; Ouyang et al. 2017). In (7), u and v are generally normalized between 0 and 1. The coordinates of $P(u,v)$ can be obtained as follows (Ugur et al. 2009):

$$(u,v) = r\cos(2\pi u)\sin(\pi v)$$

$$y(u,v) = r\sin(2\pi u)\sin(\pi v) \tag{8}$$

$$z(u,v) = r\cos(2\pi u)$$

For the spherical surface, each longitude meridian is exactly a half circle and the latitude parallels are usually small circles except the equator. By taking two points $P1({}_x1,{}_y1,{}_z1)$ as the vector $\overrightarrow{V_1}$ and $P2_{(}x2_{,}y2_{,}z2_{)}$ as the vector $\overrightarrow{V_2}$ on the spherical surface, the shortest distance between P_1 and P_2 is along the minor arc of a great circle (Ugur et al. 2009; Chen et al. 2017):

$$\overrightarrow{V_1}.\overrightarrow{V_2} = \left|\overrightarrow{V_1}\right|\left|\overrightarrow{V_2}\right|\cos\theta \tag{9}$$

where θ is the smaller angle between $\overrightarrow{V_1}$ and $\overrightarrow{V_2}$. The scalar product is calculated as:

$$\overrightarrow{V_1}.\overrightarrow{V_2} = x_1x_2 + y_1y_2 + z_1z_2 \tag{10}$$

So, the shortest distance $d(P_1,P_2)$ is calculated by the following formula:

$$d\left(P_1, P_2\right) = r.arccos\left(\frac{\overrightarrow{V_1}.\overrightarrow{V_2}}{r^2}\right) \tag{11}$$

Otherwise, the angle between two points can be obtained by the following formula.

$$= \text{arc}\cos(x_1x_2 + y_1y_2 + z_1z_2) \tag{12}$$

Then, a small-angle is obtained by taking the mode.

$$\varphi = \text{mod}(\theta, 180) \tag{13}$$

After finding the small angle between the points, the arc length is obtained by the following equation.

$$d(\mathrm{P}_1, \mathrm{P}_2) = 2*\pi*r*\frac{\varphi}{360} \tag{14}$$

For a unit sphere, since $r = 1$, the equation (14) can be written as:

$$(\mathrm{P}_1, \mathrm{P}_2) = 2*\pi*\frac{\varphi}{360} \tag{15}$$

In the classical TSP, for calculating the distance matrices, the Euclidian norm is used between two points. However, for the 3D Euclidean TSP on a spherical surface, the distance between two points (P_i, P_j) is calculated by a 3D Euclidean distance on a straight line instead of an arc length (Dorigo and Gambardella 1997):

$$d(\mathrm{P}_1, \mathrm{P}_2) = \sqrt{(x_i - x_j)^2 + (y_i - y_j)^2 + (z_i - z_j)^2} \tag{16}$$

For the spherical TSP, on the other hand, Eq. (11), (14) or (15) can be used for building the distance matrix. In this study, the distance matrix is calculated using Eq.(15) for the spherical distance.

Present your perspective on the issues, controversies, problems, etc., as they relate to theme and arguments supporting your position. Compare and contrast with what has been, or is currently being done as it relates to the chapter's specific topic and the main theme of the book.

Solution Algorithms

In this paper, for solving the spherical TSP, various datasets with 20 up to 1000 data points are generated separately. 2-opt and 3-opt algorithms are widely used algorithms in the literature to improve the quality of the solution. The 2-opt algorithm developed by Croes (1958) to solve the TSP is a simple local search algorithm. The 3-opt algorithm developed by Bock (1958) is an algorithm that is very similar to the 2-opt algorithm. The 3-opt algorithm usually gives better solutions than the 2-opt algorithm. However, it is a disadvantage that the edge changing process in the 3-opt algorithm is complicated (Yildirim 2014;Karagul et al. 2016). For more detailed information, in addition to the articles of Croes (1958) and Bock (1958), many studies in the literature can be reviewed. For investigating the benefit of using the evolutional computation strategies, GA, ACO, and FGA are used and then given as follows in detail.

For the post improvement method, 2-opt is chosen considering the simplicity. The post improvement is used only for GA and ACO.

Genetic Algorithms

GA is a population-based meta-heuristic and evolutionary search and optimization method inspired by Darwin's survival of the fittest principle of the natural selection of evolution. Holland has shown that GA can develop better individuals by manipulating individuals within a population through selection, crossover, and mutation rather than by improving the learning ability of a single individual. The structure of a GA was then represented by Goldberg (1989). The schema theory of GA had shown that it guarantees the optimal solution in infinite time. Thus, it is mainly used for obtaining approximate solutions. GA is used in may industrial engineering and optimization problems such as vehicle routing, facility layout planning, job scheduling, circuit design (Gen and Cheng 1997; Yu et al. 2011).

A basic example for solving a spherical-TSP is given with 6 points (cities) for the GA implementation.The coordinates, the distance matrix, GA parameters and initial population for the spherical TSP are given in Table 1-4, respectively.

Here, partially-mapped crossover operator (PMX), inversion mutation operator and roulette wheel selection method are selected as the setup parameters of GA. The minimum total distance is to be assigned as the cost. Two chromosomes are selected by selection from within the population and the crossover operation is applied. Let's assume that the following chromosomes have been selected.

Chromosome 1: 1-4-6-5-3-2-1
Chromosome 2: 1-2-6-5-4-3-1

If PMX is used as the crossover method, the newly formed chromosomes and their costs are as follows.

Offspring 1: 1-6-2-3-5-4-1 *Cost*: 9.50
Offspring 2: 1-2-4-6-3-5-1 *Cost*: 9.92

A number between [0;1] is randomly generated. If the number generated is larger than the mutation rate, mutation is performed. If the inversion mutation is applied to offspring 2:

Offspring 2: 1-2-4-6-3-5-1 *Cost:* 9.92
Mutationed Offspring: 1-2-3-6-4-5-1 *Cost:* 9.14

Table 1. Coordinates for the sample spherical TSP

Points	x	y	z
1	0.05092	0.84898	- 0.52596
2	0.09551	- 0.35749	0.92902
3	- 0.13171	- 0.07618	0.98836
4	- 0.13414	- 0.38473	0.91323
5	- 0.38628	0.42939	- 0.81634
6	0.84720	0.51143	- 0.14384

Table 2. Distance matrix for the sample spherical TSP

	1	2	3	4	5	6
1	0.00	2.48	2.20	1.73	0.69	0.69
2	2.48	0.00	0.37	0.80	2.82	1.81
3	2.20	0.37	0.00	0.47	2.48	1.87
4	1.73	0.80	0.47	0.00	2.13	1.62
5	0.69	2.82	2.48	2.13	0.00	1.56
6	0.69	1.81	1.87	1.62	1.56	0.00

Table 3. GA parameter for the sample spherical TSP

Parameter	Value
Population size	10
Crossover method	Partially-mapped
Mutation method	Inversion
Mutation rate	0.01
Selection method	Roulette Wheel
Iteration number	100

Then, the produced offsprings are compared to the worst chromosomes in the population and replaced if the costs of the produced chromosomes are better. After the crossover and mutation operations, the updated generation for the initial population is given as follows with the chromosome information and costs in Table 5. Then, GA is run until reaching the iteration number which is set as 100 for this example.

Table 4. Initial population and solutions for the sample spherical TSP

No	Chromosome	Cost
1	1-3-6-5-4-2-1	11.03
2	1-2-6-5-3-4-1	10.52
3	1-6-2-3-5-4-1	9.50
4	1-3-4-6-2-5-1	9.60
5	1-5-2-3-4-6-1	6.94
6	1-2-4-6-3-5-1	9.92
7	1-5-4-2-3-6-1	6.83
8	1-4-6-5-3-2-1	10.23
9	1-5-2-3-6-4-1	9.09
10	1-2-6-5-4-3-1	10.64

Hybrid Fluid Genetic Algorithms

The performance of GA is determined by its parameters such as population size, crossover, and mutation rates. The fluid genetic algorithms (FGA) alters some of GA's existing parts and adds new features and concepts to enhance its capabilities. It is shown that FGA is faster, more accurate, more reliable and more adaptable than the standard GA.In the conventional GA, each chromosome is mapped to a unique individual. On the other hand, FGA is closer to the genetic world and biological GA by having each chromosome mapped to many different individuals.This relation between the concept of chromosome and individual in FGA is obtained by a new concept called as born-an-individual function (Jafari-Marandi and Smith 2017)

In the literature, a hybrid solution method based on Fluid Genetic Algorithm, Nearest Neighbor and 2-Opt methods for the solution of TSP is given and called hybrid FGA (HFGA) (Sahin and Karagul, 2018). In this study, HFGA is also used for comparison purposes.

Table 5. The updated initial population and its solutions for the sample spherical TSP

No	Chromosome	Cost
1	*1-6-2-3-5-4-1*	*9.50*
2	*1-2-3-6-4-5-1*	*9.14*
3	1-6-2-3-5-4-1	9.50
4	1-3-4-6-2-5-1	9.60
5	1-5-2-3-4-6-1	6.94
6	1-2-4-6-3-5-1	9.92
7	1-5-4-2-3-6-1	6.83
8	1-4-6-5-3-2-1	10.23
9	1-5-2-3-6-4-1	9.09
10	1-2-6-5-4-3-1	10.64

Ant Colony Optimization Algorithm

The ant colony optimization (ACO) algorithm also called ant systems, is a population-based metaheuristic algorithm that has been effectively applied to combinatorial optimization such as TSP (Dorigo and Gambardella1997; Stüzle andHoos1997; Dorigo et al. 2006; Yang et al. 2008; Mavrovouniotis and Yang 2011). The ACO algorithm mimics the real ant colonies and their behaviors. In the ACO algorithm, artificially generated ants are randomly allocated at the n-cities of the TSP. An artificial pheromone trail model is used by the artificial ants to select the next city to visit. In biology, the ants are known to have no memory. However, in the ACO algorithm, the artificial ants have a memory and have some knowledge about their distance to neighbor cities. Thus, the artificial ants do not only use the pheromone trail but also the distance between two cities and can prefer away with respect to the distance. Also, in order to avoid repeated visits, artificial ants can store the information of the visited cities. By modeling this information, the possibility of city j to be visited after city i, selected by ant k, p_{ij}^{k}, can be written as (Yang et al. 2008):

$$p_{ij}^{k} = \begin{cases} \dfrac{\left[\tau_{ij}\right]^{\alpha} \cdot \left[\eta_{ij}\right]^{\beta}}{\sum_{s \in allowed_k} \left[\tau_{is}\right]^{\alpha} \cdot \left[\eta_{is}\right]^{\beta}} & j \in allowed_k \\ 0 & otherwise \end{cases} \tag{17}$$

where τ_{ij} is the amount of pheromone trail between cities *i* and *j*, α is the parameter to adjust the effect of τ_{ij}, η_{ij} is the visibility of city *j* from city *i*, and $\beta$is the parameter to adjust the effect of η_{ij} in ACO algorithm.

SOLUTIONS AND RECOMMENDATIONS

Parameter Settings for GA

For the parameter settings of GA, the population size, mutation rate and a number of epochs are fixed as 10, 0.01 and 100, respectively. Also, for a fair comparison, the pseudo-random generator is fixed in order to simulate the same initial population for each simulation. The crossover, mutation, and selection methods are analyzed. The partially-mapped (PMX) and order-based (OBX) crossover operators are compared in Table 6 as alternative crossover methods. It is seen from Table 6 that PMX has better cost function minimization and solution time performances. Swap, inversion, insertion and scramble mutation operators and randomly, roulette wheel, rank, and tournament selection operators are compared with each other using 20, 10 and 5 comparison tests, in Table 7 and Table 8, respectively. Consequently, the results are evaluated by minimum, maximum and average values with consumed time.

Table 6. Parameter settings for the GA crossover operators

#Run	Partially-mapped (PMX)		Order-based (OBX)	
	Cost	Time	Cost	Time
1	15012.87	166.00	15210.64	174.23
2	15091.39	152.15	14260.49	181.64
3	15685.57	163.65	15079.71	170.67
4	14869.45	145.49	15516.22	187.04
5	14345.27	146.62	14823.49	193.37
6	15490.26	145.85	15703.17	177.97
7	14919.74	152.65	14673.29	158.46
8	13363.07	150.45	14903.72	164.71
9	13180.11	155.11	14581.51	204.14
10	14471.89	148.25	14695.81	168.13
11	14008.81	146.50	15079.92	154.89
12	14946.55	146.61	14712.92	154.55
13	14984.57	145.09	15104.89	160.93
14	15766.68	147.66	16752.36	158.84
15	13217.89	146.26	16562.38	167.18
16	15218.19	145.49	16074.69	174.63
17	14716.30	146.14	14990.61	179.32
18	15181.35	145.68	15488.20	198.58
19	15251.24	146.25	14706.94	178.89
20	13793.77	160.05	16211.17	221.19
Min	**13180.12**	**145.09**	14260.50	154.55
Max	**15766.69**	**166.00**	16752.36	221.19
Avg	**14675.75**	**150.10**	15256.61	176.47

According to Table 6-8, for solving the spherical TSP with GA, the partially-mapped (PMX) crossover operator; inversion, insertion, and scramble mutation operator; and randomly, roulette wheel, rank, and tournament selection operators are preferred to run as shown in Table 9.

Parameter Settings for HFGA

The parameter settings of the HFGA is directly taken from Sahin and Karagul (2019) and given as in Table 10.

Table 7. Parameter settings to GA mutation operators

#Run	Swap		Inversion		Insertion		Scramble	
	Cost	Time	Cost	Time	Cost	Time	Cost	Time
1	14372.96	85.12	15130.59	81.36	14557.06	85.22	16090.43	89.63
2	15889.69	86.49	14158.82	80.84	15113.31	85.59	15670.30	90.68
3	15803.20	80.98	14923.77	81.77	14189.18	84.46	14543.91	90.40
4	15065.72	91.49	14778.54	81.96	15835.08	85.51	16100.26	92.18
5	14679.16	90.95	14227.43	87.19	13954.53	85.01	13726.44	90.65
6	16826.85	91.17	14588.09	84.97	14300.66	87.24	14759.96	91.85
7	14006.62	91.64	16167.29	81.86	15480.05	94.19	13914.78	91.53
8	14526.72	90.30	15186.22	82.04	14455.74	94.54	14225.29	91.44
9	15337.61	81.96	14861.96	81.27	14524.73	94.89	16440.24	90.90
10	15911.46	81.20	15964.72	80.85	14439.17	95.41	15769.35	87.98
Min	14006.62	80.98	14158.82	**80.84**	13954.53	84.46	**13726.44**	87.98
Max	16826.85	91.64	16167.29	**87.19**	**15835.08**	95.41	16440.24	92.18
Avg	15242.00	87.13	14998.74	**82.41**	**14684.95**	89.21	15124.10	90.72

Table 8. Parameter settings to GA selection operators

	Randomly		Roulette Wheel		Rank		Tournament	
#Run	Cost	Time	Cost	Time	Cost	Time	Cost	Time
1	14244.97	71.53	15409.94	140.64	13391.83	223.52	14628.39	85.52
2	13632.92	65.92	16289.21	138.34	13834.74	240.64	16257.21	90.78
3	15122.25	65.93	14193.03	144.43	15355.01	233.23	13457.66	86.91
4	15879.27	72.82	16427.92	147.55	15300.95	241.63	14804.56	90.27
5	15462.83	69.31	13636.25	145.04	14937.53	195.52	14511.84	90.65
Min	13632.92	**65.92**	13636.25	138.34	**13391.83**	195.52	13457.66	85.52
Max	15879.27	**72.82**	16427.93	147.55	**15355.01**	241.63	16257.21	90.78
Avg	15195.31	**69.36**	**14864.27**	144.58	14868.07	221.51	14497.78	88.83

Table 9. Selected parameters for GA

Parameter	Value
Population size	10
Crossover method	PMX
Mutation method	Reverse
Mutation rate	0.01
Selection method	Roulette Wheel
Iteration number	100

Table 10. Parameter settings for HFGA

Parameter	Value
Maximum iteration number	100
Population size	50
Crossover rate	0.50
Individual learning rate (ILR)	0.15
Global learning rate (GLR)	0.15
Diversity rate (DR)	0.70

Parameter Settings for ACO

Similarly, numerous tests are examined and summarized for α, β parameters of the ACO algorithm detailed in Section 3.2. A grid search is executed by 30 tests for α parameter level between [1.0, 3.9] with increasing step 0.1 and for each test, the model is run 10 times and the results are given in Table 11. About the βparameter, 30 tests are executed between [2.0, 4.9] with increasing step 0.1 and for each test, the model is run 10 times, and the results are given in Table 12. As a result, α parameter is set to 1.0, and β parameter is set to 2.4 of ACO after the analysis.

Numerical Results

For the numerical analysis, 20 to 1000 vertices Spherical TSP models are generated and analyzed with determined parameter values for each algorithm. The simulations are performed for random initial population and for each model 10 runs are executed. As the simulation platform, i7 CPU and 4 GB RAM hardware with Linux OS and MATLAB® 8.2 software package is used.

In Tables 13 and 14, the performance analysis and simulation times of GA, ACO and FGA as an average of 10 runs are listed, respectively. It is seen that the performance of FGA is superior for every simulation obtained with different numbers of vertices. In terms of performance, ACO is slightly better than GA. In Figure 1, the performance of the methods given in Table 13, are normalized by taking the performance of the HFGA as 1 for each problem with different number of nodes, in order to visualize the comparisons. The superiority of the performance of HFGA can be more clearly seen from Figure 1.

The simulation time analysis of the related methods is given in Table 14. It is seen that the simulation time of ACO is very long compared to GA and FGA. Therefore, for real-time or industrial applications, the solution times of ACO are unfeasible especially as the number of vertices increase. According to the numerical results, the FGA algorithm has better simulation time results than GA and ACO algorithms for this dataset. In Figure 2, as the ACO simulation times have large magnitude, only the normalized simulation times of GA and FGA are given. It is seen that FGA also requires less computational time than GA for solving the spherical TSP instances. Then, in Figure 3, two sample spherical graphics are illustrated with 40 and 90 points models.

Table 11. α parameter settings for ACO algorithm

α	Cost			Time		
	Min	**Max**	**Avg**	**Min**	**Max**	**Avg**
1.0	8031.41	8207.60	8135.98	20.19	24.45	22.03
1.1	8002.38	8182.19	8091.80	18.26	21.17	19.71
1.2	8010.11	8207.60	8164.58	18.29	20.58	19.08
1.3	7913.11	8182.19	8116.19	17.77	21.67	18.70
1.4	7937.18	8207.60	8123.72	18.10	20.20	18.71
1.5	7918.53	8182.48	8081.30	18.65	28.61	23.07
1.6	7978.18	8207.60	8096.00	19.99	22.44	20.66
1.7	8008.33	8207.60	8117.03	18.36	26.76	22.61
1.8	8008.33	8182.19	8118.80	17.94	20.44	18.87
1.9	7932.02	8182.19	8129.19	18.03	20.00	18.94
2.0	**7780.88**	8186.19	8043.11	18.26	18.93	18.64
2.1	7927.27	8311.06	8126.84	18.11	26.67	20.55
2.2	7890.97	8182.19	8036.10	17.78	20.90	18.80
2.3	8008.33	8185.42	8150.05	17.70	19.79	18.85
2.4	7986.28	8182.19	8122.79	17.61	18.75	18.16
2.5	8020.02	8182.19	8137.20	17.85	21.78	18.63
2.6	7991.24	8182.19	8120.97	20.84	21.79	21.04
2.7	8002.62	8182.48	8122.50	20.86	24.20	21.31
2.8	7932.72	8182.19	8107.92	20.85	22.73	21.29
2.9	7890.97	8200.12	8049.98	20.00	22.52	21.50
3.0	7920.13	8258.52	8119.15	20.23	25.59	22.41
3.1	7894.73	8182.19	8110.25	17.71	24.81	19.67
3.2	8012.63	8207.60	8154.29	17.71	**17.87**	**17.81**
3.3	7840.82	8182.19	8063.86	17.78	18.69	18.00
3.4	7913.94	8182.19	8121.39	**17.60**	18.22	17.95
3.5	7991.24	8182.19	8090.04	17.75	25.17	20.21
3.6	8150.98	8182.19	8177.15	19.81	20.73	20.01
3.7	7890.97	**8150.16**	**8024.81**	19.76	20.12	19.96
3.8	8020.34	8182.19	8126.71	19.84	22.09	20.31
3.9	7920.13	8207.60	8121.70	18.31	21.92	19.69

Table 12. β parameter settings for ACO algorithm

β	Cost			Time		
	Min	**Max**	**Avg**	**Min**	**Max**	**Avg**
2.0	7961.86	**8182.19**	8131.03	17.57	18.24	17.83
2.1	8004.85	8207.60	8113.32	17.59	18.13	17.73
2.2	7971.86	**8182.19**	8076.35	17.53	18.20	**17.68**
2.3	7913.94	8271.76	8109.52	17.56	19.46	18.04
2.4	7927.27	8207.60	8128.33	17.56	20.78	19.31
2.5	8046.22	8207.60	8145.78	19.74	22.38	20.66
2.6	7890.97	**8182.19**	8084.13	19.72	20.32	19.91
2.7	7887.35	8204.92	8091.81	19.62	20.31	19.96
2.8	8035.28	8223.02	8110.18	19.62	22.38	20.15
2.9	8041.76	8207.60	8118.37	19.16	25.93	21.74
3.0	7970.70	**8182.19**	8084.14	17.72	20.29	18.93
3.1	7992.50	**8182.19**	8066.04	17.45	19.91	17.98
3.2	7890.97	**8182.19**	8038.23	17.46	**17.98**	17.83
3.3	8002.38	**8182.19**	8130.37	**17.41**	20.30	18.22
3.4	**7831.00**	8207.60	8089.53	19.48	26.49	21.47
3.5	7951.34	8207.60	8122.67	17.78	24.47	20.35
3.6	7958.68	**8182.19**	8086.86	19.34	24.96	21.48
3.7	7966.92	8207.60	8137.03	18.77	19.33	19.17
3.8	8015.73	8207.60	8145.51	18.62	22.16	19.49
3.9	7970.70	**8182.19**	8145.16	18.65	30.69	21.60
4.0	8008.33	8207.60	8134.18	19.69	22.99	20.58
4.1	8018.56	**8182.19**	8133.12	19.70	20.85	20.06
4.2	7890.97	**8182.19**	8116.41	19.63	23.19	20.25
4.3	7918.23	**8182.19**	8075.37	19.64	20.17	19.89
4.4	7869.60	**8182.19**	8108.45	19.71	23.00	20.58
4.5	7913.94	**8182.19**	8072.23	19.63	24.337848	21.178541
4.6	8008.33	8185.42	8126.57	19.71	21.918757	20.351301
4.7	7893.17	8207.60	8081.27	19.68	22.950568	20.407498
4.8	7890.97	**8182.19**	8070.75	19.73	22.408816	20.498541
4.9	7892.87	**8182.19**	8108.87	19.71	22.445328	20.552460

Table 13. Performance analysis of the algorithms

#OfNode	**20**	**50**	**70**	**90**	**100**
Average	**Sol**	**Sol**	**Sol**	**Sol**	**Sol**
GA	11.39	17.46	21.83	23.55	25.76
ACO	11.31	16.93	21.41	22.95	25.60
FGA	**11.25**	**16.75**	**21.41**	**22.65**	**25.30**
#OfNode	**200**	**300**	**400**	**500**	**600**
Average	**Sol**	**Sol**	**Sol**	**Sol**	**Sol**
GA	34.07	43.53	49.99	57.93	61.83
ACO	33.81	43.07	49.93	57.45	61.47
FGA	**33.31**	**42.60**	**49.19**	**56.88**	**60.96**
#OfNode	**700**	**800**	**900**	**1000**	
Average	**Sol**	**Sol**	**Sol**	**Sol**	
GA	66.40	71.39	76.32	79.09	
ACO	66.26	71.49	76.26	78.66	
FGA	**65.48**	**70.38**	**75.37**	**78.01**	

Figure 1. Normalized performance of the methods

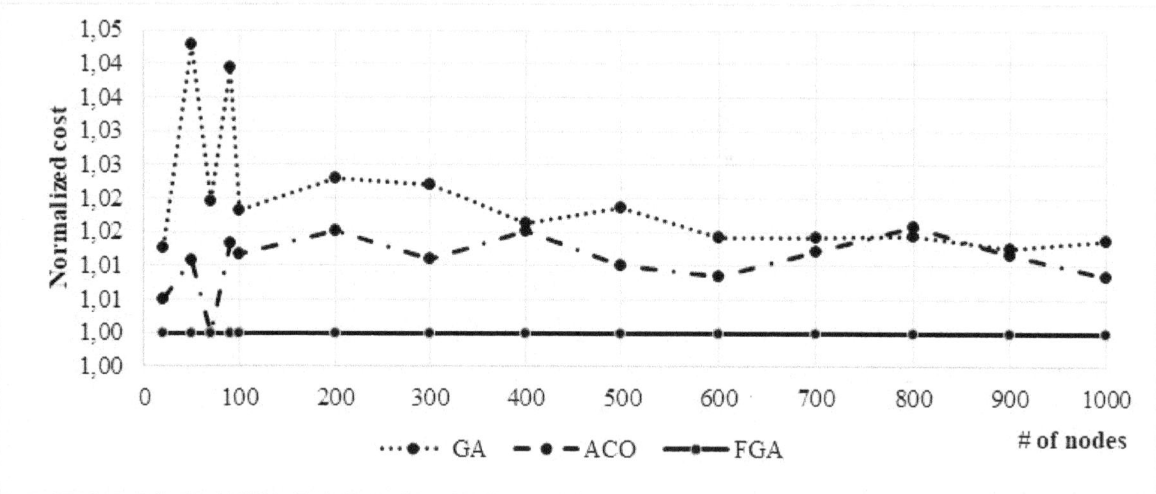

Table 14. Simulation time analysis of the algorithms

#OfNode	**20**	**50**	**70**	**90**	**100**
Average	**Time (sec)**	**Time (sec)**	**Time (sec)**	**Time (sec)**	**Time (sec)**
GA	0.03	0.17	0.32	1.05	1.17
ACO	1.32	7.97	15.44	25.82	32.03
FGA	**0.01**	**0.01**	**0.03**	**0.07**	**0.08**
#OfNode	**200**	**300**	**400**	**500**	**600**
Average	**Time (sec)**	**Time (sec)**	**Time (sec)**	**Time (sec)**	**Time (sec)**
GA	9.52	39.03	93.09	233.27	439.58
ACO	140.05	343.98	687.81	1208.99	1907.98
FGA	**0.80**	**3.00**	**11.84**	**15.82**	**91.00**
#OfNode	**700**	**800**	**900**	**1000**	
Average	**Time (sec)**	**Time (sec)**	**Time (sec)**	**Time (sec)**	
GA	780.54	1182.47	1850.36	2790.54	
ACO	2905.55	3989.22	5540.41	7166.67	
FGA	**162.37**	**369.37**	**645.26**	**966.55**	

Figure 2. Normalized performance of GA and FGA methods

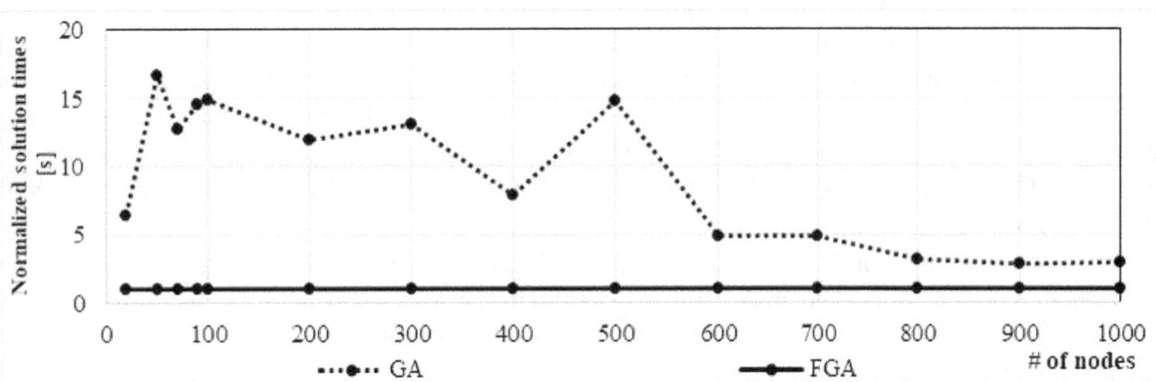

Figure 3. (a) 40 points and (b) 90 points dataset's unit sphere solutions

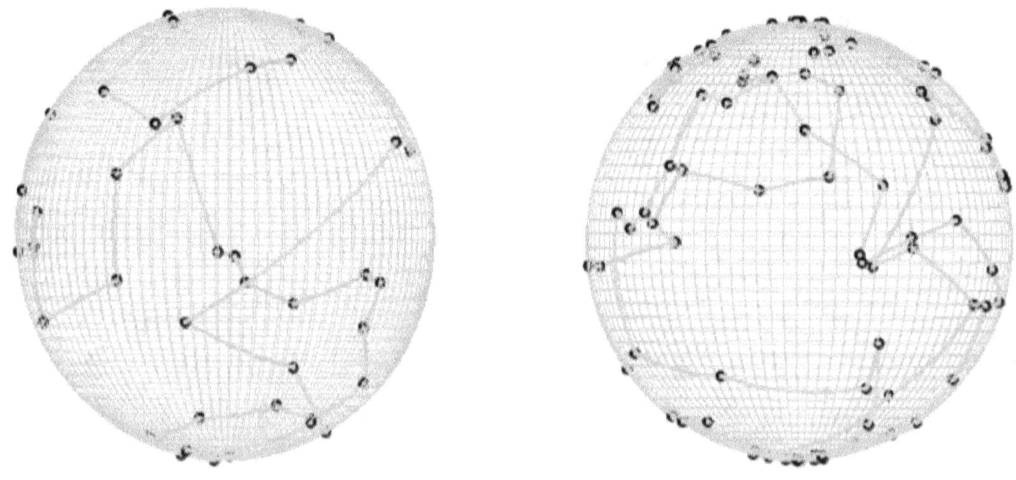

(a) (b)

CONCLUSION AND FUTURE RESEARCH

In this study, approximate algorithms based on heuristic methods are considered for the TSP on the spherical surface. Firstly, 28 test instances were produced on the unit sphere.

Including 20 to 1000 points, a spherical TSP dataset is generated and analyzed with determined parameter values from the parameter settings for each algorithm. In the parameter settings for the GA, PMX and OBX crossover operators, swap, inversion, insertion, and scramble mutation operators and randomly, roulette wheel, rank, tournament selection methods. The α, and β parameters of the ACO algorithm are evaluated with parameter analysis in order to obtain the values of simulation parameters. The FGA parameters are not adjusted for the given spherical TSP problem and directly taken from the previous study. Then, the spherical TSP is solved using the genetic algorithms (GA), ant colony optimization (ACO) and HFGA with the obtained parameters. The simulations have shown the superiority of the HFGA with respect to GA and ACO.

The TSP application of the FGA algorithm, one of which is stated as HFGA, is a recent research area and thus the literature is not mature yet. Also, from the authors' knowledge, in this study, the FGA is firstly applied for the solution of the 3D TSP. The results have shown that FGA is a state-of-the-art method for TSP, VRP and CVRP solutions.

As further research, the obtained results for the spherical surface may be beneficial in the studies carried out with unmanned aerial vehicles and drones. These tools are used extensively in areas such as reconnaissance, surveillance and city logistics.

REFERENCES

Bock, F. (1958). An algorithm for solving travelling-salesman and related network optimization problems. *Operations Research*, *6*(6), 897–897.

Chen, X., Zhou, Y., Tang, Z., & Luo, Q. (2017). A hybrid algorithm combining glowworm swarm optimization and complete 2-opt algorithm for spherical travelling salesman problems. *Applied Soft Computing*, *58*, 104–114. doi:10.1016/j.asoc.2017.04.057

Chowdhury, A., Ghosh, A., Sinha, S., Das, S., & Ghosh, A. (2013). A novel genetic algorithm to solve travelling salesman problem and blocking flow shop scheduling problem. *International Journal of Bio-inspired Computation*, *5*(5), 303–314. doi:10.1504/IJBIC.2013.057172

Chu, S. C., Roddick, J. F., & Pan, J. S. (2004). Ant colony system with communication strategies. *Information Sciences*, *167*(1-4), 63–76. doi:10.1016/j.ins.2003.10.013

Croes, G. A. (1958). A method for solving traveling-salesman problems. *Operations Research*, *6*(6), 791–812. doi:10.1287/opre.6.6.791

Dantzig, G., Fulkerson, R., & Johnson, S. (1954). Solution of a large-scale traveling-salesman problem. *Journal of the Operations Research Society of America*, *2*(4), 393–410. doi:10.1287/opre.2.4.393

Dorigo, M., Birattari, M., & Stutzle, T. (2006). Artificial ants as a computational intelligence technique. *IEEE Computational Intelligence Magazine*, *1*, 28–39. doi:10.1109/MCI.2006.329691

Dorigo, M., & Gambardella, L. M. (1997). Ant colonies for the travelling salesman problem. *Bio Systems*, *43*(2), 73–81. doi:10.1016/S0303-2647(97)01708-5 PMID:9231906

Fiechter, C. N. (1994). A parallel tabu search algorithm for large traveling salesman problems. *Discrete Applied Mathematics*, *51*(3), 243–267. doi:10.1016/0166-218X(92)00033-I

Freisleben, B., & Merz, P. (1996, May). A genetic local search algorithm for solving symmetric and asymmetric traveling salesman problems. In *Proceedings of IEEE international conference on evolutionary computation* (pp. 616-621). IEEE. 10.1109/ICEC.1996.542671

Gen, M., & Cheng, R. (1997). *Genetic algorithms and engineering design*. John Wiley&Sons.

Goldbarg, E. F., Goldbarg, M. C., & de Souza, G. R. (2008). Particle swarm optimization algorithm for the traveling salesman problem. In *Traveling Salesman Problem*. IntechOpen. doi:10.5772/5580

Goldberg, D. E. (1989). *Genetic algorithms in search optimization and machine learning*. Addison-Wesley Publishing Company.

Gülcü, Ş., Mahi, M., Baykan, Ö. K., & Kodaz, H. (2018). A parallel cooperative hybrid method based on ant colony optimization and 3-Opt algorithm for solving traveling salesman problem. *Soft Computing*, *22*(5), 1669–1685. doi:10.100700500-016-2432-3

Holland, J. H. (1975). *Adaptation in natural and artificial systems: an introductory analysis with applications to biology, control, and artificial intelligence*. University of Michigan Press.

Jafari-Marandi, R., & Smith, B. K. (2017). Fluid genetic algorithm (FGA). *Journal of Computational Design and Engineering*, *4*(2), 158–167. doi:10.1016/j.jcde.2017.03.001

Jati, G. K. (2011, September). Evolutionary discrete firefly algorithm for travelling salesman problem. In *International Conference on Adaptive and Intelligent Systems* (pp. 393-403). Springer. 10.1007/978-3-642-23857-4_38

Karagul, K., Aydemir, E., & Tokat, S. (2016). Using 2-Opt based evolution strategy for travelling salesman problem. *An International Journal of Optimization and Control: Theories & Applications*, *6*(2), 103–113. doi:10.11121/ijocta.01.2016.00268

Karp, R. M. (1972). Reducibility among Combinatorial Problems. In R. E. Miller, J. W. Thatcher, & J. D. Bohlinger (Eds.), *Complexity of Computer Computation* (pp. 85–103). Springer. doi:10.1007/978-1-4684-2001-2_9

Karp, R. M. (1972). Reducibility among combinatorial problems. In *Complexity of computer computations* (pp. 85–103). Springer. doi:10.1007/978-1-4684-2001-2_9

Laporte, G. (1992). The traveling salesman problem: An overview of exact and approximate algorithms. *European Journal of Operational Research*, *59*(2), 231–247. doi:10.1016/0377-2217(92)90138-Y

Larranaga, P., Kuijpers, C. M. H., Murga, R. H., Inza, I., & Dizdarevic, S. (1999). Genetic algorithms for the travelling salesman problem: A review of representations and operators. *Artificial Intelligence Review*, *13*(2), 129–170. doi:10.1023/A:1006529012972

Mavrovouniotis, M., & Yang, S. (2011). A memetic ant colony optimization algorithm for the dynamic travelling salesman problem. *Soft Computing*, *15*(7), 1405–1425. doi:10.100700500-010-0680-1

Menger, K. (1932). Das botenproblem. *Ergebnisse eines mathematischen kolloquiums, 2*, 11-12.

Montgomery, J., & Randall, M. (2003). The accumulated experience ant colony for the traveling salesman problem. *International Journal of Computational Intelligence and Applications*, *3*(02), 189–198. doi:10.1142/S1469026803000938

Ouaarab, A., Ahiod, B., & Yang, X. S. (2014). Discrete cuckoo search algorithm for the travelling salesman problem. *Neural Computing & Applications*, *24*(7-8), 1659–1669. doi:10.100700521-013-1402-2

Ouyang, X., Zhou, Y., Luo, Q., & Chen, H. (2013). A novel discrete cuckoo search algorithm for spherical traveling salesman problem. *Applied Mathematics & Information Sciences*, *7*(2), 777–784. doi:10.12785/amis/070248

Pan, G., Li, K., Ouyang, A., & Li, K. (2016). Hybrid immune algorithm based on greedy algorithm and delete-cross operator for solving TSP. *Soft Computing*, *20*(2), 555–566. doi:10.100700500-014-1522-3

Papadimitriou, C. H., & Steiglitz, K. (1982). *Combinatorial optimization* (Vol. 24). Prentice Hall.

Robinson, J. (1949). On the Hamiltonian game (a traveling salesman problem). RAND Project (No. RAND/RM-303).

Sahin, Y., & Karagul, K. (2019). Solving travelling salesman problem using hybrid fluid genetic algorithm (HFGA). *Pamukkale University Journal of Engineering Sciences*, *25*(1), 106–114. doi:10.5505/pajes.2018.81084

Stiitzle, T., & Hoos, H. (1997). The MAX-MIN ant system and local search for the traveling salesman problem. In *Proceedings of IEEE international conference on evolutionary computation* (pp. 309-314). Indianapolis, IN: IEEE. 10.1109/ICEC.1997.592327

Strąk, Ł., Skinderowicz, R., & Boryczka, U. (2018). Adjustability of a discrete particle swarm optimization for the dynamic TSP. *Soft Computing*, *22*(22), 7633–7648. doi:10.100700500-017-2738-9

Sureja, N. (2012). New inspirations in nature: A survey. *International Journal of Computer Applications & Information Technology*, *1*(3), 21–24.

Tsai, H. K., Yang, J. M., Tsai, Y. F., & Kao, C. Y. (2004). Some issues of designing genetic algorithms for traveling salesman problems. *Soft Computing*, *8*(10), 689–697. doi:10.100700500-003-0317-8

Uğur, A., Korukoğlu, S., Çalıskan, A., Cinsdikici, M., & Alp, A. (2009). Genetic algorithm based solution for TSP on a sphere. *Mathematical and Computational Applications, 14*(3), 219-228.

Wang, K. P., Huang, L., Zhou, C. G., & Pang, W. (2003). Particle swarm optimization for traveling salesman problem. In *Proceedings of the 2003 international conference on machine learning and cybernetics* (Vol. 3, pp. 1583-1585). Xi'an, China: IEEE. 10.1109/ICMLC.2003.1259748

Wang, Y., Tian, D., & Li, Y. (2013). An improved simulated annealing algorithm for traveling salesman problem. In *Proceedings of the 2012 International Conference on Information Technology and Software Engineering* (pp. 525-532). Springer. 10.1007/978-3-642-34522-7_56

Yang, J., Shi, X., Marchese, M., & Liang, Y. (2008). An ant colony optimization method for generalized TSP problem. *Progress in Natural Science*, *18*(11), 1417–1422. doi:10.1016/j.pnsc.2008.03.028

Yu, Y., Chen, Y., & Li, T. (2011, April). A new design of genetic algorithm for solving TSP. In *2011 Fourth International Joint Conference on Computational Sciences and Optimization* (pp. 309-313). IEEE. 10.1109/CSO.2011.46

Chapter 6
Routing Searching Algorithm for Combined Freight Transport Trips in South East Europe Region

Panagiotis Otapasidis
University of the Aegean, Greece

Christina Arampantzi
University of the Aegean, Greece

Vasileios Zeimpekis
University of the Aegean, Greece

ABSTRACT

In this chapter, the authors propose a mixed integer linear programming model (MILP) to address the shortest path problem when trans-shipments exist on combined freight transport operations. The objective is to minimize trip duration, cost, and CO2 emissions. For solving this problem, we propose an extension of Dijkstra's algorithm that identifies the optimal way of cargo movement with trans-shipment. We applied the proposed algorithm in real-life data by embracing three different types of transportation modes, namely road, rail, and inland waterways (IWW), and by testing it in three different cases where (1) goods should cross 2 or 3 South East Europe (SEE) countries, (2) goods should cover a distance more or less than 500 km, and (3) goods with origin the Port of Piraeus and destination four capital cities of SEE. The results show that the use of road transport is the most favorable solution if trip duration is the main criterion, whereas combined transport is selected when minimization of cost and CO2 emissions is the objective.

DOI: 10.4018/978-1-7998-5442-5.ch006

1. INTRODUCTION

Combined freight transport has received a large interest in recent years as part of a possible solution for a sustainable and efficient transport system. The idea behind intermodal transport is to utilize the strengths of different transport modes in one integrated transport chain. The road network has the advantage of being able to access almost any location and also of being very flexible, while rail and sea networks have the ability to transport goods long distances at a low cost. A combination of the two networks could, thus, reduce the cost of transport.

A combined freight transport network consists of: (a) an origin and a destination, (b) certain modes of transport and, (c) terminals. Usually, the first and the last leg of the transportation is executed via road transportation whereas the long-haul part is accomplished via rail or sea. Usually, a terminal (i.e. port, railway station) exists at the beginning and the end of the long-haul trip. These terminals link the first and the last leg of the trip by transferring the goods between them. To make the transfer between the modes more efficient, the goods are carried in standardized load carriers such as containers and swap-bodies.

Although combined freight transport is adopted in many cases when cargo is moved from a country to another, yet, there has been a lack of tools to evaluate the potential in combined transport and support the design of a smart, sustainable and integrated freight trip. To this end, this chapter proposes an innovative algorithm that proposes optimal selection of transport modes in trips within the South East Europe region by taking into consideration trip cost, duration and CO_2 emissions. We introduce the Shortest Path Trans-Shipment Problem (SPTP) which is based on mixed-integer linear programming model and for solving this problem, we propose an extension of Dijkstra's algorithm that identifies the optimal way of cargo movement with trans-shipment. We apply the proposed algorithm in real-life data by embracing three different types of transportation mode, namely road, rail, and Inland Waterways (IWW). We also present results from three different cases where: (a) goods should cross 2 or 3 South East Europe (SEE) countries, (b) goods should cover a distance more or less than 500 km, and (c) goods with origin the Port of Piraeus and destination four capital cities of SEE. All primary data used in the cases above have been acquired from the European funded project "Green Intermodal Freight Transport (GIFT)".

The remainder of the chapter is organized as follows. Section 2 presents the Pan-European transport corridors and the SEE transport networking focusing on corridors IV, V, VII and X. Subsequently, Section 3 describes the modelling of the Shortest Path Trans-Shipment Problem (SPTP), whereas Section 4 shows the results obtained from the three cases described above. Section 5 discusses the findings of each case. The chapter concludes with Section 6 with implications from this research as well as a future research agenda for further testing and evaluation of the proposed algorithm.

2. PAN-EUROPEAN TRANSPORT CORRIDORS (PECs) AND THE SEE NETWORK

The Corridor concept is part of the Pan-European transport infrastructure concept, which has been developed about three decades ago along the three Pan-European Transport Conferences, in 1991 at Prague, in 1994 at Crete, and in 1997 at Helsinki respectively (European Commission, 2001). The Pan-European Transport Corridors (PECs) have been developed for the benefit of the trade and social relations between European countries. Figure 1 shows the PECs of the South East Europe (SEE) region, which embrace four corridors (i.e. Corridors IV, V, VII, and X), which cross almost every country of the SEE.

Figure 1. Pan-European Corridors IV, V, VII and X

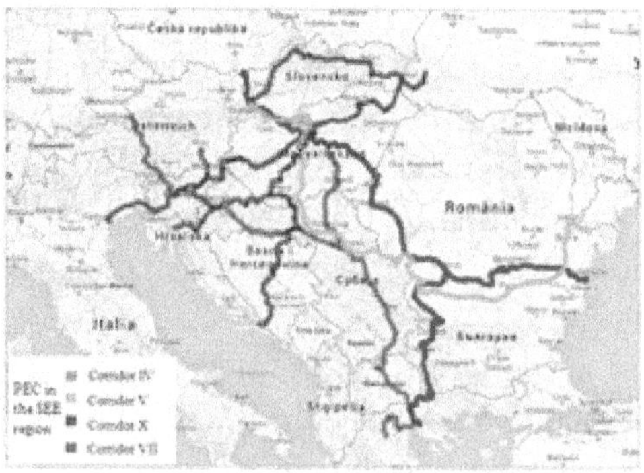

Some further details about these four corridors are presented as follows (European Commission, 1998; Dogan, 2005):

- Corridor IV is a multimodal Northwest-Southeast link, running from Germany to Hungary. Corridor IV consists of 4340 km of railways, 3640 km of roads, 10 airports and 8 sea and/or river ports.
- Corridor V is a multimodal link starting from Italy and extends across Slovenia, Hungary, and Ukraine. The components of the Corridor are 3270 km of railways, 2850 km of roads, and 3 sea and/or river ports.
- Corridor VII (the Danube River) is a waterway route on the Danube from Germany to the Black Sea. The Danube is the second longest river in Europe and links Western and Eastern Europe through the Rhine and the Main, via the Rhine-Main-Danube canal. It is approximately 2850 km length, of which 2415 are used for navigation, and comprises 44 sea and/or river ports. The function of inland ports is not limited to inland shipping. Most of the ports on the Danube have rail and road connections as well.
- Corridor X is a multimodal Northwest-Southwest link, connecting Austria with Greece. The Corridor has four branches, Branch A is Graz (Austria), Maribor (Slovenia), Zagreb (Croatia); Branch B is Budapest (Hungary), Novi Sad, Belgrade (Serbia); Branch C is Nis (Serbia), Sofia (Bulgaria), and further along Corridor IV; Branch D is Veles, Bitola (Serbia), Florina, Kozani- via Egnatia and Igoumenitsa (Greece). It consists of 2528 km of railways, 2300 km of roads, and 4 sea and/or river ports.

Figure 2 shows the index of inland freight transport volume relative to Gross Domestic Product (GDP) for certain SEE countries. This indicator is defined as the ratio between tn-km (road, rail, inland waterways) and GDP (in constant prices). Rail and inland waterways transport are based on movements on national territory, regardless of the nationality of the vehicle or vessel. Road transport is based on all movements of vehicles registered in the reporting country. As it can be seen, Bulgaria, Hungary, and Slovenia have a much higher volume of freight on inland modes than EU-28 and this is due to the increasing need for freight movement between Central and South East Europe countries (Eurostat, 2019).

Figure 2. Volume of inland freight transport, for certain SEE countries, relative to GDP (Eurostat, 2019)

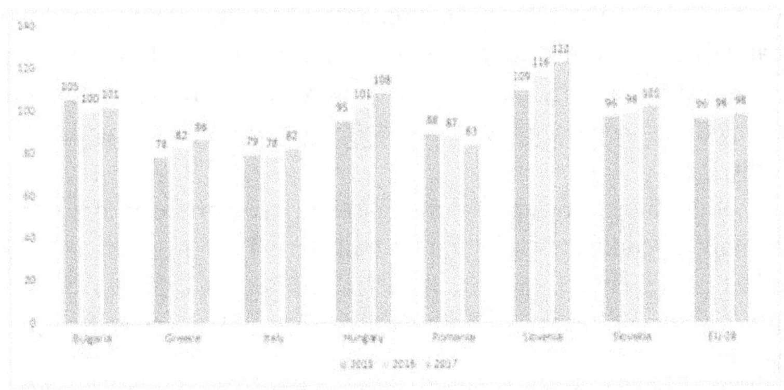

Additioanal statistial data that deal with the inland modal split of certain SEE countries are depicted in Figure 3. It is apparent that in the EU – 28 the average share of road transport between years 2015 – 2017 reaches 76% while rail transport is 17.7% and inland waterway 6.3% of the total freight transport market.

Figure 3. Inland modal split of freight transport (Eurostat, 2019)

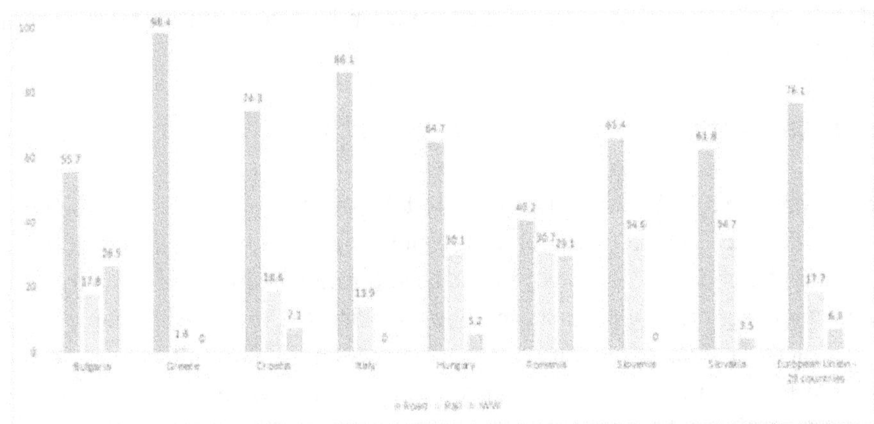

An interesting finding from Figure 3 is that most of the SEE countries (with highest to be Greece) make use of road transportation. Its dominance in SEE countries may be attributed to the lower requirements for infrastructure, flexibility, and legal framework. However, road transport exhibits significant weaknesses contributing to considerable CO_2 emissions, accidents, increased noise level, road congestion and wear (Calthrop and Proost, 1998). According to the EU Transport statistics (2019), the contribution of road transport to the GHG Emissions chart reaches 71.7% whereas according to the World Health Organization (2018), road traffic accidents occupy the third position among the ten leading causes of mortality.

EU policy makers have suggested already that in order to relieve the pressure on the roads, transport modes have to be combined so as to reduce congestion, environmental impact, improve safety, reduce economic impact (due to fatalities and environmental harm), while at the same time meet modern demands for reliability, speed and safety. Over the last two decades, the EU supports this concept by promoting co-modality (mainly under the TEN-T program) as a means for safer and greener approach to freight transport. EU adopted also recently a comprehensive strategy (Transport 2050) for a competitive transport system that will increase mobility, remove major barriers in key areas, and reduce Europe's CO_2 emissions in transport by 60% by 2050. Subsequently, many transportation companies have also started to adopt new methods to reduce freight costs (mainly through co-modality) and become more environmental friendly.

The strengthening of combined transport through green transport corridors may be an effective approach to environmentally friendly and safer freight transport. To this end, in the following section, we present a mathematical model called Shortest Path Transshipment Problem (SPTP) that we have developed for combined freight trip selection in the SEE region, which is based on a mixed-integer linear programming model.

3. THE SHORTEST PATH TRANS-SHIPMENT PROBLEM (SPTP)

This section is divided in two parts. Initially, we present a mathematical model called Shortest Path Trans-Shipment Problem (SPTP) that we use in order to describe the selection of combined trips in the SEE transport network. The model we propose is based on a mixed-integer linear programming model. Subsequently, we describe an extension to Dijkstra's algorithm which presents the optimization method adopted to solve the shortest path problem trying to find optimal freight transportation routes to South East Europe (SEE) transport network.

3.1 Mathematical Model

In this section, we provide a MILP formulation to model a shortest path problem when trans-shipments exist on middle nodes. We decided to model this problem since in combined transportation we usually have at least one middle node (i.e. terminal) that is used for modal shift (e.g. from truck to train).

Initially, we set an origin and a destination node choosing the preferred variable for minimization, which could be either, trip cost, trip duration, or CO_2 emissions. The result of the Shortest Path Trans-Shipment Problem (SPTP) reveals the optimal route, the nodes where trans-shipment(s) should take place and the transportation mode at each arc of the network. The notations used to formulate the SPTP problem are listed below:

Notation:

N: Set of nodes

M: Set of modes

C_{ijm}: Transportation cost from node $i \in n$ to node $j \in N$ by mode $m \in M$ $\left(i, j, m\right) \in \overline{A}$

c_i: Transshipment cost at node $i \in N$

x_{ijm}: Binary variable for the indication of traverse of an arc (i,j) by mode $m \in M$

h_i: Binary variable for the occurrence of transshipment at node $i \in N$

In order to formulate the problem, we let N to be the set of nodes (i.e. seaports or inland terminals) which can provide freight transportation services when containers are transported (which is the case we study). Let M the set of modes i.e. truck, train, or barge in case where inland waterway transport is used.

To formally describe the SPTP, consider a graph $G(N,A)$, which is consisted of nodes $N= \{s, s+1, \ldots, t-1, t\}$, where s is the starting node and t is the target node, that represent the nodes of the freight transportation route, and arcs

$$A=\{(i,j)\,|\,i=s, j=s+1\}\cup\{(i,j)|i\in N\setminus\{i=s,t\}, j\in N\setminus\{j=s,t\}\cup\{(i,j)\,|\,i=t-1, j=t\}.$$

The arc set represents the pre – haulage, main – haulage, and the post – haulage of cargo transportation. Pre – haulage refers to the connection between starting node s and node $s+1$. This part is covered usually by truck. Main – haulage means that freight is transported anywhere between $s+1$ and $t-1$, (any mode of transport can be used) and final part of the trip is called post – haulage and it connects $t-1$ with t which is the final destination of the route (again truck is usually used in this final part).

To provide a better declaration of the model, we created the $\bar{A}$ set of arcs that we divided it into three different parts. The first part

$$\{(i,j,m)\,|\,i\in N\setminus\{s,t-1\},(i,j)\in A, m\in M\}$$

declares that any mode of transport can be selected during the main haulage transportation. The second part

$$\{(i,j,m)\,|\,i\in\{s,t-1\},(i,j)\in A, m=1\}$$

defines that from node $s\in N$, which could be a depot connected with a container terminal that exists to the SEE network, only road transport is feasible for covering this pre-haulage distance, while the same occurs for the post-haulage, where road transport is used from the last existing terminal of our network (i.e. $t-1\in N$) to the final destination (depot) $t\in N$. A depot is a node that can only be accessed by road transport. However, freight terminals of the SEE network may accommodate multiple modes of transport (i.e. truck, train) based on the infrastructure and the equipment they operate. For instance, inland ports could serve road, rail, and inland waterway facilities. The last part of this set $\{(i,j,m)\,|\,j=t,(i,j)\in A, m=1\}$ defines that cargo can travel from $t-1\in N$ to the final destination $t\in N$ only by road transport.

To model such constraints, we define a binary variable $x_{ijm},(i,j,m)\in\bar{A}$ that receive the value of 1 if the arc (i,j) is traversed by mode m and 0 otherwise. In addition, we defined another binary variable h_i, $i\in N$ which is equal to 1 in case where a trans-shipment takes places and 0 otherwise.

Furthermore, let C_{ijm} be the transportation cost from node $i\in N$ to node $j\in N$ by mode $\in M$ $(i,j,m)\in\bar{A}$. The transportation cost depends on the distance between the terminals, the weight of the cargo, and the local prices for a tn-km of the starting node. We also take under consideration the parameter c_i to represent the average cost of trans-shipment (loading/unloading of container) from/to one mode to another in freight terminals.

According to the model formulation of the SPTP, the objective function aims to minimize the cost of freight transportation service and consists of two terms: (a) the cost of transportation, (b) the cost of trans-shipment. More specifically, the first term refers to the pre, main, and post – haulage transportation. Decision variable x_{ijm} ensures that only the cost from the travelled arcs will be counted to the total cost. The second term refers to the cost of trans-shipment in case cargo changes modes of transport at a terminal multiplied with the decision variable h_i which is responsible to consider only the nodes which participate in this process. The SPTP is formulated as follows:

$$min \sum_{(i,j,m)\in \bar{A}} C_{ijm} x_{ijm} + \sum_{i\in N} c_i h_i \quad (1)$$

$$\sum_{(k,i,m)\in \bar{A}} x_{kim} - \sum_{(i,j,m)\in \bar{A}} x_{ijm} = 0, i\in N\backslash\{s,t\} \quad (2)$$

$$x_{i(i+1)m} = x_{(j-1)jm} = 1, i=s, j=t, m=1 \quad (3)$$

$$\sum_{(k,i,m)\in \bar{A}} x_{kim} - \sum_{(i,j,m)\in \bar{A}} x_{ijm} \leq h_i, i\in N\backslash\{s,t\}, m\in M \quad (4)$$

$$h_i=0, i\in\{s,t\} \quad (5)$$

$$x_{ijm}\in\{0,1\}, (i,j,m)\in \bar{A} \quad (6)$$

$$h_i\in\{0,1\}, i\in N \quad (7)$$

Constraint (2) ensures that the already loaded mode of transport arrives at a middle node, from which has to depart to the next node. In case of the source/starting node (i.e. $s\in N$) and target node (i.e. $t\in N$) Constraint (3) ensures that the transportation mode only departs and arrives respectively from/to this node. Both of these constraints ensure the continuation of the path from the origin to the destination. What is interesting in the SPTP is the occurrence of a trans-shipment at the middle nodes (Constraint 4). This constraint provides the ability of trans-shipment during the cargo movement. Finally, we assume that at the initial and final node, trans-shipment is not feasible (Constraint 5).

3.2 Proposed Algorithm for Solving the SPTP

We propose an extension to Dijsktra's algorithm to solve the SPTP. Dijkstra's algorithm is a graph search algorithm that solves the single-source shortest path problem for a weighted and directed graph with non-negative edge path cost, producing a shortest path (Dijkstra, 1959; Cormen *et al.,* 2001).

The proposed Dijkstra's algorithm identifies the path with the minimum transportation cost, time, and CO_2 emissions. A step by step description of how the proposed algorithm operates follows.

1. Initially the variable that will be used for identifying the shortest path is set, i.e. finding the path with the minimum distance, CO_2 emissions, time, or cost
2. Then the weight and the volume of the freight are inserted by the user. The latter are used so as to identify the optimal number and type of container(s) that will be used.
3. Afterwards the proposed algorithm reads a matrix filled in with data from the network, (i.e. distance, time, cost, and CO_2 emissions) for both arcs and nodes
4. The user chooses the origin country and hub as well as the destination country, and hub.
5. Furthermore, the user inserts the distance between the pick-up point and the origin hub. The same applies to the final destination point with the final destination hub. These distances are very important for the algorithm so as to calculate the equivalent trip duration, cost, and CO_2 emissions for the initial and final leg of the total trip (i.e. the first and last mile).
6. By inserting all the necessary input data, the algorithm identifies the best path by trying to improve the distance values following the steps described below:

Step 1: Assign to every node a distance value: set it to zero for the origin node and to infinity for all other nodes.

Step 2: Mark all nodes as unvisited (temporary) and designate the initial node as current.

Step 3: For current node, consider all its unvisited neighbors and calculate their distances. For example, if current node A has a distance of 5 and an edge connecting it with node B has a length of 4, the distance to B through A will be 5+4=9. If this distance is less than the previously recorded distance, then the distance is overwritten. For the case where the main variables are cost, time, or CO_2 emissions then add to this calculation not only the value of the arc, but also the value of the current node. For illustrating this case a typical example follows. If current node A has a cost of 4 and an edge connecting cost is 3, the cost to B through A will be 4+3=7; otherwise if there is a trans-shipment, i.e. from truck to train, a node cost which is equal to 1 will be added thus the final cost to B through A will be 4+3+1=8. In some cases, more than one arc may exist between a pair of hubs due to the different transportation mode that current network supports.

Step 4: Considering all neighbors of the current node, mark it as visited (permanent). A visited node will not be checked again; its distance recorder is final and minimal.

Step 5: If all nodes have been visited, stop. Otherwise, set the unvisited node with the smallest distance (from the source node, considering all nodes in the graph) as the next temporary current node and continue from Step 3.

7. After finding the shortest path from the origin node to all other nodes, the algorithm creates the one it is chosen from the source node to the target node and displays it. Then, it calculates the total value of each variable (distance, cost, time, CO_2) and presents the final result.

Figure 4. a) Average values of crossing 2 counties tests; b)Modal split of crossing 2 countries

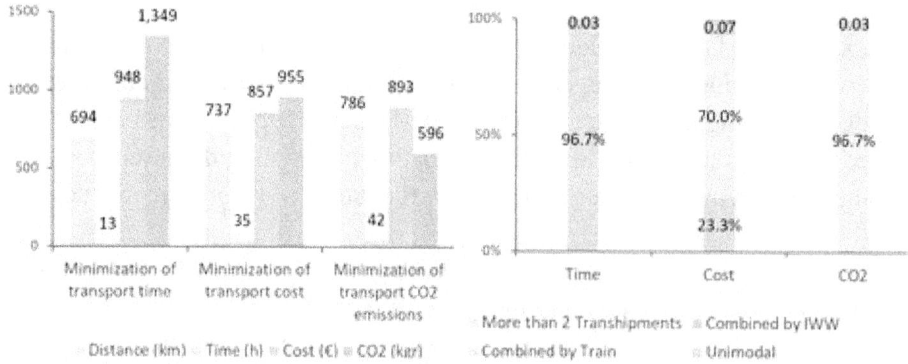

4. TESTING THE PROPOSED ALGORITHM IN THE SEE TRANSPORT NETWORK

In order to test the aforementioned Dijkstra's algorithm, a transport network of countries from the SEE region has been selected (i.e. Greece, Bulgaria, Romania, Slovenia, and Hungary). It must be mentioned however that due to lack of information (e.g. trip duration) some assumptions were made. By taking into consideration these assumptions (see Appendix) we tested the proposed algorithm in three different cases. The cases that we examined are as follows:

Case 1: Freight trips that cross either 2 or 3 countries,
Case 2: Freight trips where the total running distance is less or more than 500 km,
Case 3: Freight trips where the origin hub is Piraeus Port and the destination is 4 capital cities in the selected SEE countries.

The basic purpose of choosing these categories of tests was to study and examine cases of combined transport which usually occur in the SEE region. After performing the tests, we would be able to analyze the results concerning optimized routes in the SEE area and make a series of recommendations about the most effective ways for transporting goods in this area.

Case 1: Results for Freight Trips Crossing 2 or 3 Countries

We applied the proposed algorithm in real-life data, and we identified the optimal transportation mode for cases where goods should cross 2 or 3 countries respectively. In addition, we investigated the best option for cargo movement in terms of time, cost, and CO_2 emissions. Our test cases include several combinations of origin and destination points in SEE area for various scenarios.

Each sub-case (sub-case 1.1: crossing of 2 countries, sub-case 1.2: crossing of 3 countries) has 30 tests for each independent variable (i.e. time, cost, CO_2 emissions). In the list below an explanation of the variables follows:

- Time refers to the trip duration of a container from the origin to destination hub, loading, unloading, and the waiting time staying at the buffer area to change mode.

- Cost refers to the transportation cost from origin to destination hub. It includes also the transshipment cost at every hub that container changes mode.
- CO_2 refers to the carbon dioxide emissions that each mode of transport emits. It includes also the emissions that are produced by the lifting machines.

The average results from the 30 tests concerning sub-case 1.1 (i.e. where cargo is transported while crossing 2 countries) are shown in Figure 4a and Figure 4b.

As Figure 4a illustrates, for the case where minimization of transportation time is the objective, the results show that the average trip duration is 13 hours. Comparing the time value between the minimization of time and CO_2 emissions, a difference of 29 hours is presented, which means that more than an extra day is added to the freight trip. In case the forwarder company prefers to minimize the transportation cost (via combined transport) could save at the best case 91 €. In case of minimization of CO_2 emissions, the forwarder company could decrease at the best air pollution by 753 kg CO_2.

By taking into consideration Figure 4b we can identify the mode of transport which is the most preferable at each case. More specifically, when the objective is the minimization of transport time, then the unimodal transport option (i.e. by truck), is favored in almost all cases. On the other hand, when the objective function is taking into account CO_2 emissions, we may observe that almost all routes are executed by combined transport, where the transport mode from the long-haul part is train. Now in case where minimization of transportation cost is the objective, the results show that the combined transportation via train is the most profitable option.

The other sub-case that has been investigated is the crossing of 3 countries. Figure 5a provides information about the average values of distance, time, cost, and CO_2 emissions of the selected 30 different tests where a full-load container has to cross 3 countries.

Figure 5. a) Average values of crossing 3 counties tests; b)Modal split of crossing 3 countries

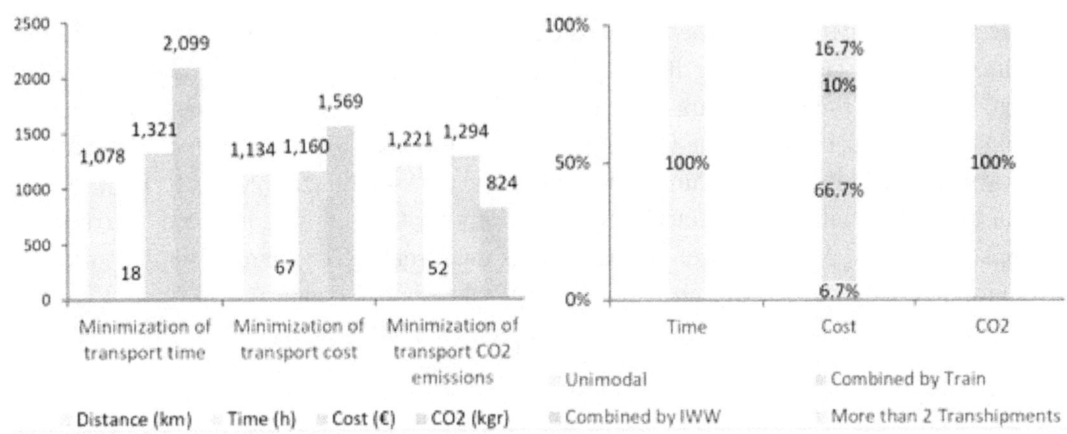

As far as the minimization of transport time is concerned, we observe that the typical trip duration is 2 days less than the case when we tried to minimize the cost. As in all cases, going faster costs more, has higher output of CO_2 emissions, and the disadvantage to cutting the trip time by 2 days is the increased amount of cost by 161€ and more than double the CO_2 emissions. There are many cases that a container has to be transported through three different countries. That means that there is an increased chance of crossing the Danube (Corridor VII). Mostly this probability is higher when the shipper specifically looks for the minimum transportation cost as it seems in Figure 5a. Inland waterway is the most cost competitive mode but due to the natural flow of the water in the Danube and the narrows passes of some areas, it becomes too slow. The pace of the barge is a factor which also varies due to multiple unpredictable factors (i.e. downstream or upstream of the river, river traffic, and depth and width of the river especially along the Romanian-Bulgarian part of the Danube). As a result, more often than not using the Danube is not a first choice, when the main variable in the objective function is time, although a good solution for large volume of freight and for products which they do not have short expiration date.

Figure 6. a) Average values when distance is less than 500 km; b)Modal split when the distance is less than 500 km

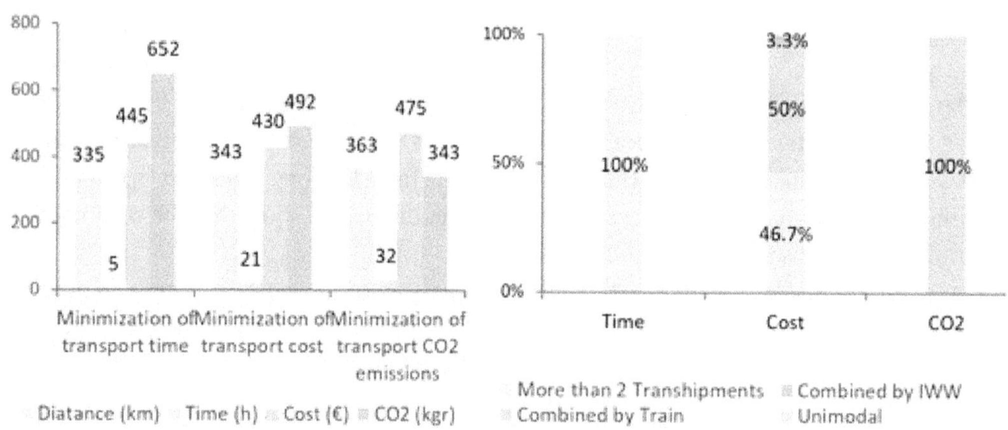

As it clearly seems in Figure 5b when we consider only time for minimization again truck (unimodal) is the optimal solution. This is due to the lack of transshipments along the route from the origin, to the destination. Moreover, the distance traveled by truck is shorter due to the efficiency of roadways, in comparison to Railways. However, train is still the most environmentally friendly mode of transport and it was always selected by the algorithm when the objective for minimization was CO_2 emissions. The majority of the SEE countries use electrified trains, except for Greece, Bulgaria, and Romania that use diesel trains, which still produce fewer CO_2 emissions than trucks. The share of modal split is mostly divided to rail and barge when minimization of cost is taken into account. As we see in Figure 5b rail transport is favored at 66.7%, IWW at 10%, and with a combination of use rail, IWW and truck being at 16.7%. The reason that these two (rail and IWW) have the majority of use, is because they are more cost effective at big distances.

Case 2: Selection of Transport Mode for Trip Distances Less or More Than 500 km

In this section we investigate two cases (sub-case 2.1 includes scenarios of travelling distance less than 500 km, whereas sub-case 2.2 more than 500 km). The cut-off point of 500 km is a crucial distance limit and plays an important role in the selection of transportation mode (i.e. road or combined transport) as it finally affects the choice of the shipper (Lammgard, 2012; EU Statistical Pocketbook, 2019) We tested the proposed algorithm in 30 cases for each sub-case in order to identify the optimal option of moving containerized freight. The results of sub-case 2.1, where the distance is less than 500 km are shown in Figure 6a and Figure 6b. All the above routes were selected randomly as there are many trips that could satisfy the term of less than 500 km (average distance equals to 347 km). While cost is usually the most important parameter for selecting a transportation mode, many customers also take into consideration speed, door-to-door services, minimization of loss or damage, and variation of delivery time. In most of sub-case 2.1 instances, cargo is transported either in the same country, e.g. from Verona to Trieste, or passing the origin's country border for travelling to the bordering country, e.g. from Thessaloniki to Sofia.

More specifically, in Figure 6a above when taking into consideration only the minimization of transport time, we see that time value is low but we have a dramatic increase of CO_2 emissions. Secondly when concentrating on the minimization of transport cost, trip duration is quadrupled while cost and CO_2 emissions are more balanced. Finally, when minimizing CO_2 emissions trip duration is exponentially increased and cost is the higher of the two previous minimizations.

But countries with developed rail lines and rail network have better market share even in short distances. That is apparent in Figure 6b at the case where cost is the input variable at the minimization function. As it seems the percentage of modal split is divided in half. The second half which belongs to rail might be a result of good equipment and infrastructure (i.e. not obsolete railcars, existence of intermodal container terminals) in rail networks such as Slovenia and Italy but also in some other SEE countries. The rest 46.7% steps from the low-cost road transport services in countries such Bulgaria and Romania or from the cases where the middle haulage distance is very short, e.g. 100 km.

In the other two cases, of minimization of time and CO_2 emissions, the choice of transportation mode seems to be easier, as all the routes are implemented entirely either by truck or train respectively. In countries where the environmental concerns are a predominant factor all routes are designed by train, while road transport is again the fastest way of freight transport. Not forgetting to mention, in such short distances inland waterway has no market share, apart from the cases that the cargo is transported between two inland ports which constitute the origin and destination point. From the other side the longer the distance is, the more likely a combined transport chain will be used. Usually distances over 500 km require combined transportation instead of unimodal (Rodrigue, 2013). So sub-case 2.2 includes routes where all the distances are above 500 km, which means that freight may pass through at the most 6 countries to reach to its final destination. The longest trip which can be created in SEE network, originates in Italy with the final destination point in Greece. In this case cargo passes from Italy, Slovenia, Hungary, Romania, Bulgaria, and Greece and in Budapest it changes from Corridor IV to Corridor V. The results of sub-case 2.2, where the distance is more than 500 km are shown in Figure 7a and Figure 7b. All the above routes were selected randomly as there are many trips that could satisfy the term of more than 500 km (average distance equals to 1270 km).

There are many instances where freight should be delivered in the shortest time possible. Such freight could have experation dates or a specific demand which should follow customer's needs for early delivery. For this reason, we have minimized transportation time, and this is seen in the first set of results, where averaged minimized time value was 23 hours, traveling a distance of 1190 km. This value is much less than the highest average value of time which can be seen in the second set of results of Figure 7a. The difference between these two values could be translated to 3 ½ days.

Figure 7. a) Average values when distance is more than 500 km; b) Modal split when the distance is more than 500 km

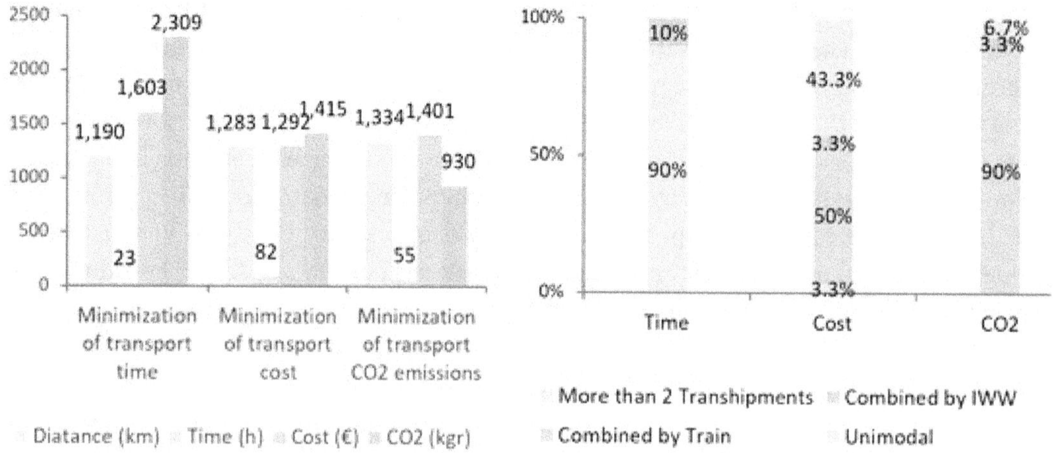

In contrast to time, we also applied the shortest path problem having cost as the input variable. As can be seen in the second set of results of the above Figure 7a the minimum value of cost is 1292 €, which is ¼ less travelling same distances along this region. The reason for the reduction of cost comes from the low cost of inland waterway which is usually chosen in long distances. In most of the cases examined in sub-case 2.2 the cargo is transshipped from truck to train then to barge, which would mean that at least 3 transshipments are occurring. Despite the amount of transshipments occurring, the transportation cost is lower, as the cost of transporting a TEU by IWW is 2.5 times less than the road and 1.5 times less than rail transport. In addition, we tried to minimize CO_2 emissions, with the result being that minimized average value was 930 kg of emissions.

When comparing the results of sub-case 2.1 and sub-case 2.2 we see that the difference between the minimum value and the maximum value gives us an increasing higher output of emissions. For example, in sub-case 2.1, the minimum value was 343 kg and the maximum was 652 kg, consistently being almost double, while in sub-case 2.2 the minimum value is 930 kg with the maximum being 2309 kg which is increased to almost 2 ½ times more.

Table 1. Capital cities and container terminals

Capital City	Container Terminal
Athens	Piraeus Container Terminal
Bucharest	Bucharest Intermodal Terminal
Budapest	Bilk Kombiterminal
Sofia	Yana Sofia Intermodal Terminal
Ljubljana	Ljubljana Moste

Figure 8. Results of choosing the optimal way of transport

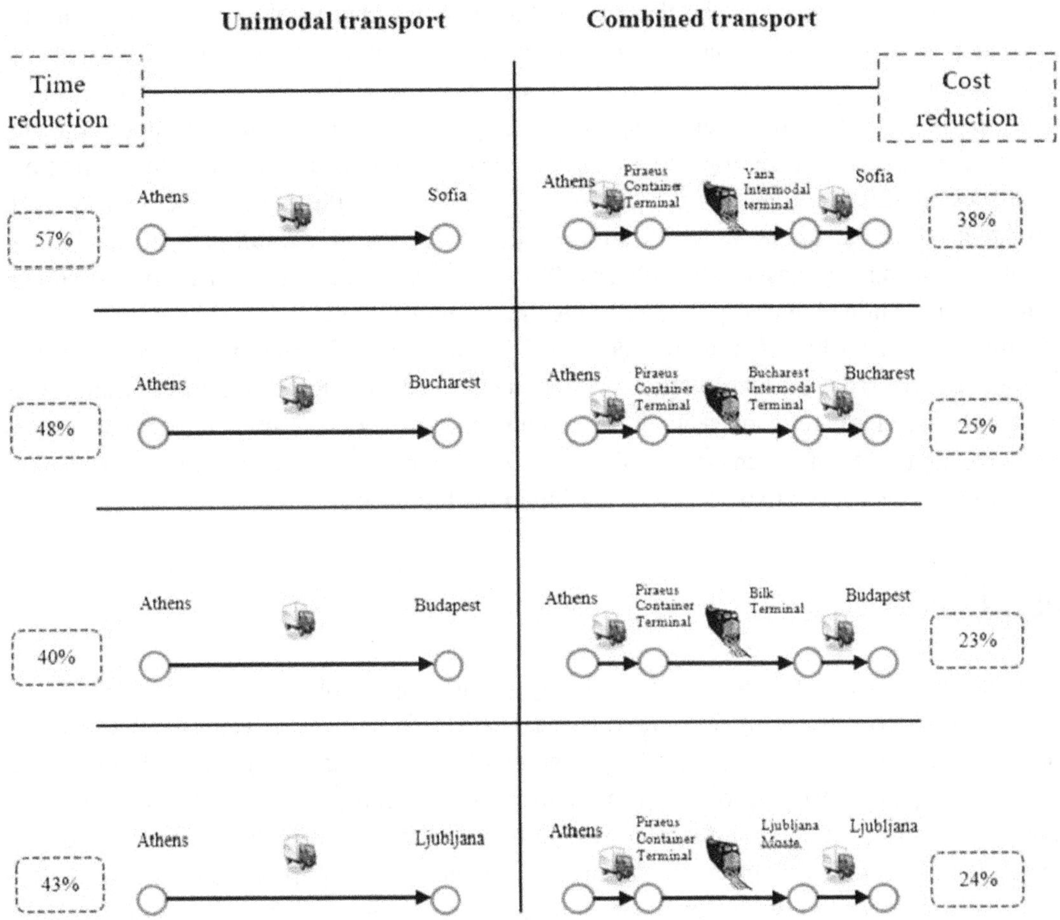

As we see in the Figure 7b above, at the minimization of transport time unimodal transport is used in most cases. Despite avoiding transshipments at a hub reduces the total transportation time even in long distances. But when CO_2 emissions are the dominating factor, diesel engine trucks will always be replaced by electrified engine trains. That can be seen in the case of minimization of CO_2 emissions when 90% of the routes were implemented by train. As mentioned before the relative unit cost of road transport is higher than both rail and inland waterway transport. For this reason, the transportation mode choice is divided between train and barge, allowing truck to have only a small percentage in long distances.

Case 3: Selecting the Transport Mode From Athens to 4 Capital Cities of SEE

The Dijkstra's algorithm was used in this case to investigate the best mode of transport (i.e. unimodal vs. combined) by taking into consideration as point of origin Piraeus port of Athens and destination point 4 capital cities of the South East Europe area (Sofia, Bucharest, Budapest, Ljubljana).

The parameters of testing were as follows: Transportation of a TEU, loaded with 25 tons of freight, from a depot around 20 km away from the Piraeus port to a warehouse which is almost 20 km away from the main cargo terminals of the these 4 capital cities. For these tests we take into consideration only road and rail transportation. Table 1 shows the names of the terminals which were used in this case.

In first instance, the scope was to minimize the trip duration to these destinations. The results from the tests revealed that the unimodal transport is the optimal solution for door-to-door delivery with minimum delays. When considering only transportation cost combined transport is the optimum option for all of these destination points. It is important to mention that, the most cost-effective freight trip is from Athens to Sofia, where we see the combination of both road and rail presents the highest cost reduction, decreasing the transportation cost by 38% in contrast to unimodal transport. The comparison of unimodal and multimodal transport is presented in Figure 8 below.

Aiming simultaneously to see the reduction of CO_2 emissions of freight transport, combination of road and rail underlies the optimal solution for every trip from Athens to 4 destinations. Using combined transport CO_2 emissions are reduced about 58% compared to unimodal transport. Moreover, in instance of transporting cargo from Athens to Ljubljana, choice of combined road and rail reduces the amount of CO_2 emissions even more to 61% in relationship to road transport.

5. DISCUSSION

Transport is one of the major sectors of the economy that are responsible for an even growing of percentage of total emissions, and mostly for CO_2 emissions (ECMT, 2005). On the Danube River, most barges still use diesel engines, which produce a large amount of emissions triggering a need to find solutions to improve barge transport performances (Wiegmans and Konings, 2007). In addition, rail services often use diesel trains to run the last kilometers. Moreover, in countries such as Romania and Bulgaria, whole networks do not support electrified trains and in Greece only a part of the main route from Thessaloniki's port to Athens railway station supports electrified trains. But even if local truck emissions, whether NO_x or SO_2, have been reduced significantly over the last two decades by introducing cleaner engines, road transport is responsible, in most cases, for more CO_2 emissions than combined transport (Reis *et al.*, 2013).

As it is shown in Cases 1, 2, 3 above, unimodal transport is indeed the fastest way to transport a container as long as there are no transshipments on the hubs. However, it is not an environmental friendly option and it is the least cost-efficient mode. Furthermore, the emissions it produces are much more than those of the other two modes (i.e. train and barge), especially on long distances. For this reason, in long distances combined transport should be the preferred environmental and cost-effective solution and in all cases for freight transportation.

Since transportation cost in most cases is the most important factor for selecting a transportation mode, we analyzed the case of combined transportation cost, and more specifically the case where rail transport is used for the middle-travelled distance. As it is presented in Figure 9 for the case of road and rail transportation, the cost increases while distance increases. The data of the graph below are taken from the sub-cases 1.1, 1.2, 2.1, and 2.2 respectively. However, if we take into consideration the key performance indicator of euro per kilometer (€/km) then it is obvious that the longer the distance is the less the value of this KPI for rail transport is (e.g. 1.3 €/km when distance is 350 km while this value is lower, i.e. 1.04 €/km, when distance is 140 0km). In addition, freight transport KPI's cost within the distance of 500 km is balanced between road and rail transport and this value is close to 1.3 €/km. From the other side, the longer the trip is the difference between the KPI's cost for rail and truck becomes larger, with the KPI of road transport gradually increases close to 1.3 €/km while rail transport's KPI decreases slightly close to 1.05 €/km.

Figure 9. Cost of rail transport per km for 25 tons of freight

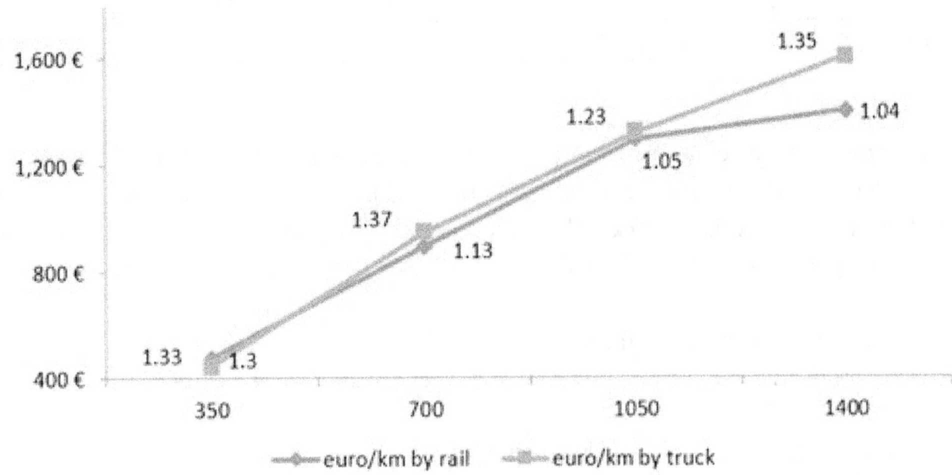

Moreover, Table 2 provides another evidence that shows why combined transport may finally be the best way for long distance transportation. After examining the results of sub-cases 1.1, 1.2, 2.1, and 2.2 we divided all the average values of distance, time, cost, and CO_2 emissions of each sub-case focusing on the ones that were executed by truck by the respective average values of rail transport. We calculated the ratio between rail and road and the outcomes of these calculations are shown in Table 2:

- rail distances are 12% longer than these of road,
- trip duration is 264% higher when combined transport is chosen due to longer waiting times at a node (i.e. transshipment time)
- transportation average cost is 4% less than truck, but as it was mentioned before distance affects this cost, and
- there is a great reduction of CO_2 emissions which is 56% less than truck.

Table 2. Ratio of train average values according to truck average values

Cases 1 and 2	Ratio Train Avg. Values/Truck Avg. Values			
	Distance	Time	Cost	CO_2
2 counties	1.13	3.36	0.94	0.44
3 countries	1.13	2.88	0.98	0.39
Distance<500 km	1.09	5.94	1.06	0.53
Distance>500 km	1.12	2.4	0.87	0.40
Average Ratio	**1.12**	**3.64**	**0.96**	**0.44**

For distances over 500 km, combined transport is always the best mode (as well as for distances between 250-500 km under certain circumstances) (Lammgard, 2012). If combined transport achieves to reduce the costs in shorter distances than 500 km then the share of it in each country's market will increase.

Combined transport, whether by rail or IWW, has the below advantages:

- Relief of the road network from the daily traffic
- Transfer of goods to a safer, more environment-friendly and more weather-independent transport modes
- Better sharing of transport volumes between modes
- Co-operative activity combining the advantages of road (flexibility, speed) and rail (more economical, mass transport on longer distances)
- Competitiveness under given circumstances (i.e. existing rail network, electrified trains, connectivity to inland ports)
- Savings in variable costs of a truck (fuel, tires, maintenance)
- Reduction in personnel (drivers, driving hours, night work)
- Less damage claims concerning transported goods.

All the above advantages explain that combined transport is a good choice in south-east Europe and still there are many opportunities exist for this section to be improved and sustainable developed.

In addition, the relationship between the average internal costs (e.g. cost of ownership, insurance, repair and maintenance, labor, energy, taxes, and toll/fees) of both networks, unimodal or combined, may partially explain the current split between truck and train (Janic, 2007). The operational cost of road transport is lower than the operational cost of the combined transport network over short, medium, and even some long-distances markets, which, in combination with other market and regulatory factors lead to lower prices. This attracts more voluminous and price sensitive commodities over such short distances.

6. CONCLUSION

This paper addressed a mathematical formulation model, the Shortest Path Trans-Shipment Problem (SPTP) which is based on mixed-integer linear programming model. The SPTP can be used to minimize cost, trip duration, or CO_2 emissions in freight transport operations.

We proposed and extension to Dijkstra's algorithm that was applied to three different transportation modes that consider trans-shipments when needed. The proposed algorithm provides the best route as well as the optimal choice of transportation mode according to the parameter to be minimized (i.e. cost, trip duration or CO_2 emissions).

The results of all tests showed that road transport is by far the fastest mode for completing a trip. It avoids all trans-shipments and it is also flexible since it can reach any destination (drop point) as road network is more developed. On the other side, combined freight transport, either by train or barge, is more cost effective and produces less than half CO_2 emissions when compared to road transportation. Combined freight transport constitutes the best way to minimize costs and take advantages of the large amount of freight that can be transported simultaneously. Thus, it can be concluded that combined transport is the preferred mode of cargo movement in medium and long distances while truck is the dominant mode in short distances.

Future research on the SPTP problem, may include the following: (a) an extended transport network (e.g. multiple road, rail and IWW arcs, additional inland terminals/ ports/ seaports), (b) more multi-criteria methods may provide better analysis to this shortest path problem, supporting decision makers to better analyze freight trips and make route scheduling according to current market demands, (c) shortest path problems are mostly taken into consideration distance or cost, however a recent trend is to investigate also the environmental impact of transport operations and at the same time minimize cost and trip duration. For this reason, goal programming model can be applied on the existing or an extended transportation network.

ACKNOWLEDGMENT

This research has been financed by the European Commission (SEE Programme) under the project entitled "Green Intermodal Freight Transport-GIFT" (SEE/C/0003/3.3/X).

REFERENCES

Calthrop, E., & Proost, S. (1998). Road Transport Externalities. *Environmental and Resource Economics, 11*(3/4), 335–348. doi:10.1023/A:1008267917001

Cormen, T., Leiserson, C., Rivest, R., & Stein, C. (2001). *Introduction to Algorithms.* MIT Press and McGraw-Hill.

Dijkstra, E. (1959). A note on two problems in connexion with graphs. *Numerische Mathematik, 1*(1), 269–271. doi:10.1007/BF01386390

Dogan, T. (2005). *An analysis of the Pan-European transport network* (MSc thesis). School of Social Sciences, Middle East Technical University, Turkey.

ECMT. (2005). Strengthening inland waterway transport, Pan-European co-operation for progress. Proceedings of Pan-European Co-operation towards Strong Inland Waterway Transport: On the Move.

European Commission (EC). (1998). Status of the Pan-European Transport Corridors and Transport Areas. TINA office, Vienna, Austria: EC.

European Commission Statistical Pocketbook. (2019). *Mobility and Transportation. Publication Office of the European Union.*

European Union (EU). (2001). Commission underlines the importance of Pan-European Corridors. IP/01/522, Brussels, Belgium: EU.

GIFT. (2013). *GIFT Green Freight Transport Observatory: System design and platform manual. Deliverable 5.2.1, Green Intermodal Freight Transport project.* SEE Programme.

Janic, M. (2007). Modeling the full costs of an intermodal and road freight transport network. *Transportation Research Part D, Transport and Environment, 12*(1), 33–44. doi:10.1016/j.trd.2006.10.004

Lammgard, C. (2012). Intermodal train services: A business challenge and a measure for decarbonisation for logistics service providers. *Research in Transportation Business & Management, 5*, 48–56. doi:10.1016/j.rtbm.2012.11.001

Reis, V., Meier, J., Pace, G., & Palacin, R. (2013). Rail and multi-modal transport. *Research in Transportation Economics, 41*(1), 17–30. doi:10.1016/j.retrec.2012.10.005

Rizet, C., Cruz, C., & Mbacke, M. (2012). Reducing freight transport CO_2 emissions by increasing the load factor. *Procedia: Social and Behavioral Sciences, 48*, 184–195. doi:10.1016/j.sbspro.2012.06.999

Rodrigue, J.-P., Comtois, C., & Slack, B. (2013). *The geography of transport systems* (3rd ed.). Routledge. doi:10.4324/9780203371183

Shin, K., & Cheong, J. (2011). Estimating transportation-related greenhouse gas emissions in the port of Busan, S. Korea. *Asian Journal of Atmospheric Environment, 5*(1), 41–46. doi:10.5572/ajae.2011.5.1.041

Wiegmans, B., & Konings, R. (2007). Strategies and innovations to improve the performance of barge transport. *European Journal of Transport and Infrastructure Research, 7*(2), 145–162.

Winfried, S. (2008). *Securing the product in the container, Container Handbook.* GDV.

World Health Organization. (2018). *Global status report on road safety.* Retrieved from:https://www.who.int/publications-detail/global-status-report-on-road-safety-2018

Table A1. Parameters and values inserted in the EcoTransit tool

Parameters	Value inserted
Amount of freight	1 ton
Type of freight	Average goods
Origin and Destination	City District
Main transportation mode	Truck, Train, Inland ship

APPENDIX

The transport network that was taken into consideration is located in the SEE area. This network consists of 36 hubs, 11 seaports, and 14 inland waterway ports. Most arcs are located on the Pan-European Corridors IV, V, and VII respectively. For most of the hubs, seaports and IWW ports, operational data were available. However, in some case there was lack of information and specific assumptions were made. The latter are presented as follows.

Type of Containers

We assume that every type of cargo whether palletized, bulk or boxed is transported by containers. The types of containers we use are 20 and 40 feet equivalent long. The length of each container is standardized due to the global standards of the trucks, railcars, and ships. Despite the length, weight and volume is not always the same. There are different manufacturers which supply various types of containers in the market with different weight and volume. In our case it is assumed that the maximum cargo weight is 21.67 and 26.48 tons for a 20' and 40' container respectively. The maximum volume is 33 cubic centimeters (m^3) for a 20' and 67 m^3 for a 40' (Winfried, 2008).

Defining Arcs

The definition of arcs was based on a SEE transport network that was designed for the purpose of this paper. All data about distances and emissions between a pair of nodes were acquired by the EcoTransit tool. The parameters described below were set in the EcoTransit tool for the collection of the aforementioned data.

The calculation of emissions assumed EURO V truck technology engines. After 2009 all commercial vehicles, whether light weight vehicles or heavy trucks, had to be manufactured according to Regulation (EC) No 715/2007 with Euro V technology engines, which are more environmental friendly than previous editions. Furthermore trains are either equipped with diesel and not only electrified engines due to some parts of the network which do not support electrification. Both trucks and trains are consider to have a 60% loading factor, whereas inland barge (>class V or Euro ship) a loading factor of 65%. The loading factor depends on the type of freight moved (i.e. density and weight) as well as on the way a truck is loaded (Rizet *et al.,* 2012).

Figure 10. Ratio of train average values according to truck average values

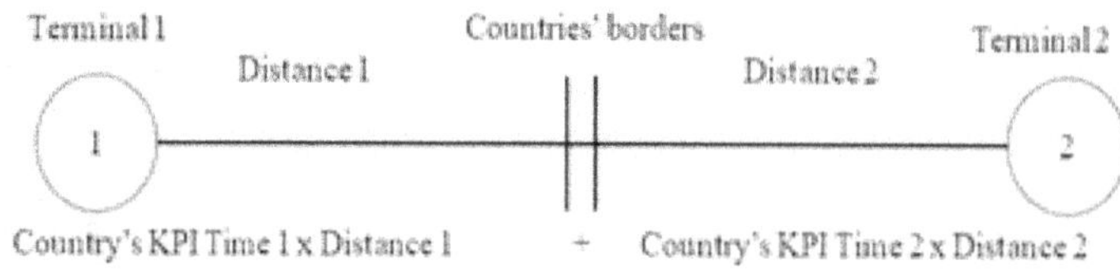

Trip duration information was collected by using primary data from the EU-funded project "GIFT". The calculation was made by multiplying the distance of the arc with each country's key performance indicator (KPI). This KPI was calculated for every country and represents the product of the average value of all the divisions of any arcs' length with the average transport time of goods in these arcs. For the cases where a country border exists between two nodes, a different approach was adopted for trip duration calculation. More specifically, it is a summary of two values. The first value is the product of the distance between the origin node and borders with the value of time's KPI of the origin country and the second one is the product of distance between the borders and the destination point with the value of time's KPI of the destination country. This calculation is shown graphically in Figure 10.

For the cost of the trip, necessary data was also collected by the GIFT project. The calculation was made by multiplying the distance of the arc between a pair of nodes with each country's KPIs and the result of this product multiplied with the tonnage of the freight. This KPI was also calculated for every country and represents the average value of transport cost of each arc. The calculation of cost was made always by adopting the same methodology even if there are country borders between two hubs. The only difference is between the two directions where the KPI depends on the country where the starting hub belongs. Figure 11 shows graphically the calculation methodology.

Another assumption was made to simulate the real way cargo is transported currently. More specifically, we assumed that the leg between the origin and the first intermediary hub as well as the leg between the last intermediary hub and the destination is served by road transport (i.e. truck) due to its flexibility and its ability to reach every place (e.g. distribution centers and warehouses). Hence, the distance between the initial and final node from their closest hub should be defined. The values of the initial and final leg of the network were added to the final results of the shortest path problem and they did not affect the path which was produced by the algorithm.

Furthermore, as there was lack of information in some cases about the physical connectivity of each hub, it was assumed that all container terminals have both road and rail connections. As far as ports were concerned, it was assumed that they are served additionally to road and rail by ship/barges as well.

Figure 11. Cost calculation in each arc

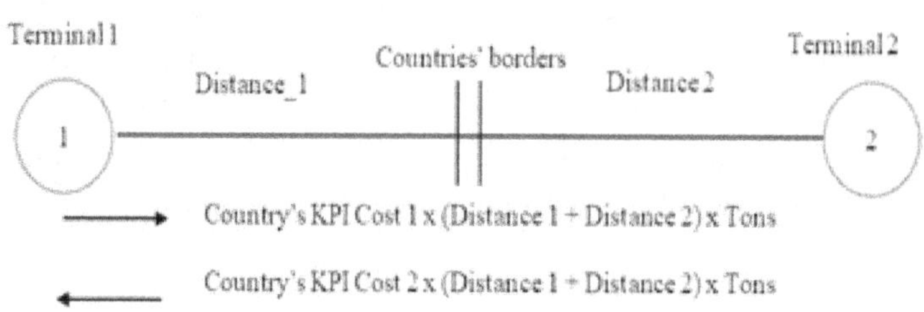

Table A2. Assumptions for arcs and nodes

Arc assumptions	Node assumptions
Data about Serbia, Slovakia, and Croatia were calculated by using Corridor's V values (GIFT, 2013)	Node time per transshipment of all TEUs is defined to 12 hours
Distances and CO_2 emissions were calculated by EcoTransit tool	Node cost per transshipment per TEU is defined to 60€ (GIFT, 2013)
Time and cost of each arc was calculated by using data (GIFT, 2013)	Node CO_2 emissions per transshipment per TEU is defined to 6 kg of emissions (Shin and Cheong, 2011)
Every location (hub) at a radius of 20 km from the center of a city is treated as being in the city center	Every node provides the facility of handling a container i.e. lifting it and storing it
Every arc that is not situated in one of these three Corridors, is calculated by using data from its nearest Corridor	There are both road and rail physical connections to each node either sea/IWW port or container terminal

Distribution centers or warehouses which are usually the initial and final nodes are served by truck. Apart from mode connectivity it was also assumed that every node has the ability to provide facilities of handling a container e.g. lifting (i.e. loading and unloading) and storing it. Last but not least, we assumed that every location (hub) at a radius of 20 km from the center of a city is treated as being in the city center. For all cities which are not crossed by a corridor, their data was calculated by using data from their nearest corridors or by the corridor, which crosses that country.

Assumptions for Nodes and Arcs

Data collected from the EU-funded project "GIFT" provided information about the cost of handling a container (i.e. loading or unloading) and storing it temporarily on the buffer zone. Buffer zone is a place that containers are left while the assigned ship, train, or truck, are available to be loaded. In buffer zones the storing time is less than 24 hours (Rodrigue *et al.*, 2013). The cost of this process was set at 60 Euros per TEU per transshipment activity (i.e. unloading, storing, and loading). Additionally, another reason which led to this cost assumption was that the nodes taken into account in the "GIFT" project were less than 1/3 of the final network nodes.

As it was mentioned previously, for the case of combined transport the initial and final leg are always operated by truck whereas the rest of the haulage is served by rail, or inland waterway transport respectively. The algorithm provides the ability, the user to manually insert the distance of the leg between the origin and the first intermediary hub as well as the leg between the last intermediary hub and the destination. However, in order to keep a uniform testing environment, these distances were set to 50 km each. To this end, we assumed that there is a certain container's transshipment time in each node. This time is calculated as the average waiting time of a container in the buffer zone which is equal to 12 hours. In addition, the container handling equipment in the storage area consists of lifting machines such as reach stackers and cranes which are usually operating using diesel engines for better performances compared to electrified engines. According to Shin and Cheong (2011), CO_2 emissions emitted by the handling of a container with one of these two machines are about 6 Kg per TEU for both loading and unloading process. All the above assumptions which were made either for arcs or nodes are presented in Table A2 below.

Chapter 7
Trigonometric Grey Prediction Method for Turkey's Electricity Consumption Prediction

Adem Tuzemen
https://orcid.org/0000-0002-5786-2686
Tokat Gaziosmanpaşa Üniversitesi, Turkey

ABSTRACT

Industry and technology continue to develop rapidly in today's world. The indisputable most important source of this development, energy is among the indispensables of daily life. Since it is one of the determining factors for the country's economy, the future forecast of electricity demand means calculating the future steps. Based on this, to forecast Turkey's electricity demand, it was benefited from grey model (GM) and trigonometric GM (TGM) techniques. The data set includes annual electricity consumption for the period 1970-2018. The performances of the methods determined were compared based on the forecast evaluation criteria (MSE, MAD, MAPE, and RMSE). Short-term forecasting analysis was carried out by determining the method that gives these values to a minimum. In the future forecast, it has been determined that electricity consumption will increase continuously.

INTRODUCTION

One of the main building blocks of economic development is energy. Electric energy is the most flexible in energy items. For this reason, it is in constant economic activities and daily life. Projected demand for energy, especially rapidly growing demand for energy is an important part of energy policy for a developing country such as Turkey. With the ongoing development level of electricity consumption has increased in recent years, as well as other goods and services, it has also increased in Turkey. Forecasting electrical energy consumption accurately and reliably is of great importance to ensure the continuation of industrialization and long-term stable energy policies as well as increasing demand for electrical energy (Hamzacebi and Es, 2014, pp. 166). Turkey in (Organisation for Economic Co-operation and Development) OECD countries over the past 15 years has become the case in countries where energy demand

DOI: 10.4018/978-1-7998-5442-5.ch007

growth occurs fastest. As a result of the rapidly increasing energy demand, especially oil and natural gas, Turkey's dependence on energy imports is increasing. However, Turkey can be fulfilled by approximately 26% of the total energy demand from its resources (www.mfa.gov.tr/turkeys-energy-strategy.en.mfa). Gross electricity consumption in Turkey was 296.7 billion KWh in 2017 and at the end of June 2018; it was realized as 147.2 billion kWh. In 2023, electricity consumption is expected to increase by 5.5% to 357.4 TWh (www.enerji.gov.tr/en-US/Pages/Electricity). Energy forecast studies in Turkey are carried out officially by the Ministry of Energy and Natural Resources by using Model for Analysis of Energy Demand (MAED) (http://www.mfa.gov.tr/turkeys-energy-strategy.en.mfa). MAED is a simulation model designed to assess medium and long-term energy demand in a country. Experimental information is needed in addition to the versatile input data to run the model (Akay and Atak, 2007, pp. 1672-1673). Keeping the supply and demand of electricity under control will be realized with a correct prediction.

Prediction is a necessary prerequisite for most operational activities. It is not possible to plan the number of expected activities without predicting the future to estimate the resources to be designed, planned and controlled (Lewis, 2012). Demand forecasting is important not only in energy but also in other areas. People will want to see ahead in planning, cost analysis, import-export situations, and sales. Demand forecasting stands out at this point. Demand forecasting is primarily important for planning. Although it is important in many sectors and subjects, it is also important for electricity. Because electricity is an energy that cannot be stored. When the electricity forecast is less than the actual consumption, power cuts are made. Also, electricity savings will become mandatory in that company. When analyzed on a country basis, a significant change will occur in the spacious levels of the people and the growth of the economy. If the electricity estimate is more than the actual consumption, an extra electricity fee is paid. Thus, the usage of electricity consumption fees paid in vain will be prevented from being used elsewhere. In other words, the more realistic the amount of consumption to be estimated, the better budgeting (Taylor, 2003, pp. 800-802).

Electricity consumption shows regional, seasonal and momentary fluctuations. During the year, up to 200% differences can occur between the lowest consumption and the highest consumption or between the lowest and the highest load on the same day. Due to the excessive variability of electricity demand and electricity as a source of energy that cannot be stored, electricity supply must be made continuously and without interruption and the demand can be met instantly. In case of failure to do so, as seen in the years 1971-1983 in Turkey, to avoid the mandatory deductions and savings applications it is next on the agenda. Factors affecting electricity demand are not limited to the economic growth rate. Factors determined to affect demand are used as 'inputs' in electricity demand forecasting models depending on their degree of impact (Gultekin, 2009).

- GDP,
- Sectoral added values,
- Per capita income,
- Population and demographic changes,
- Number of households and average household size,
- Percentage of multi-room housing and the rate of increase in house ownership,
- Urbanization rate,
- City and village income,
- Electric household and village ratio,
- Population utilizing electricity,

- Employment data,
- Technological developments and the widespread use of electrical work tools,
- Changes in the number of electrical appliances per person,
- Prices of electrical appliances and their related substitutes,
- Electricity price,
- Prices of alternative energy sources,
- Seasonal changes and climatic conditions,
- Geographical features of countries,
- Time.

To forecast the scientific environment of Turkey's electricity demand and supply, forecasting methods have been widely studied. These methods are potential forecasting tools. The most important point here is which method can be effective on which datasets. Serial behaviors (seasonal, cyclical, trending or horizontal) in the data set should be read well and appropriate prediction models should be established. An example implementation, to forecast annual net electricity consumption in Turkey, GM, Trigonometric GM techniques was determined. Models based on these methods were compared based on forecast evaluation criteria, and the model with the best performance was selected and future predictive analysis was performed. As a result, suggestions were interpreted and suggestions were tried to be made.

After the introduction part of this study, a literature review was made in the first part. In the second part, methods determined for estimation analysis are defined. In the third part, the future model has been made with the best model by including the implementation part. Finally, in the last section, the conclusion and suggestions were included.

LITERATURE REVIEW

Forecasting the demand or supply of electricity is a common subject in both national and international literature. In this section, a literature review has been made on this subject as follows.

Pi et al. (2010) in their study, they used the GM (1,1) model to estimate the electricity demand of China. The data set they used is China's annual electricity demand for the period 1984-2006. After the model was established, they carried out forecasting analysis for the period 2007-2015.

Lei and Feng (2012) forecasted in the light of data obtained from power plants located in Nordpool, California and Ontario regions using GM (1,1) and Least Squares (OLS) based Particle Swarm Optimization (PSO) model. The model established with PSO has been more meaningful.

Zhao and Magoulès (2012) modeled Artificial Neural Networks (ANN), GM (1,1), Support Vector Machines (DVM) methods to estimate residential energy demand. As a result, it was found that an artificial algorithm-based ANN model made more consistent estimates.

Wang et al. (2018) used observation data in the quarter of 2000-2016 to estimate China's electricity demand in their study. As a method; They used Seasonal GM (SGM), GM, GM-based PSO, Adaptive Parameter Learning Mechanism-Based Seasonal Fluctuation GM (APL-SFGM (1.1)). Since it gives results, electricity estimation has been made for the 2017-2020 period.

Zhou et al. (2006) in their study, they forecasted the annual and quarterly electricity demand of China by modeling the trigonometric GM (TGM) and ARIMA methods, which are among the basic estimation methods determined for this study. Models are compared based on MAPE, MAD, and MSE. With the TGM, which is determined as the best method, forecasting analysis was carried out for the future.

Che et al. (2012), in their study, they created Support Vector Regression and a fuzzy model based on this method to estimate electricity demand. The results of the analysis are compared based on MAPE. The fuzzy model gave the best result.

Wang et al. (2018) made consumption estimation for Liaoning, InnerMongolia, Beijing and Shanghai regions of China in the light of annual electricity consumption data of the 2000-2014 periods. To this end, they used a new hybrid GM modeled with multipurpose Ant Lion Optimization.

Yao et al. (2003) in their study, they used GM to estimate Taiwan's electricity demand.

Yao and Chi (2004) conducted a forecast analysis with Taguchi-based GM using an exemplary electricity demand dataset.

Bianco et al. (2010) have developed Holt-Winters and Trigonometric GM with Rolling Mechanism to forecast non-residential electricity consumption in Romania. As a result of the analysis, the models established by both methods gave close results.

Hsu and Chen (2003) developed an ANN-based GM to estimate Taiwan's energy demand. As a result, they have achieved very successful results.

Wang et al. (2018) used the Echo State Network Improved by Differential Evolution Algorithm to forecast the electrical energy consumption of Zhengzhou city of China in this study.

Xie et al. (2015) used GM and Markov models to estimate China's energy production and consumption. They predicted that the increasing trend will continue with both methods.

Kandananond (2011) used ARIMA, ANN and Multiple Linear Regression (MLR) methods to estimate Thailand's demand for electricity to plan its electricity consumption. As a result of the model comparison made based on MAPE, ARIMA and MLR models, which predicted with fewer errors, performed much better than the ANN model.

Lee and Tong (2011) forecasted energy consumption using a Gray model and Linear Regression (LR) developed based on Genetic programming. The models are compared based on MAPE and have been tested as GM developed as the best prediction model.

Wang et al. (2018) used 3 GM-based models (single-linear, hybrid-linear, and non-linear) to estimate the energy demand of China and India in this study. Estimates made with each model were analyzed based on MAPE. The method with the least error was determined as hybrid-linear GM. In the future forecast made with this model, it has been tested that the demand formed will gradually increase.

Debnath and Mourshed (2018) introduced an extensive literature review and introduced the models used to predict the energy demand of each region globally in the light of studies conducted. They found that China, which is the most used method, is also prominent in forecasting energy demand regionally.

Liu et al. (1991) used ANN and an econometric model to estimate Singapore's electricity demand in this study. As a result of the analysis, it was determined that the well-trained ANN model predicted the future electricity demand better.

Li et al. (2012) in their study established Adaptive GM (AGM), Support Vector Regression (SVR) and Backpropagation Neural Network models to estimate the electricity demand of some big countries. The models were compared based on MAPE and have been identified as the best performing model as AGM.

Hamzacebi and Es (2014) In this study, they established a basic GM by using annual data to forecasting the future of Turkey's annual electricity demand.

Feng et al. (2012) in their study, GM created the 1998-2006 period statistics to estimate the energy demand of China. As a result of the analysis, it has been determined that the demand for energy will increase continuously.

El-Fouly et al. (2006) present a sample GM for the forecast for the production of electricity from wind energy in this study.

Amber et al. (2018) used 5 different methods (MLR, Genetic Programming, Deep Neural Networks, Support Vector Machine, and ANN) to forecast the electricity demand of households in London. As a result of the analysis, the best method based on MAPE was determined as ANN.

Zhang et al. (2018) used a hybrid model developed with empirical mode decomposition to forecast the short-term electricity demand of Australia and New York, the Wavelet Neural Network model optimized with ARIMA and Fruit Fly Optimization Algorithm. When the installed models are compared based on MAPE, it is determined that the hybrid model shows a better prediction performance.

Liv et al. (2013) developed a hybrid GM based on the Taylor approach and Markov chain to predict Japan's thermal electricity production. Also, at the end of this study, they made suggestions that the model they developed may be suitable for new studies.

Kumar and Jain (2010) used GM with Rotational Mechanism, Singular Spectrum Analysis, and Grey Markov model to forecast natural gas consumption in India.

Wu et al. (2018), in their study, developed GM and Multivariate GM to forecast the electricity demand of the Shandong region of China. The newly developed GM has a better prediction performance compared to the traditional model, and its future forecast has been made.

You can also take a look at other GM-based studies on the forecast of electricity demand (Ayvaz and Kusakci, 2017; Akay and Atak, 2007; Ding et al., 2018; Kaytez et al., 2015; Pao et al., 2012; Bahrami et al., 2014; Ma et al., 2013; Ma et al., 2019; Sun and Yan, 2008; Yin et al., 2010; Niu et al., 2007; Jiang et al., 2014).

METHODOLOGY

In this study, two different methods were firstly determined to estimate the electricity demand. These are Grey Model (GM) and Trigonometric GM techniques. The methods identified in this section are described.

1. Grey Prediction Method

The Grey model was developed by Deng (1982). Grey system theory has been successfully applied in many areas; management, economy, engineering, finance, etc. There are generally three types of systems (White, Black, and Gray). When system information is completely open, it is called a white system. On the other hand, when the information of a system is not completely known, it is called a black system. Finally, if the information of a system is partially known, it is called a Grey System. The Grey prediction model adopts the basic part of the Grey system theory. The GM (1,1) Grey prediction model can be used in situations with relatively little data and can use a first-order differential equation to characterize an unknown system. Thus, GM (1,1), the Grey prediction model is suitable for predicting the competitive environment where decision-makers can only reference limited historical data (Pi et al., 2010, pp. 1519).

Grey System Theory is used as a preferred solution theory to obtain meaningful results when it comes to missing information and uncertainty. Also, the system has many advantages, besides giving the system a chance to apply in various fields, it also enables the formation of different sub-methods connected to the system. The application areas where GM is most commonly used are as follows (Aydemir, Fevzi & Gültekin, 2013, pp. 187-200); Industrial systems, Management and Organization, Social systems, Education systems, Environmental and Ecological Systems, Traffic application systems, Economic analysis, Military systems.

Research areas of the Grey System Theory are widely researched. In general, this system is used in five different areas. These are Grey Prediction, Grey Modeling, Grey Decision Making, Grey Relation Analysis, and Grey Control.

- **Grey Prediction**; is a prediction method that can produce satisfactory results when sufficient information about system behavior is not available and little information is available. It appears as a prediction model for both the forecast analysis period and the next forecast period (Köse, Erol & Temiz, 2010, pp. 299).
- **Grey Modeling**; It is the conversion of separation equations into differentiable equations with the help of an artificial concept. In cases of uncertainty and lack of data, situational analysis is based on an understanding and interpretation of the system with estimation and decision making. Three model types emerge here (Tseng, 2001, pp. 291-302; Köse et al., 2010, pp. 299). These;
    1. **GM (1,1):** Time series prediction model that includes a group of differentiable equations The grey model formed by first-order derivative equations with one variable is called the GM (1,1) model.
    2. **GM (1, *N*):** The model with first-order differentiable equations with *N* variables is called GM (1, *N*). Dynamic factor analysis is widely used in targeted studies.
    3. **GM (0, *N*):** The model with *N* variables and equations that do not require derivative information is called the GM (0, *N*) model. It is widely used in the multidimensional correlation analysis.
- **Grey Decision Making**; The most appropriate one of the different strategies that have different results is used to reach the planned goals. The point that distinguishes Grey decision making from the traditional decision definition is that the models used in decision making contain Grey parameters or the decision model is integrated with some Grey models. In other words, it is the use of GM (1,1) integrated with other methods in the solution of the decision problem (Liu & Lin, 2006, pp. 11-21).
- **Grey Relationship Analysis**; Grey relationship refers to the measurement of the relationship between two main or two sub-systems within a given system (Feng and Wang, 2000, pp. 6). The similarities or differences between the analyzed elements are called Grey relationships. System; economic, ecological, social and similar systems cover many factors. Some interrelations between these factors determine the development and direction of the system (Liu & Lin, 2006, pp.11-21).
- **Grey Control**; Greycontrol means control of systems with Grey parameter values. Using the information that the system has, the behaviors that the system may show in the future can be revealed. When predictive information about the future is reached, these data are used as feedback and new control mechanisms are created. The aim here is to direct the system towards the desired behaviors (Deng, 1982, pp. 288-294).

In traditional estimation methods, the data set is very important. The more series, the better the estimate is made. As mentioned above, GM is very good at analyzing the lower dataset. The main differences between GM and Traditional Forecasting methods are shown in Table 1.

Table 1. The Main Differences Between GM and Traditional Forecasting Methods

Forecasting Method	Minimum Number of Observations Required	Sample Type	Sample Range	Mathematical Requirement
Simple Exponential Function	5-10	Range	Short	Basic
Regression Analysis	10-20	Trend	Short	Medium
Random Regression	10	Any type	Long	Advanced
Box-Jenkins	50	Range	Long	Advanced
Neural Networks	Many	Range or not	Short	Advanced
Grey Prediction Method	4	Range	Long	Basic

Resource: Deng, 1982.

The general mathematical representation of the Grey prediction model is described as follows (Hamzacebi & Es, 2014, pp. 166-167).

Step 1: For the initial time sequence

$$x^{(0)} = \left\{x^{(1)}(1), x^{(1)}(2), \ldots, x^{(1)}(n)\right\} = \left(X_t^0; t = 1,2,3,\ldots,n; n \geq 4\right) \quad (1)$$

where X_t^0 is a non-negative sequence and n is the number of the data.

Step 2: When this sequence is subjected to AGO, the following sequence x^1 is obtained. It is obvious that X_k^1 is monotonically increasing.

$$x^1 = \left\{X_1^1, X_2^1, X_3^1, \ldots\ldots\ldots\ldots, X_n^1\right\} = \left(X_t^1; t = 1,2,3,\ldots,n; n \geq 4\right) \quad (2)$$

where;

$$X_k^1 = \left\{\sum_{t=1}^{k} x_t^0, t = 1,2,\ldots,n; k = 1,2,\ldots,n\right\} \quad (3)$$

Step 3: Form grey modeling by establishing a first-order grey differential equation

$$x_t^0 + aZ_t^1 = b, t = 2, \ldots, n \quad (4)$$

where;

$$Z_t^1 = \theta X_t^1 + (1-\theta) X_{t-1}^1, t = 2, \ldots, n \quad (5)$$

θ denotes a horizontal adjustment coefficient, and $0<\theta<1$. The selecting criterion of θ value is to yield the smallest forecasting error rate. In Eq. (4), a is called the development coefficient and b is called the driving coefficient. Applying least square method coefficients, [a b]T can be estimated as

$$\mathrm{A} = \begin{bmatrix} \mathrm{a} \\ \mathrm{b} \end{bmatrix} = (\mathrm{B_t B})^{-1} \mathrm{B^t Y_n} \quad (6)$$

where;

$$B = \begin{vmatrix} -Z_2^1 & 1 \\ -Z_3^1 & 1 \\ \vdots & \vdots \\ -Z_n^1 & 1 \end{vmatrix} Y = \begin{vmatrix} X_2^0 \\ X_3^0 \\ \ldots \\ X_n^0 \end{vmatrix} \quad (7)$$

Y= BA

Step 4: The whitenization processes is defined as follows

$$\frac{dX_t^1}{dt} + aX_t^1 = b \quad (8)$$

Thus, using *a* and *b* coefficients the AGO grey prediction model can be obtained

$$\hat{X}_{t+1}^1 = \left[X_1^0 - \frac{b}{a}\right] e^{-at} + \frac{b}{a}, t = 0, 1, 2, \ldots. \quad (9)$$

Step 5: Obtain the prediction value of the original sequence by IAGO operation. IAGO operation can be defined as:

$$X_{t+1}^{0} = X_{t+1}^{1} - X_{t}^{1} \tag{10}$$

If Eq. (9) is substituted in Eq. (10) then Eq. (11) is obtained as

$$\hat{X}_{t+1}^{0} = \left(1 - e^{a}\right)\left[X_{1}^{0} - \frac{b}{a}\right]e^{-at} \tag{11}$$

2. Trigonometric Grey Prediction Method

The trigonometric model is a residual modification technique developed to increase the accuracy of the Grey model. Thanks to this technique, the information contained in the series are now used to compensate for Greyestimation values, consisting of series resulting from differences between observed and predicted values. The trigonometric approach has the advantages of being quite simple compared to other correction techniques and requires very little computational effort (Bianco et al., 2010: 3588). The mathematical formula of the Trigonometric GreyEstimation Method can be defined as follows (Zhou Ang & Poh, 2006, pp. 2841-2843).

Now when the residual series are shown as follows;

$$r^{(0)}\left(k\right) = x^{0}\left(k\right) - \hat{x}^{(0)}\left(k\right), k = 2, 3, \ldots\ldots, n$$

$$r^{(0)}\left(k+1\right) = b_0 + b_1\left(k\right) + b_2 \sin\frac{2\pi k}{L} + b_3 \cos\frac{2\pi k}{L} + \varepsilon_k, k = 1, 2, 3, \ldots\ldots \tag{12}$$

The parameter L defines the period length and cycle. When it is monthly; When $L = 12$, quarterly; $L = 4$. εk gives random component. For annual electricity demand, L reflects the cyclical period of economic fluctuations and depends on the amount of historical data. When there are about 20 years of data, our recommendation is $15 \leq L \leq 30$, because such a selection can now capture the patterns of the series well according to preliminary trials.

Equation coefficients are calculated by Ordinary Least Squares (OLS) as follows.

$$\left|\hat{b}_0, \ \hat{b}_1, \ \hat{b}_2, \ \hat{b}_3\right|^{T} = \left(B^{T}B\right)^{-1} B^{T} R_n, \tag{13}$$

$$B = \begin{vmatrix} 1 & 1 & \sin\frac{2\pi}{L} & \cos\frac{2\pi}{L} \\ 1 & 2 & \sin\frac{4\pi}{L} & \cos\frac{4\pi}{L} \\ \vdots & \vdots & \vdots & \vdots \\ 1 & n-1 & \sin\frac{2(n-1)\pi}{L} & \cos\frac{2(n-1)\pi}{L} \end{vmatrix}, \tag{14}$$

$$R_n = \left| r^{(0)}(2), r^{(0)}(3), r^{(0)}(4), \ldots\ldots, r^{(0)}(n), \right|^T \tag{15}$$

The predicted residue series $\hat{r}^{(0)}(k)$ can be explained as follows;

$$\hat{r}^{(0)}(k+1) = \hat{b}_0 + \hat{b}_1(k) + \hat{b}_2 \sin\frac{2\pi k}{L} + \hat{b}_3 \cos\frac{2\pi k}{L}, k = 1,2,3,\ldots\ldots \tag{16}$$

The following equations are established for the forecasting process by method;

$$\hat{x}_{tr}^{(0)}(1) = x^0(1),$$

$$\hat{x}_{tr}^{(0)}(k) = \hat{x}^0(k) + \hat{r}^0(k), k = 2,3,4\ldots\ldots \tag{17}$$

$\hat{x}_{tr}^{(0)}(k)$; defines the estimated value. $\hat{x}^0(k)$ is the output of GM (1,1). The trigonometric residual modification model shown in Equation 10 is the Multiple Linear Regression model. Since the GM (1,1) model and the trigonometric residual modification model have two and four parameters, respectively, the total degree of freedom used by the Trigonometric Grey prediction approach is six. When the number of historical data points is more than 6, excessive degrees of freedom can now be used to estimate the variance of errors and then the prediction interval.

In multiple regression analysis, the idea of interval estimation is resolved as follows with 100(1-α) % interval estimation of the model output of $\hat{x}^0(k)$;

$$\hat{x}_{tr}^{(0)}(k) - t_{\frac{\alpha}{2},n-6} S\sqrt{1-h_{kk}} \le x^0(k) \le \hat{x}_{tr}^{(0)}(k) + t_{\frac{\alpha}{2},n-6} S\sqrt{1-h_{kk}}, k = 2,3,4\ldots. \tag{18}$$

$$S = \sqrt{\frac{\sum_{k=1}^{n}\left[\hat{x}_{tr}^{(0)}(k) - x^0(k)\right]^2}{n-6}},$$

$$h_{kk} = Y_k^T \left(B^T B\right)^{-1} Y_k,$$

$$Y_k = \left[1, k-1, \sin\frac{2\pi(k-1)}{L}, \cos\frac{2\pi(k-1)}{L}\right]^T$$

Although the prediction interval was obtained using the concept of multiple regression analysis, there is no clear cut or evidence. Since the center point of the prediction interval comes from the advanced Grey prediction approach, the prediction interval can be realized as a "*Grey prediction interval*".

3. Forecast Evaluation Criteria

Forecast evaluation criteria are the criteria used to measure the performance of the prediction made by any method. This study for the prediction of Turkey's electricity demand has been identified in the establishment of four different forecasting models. *MSE* (Mean Squared Error), *MAD* (Mean Absolute Deviation), *MAPE* (Mean Absolute Percentage Error) and *RMSE* (Roor Mean Squared Error). These criteria determine the errors of the estimation made in different versions. Therefore, these criteria are expected to give minimum values to determine the most successful model.

$$MSE = \frac{\Sigma(Y_t - Y_t')^2}{n-i} \tag{19}$$

$$MAD = \frac{\Sigma|e|}{n} \tag{20}$$

$$MAPE = \frac{\Sigma|e|/A}{n} \tag{21}$$

$$RMSE = \sqrt{\left(\frac{e^2}{n}\right)} \tag{22}$$

Table 2. Turkey's Annual Net Electricity Consumption

Year	(TWh)	Year	(TWh)	Year	(TWh)	Year	(TWh)	Year	(TWh)	Year	(TWh)	Year	(TWh)
1970	7,31	1977	17,97	1984	27,64	1991	49,28	1998	87,70	2005	130,26	2012	194,92
1971	8,29	1978	18,93	1985	29,71	1992	53,98	1999	91,20	2006	143,07	2013	198,05
1972	9,53	1979	19,63	1986	32,21	1993	59,24	2000	98,30	2007	155,14	2014	207,38
1973	10,53	1980	20,40	1987	36,70	1994	61,40	2001	97,07	2008	161,95	2015	217,31
1974	11,36	1981	22,03	1988	39,72	1995	67,39	2002	102,95	2009	156,89	2016	231,20
1975	13,49	1982	23,59	1989	43,12	1996	74,16	2003	111,77	2010	172,05	2017	249,02
1976	16,08	1983	24,47	1990	46,82	1997	81,89	2004	121,14	2011	186,10	2018	258,23

IMPLEMENTATION

About the prediction of Turkey's electricity consumption, Trigonometric Grey and Grey Model were used. To carry out the analysis, annual (1970-2018) net electricity consumption values have been determined as the data set. The specified data set was obtained from Turkey Statistical Institute (TSI). Analyzes were carried out using the estimation methods specified in this section. As a result, forecast evaluation criteria were compared to determine the method with the best estimation performance. With the best model from this comparison, the future prediction was made. The analyzes made are tried to be explained respectively in this section. 1970-2018 period, which will be used in the analysis of Turkey's annual net electricity consumption GWh as shown in Table 2 and Figure 1.

1. Prediction Analysis with GM (1,1) and TGM

θ, k and L parameters affect the performance of GM (1.1) and TGM. In the literature, these parameters are generally used as θ= 0.5 and k= 4 However, different values of these parameters may increase GM's forecast performance (1,1). Therefore, the best value of these parameters is investigated by experiments (Hamzacebi and Es, 2014, pp. 168).

After determining the parameter values that make the estimation evaluation criteria minimum, the analysis summary formed is shown in Table 3. The graphical representation is also given in Graph 2.

After forecast analysis with two models, the forecast evaluation criteria were calculated and it was determined that TGM was the best predictive model. The forecast evaluation summary is shown in Table 4.

When comparing models based on MAPE value, one of the prediction evaluation criteria, with the value of 0.0999149, the TGM method has shown a better prediction performance. Considering the percentile ranges of these criterion values, a pretty good estimation analysis has been carried out.

As a result of the application, the analysis was concluded by making a future forecast. The 2019-2023 period forecast has been made not only with TGM but also with GM. It is shown in Table 6. The graphical representation is also shown in Graph 3.

Figure 1. Annual Net Electricity Consumption in Turkey

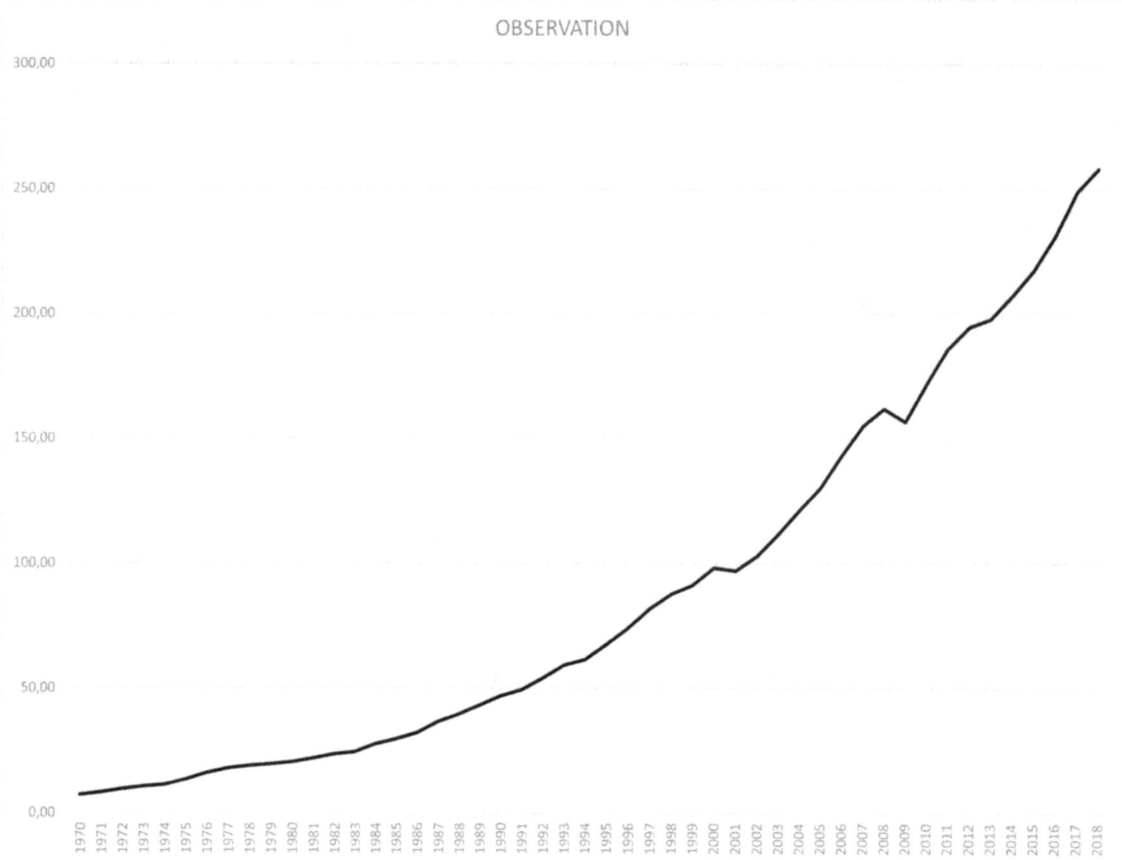

CONCLUSION AND RECOMMENDATIONS

One of the most important responsibilities of institutions and organizations that serve to provide electrical energy to end-users continuously and economically is to make the necessary investments, maintenance, and renewal by predicting future demands. The most important of these estimates is the peak demand value and the total energy demand value consumed annually on the day when the most electrical energy is taken from the system, which is defined as the peak day in the year. There are several methods to estimate these values. Although these methods cannot be separated, they all have their evaluation methods. Forecasting methods known as artificial intelligence, fuzzy logic or expert systems reveal the predictions obtained by learning from the data used as an input or by making logical inferences. Many other forecasting methods simulate the increase coefficients of data from previous years, creating forecasts for the next years.

The analysis results in a GM analysis based on TGM model estimates of net electricity consumption in Turkey that have been found to show a better performance. It has been observed that electricity consumption continued until 2019 until 2023. In the future forecast with GM, electricity consumption is constantly increasing. Unlike these methods, better estimation values may be obtained by using time series forecasting techniques, heuristic models or regression analysis.

Table 3. Forecast Analysis Summary with GM and TGM

Year Observed	value	GM (1,1)		TGM	
		Forecasted Value	RE (%)	Forecasted Value	RE (%)
1970	7,31	7,31	118,34	7,31	105,69
1971	8,29	18,10	101,97	17,05	65,13
1972	9,53	19,24	94,28	15,73	39,55
1973	10,53	20,46	91,48	14,69	22,88
1974	11,36	21,75	71,40	13,96	0,38
1975	13,49	23,12	52,90	13,54	-16,28
1976	16,08	24,59	45,47	13,46	-23,60
1977	17,97	26,14	46,78	13,73	-24,22
1978	18,93	27,79	50,49	14,35	-21,95
1979	19,63	29,55	54,00	15,32	-18,34
1980	20,40	31,41	51,60	16,66	-16,73
1981	22,03	33,40	50,54	18,34	-13,61
1982	23,59	35,51	54,30	20,38	-7,03
1983	24,47	37,75	45,23	22,74	-7,97
1984	27,64	40,14	43,63	25,43	-4,30
1985	29,71	42,67	40,85	28,43	-1,54
1986	32,21	45,37	31,44	31,71	-3,90
1987	36,70	48,23	29,10	35,27	-1,64
1988	39,72	51,28	26,44	39,07	-0,04
1989	43,12	54,52	23,80	43,10	1,11
1990	46,82	57,97	25,05	47,34	5,04
1991	49,28	61,63	21,37	51,77	4,41
1992	53,98	65,52	17,60	56,37	3,17
1993	59,24	69,66	20,62	61,12	7,50
1994	61,40	74,06	16,84	66,01	5,38
1995	67,39	78,74	12,89	71,02	2,70
1996	74,16	83,72	8,69	76,16	-0,57
1997	81,89	89,00	7,89	81,41	-1,05
1998	87,70	94,63	10,31	86,79	1,18
1999	91,20	100,61	8,82	92,28	-0,39
2000	98,30	106,96	17,15	97,91	6,82
2001	97,07	113,72	17,44	103,69	6,50
2002	102,95	120,91	15,01	109,64	3,60
2003	111,77	128,54	12,81	115,78	0,84
2004	121,14	136,67	11,54	122,17	-1,11
2005	130,26	145,30	7,97	128,82	-5,09
2006	143,07	154,48	5,87	135,79	-7,74
2007	155,14	164,24	7,82	143,13	-6,83
2008	161,95	174,62	18,33	150,89	1,43
2009	156,89	185,65	14,72	159,15	-2,38
2010	172,05	197,38	12,76	167,95	-4,68
2011	186,10	209,85	14,46	177,39	-3,80
2012	194,92	223,10	19,77	187,52	0,20
2013	198,05	237,20	21,61	198,45	1,38
2014	207,38	252,19	23,38	210,24	2,61
2015	217,31	268,12	23,29	222,99	2,42
2016	231,20	285,06	21,70	236,80	1,10
2017	249,02	303,07	24,78	251,75	3,76
2018	258,23	322,22	118,34	267,95	105,69

Figure 2. Graphical Display of the Analysis

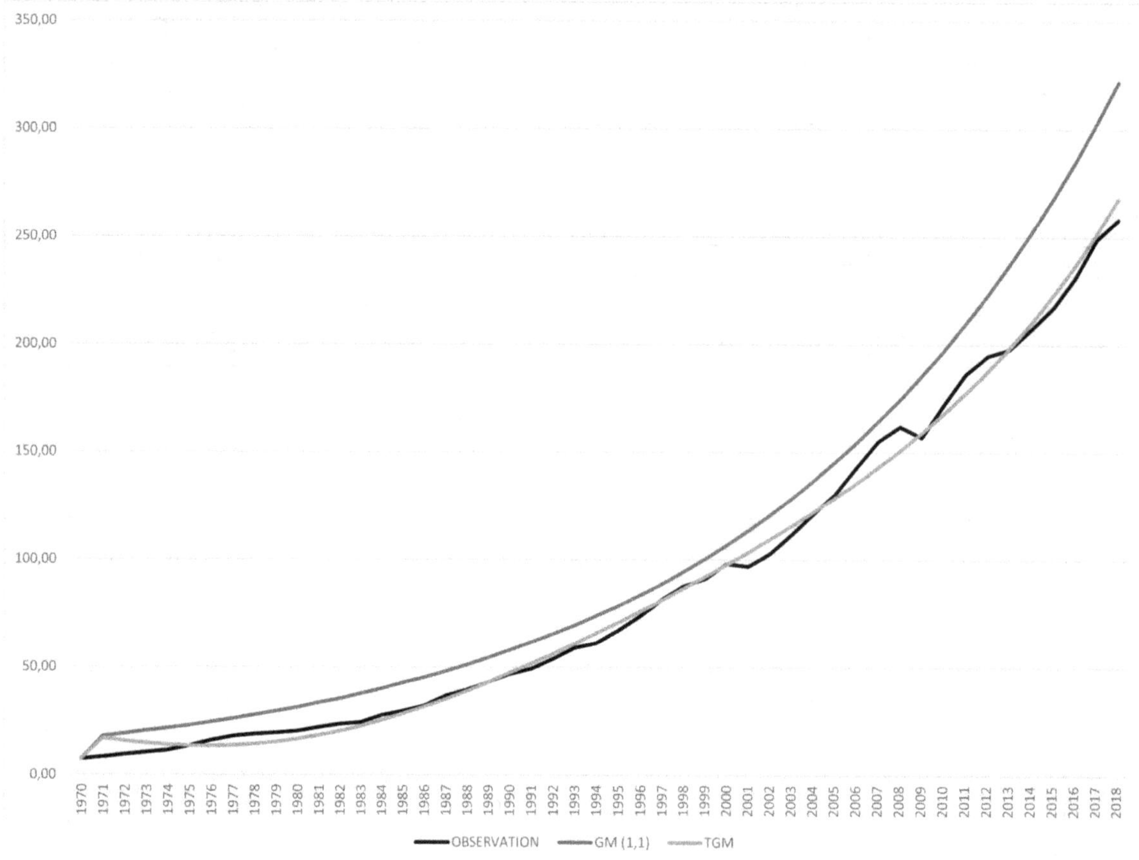

Table 4. Forecast Evaluation Summary

Model	MSE	MAD	MAPE	RMSE
GM (1,1)	490,12	17,2357179	0,3192965	22,1386886
TGM	21,71	3,59017374	0,0999149	4,65966087

Table 5. MAPE Criterion Threshold

Threshold	
Accuracy scale	**MAPE**
1st level (Perfect)	≤1
2nd level (Well)	≤ 5
3rd level (Acceptable)	≤ 10
4th level (Incapable)	> 10

Source: Liu & Lin, 2010.

Table 6. Future Prediction with Models

Year	GM (1,1)	TGM
	Forecasted Value	Forecasted Value
2019	342,57	285,50
2020	364,22	304,50
2021	387,23	325,05
2022	411,69	347,26
2023	437,70	383,44

Figure 3. Future Forecast Display with TGM

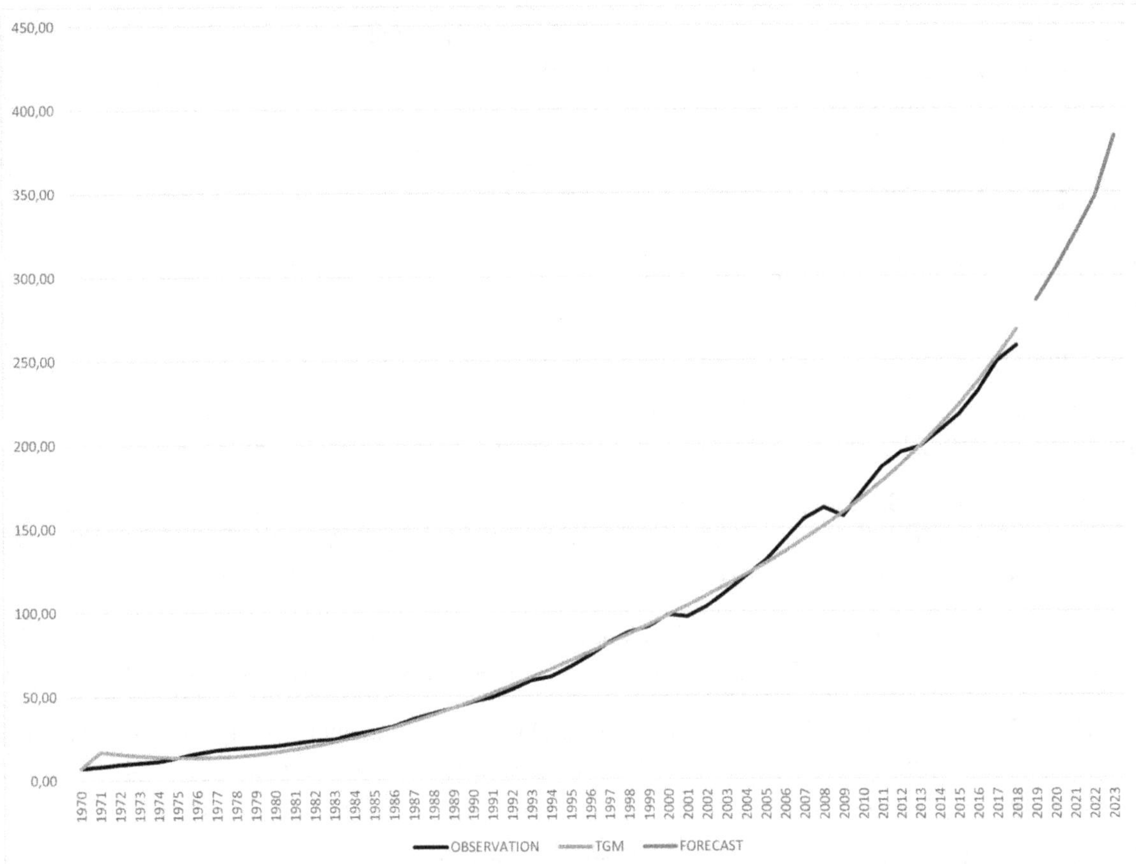

REFERENCES

Akay, D., & Atak, M. (2007). Grey prediction with rolling mechanism for electricity demand forecasting of Turkey. *Energy, 32*(9), 1670–1675. doi:10.1016/j.energy.2006.11.014

Aydemir, E., Fevzi, B., & Gültekin, Ö. (2013) Gri sistem teorisi ve uygulamaları: bilimsel yazın taraması, *Süleyman Demirel Üniversitesi İktisadi ve İdari Bilimler Fakültesi Dergisi, 18*(3), 188.

Ayvaz, B., & Kusakci, A. O. (2017). Electricity consumption forecasting for turkey with nonhomogeneous discrete grey model. *Energy Sources. Part B, Economics, Planning, and Policy, 12*(3), 260–267. doi:10.1080/15567249.2015.1089337

Bahrami, S., Hooshmand, R. A., & Parastegari, M. (2014). Short term electric load forecasting by wavelet transform and grey model ımproved by pso (particle swarm optimization) algorithm. *Energy, 72*, 434–442. doi:10.1016/j.energy.2014.05.065

Bianco, V., Manca, O., Nardini, S., & Minea, A. A. (2010). Analysis and forecasting of nonresidential electricity consumption in Romania. *Applied Energy, 87*(11), 3584–3590. doi:10.1016/j.apenergy.2010.05.018

Che, J., Wang, J., & Wang, G. (2012). An adaptive fuzzy combination model based on self-organizing map and support vector regression for electric load forecasting. *Energy, 37*(1), 657–664. doi:10.1016/j.energy.2011.10.034

Deng, J. L. (1982). Control problems of grey systems. *Systems & Control Letters, 5*, 288–294.

Deng, J. L. (1982). *Grey System Fundamental Method*. Huazhong University of Science and Technology Wuhan.

Ding, S., Hipel, K. W., & Dang, Y. G. (2018). Forecasting china's electricity consumption using a new grey prediction model. *Energy, 149*, 314–328. doi:10.1016/j.energy.2018.01.169

El-Fouly, T. H. M., El-Saadany, E. F., Salama, M. M. A., El-Fouly, T. H. M., El-Saadany, E. F., & Salama, M. M. A. (2006). Grey predictor for wind energy conversion systems output power prediction. *IEEE Transactions on Power Systems, 21*(3), 1450–1452. doi:10.1109/TPWRS.2006.879246

Feng, C. M., & Wang, R. T. (2000). Performance evaluation for airlines ıncluding the consideration financial ratios. *Journal of Air Transport Management, 6*(3), 6. doi:10.1016/S0969-6997(00)00003-X

Feng, S. J., Ma, Y. D., Song, Z. L., & Ying, J. (2012). Forecasting the energy consumption of china by the grey prediction model. *Energy Sources. Part B, Economics, Planning, and Policy, 7*(4), 376–389. doi:10.1080/15567240903330426

Gültekin, Ö. (2009). *Bursa İli Orta Dönem Elektrik Talep Tahmini*. Yüksek Lisans Tezi, Dumlupınar Üniversitesi, Fen Bilimleri Enstitüsü, Kütayha.

Hamzacebi, C., & Es, H. A. (2014). Forecasting the annual electricity consumption of turkey using an optimized grey model. *Energy, 70*, 165–171. doi:10.1016/j.energy.2014.03.105

Hsu, C. C., & Chen, C. Y. (2003). Applications of improved grey prediction model for power demand forecasting. *Energy Conversion and Management, 44*(14), 2241–2249. doi:10.1016/S0196-8904(02)00248-0

Jiang, P., Zhou, Q., Jiang, H., & Dong, Y. (2014). An optimized forecasting approach based on grey theory and cuckoo search algorithm: a case study for electricity consumption in new south wales. In *Abstract And Applied Analysis*. Hindawi.

Kandananond, K. (2011). Forecasting electricity demand in Thailand with an artificial neural network approach. *Energies*, *4*(8), 1246–1257. doi:10.3390/en4081246

Kaytez, F., Taplamacioglu, M. C., Cam, E., & Hardalac, F. (2015). Forecasting electricity consumption: A comparison of regression analysis, neural networks, and least squares support vector machines. *International Journal of Electrical Power & Energy Systems*, *67*, 431–438. doi:10.1016/j.ijepes.2014.12.036

Köse, E., Erol, S., & Temiz, İ. (2010). grey system approach for eqq models. *Journal of Engineering and Natural Sciences*, *28*, 298–309.

Lee, Y. S., & Tong, L. I. (2011). Forecasting energy consumption using a grey model ımproved by ıncorporating genetic programming. *Energy Conversion and Management*, *52*(1), 147–152. doi:10.1016/j.enconman.2010.06.053

Lei, M., & Feng, Z. (2012). A proposed grey model for short-term electricity price forecasting in competitive power markets. *International Journal of Electrical Power & Energy Systems*, *43*(1), 531–538. doi:10.1016/j.ijepes.2012.06.001

Li, G. D., Masuda, S., & Nagai, M. (2013). The prediction model for electrical power system using an improved hybrid optimization model. *International Journal of Electrical Power & Energy Systems*, *44*(1), 981–987. doi:10.1016/j.ijepes.2012.08.047

Liu, S. & Lin, Y., (2006), *Grey Information: Theory And Practical Applications*. Springer.

Liu, S., & Lin, Y. (2010). *Grey Systems: Theory And Application*. Springer-Verlag. doi:10.1007/978-3-642-13938-3

Ma, W., Zhu, X., & Wang, M. (2013). Forecasting iron ore import and consumption of China using grey model optimized by particle swarm optimization algorithm. *Resources Policy*, *38*(4), 613–620. doi:10.1016/j.resourpol.2013.09.007

Ma, X., Mei, X., Wu, W., Wu, X., & Zeng, B. (2019). A novel fractional time delayed grey model with grey wolf optimizer and ıts applications in forecasting the natural gas and coal consumption in Chongqing China. *Energy*, *178*, 487–507. doi:10.1016/j.energy.2019.04.096

Niu, D. X., Zhao, L., Zhang, B., & Wang, H. F. (2007). The application of particle swarm optimization based grey model to power load forecasting. *Chinese Journal Of Management Science*, *15*(1), 69–73.

Pao, H. T., Fu, H. C., & Tseng, C. L. (2012). Forecasting of co2 emissions, energy consumption and economic growth in china using an improved grey model. *Energy*, *40*(1), 400–409. doi:10.1016/j.energy.2012.01.037

Pi, D., Liu, J., & Qin, X. (2010). A grey prediction approach to forecasting energy demand in China. *Energy Sources. Part A, Recovery, Utilization, and Environmental Effects*, *32*(16), 1517–1528. doi:10.1080/15567030902780360

Saab, S., Badr, E., & Nasr, G. (2001). Univariate modeling and forecasting of energy consumption: The case of electricity ın Lebanon. *Energy*, *26*(1), 1–14. doi:10.1016/S0360-5442(00)00049-9

Sun, W., & Yan, Y. (2008). The model based on grey theory and pso for electricity consumption forecasting. *International Conference On Intelligent Computation Technology And Automation (Icıcta)*, 152-156. 10.1109/ICICTA.2008.40

Taylor, J. W. (2003). Short-term electricity demand forecasting using double seasonal exponential smoothing. *The Journal of the Operational Research Society*, *54*(8), 799–805. doi:10.1057/palgrave.jors.2601589

Tseng, F. M., Yu, H. C., & Tzeng, G. H. (2001). Applied hybrid grey model to forecast seasonal time series. *Technological Forecasting and Social Change*, *67*(2), 291–302. doi:10.1016/S0040-1625(99)00098-0

Wang, J., Du, P., Lu, H., Yang, W., & Niu, T. (2018). An improved grey model optimized by multi-objective ant lion optimization algorithm for annual electricity consumption forecasting. *Applied Soft Computing*, *72*, 321–337. doi:10.1016/j.asoc.2018.07.022

Wang, Z. X., Li, Q., & Pei, L. L. (2018). A seasonal gm (1, 1) model for forecasting the electricity consumption of the primary economic sectors. *Energy*, *154*, 522–534. doi:10.1016/j.energy.2018.04.155

Xie, N. M., Yuan, C. Q., & Yang, Y. J. (2015). Forecasting china's energy demand and self-sufficiency rate by grey forecasting model and Markov model. *International Journal of Electrical Power & Energy Systems*, *66*, 1–8. doi:10.1016/j.ijepes.2014.10.028

Yao, A. W., & Chi, S. C. (2004). Analysis and design of a Taguchi–grey-based electricity demand predictor for energy management systems. *Energy Conversion and Management*, *45*(7-8), 1205–1217. doi:10.1016/j.enconman.2003.08.008

Yao, A. W., Chi, S. C., & Chen, J. H. (2003). An improved grey-based approach for electricity demand forecasting. *Electric Power Systems Research*, *67*(3), 217–224. doi:10.1016/S0378-7796(03)00112-3

Yın, X., Zhou, Y., He, Y. G., & Chen, J. (2010). Power load forecasting based on adaptive particle swarm optimization grey model. *Proceedings Of The Chinese Society Of Universities For Electric Power System And İts Automation*, 4.

Zhao, H. X., & Magoulès, F. (2012). A review of the prediction of building energy consumption. *Renewable & Sustainable Energy Reviews*, *16*(6), 3586–3592. doi:10.1016/j.rser.2012.02.049

Zhou, P. A. B. W., Ang, B. W., & Poh, K. L. (2006). A trigonometric grey prediction approach to forecasting electricity demand. *Energy*, *31*(14), 2839–2847. doi:10.1016/j.energy.2005.12.002

Chapter 8
Analyzing the Efficiency of Renewable Energy Sources in Terms of Electricity Generation: The Case of Turkey

Talip Arsu
University of Aksaray, Turkey

ABSTRACT

Electricity generation, one of the renewable energy sources (RES), delivers a solution for various problems such as energy efficiency, energy supply security, reducing foreign dependency, and especially, environmental concerns. However, the solutions provided for these problems bring along the question of which RESs are produced more effectively. Therefore, in this research, RESs used for electricity generation in Turkey were analyzed by using generation data to show which one is more effective. Bi-objective multiple-criteria data envelopment analysis (BiO-MCDEA) method, a goal programming-based efficiency determination method, was used for the efficiency analysis conducted for five years between the years of 2014 and 2018. As a result of the analysis, geothermal energy came into prominence as the most effective RES for all of the years included in the solution. Geothermal energy was followed by biomass energy, wind energy, hydroelectric, and solar energy, respectively.

INTRODUCTION

Energy has great importance in terms of directing human life as well as continuous human development and human life. There is a need for secured and accessible energy supply for the sustainability of modern societies. Now, energy is an inseparable part of human life for almost every single activity such as house, transportation, industry and medical (Tiwari & Mishra, 2012, p. 3). Moreover, the rapid increase in the world population, industrialization, urbanization and the intense involvement of technological tools and equipment in human life have caused an incredible increase in energy consumption. Energy has become the basic input in various fields such as factories, ateliers, electronic household items, street lights, rail-

DOI: 10.4018/978-1-7998-5442-5.ch008

road transportation, and even electric cars. All these developments increase energy consumption every year by 4-5% (Aydın, 2013, p. 31).

Whether it is a developed or developing country, energy is essential for the functioning of all activities. It is estimated that industrial energy use in developing countries is approximately 45-50% of total commercial energy consumption. However, large-scale energy generation and energy consumption cause environmental disruption at the same time. Because most of the energy generated worldwide is still obtained from fossil sources. It caused the determination of productive energy use methods and the exploration of alternative energy sources by policymakers and industrialists (Suganthi & Samuel, 2012, p. 1224). Because the use of fossil fuels in high amounts does not only emerge as environmental problems for countries. Dependence on fossil fuels leads to other economic and social problems such as limited oil reserves, high logistics costs of solid fuel reserves and volatility in oil prices. Renewable Energy Sources (RES) are the solution to all these problems and provide politic, economic and ethical advantages (Bigerna, Bollino, & Micheli, 2015, p. 31). Also, the benefits provided by RESs for governments and societies can be summarised as below (Apergis & Payne, 2012, p. 430):

- To reduce the dependence on fossil fuels;
- To reduce the carbon emissions and environmental disruption;
- To increase the energy supply security of the region by using domestic energy sources;
- To modernize the energy infrastructure of the region;
- To minimize the macroeconomic effects of high and variable oil prices.

Energy generation technologies have a central role in social and economic development at every scale, from home and society to regional, national and international. Energy, environmental pollution and disruption, economic development and life quality can be considered among the effects of energy generation on social welfare. Today, the world is dependent on non-renewable fossil fuels, which is a major cause of pollution and climate change and will continue to be. Because of these problems and decreasing oil supply, finding renewable alternatives is becoming urgent gradually (Keyhani, Marwali, & Dai, 2010, p. 1). Within this framework, renewable energy use gains speed across the world slowly (Suganthi & Samuel, 2012, p. 1224). Moreover, RESs are separated according to criteria such as their effects on the environment, their cost, their effects on the social structure and the use of resources. Even though there is various RES which is commercialized or conceptualized, some of them can be more harmful than others for the environment. Especially hydroelectric has begun to be preferred lesser in the world, because it occupies more space than the other types of RESs, destroys agricultural estates and has larger expropriation costs. At this point, the question of which RES is more effective comes into prominence. The study designed for this purpose aims to measure which RESs used for electricity generation in Turkey is more effective by using one of the efficiencies measuring methods, Bi-Objective Multiple Criteria Data Envelopment Analysis (BiO-MCDEA).

BACKGROUND

Preindustrial energy requirements included only the warming of people and the production of basic needs products. During this period, most of the energy requirement was met by wood. Thereafter, more coal mines were reached with the rapid development of technology, and the contribution of this abundant amount of coal to the spread of steam engines also accelerated the industrial revolution process (Freris & Infield, 2008, p. 1). From the first years of the industrial revolution to the period when oil and natural gas begun to be used for industry and daily routines, coal was used as the main energy source. Together with the increase in mass-production cars and more intensive usage of cars by people in their daily lives, the importance of oil raised. Moreover, coal, oil and oil products named as fossil fuels have been used for electricity generation for a long time. Since the establishment of the first coal-fired thermal power plant with the purpose of electricity generation in New York in 1882, both coal and other fossil fuels have been used as the main electricity generation source until today. Today, electricity amount generated from coal, oil and natural gas still accounts for 65% of the total electricity generation in the world (IEA, 2019, p. 44). However, the idea that this situation is unsustainable is spreading rapidly among the citizens of the world and policymakers. The causes such as global warming, climate change, environmental disruption, air, and sea pollution have led to a decrease in trust for fossil fuels day by day.

The decrease in trust for fossil fuels and the environmental effects of its use such as emissions and climate change increased the interest of policymakers and the public in RESs. As a response to the increasing interest, the governments developed numerous policies such as renewable portfolio standards, financial incentives (tax credit, allowances, low-interest loan), net measurement policies and volunteer green power programs. Besides, the renewable energy sector and related technologies reduce not only the negative environmental effects of fossil fuels but also dependence on foreign energy sources (Apergis & Payne, 2014, p. 226). Because fossil fuels are produced only in certain regions of the world. And, it restricts the countries both in the trade of these fossil sources and in energy supply security.

Global energy demand has grown by 2.3% in 2018, which is the largest annual increase since 2010. China, the USA, and India have generated 70% of the total energy demand growth. Although the growth in renewable energy has surpassed the growths in all of the other energy types since 2010, the share of fossil fuels in the global primary energy supply is still very high (IEA, 2019, p. 38). There are also various future scenarios prepared for renewable energy. The current policy scenario in Table 1 shows the future situation of the world if any changes occur in terms of the policy. On the other hand, a sustainable development scenario shows the situation in which the sustainable energy targets that require rapid and widespread changes in all parts of the energy system are fully met (IEA, 2019, p. 23). Renewable energy supply by the year 2018, scenarios created with current policies and scenarios created according to sustainable development policies are presented in Table 1.

Table 1. Current Situation and Future Scenarios of the World Renewable Energy Supply

		Current Policies		Sustainable Development Policies	
	2018	**2030**	**2040**	**2030**	**2040**
Electricity generation (TWh)	6799	11627	15485	15434	26065
Biomass	636	1022	1256	1335	2196
Hydroelectric	4203	5171	5923	5685	6934
Wind	1265	2955	4258	4453	8295
Solar	592	2265	3658	3513	7208
Geothermal	90	161	258	282	552
Share of Total Generation	26%	33%	36%	49%	67%

Source: (IEA, 2019, p. 304)

In the scenario in which the current policies related to the RESs will continue, the foreseen growth in 2030 is 71% compared to 2018. It is also projected that the ratio of RES in total production, which was 26% in 2018, will increase to 33%. Again with the same current policies, an increase of 127% is expected in 2040 compared to 2018. On the other hand, in the scenario in which the growth for RESs will be actualized by the sustainable development policies, there will be an increase of 127% in 2030 compared to 2018, and the share of RES in total generation will increase to 49%. Likewise, if growth occurs according to sustainable development policies, there will be an increase of 283% in 2040 compared to 2018. If each RES is examined separately, it is predicted that the greatest development will be in solar energy in all scenarios. Solar energy is followed by wind and geothermal energy, respectively.

Electricity Generation From the RESs

Before the introduction of RESs, electricity generation was provided only from fossil fuels. Thanks to the nuclear power plant that produced electricity for commercial purposes in the UK in 1956 for the first time, the world met with the fuels originated from uranium. Although these sources are still generating a large part of the world's electricity, environmental damages that occurred as a result of them have been accepted by everyone, especially in recent years. Even though the whole parties related to the energy are in search of a RES which will reduce the environmental impact to zero, there is no energy source with zero environmental impact. However, RESs which have a far less environmental impact than other energy sources are inexhaustible on contrary to fossil fuel and uranium sources. If they are used wisely for proper and productive applications, they can provide reliable and sustainable energy almost permanently. Conversely, fossil fuel and uranium sources are finite and can be reduced as a result of consumption (Prakash, 2010, p. 64). In addition to these, fossil fuel and uranium used for electricity generation have been a matter of debate in various ways for many years. However, three issues, in particular, are at the center of almost all energy discussions. Firstly, even though there is a huge fossil fuel (coal, oil and natural gas) supply, it is avoided that the supply will be exhausted one day when increasing demand and limited supply are taken into consideration. Therefore, it is important to look for alternative energy sources. Secondly, global warming and climate change threads shape the relationship between energy consumption and environmental pollution. Thirdly, current tendencies in energy supply and use are still unsustainable economically, environmentally and socially (Apergis & Danuletiu, 2014, p. 578). These discussions on the environment, economy, and supply security increase the trust in RESs day by day. And, it has led energy investments to be directed mainly to the RESs. Energy investments both continue for commercialized RESs and turn into R&D investments to generate new RESs. In today's world, solar, wind, hydroelectric, biomass, geothermal and ocean energies are the RESs where electricity is generated at the level to be traded. Apart from this, R&D investments on numerous RES, such as using hydrogen as fuel, developing fuel cells from algae, continue.

Solar Energy

Solar energy is abundant and offers an important potential to mitigate climate change in the long term. There are various types of solar technology that can contribute to a series of energy services in several regions of the world. Even though the generation of solar energy still represents only a small part of the total energy consumption, solar energy technologies markets are growing rapidly. Besides, the cost of solar technologies has decreased significantly over the past 30 years. Moreover, technical developments and supportive public policies continue to offer additional cost reduction potential (Arvizu & Balaya, 2012, p. 337). Although solar energy which was commercialized in the 1980s was one of the newest RESs, it had the opportunity to become commercialized rapidly in the world. Besides, it has a promising future among the RESs.

Solar energy divides into two categories as photovoltaic and condensed solar energy. Photovoltaic systems use batteries to convert sunlight into electricity. This system does not require bright sunlight, it can generate electricity during cloudy days. Condensed solar energy makes the generation by using concentrated solar radiation technology. In general, condensed solar energy power plants are installed in uncultivated cloudless areas which are not used for agricultural purposes and have intensive sunlight (EREC, 2010). An adequate number of panels must be installed to make sufficient use of both photovoltaic and condensed solar energy. A specific area is needed for this. Especially areas with condensed solar energy power plants should be exposed to little or no shading from trees or buildings nearby. Moreover, these areas usually should be installed in an east, west or south-facing direction. Besides, cloud cover, winter season and north latitudes restrict the energy that can be generated by solar systems (Regina, 2007, p. 30).

Wind Energy

The fact that people began to benefit from wind energy dates back to very old times. From this point of view, wind energy is the oldest renewable resource. History shows us that windmills have existed in Iran, Iraq, Egypt, and China since the ancient ages. Moreover, Hammurabi, the King of Babylonia, designed a plan to irrigate the rich plain of Mesopotamia with the help of wind energy in 17th century BC (Le Gouriérès, 1982). Efforts to utilize wind energy beginning in BC evolved and were used in various fields until the 19th century. The first large wind turbine that generates electricity was installed in Cleveland, Ohio in 1888. Wind energy, which later lost its importance because of the ease of access to fossil fuels, became attractive again after the 1973 oil crisis. The most obvious example of it was seen in California in which more than 16 thousand power plants ranging from 201 to 350 kW were installed between 1981 and 1990 (Kaldellis & Zafirakis, 2011, p. 1887).

Wind energy, one of the most important RESs, is generated as solar radiation heats the ground surfaces differently. Different heating of the ground surfaces causes air temperature, pressure, and humidity differences to occur. The differences in these weather conditions cause the movement of air. And, air movement from high pressure to low pressure creates the wind (Şenel & Koç, 2015, p. 47).

Wind energy is a source which is domestic, not foreign-dependent, natural and renewable, can be supplied at the same rate in the future, does not cause acid rain and atmospheric warming, does not have CO2 emissions, does not harm natural vegetation and human health, provides fossil fuel savings, has no radioactive effects, has a swift technological development and foreign exchange-earning resource. Besides, it has an employment opportunity. In addition to all these positive contributions, it has some disadvantages such as noise, visual and aesthetic pollution, bird deaths, and the creation of interference to radio and TV receivers within an area of 2-3 km (Güler, 2005, p. 210).

Wind energy power plants can be examined within two groups as wind power plans on land and wind power plants at sea. In general, wind power plants on land are also named as wind farms. These wind power plants can generate energy ranging from 5 to 300 MW. However, other facilities have a bigger or smaller capacity. On the other hand, the technology of wind power plants at sea has some advantages along with requiring higher investment. High-quality wind sources at sea, the ability to use larger wind turbines, the ability to build power plants larger than the ones on land and reduced need for land-based transmission infrastructure can be counted as the advantages of wind energy power plants at sea (Ellabban, Abu-Rub, & Blaabjerg, 2014, p. 755).

Biomass Energy

Biofuels in solid, liquid or gas form are the only RESs that have the potential to directly substitute fossil fuels. Therefore, it is important in terms of being able to substitute oil in the transportation sector and integration with traditional fossil-based energy technologies in a short time. Especially increasing the demand for fuels needed in electricity production, space heating, and transportation sector, however, increasing the production costs of fossil fuels continuously, rising the oil prices extremely in particular, and maintaining the importance of climate change enhance the interest in bioenergy technologies (Cansın & Sohtaoğlu, 2009, p. 34).

It is important to identify biomass types and various final use areas. Biomass has historically been the main source of energy for humanity. Nowadays, it is gradually transforming from its traditional historical role into a more modern energy source. It is becoming an important component of the energy scenarios in the future. However, there is considerable disagreement about its future and potential contribution. Biomass energy is increasingly mentioned with problems such as land use, food security, biodiversity, environment, climate change, low carbon economy, sustainable development, etc (Rosillo-Calle, 2016, p. 1933). Nevertheless, the use of biomass is gaining importance to solve the energy problem due to the limited energy sources such as oil, coal and natural gas, and their effects on environmental pollution (YEGM, 2020).

Energy generation from biomass is actualized by using biogas, one of the biological fuels. Biogas, which occurs as a result of fermentation of organic-based residues in an anaerobic environment, is a colorless and odorless gas mixture which is lighter than air, burns with a bright blue flame and contains approximately 40-70% methane, 30-60% carbon dioxide, 0-3% hydrogen sulfide and very little amount of nitrogen and hydrogen depending on the composition of organic substances in its compound. 1 m3 of biogas is a fuel equivalent to 0.66 liters of diesel or 0.75 litters of gasoline (YEGM, 2020).

Sources of biological origin are primary renewable sources used for energy generation since the early days of humanity. On the other hand, the most important biological sources used for energy are woody materials in the forest ecosystem. The use of residual trunk parts from the products such as logs, poles and industrial wood obtained from tree trunks with bark, root, branches, and leaves has gained importance today to obtain energy (Karayılmazlar, Saraçoğlu, Çabuk, & Kurt, 2011, p. 64). Moreover, the plant specially grown for biomass such as corn and wheat, herbs, moss, algae in the sea, animal droppings, fertilizer, industrial wastes and all organic waste (fruit and vegetable residues) from the houses are the sources (YEGM, 2020). A wide variety of biological raw materials can be converted into fuels (biofuels) for thermal energy, electricity, and transportation by applying several different processes. While many bioenergy conversion alternatives are well-constructed and purely commercial, others are still in the early stages of development, demonstration, and commercialization (REN21, 2019, p. 71).

Geothermal Energy

Geothermal energy uses the temperature of the Earth's core as an energy source. Geothermal systems include drilling the wells into underground areas that contain water heated by the Earth's core. Hot water or steam passing through the wells can be used to generate electricity or to heat or cool a building. When geothermal energy is used for electricity, steam or hot water is used as a power to rotate turbines that generate electricity (Regina, 2007, p. 31).

The use of geothermal sources dates back to the 11th century BC. The use of geothermal sources was limited by bath and laundering in ancient ages. From the beginning of the 20th century, it has been used in residential heating, industrial heating and finally in electricity generation. Geothermal energy has been used commercially for more than 80 years due to both hundreds of MW electricity generation and direct use. In the last three decades, the speed of its use has increased even more (Fridleifsson, 2001, p. 301). Because geothermal energy has a higher capacity factor than many other power sources. Unlike wind and solar sources, which are more dependent on weather fluctuations and climate changes, geothermal sources can be used 24 hours a day, 7 days a week. Because the heat of the Earth, which is the source of geothermal energy, is permanent and unlimited (Kagel, Bates, & Gawell, 2007, p. i).

Hydroelectric Energy

Hydroelectric is a power obtained from the energy of water in motion. Running water generates an energy that can be caught by using tribunes and converted into electricity. The most common hydroelectric type is barrages. Hydroelectric is generated from the water motioning in the hydrologic cycle stimulated by solar radiation. It is the water flow in the river induced by the gravitational force which can be used for hydroelectric generation. Hydroelectric power plants have a wide range of scales from a few watts to several GW. The biggest projects are Itaipu in Brazil with 14,000 MW installed power and Three Gorges in China with 22,400 MW installed power. Both produce around 80 to 100 TW of electricity per year (Ellabban et al., 2014, p. 751).

At the moment, hydroelectric is the biggest RES used for energy generation worldwide. The hydroelectric generation has increased greatly in the last 50 years. Because of the fixed investment cost in the beginning and environmental concerns, hydroelectric energy development is still criticized. Also, hydroelectricity may cause some problems related to the need for relocation of a great number of people who live in the power plant region. Moreover, the environment is affected by hydroelectric power plants due to great engineering works (Abolhosseini, Heshmati, & Altmann, 2020, p. 4). However, it is thought that the concerns on the sustainability of hydroelectric will disappear thanks to the ever-growing sustainability regulations and criteria in the last ten years, innovative planning based on shareholder consultations and the experience obtained from scientific know-how (Kumar & Schei, 2012, p. 441). On the other hand, as it is obtained from a water source collected because of various reasons such as agricultural, domestic and industrial use, hydroelectric is attractive and clean, and it provides storage of both water and energy (Abolhosseini et al., 2020, p. 4). Besides, it is a technology that is proved, mature, foreseeable and competitive on price. Hydroelectric is among the best conversion efficiencies of all known energy sources (approximately 90% of efficiency, from water to cable) (Kumar & Schei, 2012, p. 441).

Ocean Energy

Oceans and seas, which consist of approximately 70% of 510.1 million km2, the surface of the world, cover 361.93 million km2 of the earth (Statista, 2020). This characteristic makes the oceans and seas the largest solar collectors in the world. In other words, the energy obtained from oceans and seas does not only consist of the wave or tidal energy. Oceans or seas can generate two types of energy: thermal energy from the solar heat and mechanic energy from the tide and waves. The thermal energy of the ocean or sea is used for several applications including electricity generation. Mechanic energy of the ocean or seas is quite different than their thermal energy. Even if the sun affects the whole activity of ocean and sea, tides are directed by the gravity of the moon primarily while waves are directed by the winds essentially. Consequently, tides and waves are intermittent energy sources, however, the thermal energy of the ocean or sea is quite stable (Renewable Energy World, 2020). Both mechanic and thermal ocean energy sources can generate energy in six different ways (Lewis & Estefen, 2012, p. 503);

- *Waves:* It is obtained by transferring the kinetic energy of the wind to the upper surface of the ocean and the sea.
- *Tidal range (rise and fall of the tide):* It is obtained from the gravitational forces of the earth-moon-solar system.
- *Tidal currents:* It is obtained from the water flow arising from the filling and discharge of the coastal areas as a result of the rise and fall of the tide.
- *Ocean currents:* It is obtained from the ocean and sea circulation arising from the wind effect on the surface and thermal temperature differences in the deep.
- *Ocean thermal energy conversion (OTEC):* It is obtained from the temperature differences between the solar energy stored as heat in the upper ocean or sea layers and ocean or seawater which is generally colder below 1,000 m.
- *The energy of salinity differences:* It is obtained from the salinity differences between freshwater in the river mouth and ocean or seawater.

Even if there are six different ways to generate energy from oceans or seas, energy sources other than tidal currents are either in concept or in the R&D process. Ocean energy represents the smallest part of the renewable energy market. Most of the projects are the pilot projects which are relatively small-scale and lower than 1 MW. It has reached approximately 532 MW installed power capacity at the end of the year with an additional installed power capacity of 2 MW in 2018. Two tidal barrage plants represent more than 90% of the total capacity. Together with its worldwide developments, it has become dense primarily in Europe and especially in the Scottish coasts where tidal turbine arrays were deployed in

2018. The source potential of ocean energy is huge but has not been used to a large extent despite decades (REN21, 2019, p. 91).

Solar, wind, biomass, hydroelectric, geothermal and ocean are the RESs that have already commercialized in various countries of the world. Except for these RESs, there are still several RESs which are conceptual and whose R&D process continues. Solar, wind, hydroelectric, biomass and geothermal RESs are used in Turkey. Although Turkey is a country which is being surrounded on three sides by the sea, it has not undertaken a commercial enterprise about electricity generation from the seas yet.

The Situation of RESs in Turkey and the World

The rapid economic growth of developing economies raises the question of how to meet their energy needs. Besides, several developing countries come up against the problem of providing energy supply security similar to most developed economies and ensuring an adequate supply of non-renewable resources (such as oil). Energy supply security problem occurs because the largest oil consumers do not have enough resources to meet their needs (such as the USA, Europe, Japan, and China) (Sadorsky, 2009, p. 4021). This imbalance in resources both results in an energy supply security problem for developed countries and also hits further partly weak economies of developing countries. For this reason, both developed and developing countries make huge investments in RESs.

Renewable energy has found itself a big place on a global scale in a short time in terms of human history. To increase the share of renewable energy in total energy supply, governments have tried to promote further development and adoption of renewable energy technologies. These policies were followed by the signing of the Kyoto Protocol in December 1997, when developed countries adopted emission limits for greenhouse gases (Popp, Hascic, & Medhi, 2011, p. 649). RESs, which showed a great development after this date, had at least 1 GW generation capacity (including hydroelectricity) in more than 90 countries at the end of 2018 and exceeded 10 GW capacities in at least 30 countries. China is the country that has the highest installed renewable energy generation capacity just like the previous years. China is followed by the USA, Brasilia, India, and Germany. When the capacity which is not only hydroelectric is taking into account, at least 45 countries pass 1 GW limit, and 17 countries have more than 10 GW wind, solar, biomass and geothermal energy combination. In 2018, the best countries for non-hydroelectric capacity were China, the USA, and Germany (above 100 GW). These countries were followed by India, Japan, England, Italy, Brasilia, and Spain respectively. The best countries for non-hydroelectric energy capacity per person are Iceland, Denmark, Germany, Sweden, and Finland (REN21, 2019, p. 42).

European Union (EU) countries are making great strides on RESs. For instance, an EU directive dated 2001 (Directive 2001/77/EC) provides a framework for the development of renewable energy technologies in Europe. In March 2007, EU heads of state adopted 20% of the total EU energy need as a binding target for renewable energy use by 2020 (Popp et al., 2011, p. 649). This target was succeeded before 2020, and the share of renewable energy in the total energy generated in the EU was actualized at 32.20%. In Turkey, on the other hand, the share of renewable energy in the total energy was actualized at 37.54%. However, the share of hydraulic energy, the RES which has the most environmental effects, in the total renewable energy was 67.12% in Turkey, while it was 36.56% in EU (Eurostat, 2020). In other words, although Turkey looks ahead of the EU in terms of renewable energy, it falls even further behind of EU in solar, wind, biomass and geothermal energy which have the least environmental effect.

In recent years, the development in RESs has increased together with the increase of the world population and the spread of global environmental awareness. In this respect, the recent development of all RESs is examined for both world countries and Turkey.

Solar energy converts into electrical energy by using two different methods. In 2018, the total photovoltaic installed power increased by 25% and reached 505 GW. This amount comes into prominence as a meaningful increase when it is compared with the 15 GW which was the total global installed power only ten years ago. On the other hand, concentrated solar energy installed power increased by 11% globally in 2018 (550 MW) and reached 5.5 GW levels (REN21, 2019, p. 94-107). In 2018, the amount of electricity generated from solar energy in the world increased by 28.9% to 584.6 TWh compared to 2017. Also, the generation increase in EU countries was 7.3% compared to 2017. In Turkey, the amount of electricity generated from solar energy increased from 2.9 TWh in 2017 to 7.9 TWh in 2018. In other words, the increase of 173.1% in the amount of electricity generated from solar energy in Turkey was actualized at a very higher level than both the average increase ratio of EU countries and the average increase ratio of the whole world (BP, 2019, p. 52).

Although the 51 GW increases in the wind energy global installed power capacity in 2018 were 4% less than the increases in 2017, the total capacity increased by 9% to 591 GW. 568.4 GW of wind energy installed power capacity places on land, while the rest places at sea. These capacity increases have been above 50 GW for five consecutive years. China is the leading country in terms of installed power. China is followed by the USA, Germany, India, Spain, and England respectively (REN21, 2019, p. 118-119). On the other hand, the amount of electricity generated from wind energy increased by 12.6% to 1270 TWh in 2018 compared to 2017. In EU countries, this ratio was actualized at 4.6%. In Turkey, the amount of electricity generated from the wind energy was 17.9 TWh in 2017, and it was increased to 19.8 TWh in 2018. In other words, the 10.7% increase in Turkey was higher than the average increase in EU countries and lower than the average increase in the world countries (BP, 2019, p. 52).

Biomass energy global installed power capacity was increased by 6.5% to 130 GW in 2018 from 121 GW in 2017 inferentially. On the other hand, the bioelectric generation increased by 9% to 581 TWh in 2018 compared to 532 TWh in 2017. China maintained its situation as the largest bioelectric generator country with a 14% generation increase. China was followed by the USA, Brasilia, Germany, India, England and Japan respectively (REN21, 2019, p. 73). Biomass-based electricity generation installed power in Turkey increased by 30.2% from 2017 to 2018. The installed power, which was 477.4 MW in 2017, increased to 621.9 MW in 2018. Moreover, Turkey, which shows a 25% increase in biomass-based electricity generation amount, could not manage to be at the top in the world biomass-based electricity generation league (Turkish Electricity Transmission Corporation-TETC, 2019).

The total global installed power capacity of geothermal energy was increased to 13.3 GW by using 0.5 GW new installed power capacity for electricity generation in 2018. Turkey and Indonesia became prominent as the leading countries of increasing capacity. These two countries consisted of approximately two-thirds of the new capacity. In 2018, other countries, whose capacities increased, were the USA, Iceland, New Zealand, Croatia, Philippines, and Kenya respectively. At the end of 2018, the countries which had the most geothermal energy generation capacity were the USA, Indonesia, Philippines, Turkey, New Zealand, Mexico, Italy, Iceland, Kenta and Japan (REN21, 2019, p. 80). In Turkey, the total installed power, which was 1063.7 MW in 2017, increased by 20.5% to 1282.5 MW in 2018. On the other hand, the amount of electricity generated from geothermal energy increased by 21.2 from 2017 to 2018. The amount of electricity generation obtained from geothermal energy was 6.13 TWh in 2017 and increased to 7.73 TWh in 2018 (TETC, 2019).

Hydroelectric acted similar to the previous year in 2018 in terms of the global market, capacity increase and activity density, and reached to 1.132 GW total installed power capacity with 20 GW installed power capacity increase. However, the general ranking of the countries did not change throughout the year. The first ten countries in terms of the total capacity were China, Brasilia, Canada, the United States of America, the Russian Federation, India, Norway, Turkey, Japan, and France. Moreover, these countries represented more than two-thirds of global capacity at the end of 2018. When electricity generation is taking into account, the total hydroelectric generation was 4210 TWh in the world in 2018 (REN21, 2019, p. 86). Also, the table in Turkey was not different than the world in terms of both installed power and hydroelectric generation. The installed power, which was 27.27 GW in 2017, increased by 3.7% to 28.29 GW in 2018. Likewise, the hydroelectric generation, which was 58.21 TWh in 2017, increased by 2.9% to 59.93 TWh in 2018 (TETC, 2019).

In Turkey, the investments on RES proceed swiftly, and the question about which RES is more efficient in terms of electricity generation remains among the concerns of both investments and policymakers. Therefore, in this study, efficiency research was carried out by collecting electricity generation data in which the RESs (solar, wind, biomass, geothermal and hydroelectric) were used.

Analyzing the Efficiency of Renewable Energy Sources in terms of Electricity Generation: The Case of Turkey

RESs have been mentioned in several academic studies as the focus of the interest of the researchers for many years. Especially the subject on which RES is more efficient has been discussed in academic studies. While these studies were carried out, Multiple-Criteria Decision Making (MCDM) methods were frequently used. When efficiency measurement is taking into account, Data Envelopment Analysis (DEA) comes into prominence as the most used method. It is also possible to find several studies in which RESs are evaluated and DEA and DEA-based methods are used as the method in the literature.

San Cristobal (2011) researched the efficiency of thirteen RESs with different capacities in his study in which the Multiple Criteria Data Envelopment Analysis (MCDEA) method was used by him. Likewise, Cucchiella & Gastaldi (2014) compared four different RES alternatives for the cities in Italy by using DEA. Sağlam (2018) analyzed the efficiency of seven RESs which generate electricity in the USA in his study that he used DEA. Apart from the studies in which RESs are decision-making units, there are also some studies in which the locations are examined in terms of RESs in the literature. Chien & Hu (2007) made a comparison between OECD member countries and non-OECD member countries in terms of RES by using DEA. Menegaki (2013) compared European countries in terms of renewable energy and growth by using DEA. However, in the literature research carried out, any other method excepting DEA and MCDEA was observed in the studies related to the RESs. Likewise, again in the literature research carried out, and efficiency analysis performed by using the data of electric energy that a country generated from the RESs was not found. In this study, both RES electricity generation data in Turkey was used, and the BiO-MCDEA method, which is believed to improve the discrimination power, was used as a method. From this point of view, it is thought that this study is authentic.

Table 2. Input and Output Variables

Years	RES	Inputs		Outputs	
		v_1	v_2	u_1	u_2
		MW	%	GWh	%
2014	Hydroelectric	23.643,2	34,009	40.644,7	16,131
	Geothermal	404,9	0,582	2.364,0	0,938
	Wind	3.629,7	5,221	8.520,1	3,381
	Solar	40,2	0,058	17,4	0,007
	Biomass	227,0	0,327	1.094,4	0,434
2015	Hydroelectric	25.867,8	35,364	67.145,8	25,649
	Geothermal	623,9	0,853	3.424,5	1,308
	Wind	4.503,2	6,156	11.652,5	4,451
	Solar	248,8	0,340	194,1	0,074
	Biomass	277,1	0,379	1.350,3	0,516
2016	Hydroelectric	26.681,1	33,990	67.230,9	24,500
	Geothermal	820,9	1,046	4.818,5	1,756
	Wind	5.751,3	7,327	15.517,1	5,655
	Solar	832,5	1,061	1.043,1	0,380
	Biomass	363,8	0,463	1.658,5	0,604
2017	Hydroelectric	27.273,1	32,011	58.218,5	19,584
	Geothermal	1.063,7	1,249	6.127,5	2,061
	Wind	6.516,2	7,648	17.903,8	6,023
	Solar	3.420,7	4,015	2.889,3	0,972
	Biomass	477,4	0,560	2.124,0	0,714
2018	Hydroelectric	28.291,4	31,949	59.938,4	19,665
	Geothermal	1.282,5	1,448	7.431,0	2,438
	Wind	7.005,4	7,911	19.949,2	6,545
	Solar	5.062,8	5,717	7.799,8	2,559
	Biomass	621,9	0,702	2.672,7	0,877

The Purpose and Scope of the Study

The purpose of this study is to determine the efficiency of the RESs used in electricity generation in Turkey with goal programming-based BiO-MCDEA, one of the MCDM methods. With this purpose, an efficiency analysis was performed by using capacity and electricity generation data belong to the RESs used in electricity generation in Turkey.

The Importance, Data Collection Method and Data of the Study

The RESs (hydroelectric, solar, wind, geothermal and biomass) which were used for electricity generation in Turkey between the years 2014 and 2018 are the samples of this study. Even if the RESs are varying more in the world, five of these sources are used in Turkey. The years before 2014 were excluded from the analysis because the electricity generation from solar energy started in 2014 in Turkey. The reason why the analysis was made until 2018 is that the data for 2019 has not been published yet.

In the Bio-MCDEA model in which the data of the study were analyzed, 'RES installed power amount (v1)' and 'the share of RES installed power in total installed power (v2)' were used as input, and 'RES generation amount (u1)' and 'the share of the electricity generated from the RESs in the total electricity generation (u2)' were used as output (Cucchiella & Gastaldi, 2014; Menegaki, 2013; San Cristobal, 2011). The data used in the research are given in Table 2.

The data belongs to the input and output variables of the BiO-MCDEA model created to determine the RES efficiency in Turkey between the years of 2014 and 2018 was obtained from the website of the Turkish Electricity Transmission Corporation (TETC, 2019).

Method

It is a goal programming-based efficiency determination model developed by BiO-MCDEA, Ghasemi, Ignatius, & Emrouznejad (2014) and aimed to increase the discrimination power of the MCDEA model. Bal, Örkcü, & Çelebioğlu (2010) also proposed a goal programming-based model which will eliminate discrimination power and weight distribution problems of MCDEA model. The model named as Goal Programming Data Envelopment Analysis (GPDEA) is based on the solution of unwanted deviations using equal weight. The reason why Bio-MCDEA was used in this study is that classic DEA is excluded from the model to avoid the low discrimination power disadvantages, and the solution stages of the BiO-MCDEA model are easier.

Multiple Criteria Data Envelopment Analysis (MCDEA)

Classic DEA is a non-parametric analysis frequently used for efficiency analysis in the field of social sciences in particular. The converted type of classic DEA method proposed by Charnes, Cooper, & Rhodes (1978) into linear programming form is shown below.

$$Max\, h_0 = \sum_{r=1}^{s} u_r y_{rj}$$

$$s.t. \sum_{i=1}^{m} v_i x_{ij} = 1 \qquad (1)$$

$$\sum_{r=1}^{s} u_r y_{rj} - \sum_{i=1}^{m} v_i x_{ij} \leq 0, j = 1, \ldots, n$$

$u_r \geq 0 \ \ v_i \geq 0$

Here, j, r, i, y_{rj}, x_{ij}, u_r, v_i, and h_0 respectively indicate the number of Decision Making Unit (DMU), the number of output, the number of inputs, the value of r. output for j. DMU, the value of i. input for j. DMU, the weight of r. output, the weight of i. input and the relative efficiency. In this model, h_0=1 is needed to make any DMU effective (Charnes et al., 1978, p. 432).

Classic DEA is an efficiency measurement method; however, the MCDEA model is built on inefficiencies by Li & Reeves (1999). d_0, which is limited by the interval [0, 1], can be considered as "inefficiency" measurement, and it is defined as $h_0=1-d_0$. In other words, it can be said that the smaller d_0 value means the more inefficient DMU0 (therefore, more efficient). In the method of Li & Reeves (1999), besides, to minimize d_0, the measurement of inefficiency, there are also two independent objective functions as minimizing the maximum deviation and minimizing the total of deviations. This model is as follows.

$$Min\, d_0 (veya \max h_0 = \sum_{r=1}^{s} u_r y_{rj0})$$

Min M

$$Min \sum_{j=1}^{n} d_j \tag{2}$$

$$s.t. \sum_{i=1}^{m} v_i x_{ij0} = 1$$

$$\sum_{r=1}^{s} u_r y_{rj} - \sum_{i=1}^{m} v_i x_{ij} + d_j = 0$$

$$M - d_j \geq 0 \ j=1,\ldots,n \ \ u_r, \ v_i, \ d_j \geq 0$$

MCDEA model was proposed at first as a tool to improve the discrimination power of the classic DEA model. For the resolution procedure, MCDEA was proposed as an interactive approach to solving three objectives. The first objective or criterion is to have the solution of the classic DEA within the MS-DEA solutions set. Minimax and Minisum objectives, the other two objectives, provide more restricted or slacked efficiency solutions respectively. Thanks to this model, it is proved that a wider solution is possible to obtain more reasonable input and output weights (Ghasemi et al., 2014, p. 641). Although this model solves the low discrimination power and weight distribution problems of the classic DEA, it contains the $d_0 = 1-h_0$ objective in the classic DEA model. Moreover, it wants the decision-makers to choose between three different efficiency scores instead of providing them a single efficiency score as a result of the solution of the model. BiO-MCDEA, in which these problems are eliminated, both exempts d_0 objective from the model and provides a single efficiency score for the decision-makers.

A bi-objective Multiple Criteria Data Envelopment Analysis (BiO-MCDEA)

MCDEA model consists of three independent objective functions as Min d_0, Min M and $Min\sum_j d_j$ as it is defined in Model 2. In a weighted model, three independent objective functions of the MCDEA can be turned into a single objective problem by weighting them as $w_1 d_0 + w_2 M + w_3 \sum_j d_j$. By changing the weights $w_i (i=1,2,3)$, different scores can be reached. Besides, as w_1, which is the first objective, represents the same meaning with the classic DEA model, it can be removed from the MCDEA model. Because it is proved that the discrimination power of Min M and $Min\sum_j d_j$ objectives are higher than Min d_0 objective (Li & Reeves, 1999; San Cristobal, 2011; Hatami-Marbini & Toloo, 2017). For this reason, only Min M and $Min\sum_j d_j$ objectives are weighed in the BiO-MCDEA model proposed by Ghasemi et al., (2014). The BiO-MCDEA model is shown below.

$$Min\,h = \left(w_2 M + w_3 \sum_j d_j \right)$$

$$s.t. \sum_{i=1}^{m} v_i x_{ij0} = 1 \tag{3}$$

$$\sum_{r=1}^{s} u_r y_{rj} - \sum_{i=1}^{m} v_i x_{ij} + d_j = 0$$

$M - d_j \geq 0$ $j=1,\dots,n$ $u_r \geq \varepsilon$, r=1,2,...,s $vi_{\geq}\varepsilon$, i=1,2,...,m $dj\geq_0$ j=1,2,...,n

The constraints of the BiO-MCDEA model consist of the same constraints with the MCDEA model of Li & Reeves (1999). Only *ur* and *vi* variables are constrained by the ε constant. Although Ghasemi et al. (2014) used ε=0.0001 in the examples they solved, they did not propose an approach to find a proper value for ε constant. Moreover, it is shown that the BiO-MCDEA model is still durable when ε=0 in the example solved by using a different data set. In this study, ε=0 is used as in the original of the model.

Findings

BiO-MCDEA efficiency scores of the RESs used for electricity generation in Turkey were calculated by using LINDO w32 software. The first four columns in Table 3 are the solution values that occur in the BiO-MCDEA model solution. The fifth column consists of the efficiency values obtained as a result of the BiO-MCDEA model solution. The last column refers to the ranking of the RESs according to the efficiency results that arise from the BiO-MCDEA model solution. While efficient RES is in the first place, other RESs are listed backward.

Table 3. BiO-MCDEA Model Solution Values, Efficiency Scores, and Rankings

Years	RES	Inputs		Outputs		Efficiency	Rank
		v_1	v_2	u_1	u_2		
2014	Hydroelectric	0,000042	0	0,000007	0	0,284513	4
	Geothermal	0,00247	0	0,000423	0	1	1
	Wind	0,000276	0	0,000047	0	0,400445	3
	Solar	0,024876	0	0,004261	0	0,074141	5
	Biomass	0,004405	0	0,000754	0	0,825178	2
2015	Hydroelectric	0,000039	0	0,000007	0	0,470021	3
	Geothermal	0	1,172411	0,000292	0	1	1
	Wind	0	0,162433	0,00004	0	0,4661	4
	Solar	0	2,939980	0,000732	0	0,142052	5
	Biomass	0	2,639722	0,000657	0	0,887147	2
2016	Hydroelectric	0,000037	0	0	0,017521	0,429271	4
	Geothermal	0,001218	0	0	0,569485	1	1
	Wind	0,000174	0	0	0,081284	0,459641	3
	Solar	0,001201	0	0	0,56155	0,213469	5
	Biomass	0,002749	0	0	1,285020	0,776657	2
2017	Hydroelectric	0,000037	0	0,000006	0	0,349311	4
	Geothermal	0,00094	0	0,000163	0	1	1
	Wind	0,000153	0	0,000027	0	0,483403	3
	Solar	0,000292	0	0,000051	0	0,147354	5
	Biomass	0,002095	0	0,000364	0	0,77313	2
2018	Hydroelectric	0,000035	0	0,000006	0	0,359631	4
	Geothermal	0,00078	0	0,000135	0	1	1
	Wind	0,000143	0	0,000025	0	0,49873	3
	Solar	0,000198	0	0,000034	0	0,265193	5
	Biomass	0,001608	0	0,000278	0	0,743015	2

When Table 3 is analyzed, it is seen that the only efficient RES is geothermal energy for all of the years. The rankings of other RESs according to the BiO-MCDEA model did not differ greatly for each year. Only hydroelectric ranked third in 2015, while wind energy declined to fourth. By taking the average of the efficiency values found as a result of the efficiency analysis made for five years, a single efficiency value has been reached. The results are shown in Table 4.

Table 4. BiO-MCDEA Efficiency Score Mean and Standard Deviation Values

RES	Mean	SD	Rank
Hydroelectric	0,3785494	0,072448894	4
Geothermal	1	0	1
Wind	0,4616638	0,03747669	3
Solar	0,1684418	0,073181284	5
Biomass	0,8010254	0,056430171	2

When Table 4 is analyzed, geothermal energy ranked first according to the mean of efficiency scores for the years between 2014 and 2018. Geothermal energy was followed by biomass, wind, hydroelectric and solar energies. In general, the BiO-MCDEA efficiency score rankings of the RESs did not change within the years included in the analysis. This is an indication that although RESs have improved every year, they did not reach an energy generation and installed power levels that affect the efficiency scores.

FUTURE RESEARCH DIRECTIONS

In this study, the BiO-MCDEA model carried out its mission successfully in terms of the determination of the efficiency of the RESs in Turkey. The future studies to be conducted by the use of different input and output combinations may provide different results. Moreover, different models that can be used instead of the BiO-MCDEA model or increasing the number of years included in the analysis may enable different RESs to come into prominence as effective RESs.

CONCLUSION

In recent years, all of the people and national administrations have started to see the RESs from the same perspective regardless of in which country they live. Because YEKs are technologies that reduce not only the effects of global warming and environmental disasters but also the negative effects of energy supply security and foreign dependence on fossil fuels on the national economies. From this point of view, it both reduces people's environmental concerns and provides economic value for the national administrations. However, it is still being discussed which RES has a less environmental impact or which RES is more effective in terms of economy and generation. The subject on which RES is needed to be invested in the countries which are foreign dependent on fossil fuels and has high RES potential like Turkey is at the center of all discussions. From this point of view, which RES is more efficient in Turkey was determined in this study by using generation data.

Geothermal energy and biomass energy are placed in the first two ranks for all of the years included in the analysis in this study conducted for the determination of the efficiency of RESs. The reason why geothermal and biomass energy are placed in the first two ranks for all of the years is due to the highest generation compared to the current capacity. Whereas, wind and hydroelectric energies generated less than their existing capacities. Solar energy is placed in the last rank for all of the years because solar energy is very new for Turkey. Even though both installed power and electricity generation has been increasing incrementally since 2014, when electricity generation from solar energy was started in Turkey, a complete efficiency cannot be gotten yet. The fact that electricity generation from solar energy increased by 173.1% in 2018 compared to 2017 is an indicator of future improvement in this RES type.

The rise of geothermal energy in recent years in Turkey confirms this efficiency value. In 2018, Turkey completed various geothermal energy projects and increased its installed capacity by 219 MW (21%) to 1.3 GW. Only in six years between the years of 2013 and 2018, Turkey improved its capacity more than 1 GW and ranked in the fourth line in the world in terms of cumulative geothermal energy capacity. The largest single unit completed in 2018 is Unit 2 with the installed power capacity of 65.5 MW in Kızıldere III facility which is the largest geothermal power plant (165 MW) of Turkey. Baklacı with 19.4 MW, Buharkent with 13.8 MW, 3S Kale with 25 MW and Pamukören Unit 4 with 32 MW were among the projects completed this year. Lastly, Alaşehir Unit 3 with 30 MW attended the geothermal energy generation fleet of Turkey on the date of November 3, 2018 (REN21, 2019, p. 80). These developments confirm that it is not a coincidence that the study provided effective results for all of the years included in the analysis of geothermal energy. It is thought that geothermal energy will continue its development in the following years and become one of the prominent RESs in Turkey.

REFERENCES

Abolhosseini, S., Heshmati, A., & Altmann, J. (2020). *A Review of Renewable Energy Supply and Energy Efficiency Technologies.* IZA Institute of Labor Economics: https://www.iza.org/

Apergis, N., & Danuletiu, D. C. (2014). Renewable Energy and Economic Growth: Evidence from the Sign of Panel Long-Run Causality. *International Journal of Energy Economics and Policy*, *4*(4), 578–587.

Apergis, N., & Payne, J. E. (2012). The Electricity Consumption-Growth Nexus: Renewable Versus Non-Renewable Electricity in Central America. *Energy Sources. Part B, Economics, Planning, and Policy*, *7*(4), 423–431. doi:10.1080/15567249.2011.639336

Apergis, N., & Payne, J. E. (2014). Renewable Energy, Output, CO2 Emissions, and Fossil Fuel Prices in Central America: Evidence from a Nonlinear Panel Smooth Transition Vector Error Correction Model. *Energy Economics*, *42*, 226–232. doi:10.1016/j.eneco.2014.01.003

Arvizu, D., & Balaya, P. (2012). *Direct Solar Energy. In IPCC Special Report on Renewable Energy Sources and Climate Change Mitigation Special Report of the Intergovernmental Panel on Climate Change*. Cambridge University Press.

Aydın, İ. (2013). Balıkesir'de Rüzgar Enerjisi. *Eastern Geographical Review*, *18*(29), 29–50.

Bal, H., Örkcü, H. H., & Çelebioğlu, S. (2010). Improving the Discrimination Power and Weights Dispersion in The Data Envelopment Analysis. *Computers & Operations Research*, *37*(1), 99–107. doi:10.1016/j.cor.2009.03.028

Bigerna, S., Bollino, C. A., & Micheli, S. (2015). *The Sustainability of Renewable Energy in Europe. İsviçre*. Springer International Publishing.

BP. (2019). *BP Statistical Review of World Energy 2019*. London: BP.

Cansın, Y., & Sohtaoğlu, N. H. (2009). OECD/IEA Ülkelerinin Ar-Ge Haracamalarındaki Eğilimler Kapsamında Yenilenebilir Enerji Teknolojilerindeki Gelişmelerin İncelenmesi. In *Yenilenebilir Enerji Kaynakları Sempozyumu Bildiriler Kitabı* (pp. 29-36). Diyarbakır: Elektrik Mühendisleri Odası. http://www.emo.org.tr/ekler/0e3a6c01095e487_ek.pdf

Charnes, A., Cooper, W. W., & Rhodes, E. (1978). Measuring the Efficiency of Decision Making Units. *European Journal of Operational Research*, *2*(6), 429–444. doi:10.1016/0377-2217(78)90138-8

Chien, T., & Hu, J. L. (2007). Renewable energy and macroeconomic efficiency of OECD and non-OECD economies. *Energy Policy*, *37*(7), 3606–3615. doi:10.1016/j.enpol.2006.12.033

Cucchiella, F., & Gastaldi, M. (2014). Data Envelopment Analysis to Compare Renewable Energy Efficiency in The Italian Regions. *Advanced Materials Research, 912-914*, 1607-1611. www.scientific.net/AMR.912-914.1607

Ellabban, O., Abu-Rub, H., & Blaabjerg, F. (2014). Renewable Energy Resources: Current Status, Future Prospects and Their Enabling Technology. *Renewable & Sustainable Energy Reviews*, *39*, 748–764. doi:10.1016/j.rser.2014.07.113

EREC. (2010). *Renewable Energy in Europe: Markets, Trends and Technologies. Londra*. European Renewable Energy Council.

Eurostat. (2020). *Energy from Renewable Sources*. https://ec.europa.eu/: https://ec.europa.eu/eurostat/web/energy/data/shares

Freris, L., & Infield, D. (2008). *Renewable Energy in Power Systems*. John Wiley & Sons, Ltd.

Fridleifsson, I. B. (2001). Geothermal Energy for The Benefit of The People. *Renewable & Sustainable Energy Reviews*, *5*(3), 299–312. doi:10.1016/S1364-0321(01)00002-8

Ghasemi, M. R., Ignatius, J., & Emrouznejad, A. (2014). A bi-Objective Weighted Model for Improving the Discrimination Power in MCDEA. *European Journal of Operational Research*, *233*(3), 640–650. doi:10.1016/j.ejor.2013.08.041

Güler, Ö. (2005). Dünyada ve Türkiye'de Rüzgar Enerjisi. In *TMMOB Türkiye V. Energy Conference Proceeding Book* (s. 209-2015). Ankara: UCTEA The Chamber of Electrical Engineers.

Hatami-Marbini, A., & Toloo, M. (2017). An Extended Multiple Criteria Data Envelopment Analysis Model. *Expert Systems with Applications*, *73*, 201–219. doi:10.1016/j.eswa.2016.12.030

IEA. (2019). *World Energy Outlook 2019*. International Energy Agency.

Kagel, A., Bates, D., & Gawell, K. (2007, April 1). *A Guide to Geothermal Energy and the Environment*. http://www.charleswmoore.org: http://www.charleswmoore.org/pdf/Environmental%20Guide.pdf

Kaldellis, J. K., & Zafirakis, D. (2011). The Wind Energy (R)evolution: A Short Review of a Long History. *Renewable Energy*, *36*(7), 1887–1901. doi:10.1016/j.renene.2011.01.002

Karayılmazlar, S., Saraçoğlu, N., Çabuk, Y., & Kurt, R. (2011). Biyokütlenin Türkiye'de Enerji Üretiminde Değerlendirilmesi. *Bartin Orman Fakültesi Dergisi*, *13*(19), 63–75.

Keyhani, A., Marwali, M. N., & Dai, M. (2010). *Integration of Green and Renewable Energy in Electric Power Systems*. John Wiley & Sons.

Kumar, A., & Schei, T. (2012). *Hydropower. In IPCC Special Report on Renewable Energy Sources and Climate Change Mitigation Special Report of the Intergovernmental Panel on Climate Change*. Cambridge University Press.

Le Gouriérès, D. (1982). *Wind Power Plants Theory and Design*. Pergamon Press Ltd.

Lewis, A., & Estefen, S. (2012). *Ocean Energy. In IPCC Special Report on Renewable Energy Sources and Climate Change Mitigation*. Cambridge University Press.

Li, X. B., & Reeves, G. R. (1999). A Multiple Criteria Approach to Data Envelopment Analysis. *European Journal of Operational Research*, *115*(3), 507–517. doi:10.1016/S0377-2217(98)00130-1

Menegaki, A. N. (2013). Growth and Renewable Energy in Europe: Benchmarking with Data Envelopment Analysis. *Renewable Energy*, *60*, 363–369. doi:10.1016/j.renene.2013.05.042

Popp, D., Hascic, I., & Medhi, N. (2011). Technology and The Diffusion of Renewable Energy. *Energy Economics*, *33*(4), 648–662. doi:10.1016/j.eneco.2010.08.007

Prakash, S. V. (2010). Renewable Energy And Sustainable Development: An Overview. *Birla Institute Of Technology&Science CURIE Journal*, *3*(2), 59–69.

REN21. (2019). *Renewables 2019 Global Status Report*. Paris: REN21 Renewables Now.

Regina, A. K. (2007). *Energy Supply and Renewable Resources*. Infobase Publishing.

Renewable Energy World. (2020). *Ocean Energy Tech*. www.renewableenergyworld.com: https://www.renewableenergyworld.com/types-of-renewable-energy/ocean-energy-tech/#gref

Rosillo-Calle, F. (2016). A Review of Biomass Energy – Shortcomings and Concerns. *Journal of Chemical Technology and Biotechnology (Oxford, Oxfordshire)*, *91*(7), 1933–1945. doi:10.1002/jctb.4918

Sadorsky, P. (2009). Renewable Energy Consumption and Income in Emerging Economies. *Energy Policy*, *37*(10), 4021–4028. doi:10.1016/j.enpol.2009.05.003

Sağlam, Ü. (2018). *The Efficiency Assessment of Renewable Energy Sources with Data Envelopment Analysis. Assessment of Renewable Energy Sources*.

San Cristobal, J. R. (2011). A Multi Criteria Data Envelopment Analysis Model to Evaluate the Efficiency of The Renewable Energy Technologies. *Renewable Energy*, *36*(10), 2742–2746. doi:10.1016/j.renene.2011.03.008

Şenel, M. C., & Koç, E. (2015). Dünyada ve Türkiye'de Rüzgar Enerjisi Durumu- Genel Değerlendirme. *Mühendis ve Makina*, *56*(663), 46–56.

Statista. (2020). *Surface Area of Oceans Worldwide*. https://www.statista.com/: https://www.statista.com/statistics/558499/total-area-of-global-oceans/

Suganthi, L., & Samuel, A. A. (2012). Energy Models for Demand Forecasting-A Review. *Renewable & Sustainable Energy Reviews*, *16*(2), 1223–1240. doi:10.1016/j.rser.2011.08.014

TETC. (2019). *Türkiye Elektrik Üretim-İletim İstatistikleri.* www.teias.gov.tr: https://www.teias.gov.tr/tr-TR/turkiye-elektrik-uretim-iletim-istatistikleri

Tiwari, G. N., & Mishra, R. K. (2012). *Advanced Renewable Energy Sources.* The Royal Society of Chemistry.

YEGM. (2020). *Biyokütle Enerjisi Nedir?* Yenilenebilir Enerji Genel Müdürlüğü: http://www.yegm.gov.tr/yenilenebilir/biyokutle_enerjisi.aspx

ADDITIONAL READING

Ghasemi, M. R., Ignatius, J., & Emrouznejad, A. (2014). A bi-Objective Weighted Model for Improving the Discrimination Power in MCDEA. *European Journal of Operational Research, 233*(3), 640–650. doi:10.1016/j.ejor.2013.08.041

Li-Bo, Z., & Tao, Y. (2014). The evaluation and selection of renewable energy technologies in China. *Energy Procedia, 61*, 2554–2557. doi:10.1016/j.egypro.2014.12.044

Mourmouris, J. C., & Potolias, C. (2013). A Multi-Criteria Methodology for Energy Planning and Developing Renewable Energy Sources at A Regional Level: A Case Study Thassos, Greece. *Energy Policy, 52*, 522–530. doi:10.1016/j.enpol.2012.09.074

San Cristobal, J. R. (2011). A Multi Criteria Data Envelopment Analysis Model to Evaluate the Efficiency of The Renewable Energy Technologies. *Renewable Energy, 36*(10), 2742–2746. doi:10.1016/j.renene.2011.03.008

Tasri, A., & Susilawati, A. (2014). Selection Among Renewable Energy Alternatives Based on A Fuzzy Analytic Hierarchy Process in Indonesia. *Sustainable Energy Technologies and Assessments, 7*, 34–44. doi:10.1016/j.seta.2014.02.008

Troldborg, M., Heslop, S., & Hough, R. L. (2014). Assessing the Sustainability of Renewable Energy Technologies Using Multi-Criteria Analysis: Suitability of Approach for National-Scale Assessments and Associated Uncertainties. *Renewable & Sustainable Energy Reviews, 39*, 1173–1184. doi:10.1016/j.rser.2014.07.160

Wang, J. J., Jing, Y. Y., Zhang, C. F., & Zhao, J. H. (2009). Review on Multi-Criteria Decision Analysis Aid in Sustainable Energy Decision-Making. *Renewable & Sustainable Energy Reviews, 13*(9), 2263–2278. doi:10.1016/j.rser.2009.06.021

KEY TERMS AND DEFINITIONS

Biomass Energy: A type of energy obtained from plants, animals, and organic wastes.

Fossil Fuels: Fuels that are not spread evenly around the world, can be consumed, have negative effects on the environment and contain high levels of carbon.

Geothermal Energy: A continuous energy type that uses the heat in the core of the world as a source.

Hydroelectric Energy: The most common type of renewable energy where water kinetic energy is converted into electrical energy.

Renewable Energy: Energy that will not run out as long as the sun exists and has a minimum negative impact on the environment.

Solar Energy: A type of energy that takes its power from the core of the sun and will become the main source of electricity generation in the future.

Wind Energy: A type of energy obtained from the air currents created by the temperature differences on the earth.

Chapter 9
"My Regular Pharmacy": A Study on the Evaluative Criteria Consumers Use in Selecting an Independent Community Pharmacy as Their Regular One

Niki Glaveli
University of the Aegean, Greece

ABSTRACT

The objectives of this chapter are (1) to develop an instrument that incorporates the core attributes of evaluative criteria that are used by patients/customers in selecting and repeatedly visiting a community pharmacy, (2) to analyze the proposed attributes in order to put forward valid dimensions of evaluative criteria, and (3) to assess the reliability of this instrument. The selection criteria incorporated in the final list were selected based on an extensive literature review and on experts' and customers' opinions. To collect the data, a survey was conducted in the area of Macedonia, Greece. In total, 223 questionnaires were collected. Confirmatory factor analyses revealed a four-factor structure of the proposed instrument namely: price, availability of products, staff competence, and pharmaceutical care services. Moreover, reliability was assessed through Cronbach's alpha coefficient and average variance extracted. The current study outcomes can guide pharmacists' strategic actions in boosting customer patronage behavior in the highly competitive pharmaceutical sector.

INTRODUCTION

Deregulation, cost-containment policies across EU and the broader role of pharmacies in public health promotion, therapeutic/medication management and patient counseling, are the main factors that have altered the pharmaceutical industry's landscape, heightened competition and increased the importance of adopting customer-centered strategies to attract and retain the customer (Castaldo, Grosso, Mallarini, & Rindone, 2016).

DOI: 10.4018/978-1-7998-5442-5.ch009

The deep recession of the Greek economy in the last decade (2010 onwards) has triggered even more radical reforms in the Greek pharmaceutical market with the intention to rationalize the licensing, pricing and reimbursement systems for medicines. Liberalization of the pharmacist's profession, increases in rebates to social security funds, reduction in regulated wholesale, the enforcement of electronic prescriptions and the prescription of the active substance rather than the brand, are probably the most notable changes in the sector (Vozikis, Stavropoulou, & Patrinos, 2015; Kevrekidis, Minarikova, Markos, Malovecká, & Minárik, 2018). These developments have shrunk the sales of pharmaceutical products though community pharmacies by 41.2% in the period 2009-2018 (sales dropped from 6.8 to 4.0 billion euros; Health Economics Observatory of IOBE and Hellenic Association of Pharmaceutical Companies, 2019), reduced profit margins and intensified competition amongst independent community pharmacies. The picture gets even worse if one considers the population density of pharmacies in Greece, which is the highest among EU member states. More precisely, the ratio in 2017 was 97 pharmacies per 100,000 inhabitants (1:1030), whilst the EU-28 average is 31 pharmacies per 100,000 inhabitants (1: 3226 ; Health Economics Observatory of IOBE and Hellenic Association of Pharmaceutical Companies, 2019).

Based on the above facts and data it becomes apparent that within the competitive pharmacy market environment, independent community pharmacies are required to understand the factors that influence consumers' choice of a pharmacy in order to develop the strategies and the range of quality (often niche) services required to attract, satisfy and retain their customers. Responding to this prerequisite, the objectives of this chapter are set as follows: (i) to develop an instrument that incorporates the core attributes of evaluative criteria that are used by patients/customers in selecting and repeatedly visiting a community pharmacy, (ii) to analyze the proposed attributes in order to put forward dimensions of evaluative criteria and (iii) to assess the reliability of this instrument.

BACKGROUND

Pharmacies' Expanded Role

Community pharmacies are complex organizations. *First,* they are economic entities whose survival depends on revenue/profit generation. As such, customer satisfaction and loyalty are central variables related to their viability. *Second*, they are also the places, where pharmacists undertake their professional role (Nunes, Anderson, & Martins, 2014). This role is gradually broadening and becomes more centered to the individual patient. Indeed, the traditional role assigned to the *community pharmacy* (that of *medicines supplier*) is gradually enhanced to incorporate a second one that of *patient counseling provider* (Wiedenmayer, Summers, Mackie, Gous, Edward, & Tromp, 2006) designated in this chapter as *advice provider* role. Due to their proximity and easy access, pharmacies are often the first point of contact that patients visit to seek out specialized care and counseling for minor ailments. Also, pharmacists answer questions about patients' medications and provide instructions on the effective use of medicines and their side effects, issues that can improve the outcome of medication treatment. To these two roles Nunes et al. (2014) adds a third one, that of *community health promoter*, recognizing the social aspect of a pharmacy. This aspect relates to the extent to which the pharmacy is involved in the community.

The expanded role of community pharmacies could be seen as an expected outcome of the attempt to reprofessionalize pharmacists who, by stressing the provision of care, aspire to boost the standing of their profession (Bush, Langley, & Wilson, 2009), advance professional flexibility (Goodrick & Reay,

2011), as well as earn appreciation of their role and become an effective and indispensable part of the health care system (Kevrekidis et al., 2018).

In addition, the competitive market environment requires from pharmacists to adapt their business model in order to attract new customers, maintain their present clientele and boost customer loyalty. This model needs to be redesigned on the principles of the extended role of pharmacies and on active customer-oriented management, followed by the development of an effective marketing strategy. For this purpose, pharmacists have to reconsider the source of their stream of revenues. Although prescription medications is still the main reason why people visit a pharmacy, obtaining over-the-counter (OTC) medications, food supplements and other products (e.g. cosmetics and baby care products) are also gaining ground. Most importantly these products have more stable profit margins and are bigger revenue sources (Gyaneshwari, 2015). Additionally, counseling and medical care services (e.g. measurement of blood pressure, injuries treatment, treatment of minor ailments) have been also identified as important reasons for visiting a community pharmacy (Minarikova, Malovecka, & Foltan, 2016). The latest suggests that the role of *advise provider* is gradually recognized as an important one.

Moreover, the few studies that have investigated how pharmacy users perceive pharmacies' roles (*medicines supplier, advice provider* and *community health promoter)* and react to them, provided indications that being effective in relation to all three roles is positively associated with higher levels of customer trust, satisfaction and loyalty (Wirth, Tabone, Azzopardi, Gauci, Zarb-Adami, & Serracino-Inglott, 2010; Al Arifi, 2012; Nunes et al., 2014). Thus, amongst others, the three identified roles of a pharmacy need to be reflected in the criteria that affect customers' choice of a community pharmacy as a regular one.

Research on Pharmacy Selection Criteria and Scale Development

To address the first objective of the current work, a critical review of the literature related to the *criteria* used by patients/customers in selecting a pharmacy as their regular one was conducted. The review was restricted to peer-reviewed articles published in the English language and with a good number of citations on Google Scholar. The search revealed a limited number of valuable publications (Wiendelholt, 1987; Arneson, Jacobs, Scott, & Murray,1989; Whitehead, Atkin, Krass, & Benrimoj, 1999; Villako & Raal, 2007; Bostrom, 2011; El Hajj, Salem & Mansoor, 2011; Merks, Kaźmierczak, Olszewska, & Koltowska-Häggström, 2014; Kevrekidis et al., 2018). A major conclusion that could be derived from the in-depth study of these papers (Sandberg & Aarikka-Stenroos, 2014) is that they *concentrate* mostly *on factors that affect the choice of a pharmacy for obtaining prescription medicines*, thus addressing mainly the traditional medicines supplier role of a pharmacy.

Due to the limited number of these studies and their narrow focus, further investigation was carried out to trace publications on customer/patient satisfaction with community pharmacies. Due to the identified close relationship between customer satisfaction and *customer patronage/ loyalty behavior* (e.g., Nunes et al., 2014) the pharmacies customer satisfaction literature was investigated. The relevant literature was viewed as a surrogate that could help us gain a deeper insight into the criteria used by customers to evaluate the wider range of services provided by a pharmacy. All the identified publications (MacKeigan & Larson, 1989; Marshall, Cunny, & Lawson, 1997; Kradjan, Schulz, Christensen, Stergachis, Sullivan, & Fullerton, 1999;Volume, Farris, & Kassam, 2001; Tinelli, Bond, Blenkinsopp, Jaffray, Watson, Hannaford, 2002; Larson, Rovers, & MacKeigan, 2002; Abduelkarem, Sackville, Morgan, Sackville & Hildreth, 2003; Coleman, Martin, Barber, &Patterson, 2004; Stuurman-Biez, Kokenberg, Tobi, de Boe, van Doormaal JE, & Jong-van den de Berg, 2005; Gastelurrutia, de San Vicente, Erauncetamurgil, Odriozola, & Fernandez-Llimos, 2006; Oparah & Kikanme, 2006; Traverso, Salamano, Botta, Colautti, Palchik, & Perez, 2007; Naik, Panvelkar, & Saini, 2009; Bahari & Ling, 2009;Wirth et al., 2010; Olave Quispe, Traverso, Palchik, Garcıa Bermudez, La Casa Garcıa, Perez Guerrero, & Calero, 2011; Horvat & Kos, 2011; Nunes et al., 2014; Antunes, Gomes, & Cavaco, 2015; Athavale, Banaham, Bentley, & West-Strum, 2015; Castaldo et al., 2016) *were* –once again- *studied systematically* to gain perspective (Sandberg & Aarikka-Stenroos, 2014) on the items/selection criteria utilized to measure customer satisfaction.

As expected, the pharmacies customer satisfaction literature adopted a more integrated view regarding the criteria used, an approach that is closely related to the extended role of a pharmacy. Indeed, some of the articles measured satisfaction with general services and characteristics of a community pharmacy such as: location, appearance, waiting time for prescriptions and accuracy of prescriptions, information provision, staff conscientiousness and interpersonal skills (e.g., Marshall et al., 1997; Abduelkarem et al., 2003; Gastelurrutia, de San Vicente et al., 2006; Oparah & Kikanme, 2006). Whilst, other authors incorporated satisfaction with *cognitive services* in their research (see *e.g.,* Volume et al., 2001; Kradjan et al., 2009; Naik et al., 2011). Such services were relevant to providing a combination of services including screening, patient follow-up, goal setting, medication management and support, counseling and health education.

Regarding methodology, quite different approaches seem to have been adopted for the development of the instruments used to measure either customer views on selection criteria or pharmacies customer satisfaction. Some authors derived their questionnaire's content from published papers and scales/questionnaires, as well as from guidelines relating to pharmacy services (e.g., Wiendelholt, 1987; Wirth et al., 2010; Kevrekidis et al., 2018). Others relied on experts' or their own opinions or a combination of both (e.g., Wirth et al., 2010) and a limited number consulted customers (Horvat & Kos, 2011).

Taking into consideration the previous finding, an integrated approach has been embraced regarding the development of the proposed scale in the current paper. First, after reading thoroughly the identified publications, a list of 40 items/criteria was compiled, assuring *content validity* (Wetzstein, Hartmanna, Benton jr, & Hohenstein, 2016). Items with the same meaning but different phrasing were then merged. To make the final list of criteria meaningful, items with a minimum appearance were also excluded. Lastly, items irrelevant to the Greek reality e.g. longer pharmacy opening hours (the opening hours schedule in Greece is imposed by law) were also omitted.

Second, before the actual gathering, face validity, applicability and practicality testing of the measurement scale was performed. The validation of the questionnaire was carried out by a group discussion with a panel of 10 people: 4 community pharmacists, 2 general practitioners and 4 consumers, thus strengthening further the content validity of the proposed scale as well. The panel participants were asked to run through the research instrument and recommend any modifications they considered appropriate. The above described process was completed with a discussion led by the author. This was considered necessary in order to ascertain that no significant information was omitted, the design was clear and easy to follow and establish whether the questions asked were easily understood or needed rephrasing. Particular attention was drawn on instructions. Two items: private area for consultation and loyalty card/ membership program (suggested as not common practices) were finally excluded from the list. A final list of 20 items was compiled (see Table 1and the results section). These items were expected to be organized around the following five dimensions: *price, availability of products, staff competence, pharmaceutical care services and pharmacy location and aesthetics.*

METHODOLOGY

The sampling was limited to the pharmacies located in the area of Macedonia, Greece and particularly to 7 out of its 13 administrative districts. Then, based on time and human resources available, 30 pharmacies were randomly selected. Permission was given by the owners to approach patients offsite immediately after being served at the respective pharmacy.

For 2 days of a week, in May 2019, following a plan of a random selection of 20 patients (of different gender and age) per pharmacy, respondents were approached by a trained interviewer, who asked them to take part in a study about the pharmacy they have just been served by. The first question related to whether they considered this pharmacy as their regular one (whether they prefer to make their purchases always or most of the times in the specific pharmacy). In case of a negative answer they were excluded from the survey. Then the respondents were asked to fill in their demographic characteristics, as well as to indicate to what extend the items listed in the questionnaire as criteria for the selection of the pharmacy have affected their choice of the pharmacy they had just visited as their regular one. The response rate was rather high, since 37.17% of those approached agreed to participate, resulting in 223 usable questionnaires.

The instrument used for data collection was a structured questionnaire with multiple choice and 5–point Likert scale questions anchored from: 1 = Not at all and 5= To a great extent. Thus the higher the rating the most important the criterion for the selection of the pharmacy as a regular one. The first part of the questionnaire included the multiple choice questions providing information about the demographic profile of the respondents and the second part the list of the identified selection criteria.

Regarding the demographic profile of the respondents, 151 were women and 73 men, non married was the 61.9%. The vast majority, around 85%, were under the age of 50, 54,7% held a university degree and 32,7% were employed in the private sector.

RESULTS

The statistical result from the Kaiser–Meyer–Olkin (KMO) measure of sampling adequacy was 0.918, and the Bartlett test was highly significant ($X^2 = 3082.5;0.001$) which indicated important correlations between the variables and the suitability of the factor analysis.

To put forward dimensions of evaluative criteria (second objective of the present study) exploratory factor analysis (EFA) with varimax rotation using SPSS.25 was initially performed, followed by a 2nd order confirmatory factor analysis (CFA) through LISREL 8.54. Second-order CFA can determine directly how well the super ordinate factors capture variance present in the 1st order constructs and how much variance of the 1st level order constructs occurs exclusively at that level (Arnau & Thompson, 2000).

Though, EFA the following 4 dimensions, which explained 64.207% of the variance, were identified: *price, availability of products, staff competence and pharmaceutical care services.* Three items: "convenient location", "appearance/cleanliness of the pharmacy" and "professional appearance of the pharmacy" which were expected to load to a separate dimension namely pharmacy location and aesthetics did not yield a high loading (> 0.50), thus they were excluded from CFA (see Table 1).

Means, standard deviations and CFA results are incorporated in Table 1. As indicated by EFA, the 2nd order model hypothesized four 1st order factors (consisting of 17 items) and a 2nd order one. The plausibility of the model was confirmed by the fit indexes suggested by Schumacker and Lomax (2004):

$\chi2$ =26.95, $\chi2$ /df = 2.88,GFI= 0.91, CFI=0.94, NFI = 0.92, RMSEA = 0.08, p > $\chi2$ = 0.000; p< 0.1. Also, all the items were significant, related to their hypothesized 1st order factors and their standardized loadings (standardized Lamda coefficients) were higher than 0.5. The same held true for the relationship between the 1st order factors and the 2nd order construct (see Table 3).

As regards the extracted EFA and 2nd order factors: the first factor was labeled *price* because it incorporates three items that relate to offers/discounts on non-medical products and prescription medicines, as well as payment facilities for obtaining prescription medicines. The second factor was named *availability of products* since it consists of 5 items which refer to a pharmacy's role as medicines supplier (e.g., on-the-spot availability of prescription medicines, as well as to the availability of non-medicine and alternative –medicine products).

Table 1. Confirmatory factor analysis (1st order) results, reliability coefficients and descriptives

First-order factors and items	Mean value	Sd	Std. Lamda	Alpha	AVE
I. Price					
1. Offers and/or discounts on non-medical products	2.73	1.34	0.81	0.75	0.55
2. Discounts on the prices of prescription medicines	3.36	1.37	0.78		
3. Payment facilitation for obtaining prescription medicines	3.12	1.47	0.61		
II. Availability of products					
4. Availability of a wide range of health (food) products	2.93	1.31	0.82	0.77	0.53
5. Availability of a wide range of baby care products	2.50	1.52	0.77		
6. Availability of a wide range of quality beauty products	3.14	1.34	0.70		
7. Availability of a wide range of alternative medicine products	2.43	1.41	0.68		
8. Prescription medicines are available on the spot	3.70	1.20	0.67		
III. Staff competence					
9. Guarantees respect and confidentiality	4.42	0.92	0.81	0.78	0.57
10. Provides fast and error free service	4.25	0.94	0.80		
11. I can trust the knowledge and advice I receive from the pharmacist	4.40	0.93	0.76		
12. The pharmacist cares about me, my family and community	4.38	0.89	0.74		
13. The pharmacist gives counseling on instructions to take and store medication	4.12	0.84	0.73		
14. The pharmacist answers all my questions	4.31	0.93	0.70		
15. The pharmacist and staff are friendly	4.55	0.81	0.67		
IV. Pharmaceutical care services					
16. Can offer medical care services e.g. blood pressure measurement, injections, wound dressing change, treatment of minor ailments	3.67	1.28	0.74	0.72	0.53
17. Provides products prepared by the pharmacist for e.g. treatment of minor ailments	2.71	1.37	0.72		

AVE: Average Variance Extracted

The third factor was called *staff competence*. It includes 7 items linked to the customer service abilities of the pharmacy staff such as: knowledge, expertise, friendly approach and trustworthiness), criteria closely associated with performing effectively the *advice provider* and the *community health promoter* roles of a pharmacy. Finally, the 4th factor was addressed to as *pharmaceutical care services* because of the nature of the 2 items that it incorporates ("Can offer medical care services e.g. blood pressure measurement, injections, wound dressing change, treatment of minor ailment" and "Provides products prepared by the pharmacist"). These items associate with the *advice provider role.*

The bold diagonal elements indicate the square roots of AVE. Off diagonal elements are the correlations among the first order constructs.

Table 2 presents the correlations among the 1st order constructs (off diagonal elements) as well as the square roots of the variance shared among the constructs and their items (bold diagonal elements). Since the square roots of the variance shared among the constructs and their items are higher than the correlations with the other first order constructs discriminant validity is supported (Fornell & Larcker, 1981).

Table 2. Discriminant validity of 1st order factors

1st order factors	1	2	3	4
1. Price	**0.74**			
2. Availability of products	0.41	**0.73**		
3. Staff competence	0.38	0.58	**0.76**	
4. Pharmaceutical care services	0.47	0.69	0.67	**0.73**

Furthermore, the standardized factor loading for the 1st order factors were as follows: price = 0.73, availability of products = 0.79, staff competence = 0.91 and *pharmaceutical care services* = 0.55 (see Table 3), implying that *staff competence* is the factor with the stronger association with customer behavior related to selecting a pharmacy as a regular one.

Internal reliability was assessed (third objective of the present study) with the estimation of Cronbach's alpha and average variance extracted (AVE) coefficients. The value of both measures (see Tables 1 and 3) exceed those suggested as thresholds (a> 0.70 and AVE > 0.50,) confirming that the proposed instrument is reliable (Hair, Black, Babin, & Anderson, 2010).

Table 3. Confirmatory factor analysis: 2nd order results

1st order variable: Pharmacy selection	Std. Lamda of 2nd order factors/constructs	AVE
Price	0.73	0.564
Availability of products	0.79	
Staff competence	0.91	
Pharmaceutical care services	0.55	

DISCUSSION AND IMPLICATIONS

The present study makes several contributions to the pharmaceutical literature. *First,* it adds to the limited research regarding the development of a valid and reliable instrument to help academics and practitioners understand the criteria pharmacies customers use in selecting and repeatedly visiting a pharmacy. The focus was to understand the criteria considered by customers in selecting a pharmacy as an integrated service provider and not just for obtaining prescription and OTC medicines (medicines provider role). Adopting this approach was considered critical. The reason is that the relevant papers distilled from the literature (Wiendelholt, 1987; Arneson et al., 1989; Whitehead et al., 1999; Villako & Raal, 2007; Bostrom, 2011; El Hajj et al., 2011; Merks et al., 2014; Kevrekidis et al., 2018) seemed to miss this element, providing thus not the whole picture regarding pharmacies customers choice behavior. Moreover, the limited research on pharmacies customers reactions to the extended role of a pharmacy has been considered. The relevant findings suggest that being effective in all the three identified (in the current paper) roles -those of: medicines supplier, advice provider and community health promoter- boosts customer satisfaction with and loyalty to the pharmacy. As Nunes et al. (2014) underline, being effective in relation to the *advice provider role* (associated with the competences and behavior of the pharmacist and staff) is more closely related to customer loyalty and being efficient as *medicines supplier* improves customer satisfaction more than the other two roles. So, pharmacists are advised to take into consideration multiple pharmacy attributes when they design their approach to customer service.

Second, the current research puts forward dimensions of evaluative criteria and provides understanding regarding their importance on influencing pharmacies customers selection behavior. More precisely, four dimensions emerged from EFA and were confirmed by 2nd order CFA: *price, availability of products, staff competence* and *pharmaceutical care services,* as those determining customer choice of a specific pharmacy as a regular one. Additionally, according to the standardized values of the loadings of the four lower-order factors, the *staff competence* factor is the one that appears to primarily affect customers' choice of a community pharmacy as a regular one. *Staff competence* incorporates items that relate to developing a trusting relationship between the pharmacists and their customers. As previous research also implies, the provision of customer-centered care/advice and the personal relationship (familiarity) with the pharmacist and staff could construct hedonic or emotional connection to the pharmacy, which customers prefer to mostly visit (Gavilan, Avello, & Abril, 2014). To the same direction, Castaldo et al. (2016) found that trust to a pharmacist's knowledge/professional opinion and advice is positively related to customer satisfaction. Additionally, El Hajj et al. (2011) and Minarikova et al. (2016) also consider staff competence (knowledge, skills and behavior) as one of the main reasons that lead customers' preferences towards a specific pharmacy. The vital role of pharmacies staff communication skills, as well as empathy, personality and a close relationship with the customer have also been illustrated by other relevant studies as important elements that influence the choice of a pharmacy (Antunes et al., 2015; Athavale et al., 2015). Consequently, pharmacists are recommended to invest in lifelong learning and staff training e.g. on communication skills. Also, technology and blogging can facilitate effective customer and treatment management, as well as connect the pharmacy with the local community (Retail Independent Pharmacy, 2018).

Price (payment facilitations, offers, discounts) and *product availability* were found to be equally important criteria in customer choice of a pharmacy. Although price has been underlined by earlier studies as a factor that influences customer choice of a community pharmacy (e.g. Merks et al., 2014), it seems to be more relevant in the Greek case, most likely due to the severe economic crisis and the shrinking of incomes and pensions. Another important development that may explain the significance of price as a selection criterion (for the Greek case) is possibly the legislation on: prescribing by active substance, maximum price of generics, list of non-reimbursable medicines and patients contribution in medicines expenditure (Economou, Kaitelidou, Karanikolos, & Maresso, 2017). In line with this finding, pharmacists need to consider the possibility to incorporate loyalty schemes in their pricing policies. As earlier research has illustrated, loyalty programs could be a source of competitive advantage for pharmacies (e.g., Nobre & Rodringuez, 2018). Nevertheless, a clear strategy, unfailing relationship marketing schemes, and an effective communication on the loyalty program's value proposition are essential elements in order to create differentiation advantages and loyalty towards a specific pharmacy. Technology and customer knowledge could be also helpful in boosting the effectiveness of a loyalty program (Reinert & Murray, 2017).

Regarding the importance of the factor *availability of products*, probably accessibility to a wide range of quality products (beauty/cosmetics, baby, health, alternative medicines, OTC) and on the spot availability of prescription medicines is seen as convenience and might explain the surprising finding that *location was not* supported as a selection criterion/factor by EFA. Despite the fact that location has been repeatedly reported in pharmaceutical literature as a primary pharmacy selection factor (El Hajj et al., 2011; Merks et al., 2014; Minarikova et al., 2016; Villako & Raal, 2007; Wirth et al., 2010) this evidence was not verified by the current study. Nevertheless, this point merits further investigation.

The lower importance of *pharmaceutical care services* as a factor that influences customer choice of a community pharmacy implies that this role is not yet embraced as essential by Greek customers. Nevertheless, previous research indicates that patients were more satisfied with pharmacies where pharmaceutical care was introduced and loyalty to a single pharmacy increased (see e.g. Merks et al., 2014).

Third, the validity and reliability of the proposed instrument was achieved through a rigorous process of development and validation. Initially, a critical review of the literature on pharmacies selection criteria, pharmacies customer satisfaction and pharmacies roles was conducted (theory triangulation; Bennet, 1997). The intention was to look at a situation/ phenomenon from different perspectives and through different lenses. Based on a thorough study of the identified papers, 40 items were distilled as appropriate to the choice of a pharmacy which were merged to 22. The final validation step of the questionnaire was carried out by a group discussion (led by the author) with 10 people (experts and consumers) ending up to 20 items. This process combined the approaches of other authors in the field (e.g., Wiendelholt, 1987; Wirth et al., 2010; Kevrekidis et al., 2018; Horvat & Kos, 2011) by taking into consideration the theoretical work in the specific scientific area, as well as the views of practitioners and users/customers of pharmacies. Thus, multiple sources were utilized for assuring content and face validity, as well as applicability and practicality of the research instrument.

Lastly, the analysis of the data, regarding the validity of the measurement scale and its dimensionality/structure, was performed in two steps. Initially, EFA was conducted, followed by a 2nd order CFA. The use of the 2nd order CFA has many *advantages* under explicit conditions which are met in the current study: the lower order factors are significantly correlated and the higher order factor must support theoretically justifiable relations among the lower order factors. These advantages are: First, its design imposes structure on the pattern of correlations among the 1st order factors and thus a higher-order model is more parsimonious than its lower-order equivalent (Byrne, 2005). Second, a higher-order model can assess the degree to which the 2nd order factor is responsible for this pattern of correlations among the 1st order factors. Third, a higher-order model can differentiate between the residual error related to prediction of the lower-order constructs by the 2nd order one and the measurement error related to the observed variables. Putting it simply, it can consider the unique variance associated with each 1st order factor that is not shared in common with each of the other first-order factors and partition this variance from measurement error variance.

Reliability was assessed with two indices: Cronbach's alpha and AVE. The results indicated that the proposed scale is both valid and reliable. As a result, due to its rigorous design and validation/reliability approach the proposed instrument can support researchers' and practitioners' efforts to analyze and understand pharmacies customers selection behavior.

FUTURE RESEARCH AVENUES

The questionnaire developed can be a reliable and valid instrument to assess patient/customer satisfaction with pharmaceutical care in community pharmacies in Greece. However, further research is required to check its validity and reliability in other cultures, as well as its stability in measurement models. An improvement to the study could be to document the reasons why customers who completed the questionnaires were visiting the pharmacy and analyze whether and how the demographic criteria of the customers e.g. gender, age, income, family status relate to pharmacy choice behavior.

CONCLUSION

Considering the significance of understanding consumer behavior in the pharmaceutical sector and the limited research in the field, this article presents data that refer to the evaluative criteria that customers use in selecting an independent community pharmacy as their regular one. The outcomes of this preliminary research are of importance to both academics and practitioners. For academics, the dimensions of pharmacies customers evaluative criteria are established, facilitating also the investigation of the relationships among them, as well as the more precise and accurate understanding of the reasons behind the selection of, and regular visits to, a pharmacy. With regard to practitioners, they realize that pharmacy customers use several attributes when selecting a pharmacy and understand the elements of service they need to emphasize in order to build long-term relationships with their customers.

REFERENCES

Abduelkarem, A. R., Sackville, M. A., Morgan, R. M., Sackville, M. P., & Hildreth, A. J. (2003). An assessment of the level of type 2 diabetes patients' satisfaction with community pharmacists' services. *International Journal of Pharmacy Practice*, *11*(3), 161–168. doi:10.1211/0022357021990

Al-Arifi, M. N. (2012). Patients' perception, views and satisfaction with pharmacists' role as health care provider in community pharmacy setting at Riyadh, Saudi Arabia. *Saudi Pharmaceutical Journal*, *20*(4), 323–330. doi:10.1016/j.jsps.2012.05.007 PMID:23960807

Antunes, L. O., Gomes, J. J., & Cavaco, A. M. (2015). How pharmacist–patient communication determines pharmacy loyalty? Modeling relevant factors. *Research in Social & Administrative Pharmacy*, *11*(4), 560–570. doi:10.1016/j.sapharm.2014.11.003 PMID:25577547

Arnau, R. C., & Thompson, B. (2000). Second Order Confirmatory Factor Analysis of the WAIS-III. *Assesssment*, *7*(3), 237–246. doi:10.1177/107319110000700304 PMID:11037391

Arneson, D., Jacobs, E. W., Scott, D. M., & Murray, W. J. (2011). Patronage Motives of Community Pharmacy Patrons. *Journal of Pharmaceutical Marketing & Management*, *4*(2), 3–22. doi:10.3109/J058v04n02_02

Athavale, A. S., Banaham, B. F., Bentley, J. P., & West-Strum, D. S. (2015). Antecedents and consequences of pharmacy loyalty behavior. *International Journal of Pharmaceutical and Healthcare Marketing*, *9*(1), 36–55. doi:10.1108/IJPHM-02-2013-0003

Bahari, M. B., & Ling, Y. P. (2009). Factors contributing to customer satisfaction with community pharmacies in Malaysia. *Journal of Public Health*, *18*(1), 35–41. doi:10.100710389-009-0280-7

Bennet, J. A. (1997). A case for theory triangulation. *Nursing Science Quarterly*, *10*(2), 97–106. doi:10.1177/089431849701000210 PMID:9197723

Boström, K. (2011). *Consumer behaviour of pharmacy customers. Choice of pharmacy and over-the-counter medicines* (Degree Thesis). Degree Program in International Business, Helsinki.

Briesacher, B., & Corey, R. (1997). Patient satisfaction with pharmaceutical services at independent and chain pharmacies. *American Journal of Health-System Pharmacy, 54*(5), 531–536. doi:10.1093/ajhp/54.5.531 PMID:9066860

Bush, J., Langley, C. A., & Wilson, K. A. (2009). The corporization of community pharmacy: Implications for service provision, the public health function, and pharmacy's claims to professional status in the United Kingdom. *Research in Social & Administrative Pharmacy, 5*(4), 305–331. doi:10.1016/j.sapharm.2009.01.003 PMID:19962674

Byrne, B. M. (2005). Factor Analytic Models: Viewing the Structure of an Assessment Instrument From Three Perspectives. *Journal of Personality Assessment, 85*(1), 17–32. doi:10.120715327752jpa8501_02 PMID:16083381

Castaldo, S., Grosso, M., Mallarini, E., & Rindone, M. (2016). The missing path to gain customers loyalty in pharmacy retail: The role of the store in developing satisfaction and trust. *Research in Social & Administrative Pharmacy, 12*(5), 699–712. doi:10.1016/j.sapharm.2015.10.001 PMID:27555108

Coleman, B., Martin, C., Barber, N., & Patterson, D. (2004). An evaluation of the safety and acceptability of an anticoagulation clinic in a community pharmacy setting—A pilot study. *The Pharmaceutical Journal, 273*(7328), 822–824.

Economou, C., Kaitelidou, D., Karanikolos, M., & Maresso, A. (2017). Greece: Health system review. *Health Systems in Transition, 19*(5), 1–192. PMID:29972131

El Hajj, M. S., Salem, S., & Mansoor, H. (2011). Public attitudes toward community pharmacy in Qatar: A pilot study. *Patient Preference and Adherence, 5*, 405–422. doi:10.2147/PPA.S22117 PMID:21949604

Fornell, C., & Larcker, D. F. (1981). Evaluating structural equation models with unobservable variables and measurement error. *JMR, Journal of Marketing Research, 18*(1), 39–50. doi:10.1177/002224378101800104

Gastelurrutia, M. A., de San Vicente, O. G., Erauncetamurgil, O., Odriozola, I., & Fernandez-Llimos, F. (2006). Customers' expectations and satisfaction with a pharmacy not providing advanced cognitive services. *Pharmacy World & Science, 8*(6), 374–376. doi:10.100711096-006-9049-z PMID:17120133

Gavilan, D., Avello, M., & Abril, C. (2014). Shopper marketing: A new challenge for Spanish community pharmacies. *Research in Social & Administrative Pharmacy, 10*(6), e125–e136. doi:10.1016/j.sapharm.2014.02.008 PMID:24690551

Goodrick, E., & Reay, T. (2011). Constellations of institutional logics: Changes in the professional work of pharmacists. *Work and Occupations, 38*(3), 372–416. doi:10.1177/0730888411406824

Gyaneshwari, K. (2015). A study on consumer behavior towards OTC products. *International Journal of Management and Social Science Research. RE:view, 1*(13), 134–137.

Hair, J. F., Black, W. C., Babin, B. J., & Anderson, R. E. (2010). *Multivariate Data Analysis*. Prentice Hall.

Horvat, N., & Kos, M. (2011). Slovenian pharmacy performance: A patient –centered approach to patient satisfaction survey content development. *International Journal of Pharmacy Practice*, *33*(6), 985–996. PMID:21984227

Kevrekidis, D. P., Minarikova, D., Markos, A., Ivona Malovecká, I., & Minárik, P. (2018). Community pharmacy customer segmentation based on factors influencing their selection of pharmacy and over-the-counter medicines. *Saudi Pharmaceutical Journal*, *26*(1), 33–43. doi:10.1016/j.jsps.2017.11.002 PMID:29379331

Kradjan, W. A., Schulz, R., Christensen, D. B., Stergachis, A., Sullivan, S., & Fullerton, D. S. (1999). Patients' perceived benefit from and satisfaction with asthma-related pharmacy services. *Journal of the American Pharmaceutical Association*, *39*(5), 658–666. doi:10.1016/S1086-5802(15)30350-8 PMID:10533347

Larson, L. N., Rovers, J. P., & MacKeigan, L. D. (2002). Patient satisfaction with pharmaceutical care: Update of a validated instrument. *Journal of the American Pharmaceutical Association*, *42*(1), 44–50. doi:10.1331/108658002763538062 PMID:11833515

MacKeigan, L. D., & Larson, L. N. (1989). Development and validation of an instrument to measure patient satisfaction with pharmacy services. *Medical Care*, *27*(5), 522–536. doi:10.1097/00005650-198905000-00007 PMID:2725084

Marshall, B. C., Cunny, K. A., & Lawson, K. A. (1997). HIV-positive males' satisfaction with pharmacy services. *Journal of the American Pharmaceutical Association*, *37*(1), 66–75. doi:10.1016/S1086-5802(16)30180-2 PMID:9040172

Merks, P., Kaźmierczak, J., Olszewska, A. E., & Kołtowska-Häggström, M. (2014). Comparison of factors influencing patient choice of community pharmacy in Poland and in the UK, and identification of components of pharmaceutical care. *Patient Preference and Adherence*, *8*, 715–726. doi:10.2147/PPA.S53829 PMID:24868150

Minarikova, Malovecka, I. & Foltan, V. (2016). Patient choice of pharmacy and satisfaction with pharmaceutical care – Slovak regional comparison. *Farmacia*, *64*(3), 473–480.

Naik, P. P., Saini, B., & Armour, C. (n.d.). Measurement of patient satisfaction with community pharmacy services: A review. *International Journal of Clinical Pharmacology Research*, *31*(5), 92–100.

Nobre, H., & Rodringuez, C. (2018). Loyalty programs in the pharmaceutical retail: The impact of a network loyalty program on the pharmacy loyalty. *International Journal of Business Excellence*, *14*(4), 454–469. doi:10.1504/IJBEX.2018.090312

Nunes, F. G., Anderson, J. E., & Martins, L. M. (2014). Patient reactions to community pharmacies' roles: Evidence from the Portuguese market. *Health Expectations*, *18*(6), 2853–2864. doi:10.1111/hex.12269 PMID:25229450

Olave Quispe, S. Y., Traverso, M. T., Palchik, V., Garcıa Bermudez, E., La Casa Garcıa, C., Perez Guerrero, C., & Calero, M. (2011). Validation of a patient satisfaction questionnaire for services provided in Spanish community pharmacies. *International Journal of Clinical Pharmacy*, *33*(6), 949–957. doi:10.100711096-011-9566-2 PMID:22130784

Oparah, A. C., & Kikanme, L. C. (2006). Consumer satisfaction with community pharmacies in Warri, Nigeria. *Research in Social & Administrative Pharmacy*, *2*(4), 499–511. doi:10.1016/j.sapharm.2006.02.004 PMID:17161808

Reinert, C. D., & Murray, J. A. (2017). Successful Implementation of Pharmacy Retail Store Loyalty Reward Programs. *International Journal of Applied Management and Technology*, *16*(1), 152–165. doi:10.5590/IJAMT.2017.16.1.10

Retail Independent Pharmacy. (2018). *Good medicine*. Cardinal Health. Retrieved from: https://www.readkong.com/page/good-medicine-5967197?p=1

Sandberg, B., & Aarikka-Stenroos, L. (2014). What makes it so difficult? A systematic review on barriers to radical innovation. *Industrial Marketing Management*, *43*(8), 1293–1305. doi:10.1016/j.indmarman.2014.08.003

Schumacker, R. E., & Lomax, R. G. (2004). *A beginner's guide to structural equation modeling* (2nd ed.). Lawrence Erlbaum Associates Publishers. doi:10.4324/9781410610904

Stuurman-Bieze, A.G., Kokenberg, M.E., Tobi, H., de Boer W.O., van Doormaal, J.E., & Jong-van den de Berg, L.T. (2005). Complex pharmaceutical care intervention in pulmonary care: part B patient opinion and process survey. *Pharmacy World and Science, 27*(5), 385-392.

Tinelli, M., Bond, C., Blenkinsopp, A., Jaffray, M., Watson, M., & Hannaford, P. (2007). Patient evaluation of a community pharmacy medications management service. *The Annals of Pharmacotherapy*, *41*(12), 1962–1970. doi:10.1345/aph.1K242 PMID:17971403

Traverso, M. L., Salamano, M., Botta, C., Colautti, M., Palchik, V., & Perez, B. (2011). Questionnaire to assess patient satisfaction with pharmaceutical care in Spanish language. *International Journal for Quality in Health Care*, *19*(4), 217–224. doi:10.1093/intqhc/mzm014 PMID:17545673

Villako, P., & Raal, A. (2007). A survey of Estonian consumer expectations from the the pharmacy service and a comparison with the opinions of pharmacists. *Pharmacy World & Science*, *29*(5), 546–550. doi:10.100711096-007-9102-6 PMID:17497224

Volume, C. I., Farris, K. B., Kassam, R., Cox, C. E., & Cave, A. (2001). Pharmaceutical care research and education project: Patient outcomes. *Journal of the American Pharmaceutical Association*, *4*(3), 411–420. doi:10.1016/S1086-5802(16)31255-4 PMID:11372906

Vozikis, A., Stavropoulou, L., & Patrinos, G. (2015). Community Pharmacists' Strategies in Greece: An Assessment of the Policy Environment and the Mapping of Key Players. *Health*, *7*(11), 1560–1577. doi:10.4236/health.2015.711170

Wetzstein, A., Hartmanna, E., Benton, W. C. Jr, & Hohenstein, N.-O. (2016). A systematic assessment of supplier selection literature – State-of-art and future scope. *International Journal of Production Economics*, *182*(C), 304–323. doi:10.1016/j.ijpe.2016.06.022

Whitehead, P., Atkin, P., Krass, I., & Benrimoj, S. (1999). Patient drug information and consumer choice of pharmacy. *International Journal of Pharmacy Practice*, *7*(2), 71–79. doi:10.1111/j.2042-7174.1999.tb00952.x

Wiedenmayer, K., Summers, R. S., Mackie, C., Gous, A., Everard, M., & Tromp, D. (2006). *Developing pharmacy practice: A focus on patient care.* World Health Organization Department of Medicines Policy and Standards.

Wiederholt, J. B. (1987). Development of an Instrument to Measure Evaluative Criteria that Patients Use in Selecting a Pharmacy for Obtaining Prescription Drugs. *Journal of Pharmaceutical Marketing & Management, 1*(4), 35–59. doi:10.3109/J058v01n04_04

Wirth, F., Tabone, F., Azzopardi, L. M., Gauci, M., Zarb-Adami, M., & Serracino-Inglott, A. (2011). Consumer perception of the community pharmacist and community pharmacy services in Malta. *Journal of Pharmaceutical Health Services Research, 1*(4), 189–194. doi:10.1111/j.1759-8893.2010.00034.x

ADDITIONAL READING

Desselle, S. P., Moczygemba, R. L., Zgarrick, D. P., & Alston, G. (2020). *Pharmacy Management: Essentials for All Practice Setting* (5th ed.). McGraw Hill.

Goode, J. V., Owen, J., Page, A., & Gatewood, S. (2019). Community-Based Pharmacy Practice Innovation and the Role of the Community-Based Pharmacist Practitioner in the United States. *Pharmacy (Basel, Switzerland), 7*(3), 106. https://www.ncbi.nlm.nih.gov/pmc/articles/PMC6789634/. doi:10.3390/pharmacy7030106 PMID:31382690

Health Economics Observatory of IOBE and Hellenic Association of Pharmaceutical Companies. (2019). The Pharmaceutical Market in Greece: Facts & Figures 2018. Retrieved from: http://iobe.gr/docs/research/en/RES_05_A_22042019_REP_ENG.PDF

Pharmaceutical Group of European Union (PGEU). (2019). PHARMACY 2030: A Vision for Community Pharmacy in Europe. Brussels. Retrieved from: https://www.pgeu.eu/wp-content/uploads/2019/03/Pharmacy-2030_-A-Vision-for-Community-Pharmacy-in-Europe.pdf

Smith, A. J., Scahill, S. L., Harrison, J., Carroll, T., & Medlicott, N. J. (2018). Service provision in the wake of a new funding model for community pharmacy. *BMC Health Services Research, 18*(1), 307. doi:10.118612913-018-3120-z PMID:29716610

KEY TERMS AND DEFINITIONS

Advice Provider: A pharmacy's role associated with the provision of medication advice and treatment of minor ailment.

Community Health Promoter: The extent to which the pharmacy is involved in the community.

Community Pharmacy: They are the most accessible pharmacies to the patients. They supply medicines, counsel patients, and participate in health-promotion programs. Also, they maintain links with other health professionals in primary health care.

Medicines Supplier: A pharmacy's central role associated with the provision of prescription and over-the- counter medicines.

My Regular Pharmacy: The repeatable use of a pharmacy for obtaining medicines and other type of products such as: health, beauty, baby care ones.

Reprofessionalize: Re-establish the standing of a profession.

Triangulation: The process of combining theoretical perspectives (theoretical triangulation), or data from different sources (data triangulation), or methods to conduct a study (methodological triangulation) or multiple investigators/researchers (investigator triangulation) to study a specific social phenomenon.

Chapter 10
Quality Management in Primary Dental Care:
Methods and Challenges

A. I. Syngelakis
Faculty of Dentistry, University of Athens, Greece & Department of Dentistry, European University of Cyprus, Cyprus

Maria Kamariotou
https://orcid.org/0000-0002-7873-9265
University of Macedonia, Greece

Fotis C. Kitsios
https://orcid.org/0000-0001-7269-5567
University of Macedonia, Greece

Chrystala Charalambous
European University of Cyprus, Cyprus

Argy Polychronopoulou
National and Kapodistrian University of Athens, Greece

ABSTRACT

In dental care services, quality is an important factor that affects decision making, the planning of health strategies and policies, the cost of health services, and the evaluation of them. The evaluation of quality in dental services using the assessment methods that are used in other services of primary healthcare is difficult due to the special characteristics of dentistry. However, the improvement and the evaluation of primary oral healthcare services is a complicated issue because it involves many factors that affect it. Therefore, the purpose of this chapter is to provide a complete overview of the literature using Webster and Watson's methodology. Fifty peer-reviewed papers were analyzed and the results of this review revealed that the number of publications in this domain has increased in the last decade, and there is a need to foster research (especially empirical) in this field.

DOI: 10.4018/978-1-7998-5442-5.ch010

INTRODUCTION

In dental care services, the quality is an important factor that affects decision making, the planning of health strategies and policies, the cost of health services and the evaluation of them (Delias et al., 2015; Kitsios et al., 2019; Manolitzas et al., 2018; Sindakis and Kitsios, 2016; Stefanakakis et al., 2017). However, quality in primary dental care often means different things to dental practitioners, policy makers and patients. Nevertheless, the implementation of evaluation processes as well as the improvement of health services is required for the successful implementation of dental interventions (Syngelakis and Kitsios, 2017; Syngelakis et al., 2019; Syngelakis et al., 2013; Syngelakis et al., 2012). According to Campbell et al. (2000), the key dimensions of health care quality are access, effectiveness and efficiency. Following this approach, main dimensions of primary dental care quality include access, effectiveness and efficiency, as well as appropriateness, availability, continuity, patient-centredness, efficacy, respect and caring, safety and timeliness. Key determinants of primary dental care quality include skills of dentists, patient's satisfaction, patient – dentist interpersonal relationship, dentist's satisfaction, patient's safety, trust and accountability of dentist, cost of services, use of modern technology and equipment, referral to a specialist when required, evidence – based decision making, patient's involvement in decision making and emphasis on prevention (Bader, 2009; Campbell and Tickle, 2013; Chu et al., 2001; Evans et al., 2001; Gonzalez et al., 2006; Tinanoff, 2012).

Therefore, the evaluation of quality in dental services using the assessment methods that are used in other services of primary healthcare is difficult due to the special characteristics of dentistry (Bader, 2009; Dental Quality Alliance, 2019; Tickle et al., 2015). In recent years, researchers have developed some methods in order to evaluate dental quality such as ISO, EFQM, Visitatie, TQM, EPA, benchmarking, Pay for Performance, peer review and Decision Support Systems for clinical services (Lavell et al., 2004; Tinanoff, 2012; Yamalik, 2007a,b; Voinea-Griffin et al 2010a,b). The main goal of these methods is to support patients' satisfaction and safety, improve healthcare processes and upgrade overall patient's experience. However, the improvement and the evaluation of primary oral healthcare services is a complicated issue because it involves many factors (e.g. political, legal, economic and social environment) that affect it.

Previous researchers have examined the quality assessment methods in primary dental care and the limitations of these methods. Furthermore, they evaluated the quality indicators in primary healthcare and they discussed the specific aspects of dental services and the dimensions of quality in dentistry focusing on primary dental care (Byrne et al., 2019; Mills et al., 2014; Righolt et al., 2019; Saleh et al., 2015; Samartzis and Talias, 2019; Scambler et al., 2016; Simou et al., 2013). Therefore, the purpose of this chapter is to provide a complete overview of the literature from a different perspective, not implemented in existing analyses and, thus, it is a good starting point for further research. More specifically, this chapter answers the following questions: How many papers are published each year? Which journals have published NSD peer-reviewed publications and which have published the highest number of publications? Which topics have been studied? What areas for future research can be identified?

This chapter provides an overview of the current state of the field and it can help researchers to expand it and to position their study. The evaluation of health quality is an immature field, so it could inspire researchers for empirical investigation. Managers can assist to understand the characteristics of health services, the factors which affect their quality as well as quality in dentistry. It is anticipated that the effective development of health services will be increasingly important in next years as a result of rapid developments in new technologies and changing customer needs or preferences. The usefulness of this chapter is important for researchers who are interested in studying to expand knowledge in this area and they can use this review to position their research. This review may be of interest also to academics who are already studying the evaluation of quality in dental services or health-related scientific areas, or researchers who have been introduced to the field but they are interested in examining more specific insights into where current research topics in this review can be located, and how they may contribute to them.

The structure of the rest of the paper is as follows: Section 2 analyzes the methodology used for conducting the literature review. Then, the results of the analysis of the papers are provided in Section 3. Finally, the conclusion and suggestions for future research conclude the paper.

BACKGROUND

As it has already been stated, the aim of this paper is to examine the current state of quality assessment methods in primary dental care. Studies were identified using a three-phased literature review methodology, which was suggested by Webster and Watson (2002), and has been previously used in the field of strategic management and Information Systems (Kamariotou and Kitsios, 2017; Kitsios and Kamariotou, 2019a; b; c; 2016a; b; Kitsios et al., 2020; 2019). First, a search of the extant literature reviews was done to select the databases and keywords of the basic search. Then, the backward search was implemented to examine the references of the selected papers and finally, the forward search to examine the citations of the selected papers in order to increase their amount. After the selection of the papers, these were classified according to their content.

The existing literature review papers from 2013 to 2019 are presented to place the current literature review alongside the existing knowledge about the field of quality evaluation on the primary health care and to examine the previous knowledge of this area, as well as to discuss the identified research questions based on the findings of previous studies. Additionally, previous literature reviews give an overview of the literature review methodologies used by researchers and highlight their importance and gaps in their implementation. Table 1 presents a summary of the existing literature reviews on this field.

Simou et al., (2013) conducted a systematic literature review in order to study implemented quality projects used for evaluation of quality in primary health care services. They made comparisons between indicators and they identified problems and actions for alignment of quality projects on primary health care.

Next, Mills et al., (2014) conducted a systematic literature review in order to identify the features of patient-centred care relevant to dentistry and ascertain the current research evidence base underpinning its use as a measure of quality within general dental practice. Authors concluded that the current literature in this field is limited and further research is necessary to understand the important features of patient-centred care in dentistry and patients' views should be central to this research.

Later, Saleh et al., (2015) conducted a comprehensive synthesis of studies addressing quality of care in the primary healthcare sector of the Eastern Mediterranean Region. They analyzed 159 papers in order to address the gaps to generate contextualized information and knowledge that ensures relevance and targeted high-impact interventions.

Scambler et al., (2016) conducted a literature review in order to explore how patient-centred care is defined in the dental literature. They analyzed 28 papers and they concluded that the concept of patient-centred care is neither clearly understood nor empirically and systematically assessed in dental settings. Whilst most authors seem to suggest that patient-centred care is about delivering care that is humane, involving good communication and shared decision-making, there is no work assessing these concepts empirically or relating them to practical outcomes.

Table 1. Previous literature reviews

Authors	Year	Methodology	Results
Byrne et al.	2019	5 databases using keywords regarding to "quality", "performance", "satisfaction", "matrix", "measure", "guidelines", "index", "standards", "control", "improvement" and "assessment"	identification of measures used to assess quality in primary care dentistry and classification according to which dimensions of quality they attempt to measure
Righolt et al.	2019	3 databases using keywords regarding to "Quality of Health Care", "indicators", "assessment", "oral health", "measure", "performance", "dental care", "oral health care"	overview of existing quality measures in the field of oral health care and evaluation of the scientific soundness and applicability of these quality measures.
Samartzis and Talias	2019	3 databases using keywords regarding to "mental", "health", "quality" and "indicators"	8 dimensions of indicators in order to evaluate the quality in mental health services
Scambler et al.	2016	4 databases using keywords regarding to "patient-centred care", "empower", "shared decision-making", "dentistry", "dental" and "oral health"	definitions of patient-centred care in the dental literature
Saleh et al.	2015	3 databases using keywords regarding to "quality", "quality of healthcare", "quality of care", "total quality management", "clinical performance assessment", "clinical competence", "guideline adherence", "performance measurement", "outcome assessment/ measurement", "process assessment/ measurement", "quality assurance", "quality improvement" and "quality indicators"	a comprehensive synthesis of studies addressing quality of care in the primary healthcare sector of the Eastern Mediterranean Region
Mills et al.	2014	11 databases using keywords relating to "patient-centred care", "person centred care", "person focussed care", "oral", "dental" and "dentistry"	identification of the features of patient-centred care relevant to dentistry and its use as a measure of quality within general dental practice
Simou et al.	2013	2 databases using keywords regarding to "quality indicators", "primary healthcare", "assess" and "evaluate"	evaluation of quality in primary health care services

Byrne et al., (2019) conducted a systematic literature review in order to identify measures used to assess quality in primary care dentistry and categorize them according to which dimensions of quality they attempt to measure. Authors concluded that there has been a proliferation in the development of dental quality measures in recent years. However, this development has not been guided by a clear understanding of the meaning of quality. In this view, Righolt et al., (2019) provided an overview of existing quality measures in the field of oral health care and evaluated the scientific soundness and applicability of these quality measures. The results of their review provided potential opportunities including the piloting and testing of quality measures and the establishment of suitable information systems that allow the provision of transparent routine feedback on the quality of oral health care.

Finally, Samartzis and Talias (2019) conducted a narrative review study in order to describe the methodology for qualitative and quantitative evaluation and improvement of the mental health service system. They provided criteria, indicators, and methodology for evaluating and improving the quality of mental health services and the related qualitative and quantitative indicators. These indicators are classified to eight dimensions of quality assessment: suitability of services, accessibility of patients to services, acceptance of services by patients, ability of healthcare professionals to provide services, efficiency of health professionals and providers, continuity of service over time (ensuring therapeutic continuity), efficiency of health professionals and services and safety.

The search was done in Scopus and Science Direct using the following keywords "quality indicators", "assessment OR evaluation", and "dentistry OR dental OR primary healthcare". Papers published in academic journals, book chapters and proceedings of conferences. Furthermore, there were no limitations concerning the publication years. The searched papers belonged to fields of business management and healthcare. Finally, the published papers were only in English.

The total number of articles which were gathered using keywords in all databases was 622. Using exclusion criteria in databases, papers were reduced to 400. Then, examining their titles, 120 articles were found relevant with the purpose of this paper. Next, scanning their abstract, 96 were accepted. A number of 10 studies were rejected because their full text was not accessible. Consequently, 86 articles were selected according to their full text. Duplicate articles were removed and 46 articles were addressed (Figure 1).

Figure 1. Article selection process I

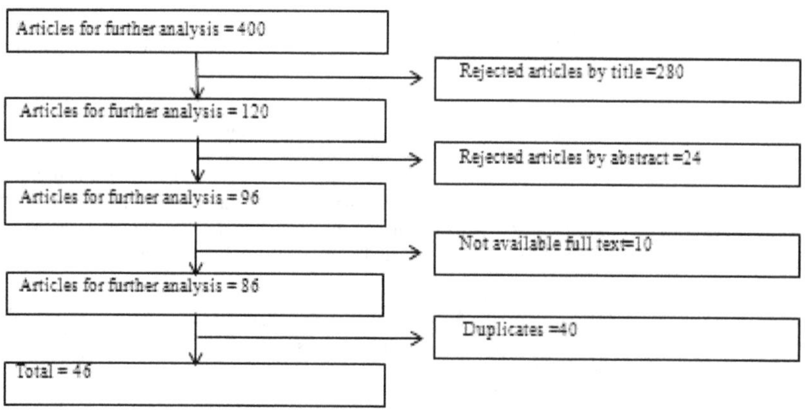

In the current amount of papers, more 4 papers were added from the backward search and forward search and thus revealed a total of 50 articles (Figure 2). Search was finished when it came to common articles from all databases and using different combinations of keywords. It was then that it was decided that the critical mass of related publications was collected (Webster and Watson, 2002).

Figure 2. Article selection process II

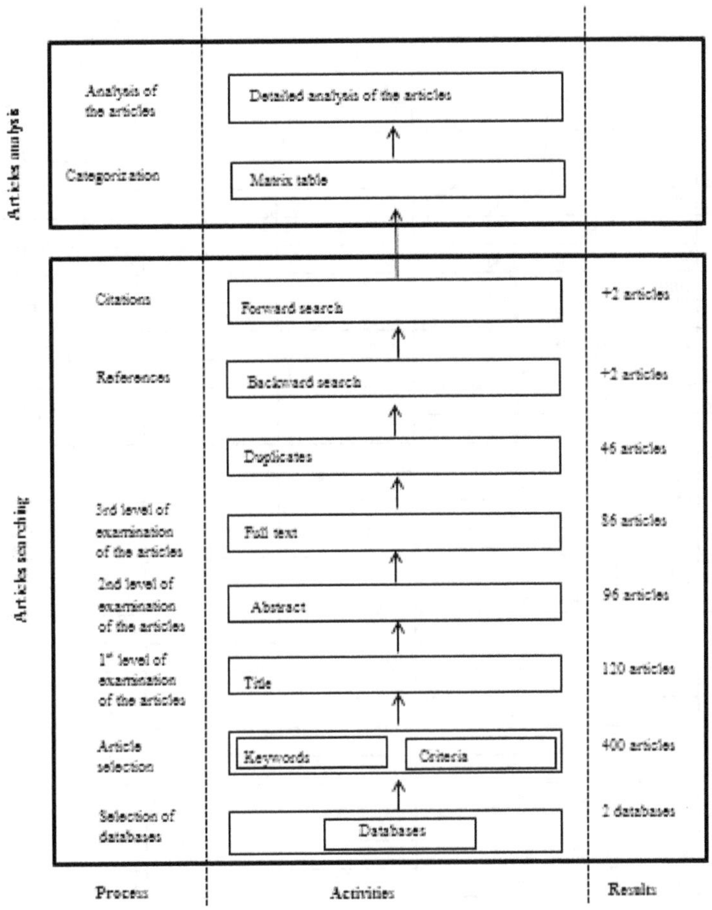

MAIN FOCUS OF THE CHAPTER

Fifty papers were analyzed based on a classification framework. These articles were analyzed on 5 broad dimensions which will provide a better understanding of the integration between quality improvement in dental care and patients' satisfaction based on their perceptions, and will also help future researchers to expand the knowledge in this field. Papers were classified based on publication year, journals and publishers, authors, keywords used, research method adopted, quality indicators, dimensions of quality and methods of quality improvement. Table 2 presents the classification of papers based on their concept.

Based on the classification of papers that was presented in Table 2, Figure 3 shows the percentage of articles per concept. The majority of papers (38.00%) refer to quality indicators. 36.00% of papers are related to factors that affect quality improvement. 28% of papers refer to service use and patients' perceptions.

Figure 4 shows that 66.00% of the papers were classified to quantitative surveys and 12% of studies were qualitative surveys whereas 20% of studies were theoretical. 2% of papers used observations.

VOSviewer was used to indicate the most frequently used keywords of the 50 papers and the relationships among them. The heat map which is presented in Figure 5 shows the relationship among keywords.

Table 2. Concept-matrix table

				Concepts				
No.	Authors	Year	Method	Methods of quality improvement	Factors for quality improvement	Quality indicators	Patient satisfaction	Service use and patients' perceptions
1.	Torppa-Saarinen et al.	2019	Quantitative Survey					x
2.	Obadan-Udoh et al.	2019	Theoretical			x		
3.	Top et al.	2019	Quantitative Survey		x	x		x
4.	Kahar et al.	2019	Quantitative Survey					x
5.	Byrne & Tickle	2019	Theoretical		x	x		
6.	Bukhari	2019	Quantitative Survey			x		
7.	Pérez-Rivas et al.	2019	Qualitative survey			x		
8.	Kovács et al.	2019	Quantitative Survey			x		
9.	Millá-Perseguer et al.	2019	Quantitative Survey		x			
10.	Yeon et al.	2019	Quantitative Survey				x	x
11.	Renggli et al.	2019	Quantitative Survey	x				
12.	Niaz et al.	2019	Quantitative Survey			x		
13.	Kumar & Chacko	2018	Quantitative Survey	x				
14.	Torppa-Saarinen et al.	2018	Quantitative Survey					x
15.	Quercioli et al.	2018	Quantitative Survey					x
16.	Limpuangthip et al.	2018	Quantitative Survey		x		x	x
17.	Contreras-Ramírez et al.	2018	Quantitative Survey		x		x	
18.	Heale et al.	2018	Quantitative Survey	x				
19.	Pavlič et al.	2018	Quantitative Survey			x		
20.	Mubeen et al.	2017	Quantitative Survey		x			
21.	Neumann et al.	2017	Quantitative Survey		x			
22.	Neves et al.	2017	Quantitative Survey			x		
23.	Glurich et al.	2017	Theoretical	x				
24.	Masood et al.	2017	Quantitative Survey					x
25.	Mathe et al.	2015	Quantitative survey					x
26.	Harris et al.	2015	Qualitative survey	x				
27.	Puszka et al.	2015	Qualitative survey					x
28.	Basto-Pereira et al.	2015	Qualitative survey			x		
29.	Waraich et al.	2015	Quantitative Survey			x		
30.	ni Riain et al.	2015	Qualitative survey			x		
31.	Orueta et al.	2015	Quantitative Survey			x		
32.	De Clercq	2014	Quantitative Survey			x		
33.	Campbell & Tickle	2013	Theoretical	x	x			
34.	Alhassan et al.	2013	Quantitative Survey		x			
35.	Heard et al.	2013	Quantitative Survey		x			x
36.	Tickle & Campbell	2013	Theoretical	x				
37.	Mills et al.	2013	Theoretical	x				
38.	John et al.	2011	Quantitative Survey		x		x	x
39.	Zhao et al.	2011	Quantitative Survey		x			
40.	Wong et al.	2011	Qualitative survey			x		
41.	Harris et al.	2011	Quantitative Survey			x		
42.	Mills & Batchelor	2011	Theoretical		x	x		
43.	Bader	2009	Theoretical	x				
44.	Marino et al.	2008	Quantitative Survey					x
45.	Yamalik	2007	Theoretical			x		
46.	Fenlon & Sherriff	2004	Quantitative Survey		x		x	
47.	Lionis et al.	2004	Quantitative Survey	x	x		x	
48.	Karydis et al.	2001	Quantitative Survey		x			x
49.	Marcus & Spolsky	1998	Quantitative Survey		x	x		
50.	Harteloh & Verheggen	1994	Theoretical		x			

Figure 3. Percentage of articles per concept

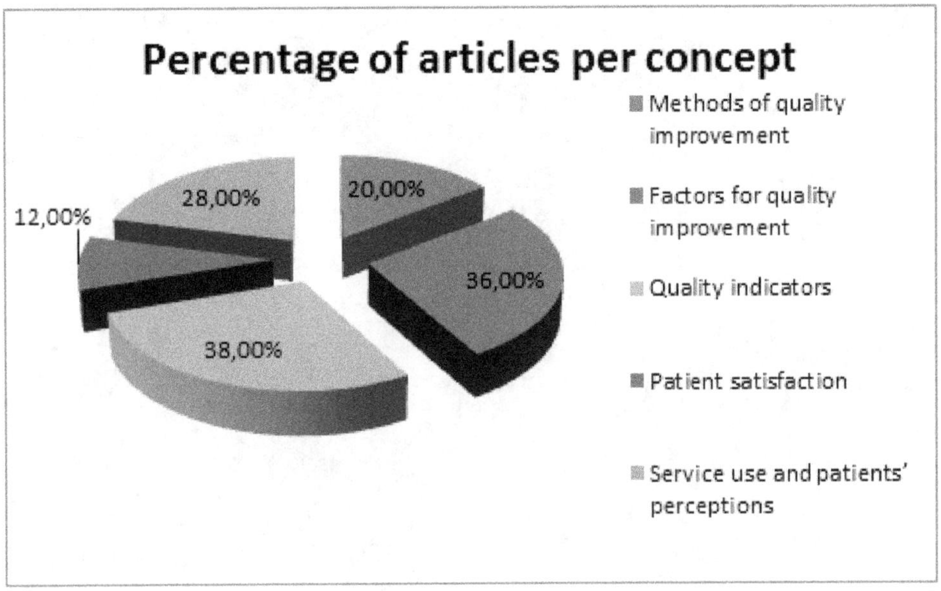

Figure 4. Percentage of articles per method

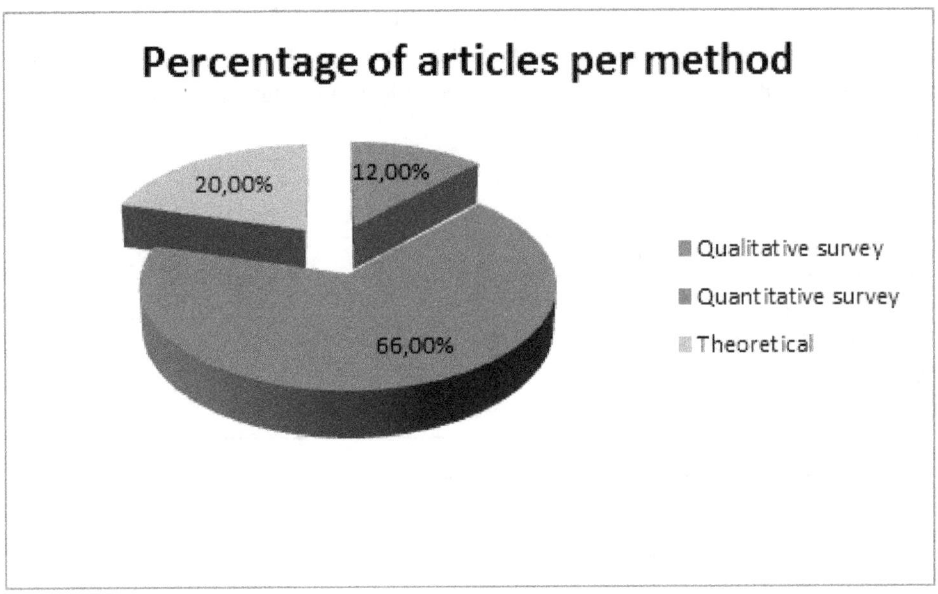

The most frequently keywords are "quality", "life" and "primary healthcare". This figure shows that there are 3 clusters among the keywords. The first cluster refers to the measurement of quality. The second cluster combines the term of quality with the assessment, the effects in oral health, the association between health quality and quality of life and patients' expectations from health services. The third cluster includes the keywords 'dental care" and "primary healthcare".

Figure 5. Frequency of keywords

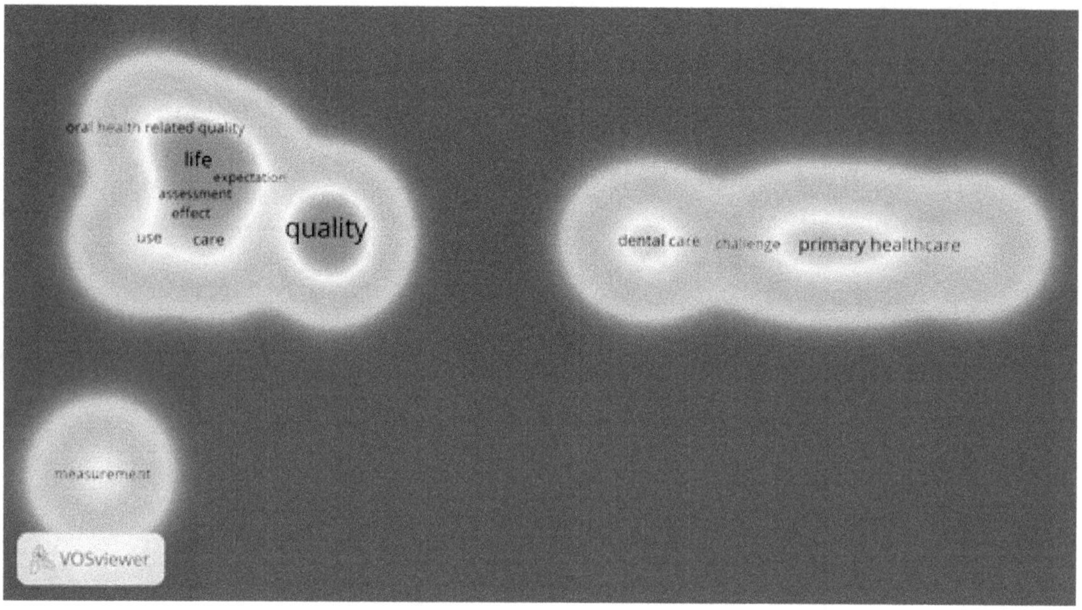

Table 3 presents the journals that have published the majority of papers. The British dental journal has published 7 papers, The journal of Community dentistry and oral epidemiology and the Journal of dentistry have published 4 papers each one. Then, the International journal for quality in health care and the Journal of the American Dental Association have published 3 papers. Finally, the Health Policy journal and the Journal of Oral Biology and Craniofacial Research have published 2 papers.

Table 3. Articles per journal

Journal	Number of papers
British dental journal	7
Community dentistry and oral epidemiology	4
Journal of dentistry	4
International journal for quality in health care	3
The Journal of the American Dental Association	3
Health Policy	2
Journal of Oral Biology and Craniofacial Research	2

Although researchers in the area of quality management in dental care conducted studies two decades ago, the majority of the papers have only been published in the last three years. Figure 6 presents the number of papers published each year. Especially, in the early 2000s, the awareness of quality management into primary healthcare was found to be very low. The strong practice of quality evaluation came into existence around 2011, when researchers realized the significance of the integration between quality and dental care and started examined indicators that affect the assessment of quality. Such a finding highlights both the importance of the field and its continuous development. Figure 7 presents a clear increasing direction the last three years.

Figure 6. Number of articles per year

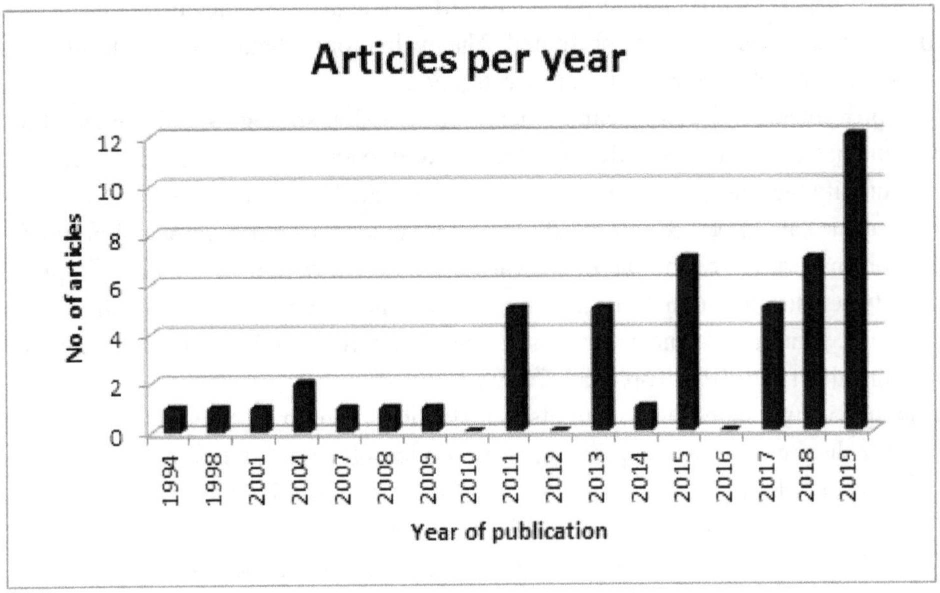

Figure 7. Distribution of articles per year

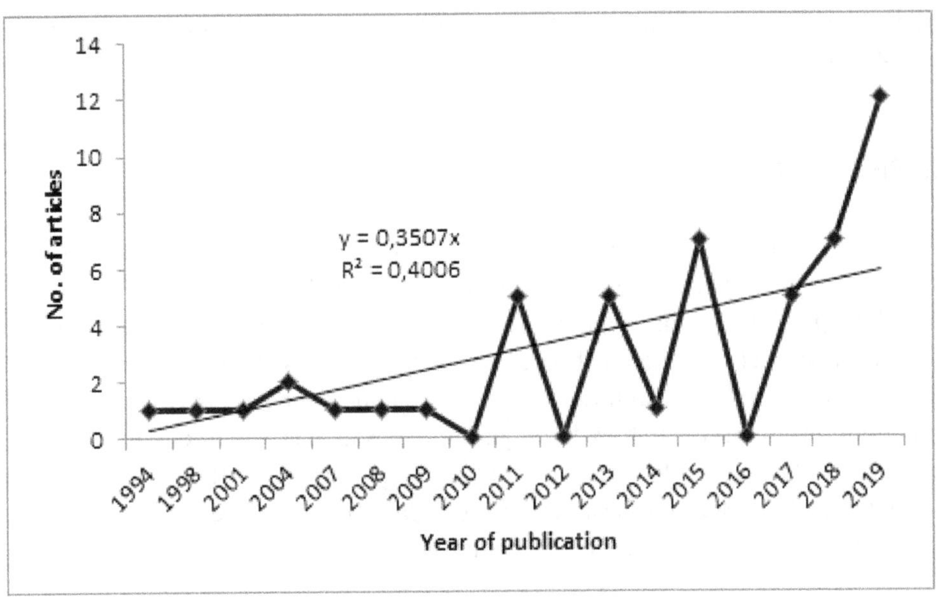

SOLUTIONS AND RECOMMENDATIONS

Previous researchers have examined the quality assessment methods in primary dental care and the limitations of these methods. Furthermore, they evaluated the quality indicators in primary healthcare and they discussed the specific aspects of dental services and the dimensions of quality in dentistry focusing on primary dental care (Byrne et al., 2019; Mills et al., 2014; Righolt et al., 2019; Saleh et al., 2015; Samartzis and Talias, 2019; Scambler et al., 2016; Simou et al., 2013; Syngelakis et al., 2013; Syngelakis et al., 2012). As this chapter presents a different analysis of the current literature, some recommendations for future research are presented. The evaluation of health quality is an immature field, so it could inspire researchers for empirical investigation.

Quality improvement in dentistry can be achieved by using strategies and policies for evaluation. Stakeholders and policy-makers should be involved in order to identify goals for quality improvement and indicators for quality measurement (Campbell and Tickle, 2013; Harteloh, 2003). Results from surveys in different countries are expected to provide important evidence-based information to health managers and decision-makers in health planning regarding to health policies for quality improvement. The comparison between different indicators in dental quality and policies will help health policy-makers to formulate strategic plans and define objectives in order to improve quality in primary health care (Neves et al., 2017; ni Riain et al., 2015; Top et al., 2019).

Dental quality is associated with patients' satisfaction (Fenlon & Sherriff, 2004; Karydis et al., 2001). Access to health services, improvement of service planning and perceived use of service are important actions in order to increase patients' satisfaction and oral-health quality (Torppa-Saarinen et al., 2019). As customer satisfaction is being recognized to play an important role in the success of any service provided in today's competitive environment, researchers can examine the factors that affect the assessment of quality in primary healthcare. Furthermore, dental care providers need to plan and direct resources effectively in order to increase patients' satisfaction based on patients' expectations about responsiveness, reliability, empathy and effective treatment (John et al., 2011; Karydis et al., 2001). What dimensions of quality affect patients' satisfaction? What do patients think about the perceived use of health services and how can these perceptions affect dental quality? (Top et al., 2019). Dentists require an appropriate set of indicators to demonstrate that a service is continually improving (Byrne and Tickle, 2019; Voinea-Griffin et al., 2010a; b).

FUTURE RESEARCH DIRECTIONS

Future researchers could explore the differences among the indicators in dental care and conduct quantitative surveys in order to collect data and check the reliability of the theory developed, discussing the challenges and the problems of this integration (e.g. feasibility of implementing quality dental indicators, dentists' reaction and cost). From a theoretical perspective, this chapter is a bibliometric study that provides a macro picture of a research field, its evolution and connections among studies, in order to be a starting point for future research by highlighting issues among health quality evaluation research domains, while providing a complete overview of the literature from a different perspective, not implemented in existing analyses. Future researchers could expand this literature review and provide different bibliometric analyses such as co-author or co-citation.

CONCLUSION

The purpose of this chapter was to provide a complete overview of the literature from a different perspective, not implemented in existing analyses and, thus, it is a good starting point for further research. The chapter was based upon an analysis of 50 papers, derived from databases and categorized according to the main issues of this emerging research area. This chapter has some limitations that are described below. First, criteria for searching articles in databases include specific keywords in the title and abstract of the papers. There probably exist papers which lack these keywords in the title but still focus on the field of quality evaluation in dental care. Another limitation is that only English papers were searched which may skip other publications in other languages. Thus, the use of different journals or papers from other sources could possibly lead to different findings regarding the most cited papers, or the most active researchers or institutions, and the percentage of papers per publisher.

FUNDING

This research received no specific grant from any funding agency in the public, commercial, or not-for-profit sectors.

REFERENCES

Alhassan, R. K., Spieker, N., van Ostenberg, P., Ogink, A., Nketiah-Amponsah, E., & de Wit, T. F. R. (2013). Association between health worker motivation and healthcare quality efforts in Ghana. *Human Resources for Health*, *11*(1), 1–11. doi:10.1186/1478-4491-11-37 PMID:23945073

Bader, J. D. (2009). Challenges in quality assessment of dental care. *The Journal of the American Dental Association*, *140*(12), 1456–1464. doi:10.14219/jada.archive.2009.0084 PMID:19955057

Basto-Pereira, M., Furtado, S. I. F., Silva, R. J. P., Fachado González, F., Vara Fernandes, T. M., Correia de Sousa, J., & Yaphe, J. (2015). Performance indicators for clinical practice management in primary care in Portugal: Consensus from a Delphi study. *The European Journal of General Practice*, *21*(1), 52–57. doi:10.3109/13814788.2014.907268 PMID:24845034

Bukhari, O. M. (2019). (in press). Dental Caries Experience and Oral Health Related Quality of Life in Working Adults. *The Saudi Dental Journal*. Advance online publication. doi:10.1016/j.sdentj.2019.11.003

Byrne, M., & Tickle, M. (2019). Conceptualising a framework for improving quality in primary dental care. *British Dental Journal*, *227*(10), 865–868. doi:10.103841415-019-0900-8 PMID:31758120

Byrne, M. J., Tickle, M., Glenny, A. M., Campbell, S., Goodwin, T., & O'Malley, L. (2019). A systematic review of quality measures used in primary care dentistry. *International Dental Journal*, *69*(4), 252–264. doi:10.1111/idj.12453 PMID:30430563

Campbell, S., & Tickle, M. (2013). How do we improve quality in primary dental care? *British Dental Journal*, *215*(5), 239–243. doi:10.1038j.bdj.2013.831 PMID:24029991

Campbell, S., & Tickle, M. (2013). What is quality primary dental care? *British Dental Journal, 215*(3), 135–139. doi:10.1038j.bdj.2013.740 PMID:23928610

Campbell, S. M., Roland, M. O., & Buetow, S. A. (2000). Defining quality of care. *Social Science & Medicine, 51*(11), 1611–1625. doi:10.1016/S0277-9536(00)00057-5 PMID:11072882

Chu, C. H., Yeung, C. Y. Y. J., & Lo, E. C. M. (2001). Monitoring patient satisfaction with university dental services under two fee-paying systems. *Community Dentistry and Oral Epidemiology, 29*(5), 390–398. doi:10.1034/j.1600-0528.2001.290509.x PMID:11553112

Contreras-Ramírez, R., Davis-Toledo, G., & Nuñez-Franz, L. (2018). Patient satisfaction and quality of emergency dental care in Chilean public health services. *Journal of Oral Research, 7*(6), 236–243. doi:10.17126/joralres.2018.057

De Clercq, E., Van Casteren, V., Bossuyt, N., Goderis, G., & Moreels, S. (2014). EPR-based, quality-related process parameters: A nationwide assessment. *Studies in Health Technology and Informatics*, 121–125. PMID:25160158

Delias, P., Doumpos, M., Grigoroudis, E., Manolitzas, P., & Matsatsinis, N. (2015). Supporting health-care management decisions via robust clustering of event logs. *Knowledge-Based Systems, 84*, 203–213. doi:10.1016/j.knosys.2015.04.012

Dental Quality Alliance. (2019). *Quality Measurement in Dentistry. A Guidebook.* Retrieved from https://www.ada.org/~/media/ADA/DQA/2019_Guidebook.pdf?la=en

Elani, H. W., Simon, L., Ticku, S., Bain, P. A., Barrow, J., & Riedy, C. A. (2018). Does providing dental services reduce overall health care costs?: A systematic review of the literature. *The Journal of the American Dental Association, 149*(8), 696–703. doi:10.1016/j.adaj.2018.03.023 PMID:29866364

Evans, D. B., Edejer, T. T. T., Lauer, J., Frenk, J., & Murray, C. J. (2001). Measuring quality: From the system to the provider. *International Journal for Quality in Health Care, 13*(6), 439–446. doi:10.1093/intqhc/13.6.439 PMID:11769745

Fenlon, M. R., & Sherriff, M. (2004). Investigation of new complete denture quality and patients' satisfaction with and use of dentures after two years. *Journal of Dentistry, 32*(4), 327–333. doi:10.1016/j.jdent.2004.01.007 PMID:15053917

Flores-Mir, C., Palmer, N. G., Northcott, H. C., Khurshed, F., & Major, P. W. (2006). Perceptions and attitudes of Canadian dentists toward digital and electronic technologies. *Journal of the Canadian Dental Association, 72*(3), 243.

Glurich, I., Nycz, G., & Acharya, A. (2017). Status update on translation of integrated primary dental-medical care delivery for management of diabetic patients. *Clinical Medicine & Research, 15*(1-2), 21–32. doi:10.3121/cmr.2017.1348 PMID:28373288

González, G. Z., Klazinga, N., ten Asbroek, G., & Delnoij, D. M. (2006). Performance indicators used to assess the quality of primary dental care. *Community Dental Health, 23*(4), 228–235. PMID:17194070

Harris, R., Bridgman, C., Ahmad, M., Bowes, L., Haley, R., Saleem, S., Singh, R., & Taylor, S. (2011). Introducing care pathway commissioning to primary dental care: Measuring performance. *British Dental Journal, 211*(11), 22–29. doi:10.1038j.bdj.2011.1005 PMID:22158196

Harris, S. B., Green, M. E., Brown, J. B., Roberts, S., Russell, G., Fournie, M., Webster-Bogaert, S., Paquette-Warren, J., Kotecha, J., Han, H., Thind, A., Stewart, M., Reichert, S., Tompkins, J. W., & Birtwhistle, R. (2015). Impact of a quality improvement program on primary healthcare in Canada: A mixed-method evaluation. *Health Policy (Amsterdam), 119*(4), 405–416. doi:10.1016/j.healthpol.2014.10.019 PMID:25476553

Harteloh, P. P. (2003). Quality systems in health care: A sociotechnical approach. *Health Policy (Amsterdam), 64*(3), 391–398. doi:10.1016/S0168-8510(02)00183-5 PMID:12745175

Harteloh, P. P., & Verheggen, F. W. (1994). Quality assurance in health care. From a traditional towards a modern approach. *Health Policy (Amsterdam), 27*(3), 261–270. doi:10.1016/0168-8510(94)90120-1 PMID:10134583

Heale, R., Wenghofer, E., James, S., & Garceau, M. L. (2018). Quality of care for patients with diabetes and mulitmorbidity registered at nurse practitioner-led clinics. *Canadian Journal of Nursing Research, 50*(1), 20–27. doi:10.1177/0844562117744137 PMID:29216748

Heard, A., Nath, D. K., & Loevinsohn, B. (2013). Contracting urban primary healthcare services in B angladesh–effect on use, efficiency, equity and quality of care. *Tropical Medicine & International Health, 18*(7), 861–870. doi:10.1111/tmi.12113 PMID:23611146

John, J., Yatim, F. M., & Mani, S. A. (2011). Measuring service quality of public dental health care facilities in Kelantan, Malaysia. *Asia-Pacific Journal of Public Health, 23*(5), 742–753. doi:10.1177/1010539509357341 PMID:20460281

Kahar, P., Chapman, C., & Gupta, J. (2019). Assessment of the Effect of Oral Health on Quality of Life and Oral-Health Indicators among ESRD Patients in Southwest Florida: A Pilot Study. *International Journal of Dentistry, 2019*, 1–10. doi:10.1155/2019/1608329 PMID:31662758

Kamariotou, M., & Kitsios, F. (2017). Information Systems Phases and Firm Performance: A conceptual Framework. In A. Kavoura, D. Sakas, & P. Tomaras (Eds.), *Strategic Innovative Marketing* (pp. 553–560). Springer. doi:10.1007/978-3-319-33865-1_67

Karydis, A., Komboli-Kodovazeniti, M., Hatzigeorgiou, D., & Panis, V. (2001). Expectations and perceptions of Greek patients regarding the quality of dental health care. *International Journal for Quality in Health Care, 13*(5), 409–416. doi:10.1093/intqhc/13.5.409 PMID:11669569

Kitsios, F., & Kamariotou, M. (2016a). Decision Support Systems and Business Strategy: A conceptual framework for Strategic Information Systems Planning. In *Proceedings of the 6th IEEE International Conference on IT Convergence and Security (ICITCS2016)* (pp. 149-153). Prague, Czech Republic: IEEE 10.1109/ICITCS.2016.7740323

Kitsios, F., & Kamariotou, M. (2016b). Critical success factors in service innovation strategies: An annotated bibliography on NSD. In *Proceedings of British Academy of Management (BAM) Conference 2016* (pp. 1-28). Academic Press.

Kitsios, F., & Kamariotou, M. (2019a). Business Strategy Modelling based on Enterprise Architecture: A State of the Art Review. *Business Process Management Journal, 25*(4), 606–624. doi:10.1108/BPMJ-05-2017-0122

Kitsios, F., & Kamariotou, M. (2019b). Mapping New Service Development: A Review and Synthesis of Literature. *Service Industries Journal.*

Kitsios, F., & Kamariotou, M. (2019c). (in press). Service innovation process digitization: Areas for exploitation and exploration. *Journal of Hospitality and Tourism Technology*. Advance online publication. doi:10.1108/JHTT-02-2019-0041

Kitsios, F., Kamariotou, M., & Manthou, V. (2019a). Hospital Information Systems Planning: Strategic IT Alignment in Healthcare. In D. P. Sakas & D. K. Nasiopoulos (Eds.), *Strategic Innovative Marketing* (pp. 203–209). Springer. doi:10.1007/978-3-030-16099-9_25

Kitsios, F., Kamariotou, M., & Talias, M. (2020). Corporate Sustainability Strategies and Decision Support Methods: A Bibliometric Analysis. *Sustainability, 12*(521), 1–21.

Kitsios, F., Stefanakakis, S., Kamariotou, M., & Dermentzoglou, L. (2019b). E-service Evaluation: User Satisfaction Measurement and Implications in Health Sector. *Computer Standards & Interfaces Journal, 63*, 16–26. doi:10.1016/j.csi.2018.11.002

Kovács, N., Pálinkás, A., Sipos, V., Nagy, A., Harsha, N., Kőrösi, L., Papp, M., Ádány, R., Varga, O., & Sándor, J. (2019). Factors Associated with Practice-Level Performance Indicators in Primary Health Care in Hungary: A Nationwide Cross-Sectional Study. *International Journal of Environmental Research and Public Health, 16*(17), 1–15. doi:10.3390/ijerph16173153 PMID:31470573

Kumar, S., & Chacko, R. K. (2018). Assessing perceptions of oral health related quality of life in dental implant patients. Experience of a tertiary care center in India. *Journal of Oral Biology and Craniofacial Research, 8*(2), 74-77.

Lavelle, C., Schroth, R., & Wiltshire, W. A. (2004). Performance measures to improve the quality of orthodontic services and control expenditures. *American Journal of Orthodontics and Dentofacial Orthopedics, 126*(4), 445–449. doi:10.1016/j.ajodo.2003.10.030 PMID:15470347

Lee, D. W., & Shin, I. S. (2018). Critical quality evaluation of network meta-analyses in dental care. *Journal of Dentistry, 75*, 7–11. doi:10.1016/j.jdent.2018.05.010 PMID:29879456

Limpuangthip, N., Somkotra, T., & Arksornnukit, M. (2018). Modified retention and stability criteria for complete denture wearers: A risk assessment tool for impaired masticatory ability and oral health-related quality of life. *The Journal of Prosthetic Dentistry, 120*(1), 43–49. doi:10.1016/j.prosdent.2017.09.010 PMID:29195820

Lionis, C., Tsiraki, M., Bardis, V., & Philalithis, A. (2004). Seeking quality improvement in primary care in Crete, Greece: The first actions. *Croatian Medical Journal, 45*(5), 599–603. PMID:15495288

Listl, S. (2019). Value-based oral health care: Moving forward with dental patient-reported outcomes. *The Journal of Evidence-Based Dental Practice, 19*(3), 255–259. doi:10.1016/j.jebdp.2019.101344 PMID:31732101

Lohr, K. N., & Harris-Wehling, J. (1991). Medicare: A Strategy For Quality Assurance, I: A Recapitulation of the Study and a Definition of Quality of Care. *QRB. Quality Review Bulletin*, *17*(1), 6–9. doi:10.1016/S0097-5990(16)30413-4 PMID:1900934

Manolitzas, P., Kostagiolas, P., Grigoroudis, E., Intas, G., & Stergiannis, P. (2018). Data on patient's satisfaction from an emergency department: Developing strategies with the Multicriteria Satisfaction Analysis. *Data in Brief*, *21*, 956–961. doi:10.1016/j.dib.2018.10.041 PMID:30426051

Marcus, M., & Spolsky, V. (1998). Concepts of quality and the provision of periodontal care: A survey. *Journal of Periodontology*, *69*(2), 228–240. doi:10.1902/jop.1998.69.2.228 PMID:9526923

Marino, R., Schofield, M., Wright, C., Calache, H., & Minichiello, V. (2008). Self-reported and clinically determined oral health status predictors for quality of life in dentate older migrant adults. *Community Dentistry and Oral Epidemiology*, *36*(1), 85–94. PMID:18205644

Masood, M., Newton, T., Bakri, N. N., Khalid, T., & Masood, Y. (2017). The relationship between oral health and oral health related quality of life among elderly people in United Kingdom. *Journal of Dentistry*, *56*, 78–83. doi:10.1016/j.jdent.2016.11.002 PMID:27825838

Mathe, N., Johnson, S. T., Wozniak, L. A., Majumdar, S. R., & Johnson, J. A. (2015). Alternation as a form of allocation for quality improvement studies in primary healthcare settings: The on-off study design. *Trials*, *16*(1), 1–10. doi:10.118613063-015-0904-x PMID:26303892

Mehrotra, D. (2018). Quality Of Care In Dental Institutes. *Journal of Oral Biology and Craniofacial Research*, *8*(2), 73. doi:10.1016/j.jobcr.2018.05.007 PMID:29892523

Millá-Perseguer, M., Guadalajara-Olmeda, N., Vivas-Consuelo, D., & Usó-Talamantes, R. (2019). Measurement of health-related quality by multimorbidity groups in primary health care. *Health and Quality of Life Outcomes*, *17*(1), 1–10. doi:10.118612955-018-1063-z PMID:30634992

Mills, I., & Batchelor, P. (2011). Quality indicators: The rationale behind their use in NHS dentistry. *British Dental Journal*, *211*(1), 11–15. doi:10.1038j.bdj.2011.521 PMID:21738180

Mills, I., Frost, J., Cooper, C., Moles, D. R., & Kay, E. (2014). Patient-centred care in general dental practice-a systematic review of the literature. *BMC Oral Health*, *14*(1), 1–13. doi:10.1186/1472-6831-14-64 PMID:24902842

Mills, I., Frost, J., Moles, D. R., & Kay, E. (2013). Patient-centred care in general dental practice: Sound sense or soundbite? *British Dental Journal*, *215*(2), 81–85. doi:10.1038j.bdj.2013.684 PMID:23887534

Mubeen, S., Patel, K., Cunningham, Z., O'Rourke, N., Pandis, N., Cobourne, M. T., & Seehra, J. (2017). Assessing the quality of dental clinical practice guidelines. *Journal of Dentistry*, *67*, 102–106. doi:10.1016/j.jdent.2017.10.003 PMID:29031996

Neumann, A., Kalenderian, E., Ramoni, R., Yansane, A., Tokede, B., Etolue, J., Vaderhobli, R., Simmons, K., Even, J., Mullins, J., Kumar, S., Bangar, S., Kookal, K., White, J., & Walji, M. (2017). Evaluating quality of dental care among patients with diabetes: Adaptation and testing of a dental quality measure in electronic health records. *The Journal of the American Dental Association*, *148*(9), 634–643. doi:10.1016/j.adaj.2017.04.017 PMID:28624074

Neves, M., do Amaral Giordani, J. M., Ferla, A. A., & Hugo, F. N. (2017). Primary care dentistry in Brazil: From prevention to comprehensive care. *The Journal of Ambulatory Care Management, 40*(2), 35–48. doi:10.1097/JAC.0000000000000186 PMID:28252501

ni Riain, A., Vahey, C., Kennedy, C., Campbell, S., & Collins, C. (2015). Roadmap for developing a national quality indicator set for general practice. *International Journal of Health Care Quality Assurance, 28*(4), 382-393.

Niaz, Q., Godman, B., Massele, A., Campbell, S., Kurdi, A., Kagoya, H. R., & Kibuule, D. (2019). Validity of World Health Organisation prescribing indicators in Namibia's primary healthcare: Findings and implications. *International Journal for Quality in Health Care, 31*(5), 338–345. doi:10.1093/intqhc/mzy172 PMID:30169688

Obadan-Udoh, E. M., Calvo, J. M., Panwar, S., Simmons, K., White, J. M., Walji, M. F., & Kalenderian, E. (2019). Unintended consequences and challenges of quality measurements in dentistry. *BMC Oral Health, 19*(1), 1–5. doi:10.118612903-019-0726-4 PMID:30823894

Orueta, J. F., García-Alvarez, A., Grandes, G., & Nuño-Solinís, R. (2015). The origin of variation in primary care process and outcome indicators: Patients, professionals, centers, and health districts. *Medicine, 94*(31), 1–10. doi:10.1097/MD.0000000000001314 PMID:26252315

Pavlič, D. R., Sever, M., Klemenc-Ketiš, Z., Švab, I., Vainieri, M., Seghieri, C., & Maksuti, A. (2018). Strength of primary care service delivery: A comparative study of European countries, Australia, New Zealand, and Canada. *Primary Health Care Research and Development, 19*(3), 277–287. doi:10.1017/S1463423617000792 PMID:29307317

Pérez-Rivas, F. J., Martín-García, Á., Sánz-Bayona, M. T., Fernández-Díaz, M. C., Barberá-Martín, A., Cárdenas-Valladolid, J., López-Palacios, S., & Rico-Blázquez, M., & Nurses at the Quality of Care Commission of the Community of Madrid, Spain. (2019). Establishing Technical Values for Nursing Diagnoses in Primary Healthcare. *International Journal of Nursing Knowledge*. PMID:31379131

Puszka, S., Nagel, T., Matthews, V., Mosca, D., Piovesan, R., Nori, A., & Bailie, R. (2015). Monitoring and assessing the quality of care for youth: Developing an audit tool using an expert consensus approach. *International Journal of Mental Health Systems, 9*(1), 1–12. doi:10.118613033-015-0019-5 PMID:26170899

Quercioli, C., Nisticò, F., Troiano, G., Maccari, M., Messina, G., Barducci, M., Carriero, G., Golinelli, D., & Nante, N. (2018). Developing a new predictor of health expenditure: Preliminary results from a primary healthcare setting. *Public Health, 163*, 121–127. doi:10.1016/j.puhe.2018.07.007 PMID:30142482

Renggli, S., Mayumana, I., Mboya, D., Charles, C., Mshana, C., Kessy, F., Glass, T. R., Pfeiffer, C., Schulze, A., Aerts, A., & Lengeler, C. (2019). Towards improved health service quality in Tanzania: Appropriateness of an electronic tool to assess quality of primary healthcare. *BMC Health Services Research, 19*(1), 1–16. PMID:30670011

Righolt, A. J., Sidorenkov, G., Faggion, C. M. Jr, Listl, S., & Duijster, D. (2019). Quality measures for dental care: A systematic review. *Community Dentistry and Oral Epidemiology, 47*(1), 12–23. doi:10.1111/cdoe.12429 PMID:30375669

Saleh, S., Alameddine, M., Mourad, Y., & Natafgi, N. (2015). Quality of care in primary health care settings in the Eastern Mediterranean region: A systematic review of the literature. *International Journal for Quality in Health Care*, *27*(2), 79–88. PMID:25574040

Samartzis, L., & Talias, M. A. (2020). Assessing and improving the quality in mental health services. *International Journal of Environmental Research and Public Health*, *17*(1), 1–31. PMID:31905840

Scambler, S., Delgado, M., & Asimakopoulou, K. (2016). Defining patient-centred care in dentistry? A systematic review of the dental literature. *British Dental Journal*, *221*(8), 477–484. doi:10.1038j.bdj.2016.777 PMID:27767160

Shimpi, N., Schwei, K., Cooper, S., Chyou, P. H., & Acharya, A. (2018). Understanding patients' oral health information needs: Findings of a survey on use of patient portals in dentistry. *The Journal of the American Dental Association*, *149*(3), 184–190. doi:10.1016/j.adaj.2017.09.051 PMID:29395012

Simou, E., Pliatsika, P., Koutsogeorgou, E., & Roumeliotou, A. (2015). Quality indicators for primary health care: A systematic literature review. *Journal of Public Health Management and Practice*, *21*(5), 8–16. doi:10.1097/PHH.0000000000000037 PMID:24374361

Sindakis, S., & Kitsios, F. (2016). Entrepreneurial Dynamics and Patient Involvement in Service Innovation: Developing a Model to Promote Growth and Sustainability in Mental Health Care. *Journal of the Knowledge Economy*, *7*(2), 545–564. doi:10.100713132-014-0228-1

Stefanakakis, S., Trevlakis, S., & Kitsios, F. (2017). Analyzing quality characteristics: The case of an e – health application. In *Proceedings of the 6th International Symposium and 28th National Conference on Operational Research* (pp. 246-251). Thessaloniki, Greece: Academic Press.

Syngelakis, A., Kitsios, F., & Kamariotou, M. (2019). Work-related Stress in Public Healthcare: A Case Study in Greece during economic crisis. In D. P. Sakas & D. K. Nasiopoulos (Eds.), *Strategic Innovative Marketing* (pp. 211–216). Springer. doi:10.1007/978-3-030-16099-9_26

Syngelakis, A., Koletsi-Kounari, H., Polychronopoulou, A., Panagiotakos, D. B., & Lionis, C. (2012). Evaluation of the staffing and the organization of the Greek NHS Primary Health Care Centres Dental Departments. *Hellenic Hospital Dentistry*, *5*, 11–20.

Syngelakis, A. I., & Kitsios, F. (2017). Evaluating occupational stress and job satisfaction in health sector during the austerity: the case of Greek NHS dentists. In *Proceedings of the 6th International Symposium and 28th National Conference on Operational Research* (pp. 252-256). Thessaloniki, Greece: Academic Press.

Syngelakis, A. I., Koletsi-Kounari, H., Polychronopoulou, A., Panagiotakos, D. B., & Lionis, C. (2013). Dentistry in Primary Health Care Centres: Attendance and access of users, professional satisfaction and security of dentists. *Stomatologia*, *70*(3), 113–123.

Tickle, M., & Campbell, S. (2013). How do we measure quality in primary dental care? *British Dental Journal*, *215*(4), 183–187. doi:10.1038j.bdj.2013.789 PMID:23969661

Tickle, M., O'malley, L., Brocklehurst, P., Glenny, A. M., Walsh, T., & Campbell, S. (2015). A national survey of the public's views on quality in dental care. *British Dental Journal, 219*(3), 1–7. doi:10.1038j.bdj.2015.595 PMID:26271885

Tinanoff, N. (2012). Potential to improve oral health care through evidence, protocols, and payment models. *Journal of Public Health Dentistry, 72*, S48–S51. doi:10.1111/j.1752-7325.2012.00325.x PMID:22433101

Top, M., Sönmez, S., & Aslan, H. (2019). An Evidence-Based Approach to Outcome Measurement in Oral and Dental Health Services: Oral Health-Related Quality of Life and Oral Health Impact. *Worldviews on Evidence-Based Nursing, 16*(5), 408–415. doi:10.1111/wvn.12398 PMID:31544360

Torppa-Saarinen, E., Suominen, A. L., Lahti, S., & Tolvanen, M. (2019). Longitudinal pathways between perceived oral health and regular service use of adult Finns. *Community Dentistry and Oral Epidemiology, 47*(5), 374–380. doi:10.1111/cdoe.12478 PMID:31179572

Torppa-Saarinen, E., Tolvanen, M., Suominen, A. L., & Lahti, S. (2018). Changes in perceived oral health in a longitudinal population-based study. *Community Dentistry and Oral Epidemiology, 46*(6), 569–575. doi:10.1111/cdoe.12393 PMID:29956841

Voinea-Griffin, A., Fellows, J. L., Rindal, D. B., Barasch, A., Gilbert, G. H., & Safford, M. M. (2010a). Pay for performance: Will dentistry follow? *BMC Oral Health, 10*(1), 1–8. doi:10.1186/1472-6831-10-9 PMID:20423526

Voinea-Griffin, A., Rindal, D. B., Fellows, J. L., Barasch, A., Gilbert, G. H., & Safford, M. M.DPBRN Collaborative Group. (2010b). Pay-for-performance in dentistry: What we know. *Journal for Healthcare Quality, 32*(1), 51–58. doi:10.1111/j.1945-1474.2009.00064.x PMID:20151592

Wanyonyi, K. L., Radford, D. R., & Gallagher, J. E. (2019). Electronic primary dental care records in research: A case study of validation and quality assurance strategies. *International Journal of Medical Informatics, 127*, 88–94. doi:10.1016/j.ijmedinf.2019.04.007 PMID:31128836

Waraich, P., Saklikar, R. S., Aubé, D., Jones, W., Haslam, D., & Hamill, K. (2010). Quality measures for primary mental healthcare: A multistakeholder, multijurisdictional Canadian consensus. *BMJ Quality & Safety, 19*(6), 519–525. doi:10.1136/qshc.2008.027839 PMID:20702436

Webster, J., & Watson, R. T. (2002). Analyzing the past to prepare for the future: Writing a literature review. *Management Information Systems Quarterly, 26*, 13–23.

Wong, S. T., Browne, A. J., Varcoe, C., Lavoie, J., Smye, V., Godwin, O., Littlejohn, D., & Tu, D. (2011). Enhancing measurement of primary health care indicators using an equity lens: An ethnographic study. *International Journal for Equity in Health, 10*(1), 1–12. doi:10.1186/1475-9276-10-38 PMID:21892956

Yamalik, N. (2007). Quality systems in dentistry. Part 1. The increasing pressure for quality and implementation of quality assurance and improvement (QA/I) models in health care. *International Dental Journal, 57*(5), 338–346. doi:10.1111/j.1875-595X.2007.tb00143.x PMID:17992920

Yamalik, N. (2007). Quality systems in dentistry* Part 2. Quality Assurance and Improvement (QA/I) tools that have implications for dentistry. *International Dental Journal*, *57*(6), 459–467. doi:10.1111/j.1875-595X.2007.tb00150.x PMID:18265780

Yeon, A. A. A., Kamaruddin, A. I., Harun, N. A., & Kusumawardani, A. (2019). Parents' satisfaction and the quality of dental care provided by dental students at International Islamic University Malaysia. *Materials Today: Proceedings*, *16*, 2341–2350.

Zhao, L., Lin, H. C., Lo, E. C. M., & Wong, M. C. M. (2011). Clinical and socio-demographic factors influencing the oral health-related quality of life of Chinese elders. *Community Dental Health*, *28*(3), 206–210. PMID:21916355

ADDITIONAL READING

Bhardwaj, A., Ramoni, R., Kalenderian, E., Neumann, A., Hebballi, N. B., White, J. M., McClellan, L., & Walji, M. F. (2016). Measuring up: Implementing a dental quality measure in the electronic health record context. *The Journal of the American Dental Association*, *147*(1), 35–40. doi:10.1016/j.adaj.2015.06.023 PMID:26562736

Elg, M., Stenberg, J., Kammerlind, P., Tullberg, S., & Olsson, J. (2011). Swedish healthcare management practices and quality improvement work: Development trends. *International Journal of Health Care Quality Assurance*, *24*(2), 101–123. doi:10.1108/09526861111105077 PMID:21456487

Herndon, J. B., & Amundson, C. W. (2019). Dental quality measures. *The Journal of the American Dental Association*, *150*(1), 4. doi:10.1016/j.adaj.2018.11.016 PMID:30611324

Herndon, J. B., Aravamudhan, K., Stephenson, R. L., Brandon, R., Ruff, J., Catalanotto, F., & Le, H. (2017). Using a stakeholder-engaged approach to develop and validate electronic clinical quality measures. *Journal of the American Medical Informatics Association*, *24*(3), 503–512. PMID:28339559

Hunt, R. J., & Ojha, D. (2017). Oral health care quality measurement and its role in dental education. *Journal of Dental Education*, *81*(12), 1395–1404. doi:10.21815/JDE.017.099 PMID:29196327

Obadan-Udoh, E., Simon, L., Etolue, J., Tokede, O., White, J., Spallek, H., Walji, M. F., & Kalenderian, E. (2017). Dental providers' perspectives on diagnosis-driven dentistry: Strategies to enhance adoption of dental diagnostic terminology. *International Journal of Environmental Research and Public Health*, *14*(7), 1–10. doi:10.3390/ijerph14070767 PMID:28703751

Roberts, M. B. (2013). Dental Quality Alliance: Measuring quality in dentistry. *Texas Dental Journal*, *130*(10), 1006–1009. PMID:24354159

Sirotich, F., Adair, C. E., Durbin, J., Lin, E., & Canning, C. (2019). Key steps for a mental health and addictions performance measurement framework for Canada. *Healthcare Management Forum*, *32*(2), 97–104. doi:10.1177/0840470418812152 PMID:30739489

Vidone, L., Hunt, R. J., & Ojha, D. (2016). An Emerging Era in Dentistry Quality Measurement. *Journal of the Massachusetts Dental Society*, *65*(3), 18–21. PMID:29847045

Ward, P. (2011). Quality measures. *British Dental Journal, 211*(8), 1–2. doi:10.1038j.bdj.2011.876 PMID:22015500

KEY TERMS AND DEFINITIONS

Decision Maker: A person who is responsible for making decisions in an organization.

Dentistry: It is a part of medicine that is responsible for the study, diagnosis, prevention, and treatment of diseases, disorders, and conditions that affect the teeth.

Evaluation: The process of assessment using specific criteria/standards.

Health Policy: It includes decisions and plans in order to achieve specific goals in healthcare.

Measurement: It can be a number to a characteristic of an object or event, and it can be compared with other objects or events.

Policy Maker: A person who is responsible for formulating policies and strategic plans.

Primary Healthcare: It includes physical, mental, and social well-being as well as person's health needs throughout their lifetime.

Quality: The standard of something as measured in comparison with other similar things.

Satisfaction: A condition that expresses person's wishes, expectations, or needs and the pleasure from this.

Service Delivery: It is the process from service design to offering from a provider to customers.

Chapter 11
SERVQUAL-Based Evaluation of Service Quality in Turkish Health Industry With Fuzzy Logic

Hasan Dinçer
Istanbul Medipol University, Turkey

Serhat Yüksel
Istanbul Medipol University, Turkey

Yaşar Gökalp
Istanbul Medipol University, Turkey

Serkan Eti
https://orcid.org/0000-0002-4791-4091
Istanbul Medipol University, Turkey

ABSTRACT

The aim of this study is to analyze the service quality in the Turkish health sector. In this context, a literature review is conducted, and 16 different criteria based on SERVQUAL are determined. Fuzzy DEMATEL method is used to determine which of these criteria are more important. The findings indicate that operating hours and quality of equipment have the highest weights. In addition to these items, experience of personnel and individual attention to patient are other important factors with respect to the service quality in Turkish health industry. Hence, it is strongly recommended that hospitals in Turkey should operate in an extended time because Turkish people prefer to go to the hospitals after working hours. It is also obvious that Turkish hospitals should make investment to quality equipment. Finally, it is a good idea for new hospitals to have qualified employee. This situation has a positive contribution to the service quality of these hospitals.

DOI: 10.4018/978-1-7998-5442-5.ch011

INTRODUCTION

Health is an important factor in all societies. In this context, people have always wanted to be healthier over the years. The main reason for this is that the quality of life increases when people are healthy. In other words, healthy individuals do not experience pain that makes them unhappy. In addition to the mentioned issue, if people are healthy, it may be more possible for them to be included in social environments. On the other hand, when individuals are healthy, they will be able to find jobs more easily. In this way, the persons concerned can have a regular income (Pansiri and Mmereki, 2010). This situation will contribute to these people to lead a happier life.

In addition to the issues mentioned, labor force participation will increase in countries with healthier individuals. In this way, an increase in the amount of production in the country will come to the benefit. It is easier for economic development to be continuous in societies with developing industry. Moreover, there will be new investments in the country thanks to the developing industry. Thanks to these new investments, new job opportunities will emerge (Shaikh et al., 2008). In summary, in a society with healthy individuals, it is possible for the country to grow economically. Therefore, countries also prefer their citizens to be healthier. Hence, they aim to implement many policies that contribute to healthy life.

However, people sometimes experience some illnesses. Naturally, these people aim to improve their quality of life by getting rid of their illnesses in a short time. In this case, the effectiveness of the health system in the country plays an important role. If the sick individuals cannot get the treatment they need in time, they are unhappy. Depending on the population density in the country, state hospitals are sometimes not sufficient to meet this need. In this case, private hospitals are established by entrepreneurs in the country. In this way, it is aimed to treat sick people in the country in a timely and correct manner.

On the other hand, after the intense opening of private hospitals, there was a serious competition in the health sector. In this case, it has become necessary to take some measures to ensure the continuity of the hospitals. In this framework, it is accepted by everyone that hospitals should satisfy their patients. Otherwise, patients will not prefer these hospitals when they need it, and they will apply to other hospitals operating in the highly competitive sector (Taner and Antony, 2006). This will make it very difficult for hospitals to continue their activities for a long time. As a result, hospitals should give importance to the customer expectations.

There are many issues that hospitals need to pay attention in order to make patients satisfied. First of all, the physical conditions of the hospital affect the choice decisions of the patients. In this context, the working hours of the hospital and the number of employees are important issues. Patients prefer to reach hospitals and doctors in different time periods. Besides the issues mentioned, the location of the hospital is also an important factor (Zarei et al., 2012). The main reason for this is that patients want to easily reach the location of the hospital. On the other hand, the quality of physical safety and equipment in the hospital can be considered as a reason for preference for patients.

In addition to physical conditions, the fact that the hospital is functional and responds quickly to problems are other important factors. In this context, the short waiting time for the results of the tests will increase the happiness of the patients. In addition, the accuracy of patient reports and the experience of hospital staff will contribute to increasing patient satisfaction (Kalaja et al., 2016). On the other hand, the communication of the hospital with the patients should also be of high quality. In this framework, the establishment of a patient communication center where patients can directly apply will contribute to achieving this goal. Furthermore, hospitals with a quality call center will also be preferred by patients.

Furthermore, safety is another issue that affects patient satisfaction. In this context, the safety of health personnel is very important. Patients prefer a hospital where security measures are taken effectively. The security of patient information is another important issue in this process (Ocampo et al., 2017). The reason for this is that hospitals have very specific information about patients. On the other hand, special attention to patients will increase the preference of hospitals. In addition to the issues mentioned, hospital costs are other issues that patients care about. Hospitals with very high treatment costs are more difficult to choose.

Considering the aforementioned issues, it is seen that there are many different factors affecting the service quality in hospitals. Hospitals need to take some precautions to improve these factors. Therefore, it is very difficult for hospitals to be able to improve all factors at the same time. In this context, it is necessary to know which factors are more important. In this process, it is essential to determine what patients care about most in hospitals. Therefore, the method used in the analysis is very important (Sam et al., 2018). Otherwise, it is very difficult to provide correct suggestions in order to improve the quality of the service.

The SERVQUAL method is a frequently preferred approach to measuring service quality (Roslan et al., 2015). According to the measurement model in question, there are 5 main elements of service quality. In this context, different factors such as physical elements of the company, reliability, responsiveness, assurance and empathy are taken into account. Therefore, it is thought that determining the criteria to be used in the measurement of service quality in hospitals according to the SERVQUAL approach will increase the quality of the analysis (Chatzoglou et al., 2014). In addition, multi-criteria decision-making methods are taken into account in identifying the most important of the different criteria. The DEMATEL method has also been used frequently by different researchers in the literature. The most important advantage of this method is that the effect-relationship matrix can be created between the criteria.

Turkey is also a country with a highly developed healthcare sector. There are many state hospitals in the country as well as many private hospitals. Therefore, this situation increases the competition in the sector. In this specified competitive environment, private hospitals need to develop some strategies to ensure their continuity. In this context, it is vital to determine whether the patients in Turkey about what the hospital. In this way, it will be possible for private hospitals to increase the satisfaction of patients. This will increase the preference of hospitals and will provide competitive advantage to these hospitals (Akbaba, 2006).

The aim of this study is to determine the issue of increasing the quality of the health sector in Turkey. In this context, studies in the literature have been examined in detail. As a result of the examination, 16 different criteria were determined by considering the dimensions of the SERVQUAL approach. After that, it is aimed to determine which of these criteria is more important. To achieve this goal, the fuzzy DEMATEL method is used. Thanks to the results to be obtained, recommendations to improve the quality of services in the health sector in Turkey could be developed. The most important contribution of this study to the literature is the service quality criteria determined by considering the SERVQUAL approach. These criteria, which are formed as a result of detailed studies of many studies, can also be taken into consideration for different sectors. On the other hand, there is no study in the health sector using the fuzzy DEMATEL method. Therefore, this study is thought to be originally methodological compared to others.

In this study, firstly, general information about service quality and the importance of this issue in the health sector was shared. After that, the results of the literature analysis were shared. Next, the stages of the fuzzy DEMATEL approach used in the analysis process are explained. In the next sections of the study, the results of the analysis made for the Turkish health sector are included. After that, information was given about solution suggestions and new issues to be considered in future studies.

LITERATURE REVIEW

A literature review on the subject is included under this title. First, the issues that take into account the service quality in different sectors will be examined. In addition, studies examining service quality in the health sector will be highlighted in another subtitle. Finally, the points obtained as a result of the literature review will be underlined.

Literature Based on Service Quality in Different Industries

The issue of improving service quality has been addressed by many different people in the literature. It is observed that these studies focus on different sectors. For example, Ocampo et al. (2017) aimed to measure the service quality of state institutions in the Philippines. In this context, they have set criteria for four different dimensions of the SERVQUAL approach. In the analysis process of this study, AHP method was taken into consideration. According to the analysis results obtained, it was determined that many different factors affect the service quality more. In this context, it has been determined that formal dressing of employees in public institutions contributes to the increase of service quality. In addition, the fact that the personnel working in these state institutions are helpful while providing service is another prominent criterion.

Roslan et al. (2015) aimed to increase the service quality of the logistics sector in Malaysia. In this study, the criteria have been determined by considering the SERVQUAL methodology. In other words, many different factors that increase the quality of service have been identified and among them, the more important ones have been tried to be identified. It is concluded that increasing the communication of the company personnel with the customers will increase the service quality. In addition to this factor, it is also stated that customers should trust the company. Otherwise, it was stated that it would not be possible to increase the service quality of logistics companies.

Similarly, Lee et al. (2019) also focused on increasing the service quality in the retail sector. 12 different criteria were selected for this purpose. According to the analysis results obtained, in order for a company to improve its service quality, it was first determined that it should find solutions to the complaints of the customers. For this, it was stated that the complaints of the customers related to the company should be determined. In this way, it will be easier to develop a solution proposal for problems. In addition, the competitive prices of products are another prominent factor. In other words, it will be very difficult for customers to choose a product that is very expensive, which will lead to a decrease in the service quality of the company.

Chatzoglou et al. (2014) is another study that examines to increase the service quality. Within this framework, criteria for 5 different dimensions based on SERVQUAL have been determined. Questions have been created by taking these criteria into consideration. A survey study was applied with these questions. As a result, it is concluded that the positive attitudes of the employees towards the customers are the most important issue in this process. On the other hand, the high level of knowledge of employees is also a factor in customers' preference for these companies. In this way, it will be possible for companies that can improve their service quality to gain competitive advantage.

Sayareh et al. (2016) also aimed to identify issues that improve service quality in the logistics industry. In this study, it is concluded that the accuracy of the service provided is the most important issue. Similarly, Ariff et al. (2012) focused on how to improve the quality of service in internet banking services. In this context, it has been determined that the connection quality in internet banking is continuous and of high quality. Limbourg et al. (2016) carried out an investigation on the logistics industry. According to the analysis results obtained, it has been determined that promotions given to customers play an important role in increasing the service quality.

Chen et al. (2009) also carried out a study to increase the quality of service in the shipping industry. It is concluded that in order to increase the service quality, customers' trust in the company should be increased. Jeeradist et al. (2016) focused on service quality in the airline sector in their study. In this study, some criteria were determined based on the TRIZ approach. As a result of the analysis, it has been determined that in cases where the knowledge of the airline personnel is high, the service quality can be increased. Akbaba (2006) tried to determine how the service quality in the hotel sector can be improved. According to the analysis results obtained, it was concluded that the hotel customers should be closely related.

Roses et al. (2009) tried to determine how to improve service quality for information technologies. In this study, it was stated that the problems should be solved in time in order to increase the service quality. Rezaei et al. (2018), on the other hand, researched ways to improve the quality of service in the airline industry. In this framework, the criterion has been determined by considering the SERVQUAL approach. According to the results of the analysis, it has been concluded that flight safety is the most important factor in service quality. In addition, Dinçer et al. (2019) also conducted an analysis of the quality of service in the financial sector. In this study, fuzzy DEMATEL and fuzzy TOPSIS methods were used. As a result, it was concluded that the cost of the service provided should be at a reasonable level.

Alam and Mondal (2019) also focused on the quality of service in the transportation sector. In this study, different criteria were determined depending on the SERVQUAL approach. AHP method was used to understand which of these factors are more important. According to the analysis results obtained, it was stated that the security issue is very important in the transportation sector. Awasthi et al. (2011) also conducted a review on the transportation sector. In this study, the fuzzy TOPSIS method was taken into consideration. As a result, it is stated that the quality of service will increase if the metro sector is clean. Maghsoodi et al. (2019) carried out a similar study for a different sector. In this study, it is concluded that the service quality will increase if the customer expectations are met correctly.

Belwal and Amireh (2018) examined the telecommunications industry in Oman. In this study, it is concluded that solving customer problems is the most important issue. Basfirinci and Mitra (2015) took this issue into consideration for the airline industry by using KANO and SERVQUAL methods. According to the analysis results obtained, reliability is determined to be the most important factor. On the other hand, Souri et al. (2018) emphasized the importance of developing green products. In addition to the studies mentioned, Sam et al. (2018) investigated the quality of service for public transport in Ghana. As a result, it was concluded that physical conditions are the most important.

Literature Regarding Service Quality in Health Industry

In this sub-title of the study, the important ones that examine the service quality in the health sector will be included. Pansiri and Mmereki (2010) aimed to measure the quality of health services in Botswana. By considering the SERVQUAL approach, criteria for service quality have been determined. In the study in question, it has been determined that physical issues are the most important. Shaikh et al. (2008) also examined the health sector in Pakistan. In this study, it was determined that the doctor's room was clean and communication with patients was important factors.

Taner and Antony (2006), quality of service in the health sector in Turkey have focused on how to increase. In this study, it was determined that hospitals should give importance to technological infrastructure in order to have high service quality. Zarei et al. (2012), on the other hand, conducted a review on the quality of service in the health sector in Iran. As a result, it was emphasized that hospitals should be clean. Kalaja et al. (2016), on the other hand, argued that efforts should be made to shorten the service period, especially in public hospitals. Ocampo et al. (2017) focused on the service quality of state hospitals in the Philippines. Similar results were obtained in this study using AHP and TOPSIS method.

Soita (2012) conducted an analysis of the quality of service in the health sector in Uganda. According to the analysis results obtained, it has been determined that the safety of laboratory results is the most important criterion. Similarly, Kilbourne et al. (2004) also emphasized that test results should be safe in order to increase their service quality. Punnakitikashem et al. (2012) stated that the relationship with patients should be improved. Yeşilada and Direktör (2010) compared public and private hospitals in their study. As a result, it was concluded that the safety of patient records and the competence of staff are important for both hospital groups.

The Results of Literature Review

The most important issue obtained as a result of the literature review is that the quality of service is very popular. It is seen that this issue has been dealt with in many sectors. On the other hand, another result obtained is that the factors stated to increase the service quality in the studies are mostly different from each other. Therefore, it is seen that performing studies on service quality on a single country will yield more accurate results. The main reason for this is that this situation may vary from country to country. In other words, by doing such studies on a country basis, the expectations of the people in that country should be determined. This study focused on the quality of service in the health sector in Turkey. In this way, it will be possible to develop country-based strategy suggestions.

AN APPLICATION ON TURKISH HEALTH INDUSTRY

SERVQUAL-Based Criteria List

The SERVQUAL model is an approach that aims to measure the service quality of companies. The main advantage of this model is that it takes into account different issues when measuring a company's service quality (Souri et al., 2018). Primarily, it gives importance to the physical properties of the company. In this context, importance is given to issues such as the physical appearance of the workplace, the quality of the materials used and the dressing of the personnel working in the company. On the other hand, reliability is another important factor in the SERVQUAL method (Maghsoodi et al., 2019). In this framework, the issue of how much customers trust the company is discussed. In addition to the issues mentioned, the time of the service provided is another consideration in the SERVQUAL approach. In addition, the issue of how confident the company employees create with the customers is also important. In addition, the company needs to pay attention to whether the special needs of customers are understood (Rezaei et al., 2018). This study examined the quality of services in the health sector in Turkey. It was selected based on the criteria for increasing the service quality in hospitals and the SERVQUAL method. In this process, the issues obtained as a result of the literature review were also taken into consideration. As a result, a total of 16 criteria were determined for three different dimensions. Details of these criteria are shared in Table 1.

Table 1. SERVQUAL-based Criteria List

Dimension	Criteria	Literature
Physical Conditions (Tangibility)	Operating hours (C1)	Ocampo et al. (2017); Roslan et al. (2015)
	Physical Security (C2)	Lee et al. (2019); Chatzoglou et al. (2014)
	Quality of Equipment (C3)	Sayareh et al. (2016); Chen et al. (2009)
	Quality of Building (C4)	Jeeradist et al. (2016); Ariff et al. (2012)
	Location (C5)	Limbourg et al. (2016); Jeeradist et al. (2016)
	Number of Personnel (C6)	Akbaba (2006); Roses et al. (2009)
Functionality and Responsiveness	Accuracy of reports (C7)	Rezaei et al. (2018); Dinçer et al. (2019)
	Waiting time of tests (C8)	Alam and Mondal (2019); Awasthi et al. (2011)
	Experience of personnel (C9)	Maghsoodi et al. (2019); Belwal and Amireh (2018)
	The quality of call center (C10)	Basfirinci and Mitra (2015); Souri et al. (2018)
	Patient Communication (C11)	Sam et al. (2018); Pansiri and Mmereki (2010)
Security and Empathy	Security of doctors (C12)	Shaikh et al. (2008); Maghsoodi et al. (2019)
	Patients security (C13)	Taner and Antony (2006); Zarei et al. (2012)
	Data security (C14)	Kalaja et al. (2016); Ocampo et al. (2017)
	Individual Attention to patient (C15)	Soita (2012); Kilbourne et al. (2004)
	Reasonable Cost (C16)	Punnakitikashem et al. (2012); Yeşilada and Direktör (2010)

As can be seen from Table 1, 16 different criteria, which are thought to affect the service quality in the health sector, have been determined. These criteria were created by taking into consideration both the literature review and the dimensions in the SERVQUAL approach. Concepts such as the length of working hours and physical security will increase patient satisfaction. Similarly, the quality of the equipment and building are other important factors. The location of the hospital and the number of employees are other physical factors that affect the service quality. On the other hand, patients want the tests to be explained on time and to be correct. Employee experience is another important issue in this process. In addition, the quality of the call center and the effective operation of the patient relations unit will also contribute to improving the quality of service. In addition, doctor and patient safety should be given importance in order to increase the service quality. The confidentiality of personal information of patients should also be considered within this framework. In addition, showing special care to patients will increase the degree of satisfaction of these patients. Finally, the treatment fees are not too high are among the expectations of customers in this process.

Fuzzy DEMATEL

The DEMATEL approach is used to identify the most important of the different alternatives. In this way, it provides convenience to the decision maker in a situation where there are too many alternatives. The biggest advantage of the DEMATEL method compared to other similar methods is that it also determines the relationship between these factors (Zhang et al., 2020). Firstly, the purpose is defined. On the other side, the second step is related to the evaluation of different criteria. Within this framework, 5 different factors are used that are "No", "Low", "Medium", "High", "Very High". In the third step, initial direct-relation fuzzy matrix ($\tilde{Z}$) is obtained by considering the equations (1) and (2).

$$\tilde{Z} = \begin{bmatrix} 0 & \tilde{z}_{12} & \cdots & \cdots & \tilde{z}_{1n} \\ \tilde{z}_{21} & 0 & \cdots & \cdots & \tilde{z}_{2n} \\ \vdots & \vdots & \ddots & \cdots & \cdots \\ \vdots & \vdots & \vdots & \ddots & \vdots \\ \tilde{z}_{n1} & \tilde{z}_{n2} & \cdots & \cdots & 0 \end{bmatrix} \tag{1}$$

$$\tilde{Z} = \frac{\tilde{Z}^1 + \tilde{Z}^2 + \tilde{Z}^3 + \ldots \tilde{Z}^n}{n} \tag{2}$$

After that, this direct relation matrix is normalized with the equations (3)-(5).

$$\tilde{X} = \begin{bmatrix} \tilde{x}_{11} & \tilde{x}_{12} & \cdots & \cdots & \tilde{x}_{1n} \\ \tilde{x}_{21} & \tilde{x}_{22} & \cdots & \cdots & \tilde{x}_{2n} \\ \vdots & \vdots & \ddots & \cdots & \cdots \\ \vdots & \vdots & \vdots & \ddots & \vdots \\ \tilde{x}_{n1} & \tilde{x}_{n2} & \cdots & \cdots & \tilde{x}_{nn} \end{bmatrix} \tag{3}$$

$$\tilde{x}_{ij} = \frac{\tilde{z}_{ij}}{r} = \left(\frac{l_{ij}}{r}, \frac{m_{ij}}{r}, \frac{u_{ij}}{r}\right) \tag{4}$$

$$r = max_{1 \le i \le n}\left(\sum_{j=1}^{n} u_{ij}\right) \tag{5}$$

Next, the total influence fuzzy matrix is generated by using the equations (6)-(12).

$$X_l=\begin{bmatrix} 0 & l'_{12} & \cdots & \cdots & l'_{1n} \\ l'_{21} & 0 & \cdots & \cdots & l'_{2n} \\ \vdots & \vdots & \ddots & \cdots & \cdots \\ \vdots & \vdots & \vdots & \ddots & \vdots \\ l'_{n1} & l'_{n2} & \cdots & \cdots & 0 \end{bmatrix}$$

$$X_m= X_m =\begin{bmatrix} 0 & m'_{12} & \cdots & \cdots & m'_{1n} \\ m'_{21} & 0 & \cdots & \cdots & m'_{2n} \\ \vdots & \vdots & \ddots & \cdots & \cdots \\ \vdots & \vdots & \vdots & \ddots & \vdots \\ m'_{n1} & m'_{n2} & \cdots & \cdots & 0 \end{bmatrix}$$

$$X_u=\begin{bmatrix} 0 & u'_{12} & \cdots & \cdots & u'_{1n} \\ u'_{21} & 0 & \cdots & \cdots & u'_{2n} \\ \vdots & \vdots & \ddots & \cdots & \cdots \\ \vdots & \vdots & \vdots & \ddots & \vdots \\ u'_{n1} & u'_{n2} & \cdots & \cdots & 0 \end{bmatrix} \tag{6}$$

$$\tilde{T} = \lim_{k\to\infty} \tilde{X} + \tilde{X}^2 + \ldots + \tilde{X}^k \tag{7}$$

$$\tilde{T} =\begin{bmatrix} \tilde{t}_{11} & \tilde{t}_{12} & \cdots & \cdots & \tilde{t}_{1n} \\ \tilde{t}_{21} & \tilde{t}_{22} & \cdots & \cdots & \tilde{t}_{2n} \\ \vdots & \vdots & \ddots & \cdots & \cdots \\ \vdots & \vdots & \vdots & \ddots & \vdots \\ \tilde{t}_{n1} & \tilde{t}_{n2} & \cdots & \cdots & \tilde{t}_{nn} \end{bmatrix} \tag{8}$$

$$\tilde{t}_{ij} = \left(l''_{ij}, m''_{ij}, u''_{ij}\right) \tag{9}$$

$$\left[l''_{ij}\right] = X_l \times \left(I - X_l\right)^{-1} \tag{10}$$

$$\left[m''_{ij}\right] = X_m \times \left(I - X_m\right)^{-1} \tag{11}$$

$$\left[u''_{ij}\right] = X_u \times \left(I - X_u\right)^{-1} \tag{12}$$

In the final stage, defuzzification is occurred. For this purpose, equations (13)-(21) are taken into consideration.

$$u_i^{max} = max u_i, l_i^{min} = \min_j l_i \tag{13}$$

$$\bigcup_{min}^{max} = u_i^{max} - l_i^{min} \tag{14}$$

$$x_{lj} = \left(l_{ij} - l_i^{min}\right) / \bigcup_{min}^{max} \tag{15}$$

$$x_{mj} = \left(m_{ij} - l_i^{min}\right) / \bigcup_{min}^{max} \tag{16}$$

$$x_{uj} = \left(u_{ij} - l_i^{min}\right) / \bigcup_{min}^{max} \tag{17}$$

$$x_j^{ls} = x_{mj} / \left(1 + x_{mj} - x_{lj}\right) \tag{18}$$

$$x_j^{rs} = x_{uj} / \left(1 + x_{uj} - x_{mj}\right) \tag{19}$$

$$x_j^{crisp} = \left[x_j^{ls}\left(1 - x_j^{ls}\right) + x_j^{rs} x_j^{rs}\right] / \left[1 - x_j^{ls} + x_j^{rs}\right] \tag{20}$$

$$f_{ij} = l_i^{min} + x_j^{crisp} \bigcup_{min}^{max} \tag{21}$$

DEMATEL methodology was also considered in many different studies. In these studies, various industries are evaluated. Table 2 gives information about some of these studies.

Table 2. Studies in the Literature with DEMATEL

Studies	Subject
Yazdi et al. (2020)	probabilistic safety analysis
Feng and Ma (2020)	service innovation in manufacturing enterprises
Jiang et al. (2020)	key performance indicators in hospital performance management
Can and Delice (2020)	human error assessment
Zhang et al. (2020)	key performance indicators
Jayakrishna et al. (2020)	innovation in an industrial symbiosis environment
Song et al. (2020)	sustainable online consumption
Chen et al. (2020)	innovative value propositions for smart product service system
Dinçer et al. (2020)	customer expectations in retail service industry
Liu et al. (2020)	critical factors in construction and demolition waste recycling
Jiao et al. (2020)	risk analysis and evaluation
Sagnak (2020)	supplier selection in the green environment
Wu et al. (2020)	improving quality function deployment analysis
Mohammed et al. (2020)	prioritization in patients with multiple chronic diseases
Solanki et al. (2020)	key factors of supply chain
Motaghifard et al. (2020)	evaluating health safety environmental performance

Analysis Results

In the first stage of the analysis, 3 decision makers made evaluations about these 16 different criteria with respect to the service quality in health industry. These experts have at least 15-year work experience. Additionally, they are also top managers. They made evaluations regarding 5 different scales that are no (N), low (L), medium (M), high (H), very high (V). These evaluations are given on the Appendix part. After that, these evaluations are converted into linguistic variables and arithmetic mean of these three evaluations are calculated. The initial direct-relation fuzzy matrix is given on Table 3.

In the next step, this matrix is normalized. Normalized matrix is also given in Table 4.

In the next aspect, defuzzification process is performed. As a result, the weighted results of these 16 criteria can be calculated. These results are given on Table 5.

Table 5 indicates that operating hour (C1) is the most significant criteria because it has the highest weight. In addition to this situation, the quality of equipment (C3) has the second greatest weight. Moreover, experience of personnel (C9) and individual attention to patient (C15) are other significant factors that affect service quality in Turkish health industry. On the other side, the quality of call center (C10), location (C5) and patient communication (C11) are the least important factors.

Table 3. Initial Direct Relation Matrix

C	C1			C2			C3			C4			C5			C6			C7			C8		
C1	.00	.00	.00	.50	.75	1.00	.17	.42	.67	.08	.33	.58	.33	.58	.83	.42	.67	.92	.50	.75	1.00	.42	.67	.83
C2	.33	.58	.75	.00	.00	.00	.08	.33	.58	.25	.50	.75	.17	.42	.67	.33	.58	.83	.08	.33	.58	.17	.42	.67
C3	.17	.33	.58	.33	.58	.83	.00	.00	.00	.42	.67	.92	.08	.25	.50	.33	.50	.67	.67	.92	1.00	.67	.92	1.00
C4	.17	.42	.67	.50	.75	.92	.33	.58	.83	.00	.00	.00	.08	.33	.58	.25	.50	.75	.00	.25	.50	.08	.33	.58
C5	.25	.50	.75	.33	.58	.75	.08	.33	.58	.17	.42	.67	.00	.00	.00	.25	.50	.75	.00	.17	.42	.00	.17	.42
C6	.33	.50	.75	.17	.33	.58	.00	.17	.42	.08	.25	.50	.08	.25	.50	.00	.00	.00	.33	.58	.83	.33	.58	.75
C7	.25	.42	.67	.08	.25	.50	.08	.25	.50	.00	.17	.42	.00	.17	.42	.25	.42	.67	.00	.00	.00	.33	.50	.67
C8	.17	.42	.67	.17	.42	.67	.25	.50	.75	.00	.17	.42	.08	.25	.50	.25	.50	.75	.17	.33	.58	.00	.00	.00
C9	.17	.42	.67	.42	.67	.92	.17	.33	.58	.08	.25	.50	.00	.17	.42	.25	.50	.75	.67	.92	1.00	.33	.58	.75
C10	.17	.33	.58	.17	.33	.58	.17	.33	.58	.08	.25	.50	.25	.42	.67	.25	.42	.67	.08	.25	.50	.08	.25	.50
C11	.08	.25	.50	.17	.33	.58	.00	.17	.42	.08	.25	.50	.08	.25	.50	.17	.33	.58	.17	.33	.58	.17	.33	.58
C12	.42	.67	.83	.58	.83	1.00	.50	.75	.92	.42	.67	.92	.33	.50	.67	.33	.50	.67	.33	.58	.83	.17	.33	.58
C13	.17	.33	.58	.25	.42	.67	.25	.42	.67	.25	.42	.67	.33	.50	.67	.33	.50	.75	.42	.58	.75	.17	.33	.58
C14	.17	.42	.67	.17	.42	.67	.33	.58	.75	.08	.33	.58	.17	.42	.67	.25	.50	.75	.25	.50	.67	.17	.42	.67
C15	.33	.58	.75	.33	.58	.75	.25	.50	.67	.17	.42	.67	.25	.50	.75	.42	.67	.83	.33	.58	.75	.33	.58	.75
C16	.25	.42	.67	.17	.33	.58	.25	.42	.67	.25	.42	.67	.33	.50	.67	.33	.50	.75	.08	.25	.50	.17	.33	.58

C	C9			C10			C11			C12			C13			C14			C15			C16		
C1	.25	.50	.75	.33	.50	.67	.58	.83	1.00	.58	.83	1.00	.58	.83	1.00	.17	.42	.67	.50	.75	.92	.33	.50	.75
C2	.08	.33	.58	.08	.33	.58	.17	.42	.67	.50	.75	.92	.58	.83	1.00	.33	.58	.83	.17	.42	.67	.08	.33	.58
C3	.25	.50	.75	.17	.42	.67	.25	.50	.75	.67	.92	1.00	.67	.92	1.00	.33	.58	.75	.33	.58	.75	.25	.50	.75
C4	.08	.33	.58	.08	.33	.58	.25	.50	.75	.25	.50	.75	.33	.58	.83	.17	.42	.67	.33	.58	.75	.17	.42	.67
C5	.17	.33	.58	.00	.17	.42	.25	.50	.75	.33	.58	.75	.25	.50	.75	.00	.17	.42	.25	.50	.75	.17	.33	.58
C6	.33	.58	.75	.25	.50	.75	.25	.50	.75	.17	.42	.67	.33	.58	.83	.17	.42	.67	.50	.75	.92	.17	.42	.67
C7	.42	.58	.75	.17	.33	.58	.17	.33	.58	.25	.42	.67	.33	.50	.75	.25	.42	.67	.33	.50	.67	.17	.33	.58
C8	.17	.33	.58	.00	.17	.42	.25	.42	.67	.17	.33	.58	.33	.50	.75	.08	.25	.50	.42	.58	.75	.25	.42	.67
C9	.00	.00	.00	.33	.58	.83	.50	.75	.92	.42	.67	.92	.58	.83	1.00	.50	.75	.92	.67	.92	1.00	.42	.67	.83
C10	.17	.33	.58	.00	.00	.00	.42	.58	.75	.17	.33	.58	.17	.33	.58	.25	.42	.67	.33	.50	.67	.33	.50	.67
C11	.08	.25	.50	.25	.42	.67	.00	.00	.00	.25	.42	.67	.33	.50	.67	.25	.42	.67	.42	.58	.75	.25	.42	.67
C12	.08	.25	.50	.17	.33	.58	.25	.42	.67	.00	.00	.00	.33	.50	.75	.25	.42	.67	.33	.50	.67	.25	.42	.67
C13	.33	.50	.67	.25	.42	.67	.33	.50	.67	.33	.50	.75	.00	.00	.00	.25	.42	.67	.33	.50	.67	.17	.33	.58
C14	.33	.58	.83	.17	.42	.67	.42	.67	.83	.33	.58	.83	.42	.67	.83	.00	.00	.00	.33	.58	.75	.25	.50	.75
C15	.42	.67	.83	.42	.67	.83	.42	.67	.83	.25	.50	.75	.25	.50	.75	.25	.50	.75	.00	.00	.00	.08	.33	.58
C16	.17	.33	.58	.33	.50	.75	.08	.25	.50	.17	.33	.58	.17	.33	.58	.17	.33	.58	.17	.33	.58	.00	.00	.00

Table 4. Normalized Direct Relation Matrix

C	C1			C2			C3			C4			C5			C6			C7			C8		
C1	.00	.00	.00	.04	.06	.08	.01	.03	.06	.01	.03	.05	.03	.05	.07	.03	.06	.08	.04	.06	.08	.03	.06	.07
C2	.03	.05	.06	.00	.00	.00	.01	.03	.05	.02	.04	.06	.01	.03	.06	.03	.05	.07	.01	.03	.05	.01	.03	.06
C3	.01	.03	.05	.03	.05	.07	.00	.00	.00	.03	.06	.08	.01	.02	.04	.03	.04	.06	.06	.08	.08	.06	.08	.08
C4	.01	.03	.06	.04	.06	.08	.03	.05	.07	.00	.00	.00	.01	.03	.05	.02	.04	.06	.00	.02	.04	.01	.03	.05
C5	.02	.04	.06	.03	.05	.06	.01	.03	.05	.01	.03	.06	.00	.00	.00	.02	.04	.06	.00	.01	.03	.00	.01	.03
C6	.03	.04	.06	.01	.03	.05	.00	.01	.03	.01	.02	.04	.01	.02	.04	.00	.00	.00	.03	.05	.07	.03	.05	.06
C7	.02	.03	.06	.01	.02	.04	.01	.02	.04	.00	.01	.03	.00	.01	.03	.02	.03	.06	.00	.00	.00	.03	.04	.06
C8	.01	.03	.06	.01	.03	.06	.02	.04	.06	.00	.01	.03	.01	.02	.04	.02	.04	.06	.01	.03	.05	.00	.00	.00
C9	.01	.03	.06	.03	.06	.08	.01	.03	.05	.01	.02	.04	.00	.01	.03	.02	.04	.06	.06	.08	.08	.03	.05	.06
C10	.01	.03	.05	.01	.03	.05	.01	.03	.05	.01	.02	.04	.02	.03	.06	.02	.03	.06	.01	.02	.04	.01	.02	.04
C11	.01	.02	.04	.01	.03	.05	.00	.01	.03	.01	.02	.04	.01	.02	.04	.01	.03	.05	.01	.03	.05	.01	.03	.05
C12	.03	.06	.07	.05	.07	.08	.04	.06	.08	.03	.06	.08	.03	.04	.06	.03	.04	.06	.03	.05	.07	.01	.03	.05
C13	.01	.03	.05	.02	.03	.06	.02	.03	.06	.02	.03	.06	.03	.04	.06	.03	.04	.06	.03	.05	.06	.01	.03	.05
C14	.01	.03	.06	.01	.03	.06	.03	.05	.06	.01	.03	.05	.01	.03	.06	.02	.04	.06	.02	.04	.06	.01	.03	.06
C15	.03	.05	.06	.03	.05	.06	.02	.04	.06	.01	.03	.06	.02	.04	.06	.03	.06	.07	.03	.05	.06	.03	.05	.06
C16	.02	.03	.06	.01	.03	.05	.02	.03	.06	.02	.03	.06	.03	.04	.06	.03	.04	.06	.01	.02	.04	.01	.03	.05

C	C9			C10			C11			C12			C13			C14			C15			C16		
C1	.02	.04	.06	.03	.04	.06	.05	.07	.08	.05	.07	.08	.05	.07	.08	.01	.03	.06	.04	.06	.08	.03	.04	.06
C2	.01	.03	.05	.01	.03	.05	.01	.03	.06	.04	.06	.08	.05	.07	.08	.03	.05	.07	.01	.03	.06	.01	.03	.05
C3	.02	.04	.06	.01	.03	.06	.02	.04	.06	.06	.08	.08	.06	.08	.08	.03	.05	.06	.03	.05	.06	.02	.04	.06
C4	.01	.03	.05	.01	.03	.05	.02	.04	.06	.02	.04	.06	.03	.05	.07	.01	.03	.06	.03	.05	.06	.01	.03	.06
C5	.01	.03	.05	.00	.01	.03	.02	.04	.06	.03	.05	.06	.02	.04	.06	.00	.01	.03	.02	.04	.06	.01	.03	.05
C6	.03	.05	.06	.02	.04	.06	.02	.04	.06	.01	.03	.06	.03	.05	.07	.01	.03	.06	.04	.06	.08	.01	.03	.06
C7	.03	.05	.06	.01	.03	.05	.01	.03	.05	.02	.03	.06	.03	.04	.06	.02	.03	.06	.03	.04	.06	.01	.03	.05
C8	.01	.03	.05	.00	.01	.03	.02	.03	.06	.01	.03	.05	.03	.04	.06	.01	.02	.04	.03	.05	.06	.02	.03	.06
C9	.00	.00	.00	.03	.05	.07	.04	.06	.08	.03	.06	.08	.05	.07	.08	.04	.06	.08	.06	.08	.08	.03	.06	.07
C10	.01	.03	.05	.00	.00	.00	.03	.05	.06	.01	.03	.05	.01	.03	.05	.02	.03	.06	.03	.04	.06	.03	.04	.06
C11	.01	.02	.04	.02	.03	.06	.00	.00	.00	.02	.03	.06	.03	.04	.06	.02	.03	.06	.03	.05	.06	.02	.03	.06
C12	.01	.02	.04	.01	.03	.05	.02	.03	.06	.00	.00	.00	.03	.04	.06	.02	.03	.06	.03	.04	.06	.02	.03	.06
C13	.03	.04	.06	.02	.03	.06	.03	.04	.06	.03	.04	.06	.00	.00	.00	.02	.03	.06	.03	.04	.06	.01	.03	.05
C14	.03	.05	.07	.01	.03	.06	.03	.06	.07	.03	.05	.07	.03	.06	.07	.00	.00	.00	.03	.05	.06	.02	.04	.06
C15	.03	.06	.07	.03	.06	.07	.03	.06	.07	.02	.04	.06	.02	.04	.06	.02	.04	.06	.00	.00	.00	.01	.03	.05
C16	.01	.03	.05	.03	.04	.06	.01	.02	.04	.01	.03	.05	.01	.03	.05	.01	.03	.05	.01	.03	.05	.00	.00	.00

SOLUTIONS AND RECOMMENDATIONS

It is identified that Turkish patients mainly focus on operating hours of the hospitals. This is mainly because of the fact that people cannot find an opportunity to go to the hospital since they work during the day. Hence, they prefer to go to the hospitals after working hours. Another important point is that Turkish patients prefer the hospital to be working in a flexible time. Therefore, it is recommended that hospitals in Turkey should operate in an extended time. It is also concluded that the quality of equipment is essential for Turkish patients. Thus, Turkish hospitals should make investment to quality equipment. This situation has a positive contribution to the service quality of these hospitals. In addition to them, experience of personnel is also another important indicator with respect to the service quality of Turkish health industry. Because patients pay attention to this situation, it is a good idea for new hospitals to have qualified employee.

Table 5. Weighted Results of 16 Different Service Quality Criteria

	Di	Ri	D-R	D+R	Weights	Rank
C1	5,581382	4,563298	1,018084	10,14468	0,070846	1
C2	4,614205	4,433458	0,180747	9,047664	0,063185	7
C3	5,400744	4,490727	0,910016	9,891471	0,069078	2
C4	4,455224	4,337555	0,117669	8,792779	0,061405	10
C5	3,890681	4,271943	-0,38126	8,162624	0,057004	15
C6	4,420544	4,493548	-0,073	8,914092	0,062252	9
C7	3,800572	4,503211	-0,70264	8,303783	0,05799	12
C8	3,822984	4,536639	-0,71365	8,359623	0,05838	11
C9	5,377439	4,327412	1,050027	9,704851	0,067775	3
C10	3,760837	4,488115	-0,72728	8,248952	0,057607	14
C11	3,585058	4,556751	-0,97169	8,141809	0,056859	16
C12	4,873963	4,403763	0,4702	9,277726	0,064792	6
C13	4,339813	4,58467	-0,24486	8,924483	0,062325	8
C14	4,781401	4,529998	0,251404	9,311399	0,065027	5
C15	5,062034	4,624083	0,437952	9,686117	0,067644	4
C16	3,829623	4,451335	-0,62171	8,280957	0,057831	13

FUTURE RESEARCH DIRECTIONS

This paper focuses on the improvement of service quality in the health sector in Turkey. In this framework, an analysis has been made taking into consideration the fuzzy DEMATEL method. In a new study, it will be appropriate to focus on different countries or country groups. The factor affecting service quality in hospitals can vary from country to country. Therefore, it is thought that the study of different countries will guide practitioners. In addition, other methods will help to look at the event from a different perspective. In this context, methods such as fuzzy AHP and fuzzy ANP can be preferred to determine which factors are more important.

CONCLUSION

In this study, factors affecting the quality of services in the health sector in Turkey has tried to determine. In this framework, similar studies in the literature have been examined in detail. In addition, different dimensions of the SERVQUAL method have been considered. Taking into account both issues, 16 different factors were selected that affect the quality of service in hospitals. Then, it is aimed to determine which of these factors are more important. On the other hand, for this purpose fuzzy DEMATEL method was used.

It is concluded that operating hour (C1) has the highest weight among all criteria. Furthermore, it is also identified that the quality of equipment (C3) is the second most significant criterion. In addition to these issues, experience of personnel (C9) and individual attention to patient (C15) are other important items that affect service quality in Turkish health industry. On the other side, the quality of call center (C10), location (C5) and patient communication (C11) are the least important factors.

The obtained results show that patients in Turkey is requested to reach a doctor at any moment. Therefore, it is important that the hospitals aiming to increase the service quality adjust the working hours accordingly. In addition, another consideration is the quality of the equipment that take care of patients in Turkey. In this framework, attention should be paid to the quality of equipment to be purchased by hospitals. In this way, it will be possible to increase the service quality. This is the biggest constraint that analysis in this study is made only for Turkey. Thus, the recommendations made in this study will be specific to the healthcare system in Turkey.

REFERENCES

Akbaba, A. (2006). Measuring service quality in the hotel industry: A study in a business hotel in Turkey. *International Journal of Hospitality Management*, *25*(2), 170–192. doi:10.1016/j.ijhm.2005.08.006

Alam, M. S., & Mondal, M. (2019). Assessment of sanitation service quality in urban slums of Khulna city based on SERVQUAL and AHP model: A case study of railway slum, Khulna, Bangladesh. *Journal of Urban Management*, *8*(1), 20–27. doi:10.1016/j.jum.2018.08.002

Ariff, M. S. M., Yun, L. O., Zakuan, N., & Jusoh, A. (2012). Examining dimensions of electronic service quality for internet banking services. *Procedia: Social and Behavioral Sciences*, *65*(3), 854–859. doi:10.1016/j.sbspro.2012.11.210

Awasthi, A., Chauhan, S. S., Omrani, H., & Panahi, A. (2011). A hybrid approach based on SERVQUAL and fuzzy TOPSIS for evaluating transportation service quality. *Computers & Industrial Engineering*, *61*(3), 637–646. doi:10.1016/j.cie.2011.04.019

Basfirinci, C., & Mitra, A. (2015). A cross cultural investigation of airlines service quality through integration of Servqual and the Kano model. *Journal of Air Transport Management*, *42*, 239–248. doi:10.1016/j.jairtraman.2014.11.005

Belwal, R., & Amireh, M. (2018). Service quality and attitudinal loyalty: Consumers' perception of two major telecommunication companies in Oman. *The Arab Economics and Business Journal*, *13*(2), 197–208. doi:10.1016/j.aebj.2018.11.003

Can, G. F., & Delice, E. K. (2020). An advanced human error assessment approach: HEART and AV-DEMATEL. *Human Factors and Ergonomics in Manufacturing & Service Industries*, *30*(1), 29–49. doi:10.1002/hfm.20819

Chatzoglou, P., Chatzoudes, D., Vraimaki, E., & Leivaditou, E. (2014). Measuring citizen satisfaction using the SERVQUAL approach: The case of the 'Hellenic post'. *Procedia Economics and Finance*, *9*, 349–360. doi:10.1016/S2212-5671(14)00036-7

Chen, K. K., Chang, C. T., & Lai, C. S. (2009). Service quality gaps of business customers in the shipping industry. *Transportation Research Part E, Logistics and Transportation Review*, *45*(1), 222–237. doi:10.1016/j.tre.2008.02.005

Chen, Z., Lu, M., Ming, X., Zhang, X., & Zhou, T. (2020). Explore and evaluate innovative value propositions for smart product service system: A novel graphics-based rough-fuzzy DEMATEL method. *Journal of Cleaner Production*, *243*, 118672. doi:10.1016/j.jclepro.2019.118672

Dinçer, H., Yüksel, S., & Martínez, L. (2019). Analysis of balanced scorecard-based SERVQUAL criteria based on hesitant decision-making approaches. *Computers & Industrial Engineering*, *131*, 1–12. doi:10.1016/j.cie.2019.03.026

Dinçer, H., Yüksel, S., & Pınarbaşı, F. (2020). Kano-Based Measurement of Customer Expectations in Retail Service Industry Using IT2 DEMATEL-QUALIFLEX. In *Handbook of Research on Positive Organizational Behavior for Improved Workplace Performance* (pp. 349–370). IGI Global. doi:10.4018/978-1-7998-0058-3.ch020

Feng, G. C., & Ma, R. (2020). Identification of the factors that influence service innovation in manufacturing enterprises by using the fuzzy DEMATEL method. *Journal of Cleaner Production*, *253*, 120002. doi:10.1016/j.jclepro.2020.120002

Jayakrishna, K., Vimal, K. E. K., Vibha, M., Jain, S., & Kulatunga, A. K. (2020). Fostering systematic eco-innovation in an industrial symbiosis environment using DEMATEL. *International Journal of Industrial and Systems Engineering*, *34*(1), 20–42. doi:10.1504/IJISE.2020.104314

Jeeradist, T., Thawesaengskulthai, N., & Sangsuwan, T. (2016). Using TRIZ to enhance passengers' perceptions of an airline's image through service quality and safety. *Journal of Air Transport Management*, *53*, 131–139. doi:10.1016/j.jairtraman.2016.02.011

Jiang, S., Shi, H., Lin, W., & Liu, H. C. (2020). A large group linguistic Z-DEMATEL approach for identifying key performance indicators in hospital performance management. *Applied Soft Computing*, *86*, 105900. doi:10.1016/j.asoc.2019.105900

Jiao, J., Wei, M., Yuan, Y., & Zhao, T. (2020). Risk Quantification and Analysis of Coupled Factors Based on the DEMATEL Model and a Bayesian Network. *Applied Sciences (Basel, Switzerland)*, *10*(1), 317. doi:10.3390/app10010317

Kalaja, R., Myshketa, R., & Scalera, F. (2016). Service quality assessment in health care sector: The case of Durres public hospital. *Procedia: Social and Behavioral Sciences*, *235*, 557–565. doi:10.1016/j.sbspro.2016.11.082

Kilbourne, W. E., Duffy, J. A., Duffy, M., & Giarchi, G. (2004). The applicability of SERVQUAL in cross-national measurements of health-care quality. *Journal of Services Marketing*, *18*(7), 524–533. doi:10.1108/08876040410561857

Lee, C. H., Zhao, X., & Lee, Y. C. (2019). Service Quality Driven Approach for Innovative Retail Service System Design and Evaluation: A Case Study. *Computers & Industrial Engineering*, *135*, 275–285. doi:10.1016/j.cie.2019.06.001

Limbourg, S., Giang, H. T. Q., & Cools, M. (2016). Logistics service quality: The case of Da Nang city. *Procedia Engineering*, *142*, 124–130. doi:10.1016/j.proeng.2016.02.022

Liu, H., Long, H., & Li, X. (2020). Identification of critical factors in construction and demolition waste recycling by the grey-DEMATEL approach: A Chinese perspective. *Environmental Science and Pollution Research International*, *27*(8), 1–19. doi:10.100711356-019-07498-5 PMID:31907809

Maghsoodi, A. I., Saghaei, A., & Hafezalkotob, A. (2019). Service quality measurement model integrating an extended SERVQUAL model and a hybrid decision support system. *European Research on Management and Business Economics*, *25*(3), 151–164. doi:10.1016/j.iedeen.2019.04.004

Mohammed, K. I., Zaidan, A. A., Zaidan, B. B., Albahri, O. S., Albahri, A. S., Alsalem, M. A., & Mohsin, A. H. (2020). Novel technique for reorganisation of opinion order to interval levels for solving several instances representing prioritisation in patients with multiple chronic diseases. *Computer Methods and Programs in Biomedicine*, *185*, 105151. doi:10.1016/j.cmpb.2019.105151 PMID:31710981

Motaghifard, A., Omidvari, M., & Kaazemi, A. (2020). Introducing a conceptual model for evaluating health safety environmental performance of residential buildings using the fuzzy decision-making approach. *Environmental Monitoring and Assessment*, *192*(1), 19. doi:10.100710661-019-8004-0 PMID:31820115

Ocampo, L., Alinsub, J., Casul, R. A., Enquig, G., Luar, M., Panuncillon, N., Bongo, M., & Ocampo, C. O. (2019). Public service quality evaluation with SERVQUAL and AHP-TOPSIS: A case of Philippine government agencies. *Socio-Economic Planning Sciences*, *68*, 100604. doi:10.1016/j.seps.2017.12.002

Ocampo, L., Alinsub, J., Casul, R. A., Enquig, G., Luar, M., Panuncillon, N., Bongo, M., & Ocampo, C. O. (2019). Public service quality evaluation with SERVQUAL and AHP-TOPSIS: A case of Philippine government agencies. *Socio-Economic Planning Sciences*, *68*, 100604. doi:10.1016/j.seps.2017.12.002

Pansiri, J., & Mmereki, R. N. (2010). Using the SERVQUAL model to evaluate the impact of public service reforms in the provision of primary health care in Botswana. *Journal of African Business*, *11*(2), 219–234. doi:10.1080/15228916.2010.509005

Punnakitikashem, P., Buavaraporn, N., Maluesri, P., & Leelartapin, K. (2012, April). Health care service quality: case example of a hospital with lean implementation. In *POMS 23 rd Annual Conference* (pp. 20-23). Academic Press.

Rezaei, J., Kothadiya, O., Tavasszy, L., & Kroesen, M. (2018). Quality assessment of airline baggage handling systems using SERVQUAL and BWM. *Tourism Management*, *66*, 85–93. doi:10.1016/j.tourman.2017.11.009

Roses, L. K., Hoppen, N., & Henrique, J. L. (2009). Management of perceptions of information technology service quality. *Journal of Business Research*, *62*(9), 876–882. doi:10.1016/j.jbusres.2008.10.005

Roslan, N. A. A., Wahab, E., & Abdullah, N. H. (2015). Service Quality: A case study of logistics sector in Iskandar Malaysia using SERVQUAL Model. *Procedia: Social and Behavioral Sciences*, *172*, 457–462. doi:10.1016/j.sbspro.2015.01.380

Sagnak, M. (2020). Selecting the Most Appropriate Supplier in the Green Environment: A Hybrid Multi-Criteria Decision-Making Application. In Handbook of Research on Industrial Applications for Improved Supply Chain Performance (pp. 101-124). IGI Global. doi:10.4018/978-1-7998-0202-0.ch005

Sam, E. F., Hamidu, O., & Daniels, S. (2018). SERVQUAL analysis of public bus transport services in Kumasi metropolis, Ghana: Core user perspectives. *Case Studies on Transport Policy*, *6*(1), 25–31. doi:10.1016/j.cstp.2017.12.004

Sayareh, J., Iranshahi, S., & Golfakhrabadi, N. (2016). Service quality evaluation and ranking of container terminal operators. *The Asian Journal of Shipping and Logistics*, *32*(4), 203–212. doi:10.1016/j.ajsl.2016.12.003

Shaikh, B. T., Mobeen, N., Azam, S. I., & Rabbani, F. (2008). *Using SERVQUAL for assessing and improving patient satisfaction at a rural health facility in Pakistan*. Academic Press.

Soita, P. W. (2012). Measuring perceived service quality using SERVQUAL: A case study of the Uganda health and fitness sector. *International Journal of Business and Social Science*, *3*(5).

Solanki, R., Darbari, J. D., Agarwal, V., & Jha, P. C. (2020). A Fuzzy Multi-criteria Decision Model for Analysis of Socio-ecological Performance Key Factors of Supply Chain. In *Soft Computing for Problem Solving* (pp. 671–685). Springer. doi:10.1007/978-981-15-0035-0_55

Song, W., Zhu, Y., & Zhao, Q. (2020). Analyzing barriers for adopting sustainable online consumption: A rough hierarchical DEMATEL method. *Computers & Industrial Engineering*, *140*, 106279. doi:10.1016/j.cie.2020.106279

Souri, M. E., Sajjadian, F., Sheikh, R., & Sana, S. S. (2018). Grey SERVQUAL method to measure consumers' attitudes towards green products-A case study of Iranian consumers of LED bulbs. *Journal of Cleaner Production*, *177*, 187–196. doi:10.1016/j.jclepro.2017.12.105

Taner, T., & Antony, J. (2006). Comparing public and private hospital care service quality in Turkey. *Leadership in Health Services*, *19*(2), 1–10. doi:10.1108/13660750610664991 PMID:16875104

Wu, S. M., You, X. Y., Liu, H. C., & Wang, L. E. (2020). Improving quality function deployment analysis with the cloud MULTIMOORA method. *International Transactions in Operational Research*, *27*(3), 1600–1621. doi:10.1111/itor.12484

Yazdi, M., Nedjati, A., Zarei, E., & Abbassi, R. (2020). A novel extension of DEMATEL approach for probabilistic safety analysis in process systems. *Safety Science*, *121*, 119–136. doi:10.1016/j.ssci.2019.09.006

Yesilada, F., & Direktör, E. (2010). Health care service quality: A comparison of public and private hospitals. *African Journal of Business Management*, *4*(6), 962.

Zarei, A., Arab, M., Froushani, A. R., Rashidian, A., & Tabatabaei, S. M. G. (2012). Service quality of private hospitals: The Iranian Patients' perspective. *BMC Health Services Research*, *12*(1), 31. doi:10.1186/1472-6963-12-31 PMID:22299830

Zhang, L., Liu, R., Jiang, S., Luo, G., & Liu, H. C. (2020, March). Identification of Key Performance Indicators for Hospital Management Using an Extended Hesitant Linguistic DEMATEL Approach. *Health Care*, *8*(1), 7. PMID:31881773

KEY TERMS AND DEFINITIONS

AHP: Analytical hierarchy process.

ANP: Analytical network process.

DEMATEL: Decision-making trial and evaluation laboratory.

SERVQUAL: It is a methodology which has some dimensions to measure the service quality of the companies.

Table A1. Evaluations of Decision Maker 1

Criteria	C1	C2	C3	C4	C5	C6	C7	C8	C9	C10	C11	C12	C13	C14	C15	C16
C1	-	H	M	M	M	H	H	V	M	V	H	V	H	M	M	H
C2	V	-	M	H	M	H	M	M	L	L	M	V	V	M	M	M
C3	H	M	-	M	L	V	V	V	M	M	H	V	V	V	V	M
C4	H	V	M	-	M	H	L	M	M	L	H	M	H	M	V	M
C5	H	V	L	M	-	H	L	L	M	L	H	V	H	L	H	M
C6	H	M	L	L	L	-	H	V	V	H	H	H	H	H	V	M
C7	H	M	L	L	L	H	-	V	V	H	H	H	H	H	V	M
C8	M	H	L	L	L	L	H	-	M	L	H	H	H	M	V	H
C9	H	H	M	L	L	H	V	V	-	H	V	H	V	V	V	V
C10	H	M	M	L	H	H	M	M	M	-	V	H	H	H	V	V
C11	M	M	L	L	L	M	H	M	L	M	-	H	V	H	V	H
C12	V	H	V	H	V	V	M	H	M	H	H	-	H	H	V	H
C13	M	M	H	H	V	H	V	H	V	H	V	H	-	H	V	M
C14	M	M	V	M	M	H	V	H	H	M	V	H	V	-	V	H
C15	V	V	V	H	H	V	V	V	V	V	V	H	H	H	-	M
C16	H	H	H	H	V	H	M	M	M	H	M	H	H	H	M	-

APPENDIX

Table A2: Evaluations of Decision Maker 2

Criteria	C1	C2	C3	C4	C5	C6	C7	C8	C9	C10	C11	C12	C13	C14	C15	C16
C1	-	H	L	L	H	M	H	L	L	N	V	H	H	L	V	N
C2	L	-	L	L	L	L	L	L	L	L	L	M	H	M	L	L
C3	N	H	-	H	N	N	V	V	M	L	L	V	V	L	L	M
C4	L	H	H	-	L	L	L	L	L	L	L	L	L	L	L	L
C5	L	L	L	L	-	L	N	N	N	N	L	L	L	N	L	N
C6	N	N	N	N	N	-	M	L	L	L	L	L	L	L	M	M
C7	N	N	N	N	N	N	-	N	N	N	N	N	N	N	N	N
C8	L	L	H	N	N	H	N	-	N	N	N	N	N	N	N	N
C9	L	H	N	N	N	L	V	L	-	L	H	H	H	H	V	M
C10	N	N	N	N	N	N	N	N	N	-	N	N	N	N	N	N
C11	N	N	N	N	N	N	N	N	N	N	-	N	N	N	N	N
C12	M	V	H	H	N	N	H	N	N	N	N	-	N	N	N	N
C13	N	N	N	N	N	N	N	N	N	N	N	N	-	N	N	N
C14	L	L	L	L	L	L	L	L	L	L	L	L	L	-	L	L
C15	L	L	L	L	L	L	L	L	L	L	L	L	L	L	-	L
C16	N	N	N	N	N	N	N	N	N	N	N	N	N	N	N	-

Table A3: Evaluations of Decision Maker 3

Criteria	C1	C2	C3	C4	C5	C6	C7	C8	C9	C10	C11	C12	C13	C14	C15	C16
C1	-	H	M	L	M	H	H	H	H	M	H	H	V	M	H	H
C2	M	-	L	M	M	H	L	M	M	M	M	H	H	H	M	L
C3	L	M	-	H	M	M	H	H	M	M	M	H	H	M	M	M
C4	L	M	M	-	L	M	L	L	L	M	M	H	H	M	M	M
C5	M	M	M	M	-	M	L	L	M	L	M	M	M	L	M	M
C6	H	M	L	M	M	-	M	M	M	M	M	L	H	L	H	L
C7	M	L	M	L	L	M	-	M	H	L	L	M	H	M	M	M
C8	M	L	M	L	M	M	L	-	M	L	M	L	H	L	H	M
C9	L	M	M	M	L	M	H	M	-	H	M	M	H	M	H	M
C10	L	M	M	M	M	M	L	L	M	-	H	L	L	M	M	M
C11	L	M	L	M	M	M	L	M	M	H	-	M	M	M	H	M
C12	M	H	M	M	M	M	M	L	L	L	M	-	H	M	M	M
C13	M	H	M	M	M	H	H	L	M	M	M	H	-	M	M	M
C14	M	M	M	L	M	M	L	L	H	M	H	H	H	-	M	M
C15	M	M	L	L	M	H	M	M	H	H	H	M	M	M	-	L
C16	M	L	M	M	M	H	L	M	M	H	L	L	L	L	M	-

Chapter 12
Economic Sustainable Health Information Systems

Angelos I. Stoumpos
Open University of Cyprus, Cyprus

Michael A. Talias
https://orcid.org/0000-0002-1802-5586
Open University of Cyprus, Cyprus

ABSTRACT

Public health systems have adopted computer health and information technology as a dynamic transformational tool both to improve real-time surveillance systems and to communicate and exchange information between different organizations. Health information systems (HIS) incorporate data collection, processing, report creation, and the use of information necessary to improve the efficiency and effectiveness of health services through better management of these services at every level. The sustainable development of these systems depends on their interoperability, the combination of the "three pillars of sustainability" (economic, social, environmental). At the global health level, there are variations between countries in terms of the application and implementation of sustainable systems. However, taking advantage of the available knowledge and technology, and with proper management of the economy, the viability of health information systems can be sustainable.

INTRODUCTION

Technological developments are increasingly raising concerns in the field of e-health about costs, privacy, security, and information. However, the main technical obstacles mainly concern the infrastructure and the issues of using standards and interoperability. Advances in e-health standards can meet the challenges posed in the health sector. Some benefits of adopting the standards are summarized as follows:

- They create the necessary interoperability between health systems.
- They prevent "locking" by a single supplier.

DOI: 10.4018/978-1-7998-5442-5.ch012

- Minimize the risks of developing new technologies.
- They reduce costs, as they facilitate the growth of competition and eliminate the need for expensive solutions.
- They address specific concerns about patient privacy, security, and identification.
- Ensure the dissemination and adoption of new IT (Information Technology) and Health Communication Technology solutions.

Therefore, perhaps one of the most critical problems facing the medical field is the lack of a unique or otherwise varied classification and coding. At this point, we should mention the production of data, and information on health is enormous. The source (administrative, medical, financial, nursing, etc.) as well as their form (images, texts, audio, biomarkers, videos, etc.) varies. This is also the reason why they need for both standardization and codification in the medical field is becoming more and more necessary so that the various subsystems and their subunits can communicate and exchange data and messages.

The use of proper coding allows the standardization and systematic storage of information (including financial information) to assist in diagnoses, to provide better and better health services ultimately, and to preserve medical history. At the same time, through coding, as in any other administrative process, we can compare the data and reach statistical and administrative conclusions, drawing up the appropriate strategy promptly.

Coding can be defined as the process of organizing both data and health information into categories, for shortening, storing, and quickly retrieving this data. Coding, therefore, offers a "common language" for interpreting and representing concepts and data, which is a crucial prerequisite for achieving semantic interoperability. In this way, not only the ambiguity and errors are reduced, but at the same time, the quality of the data is upgraded and their comparability and standardization are achieved, as well as the faster processing and search of data via computers and the production of other data correlations.

On the other hand, a set of rules, demands, conditions, and ways of representing information is defined as a standard. Because of the criteria, ambiguities, incompatibilities, and misinterpretations of the data are avoided. Some of the basic template categories are:

- Clinical data representation standards.
- Communication standards.
- Recognition standards.
- Medical technology product safety standards.
- Data security standards and medical confidentiality.
- Electronic health envelope architecture standards.
- Costing standards.

Nevertheless, it should be mentioned that without interoperability, e-health solutions cannot be achieved. The development of independent healthcare solutions and applications has led to the existence of information systems that most of the time, cannot communicate with each other. However, health as a total good requires total solutions. The European Union defines interoperability as follows: "The exchange of two or more e-health applications (e.g., electronic health records) that can exchange, understand and act based on the citizen/patient and other health-related information and knowledge between different - linguistically and culturally - clinical scientists, patients, and other bodies or organizations within and between the various health systems, in a collaborative framework."

Interoperability is a prerequisite for facilitating the efficiency of e-health distribution. It concerns the need for collaboration, information exchange, shared and further use of information to increase administrative efficiency, reduce bureaucracy for patients/citizens, and to save financial resources. The result is improved patient service delivery, as well as reduced costs for administrations and patients/citizens. The concept of interoperability consists of four primary levels:

- Technical: It refers to the interconnection of used computers and services and the facilitation of independent systems to exchange health-related information. This category includes topics such as "open" interfaces, interface services, data integration, and the use of intermediate applications, data presentation and exchange, access, and data security services.
- Semantic: Semantic interoperability not only implies the ability of systems to exchange data but also to be able to understand them. It therefore deals with the possibility of exchange, understanding, and action concerning the health information transmitted.
- Organizational: Organizational interoperability defines the necessary operational objectives and procedures for promoting collaboration between services and healthcare providers who wish to exchange information but have different internal structures and processes.
- Legal: It is related to the development of the appropriate legal framework to allow safe access and processing of data transmitted electronically to patients.

At this point, given the data so far, we could argue about the importance of interoperability. The creation of interoperable health systems contributes, among other things, to reducing costs and providing better and more efficient health services. At this time, communication between different e-health systems can bring significant benefits, both for health professionals (e.g., doctors in the same or different hospitals) and for the patients themselves, who will receive the most effective care. For this reason, interoperability is a prerequisite, but also a catalyst for the development of e-health services. The benefits of the interoperability of e-health solutions are varied:

- For patients: provision of services at the point of need, sense of security, quality of care even abroad.
- For healthcare professionals: upgraded quality and safety of care through successful collaboration between different care providers and physicians. Updated patient information and enhanced physician decision-making and diagnosis support based on experience and clinical protocols.
- For payers: reduction of spending on electronic health information systems. For example, because of the interoperability, an MRI (Magnetic Resonance Imaging) scan performed immediately at a hospital in a doctor's office or another hospital during the patient's treatment can be carried out directly, without the latter having to repeat the examination.
- For the healthcare industry: the use of a common framework for the interoperability of e-health solutions at the European level can enhance the creation of a single digital healthcare market, increasing competition and reducing costs for organizations developing such solutions.

ECONOMY, TECHNOLOGY AND KNOWLEDGE

Electronic Health Record (EHR)

Healthcare systems are becoming patient-centered, with a focus on prevention, preventive and personalized services, and the healthy lifestyle of citizens. Innovative technologies allow citizens to survive, as they enable them to manage their complete health and well-being. From healthcare organizations, the new standard implies changes in the production of services, both in terms of use and content. Some of the critical results of the anthropocentric approach are improving the lifestyle of citizens, losing weight, reducing the length of hospital stay, better access to health-related services, and improving communication between healthcare providers.

Based on the literature review, a new innovative approach is proposed to build the next generation of collaborative health information space that connects health care providers with citizens and helps them by giving them access to distributed health resources at any time and wherever required. The visible, sustainable health environment that focuses on citizens offers a means for the gradual transition from the current situation to a caring environment focused on citizens, where the latter has a participatory role in healthcare activities.

To implement this patient-centered model of care, we need new tools for extensive collaboration and support activities for the health and well-being of citizens. One of the existing tools is the Electronic Health Record (EHR). Some of its features are:

1. Each person checks their electronic envelope.
2. Contains information from the entire life of the individual, including information from all health care providers.
3. It is accessible from anywhere and at any time.
4. It is private and secure.
5. It is transparent, i.e., individuals can see who entered each piece of the transferred data and who has access to it.
6. Allows secure exchange of information throughout the health system.

The World Health Organization (WHO) states that e-health is "the leverage of information and communication technology (ICT) to connect providers, patients and governments, training and informing professionals. "The health sector, managers and consumers, the stimulation of innovation in the provision of care and the management of the health system, as well as its improvement" (Wickramasingle, Bali, Kirn, & Suomi, 2012). Therefore, e-health is a long term that includes various activities in a growing field. So, we can accept e-health is mainly activated and guided by the use of Information and Communication Technology (ICT) in healthcare with the ability to change the healthcare industry worldwide in terms of infrastructure as well as cost and services.

According to the WHO, e-health assessment can be performed when designing, developing, or operating an e-health system (Wickramasingle, Bali, Kirn, & Suomi, 2012).

E-health aims to improve traditional health care services. Therefore, the goals are to eliminate inequalities, inadequacies, poor quality, lack of health resources, and inadequate distribution. In essence, e-health aims: at efficiency, quality of care, documentation, empowerment of consumers and patients, training of physicians, expanding the scope of health care, impartiality, ethics, and cost savings.

E-health can bring to a country, either benefits or disadvantages. Briefly, we could summarize them in the following categories:

- Effects on Informatics Education. Many countries have a large number of consumers with a low level of computer education. Especially for the population in rural areas, it is difficult to learn e-health and telemedicine skills in the short term. Informatics and e-health education are largely related. However, the vast majority of young people have grown up in the information age. They have received extensive training in information technology, which they use even in their workplace. More and more young people are also starting to use computers, search online, and shop online. In the long run, young people will be the backbone of the e-health market in the future. Therefore, to develop e-health, each country needs to educate its nation and make its citizens more aware of e-health and e-consumption (Wickramasingle, Bali, Kirn, & Suomi, 2012).
- Effects on morbidity. For instance, China was a country with a high rate of morbidity, according to statistics from the Chinese Ministry of Health in 2008, where the occupancy rate was 0.684%, and the hospital bed occupancy rate rose to 74.6%. These statistics show that medical infrastructure needs more investment. The government and public hospitals are the main carriers of the medical infrastructure, and the development of information systems Health Information System (HIS), Electronic Health Record (EHR), and several other supporting infrastructures are considered sustainable. It is worth noting at this point that in the case of China, the financing of the construction of applications and e-health solutions is decided mainly by the investment policy of each government (Wickramasingle, Bali, Kirn, & Suomi, 2012).
- Impacts on cultural / social dimensions. Each country implements its treatments. In China, for example, although people accept Western medicine, they place more emphasis on traditional Chinese medicine. A personal treatment appointment with a doctor is always preferable to telemedicine. E-health should not only be used as a technological achievement or should be oriented only in the direction of information, but should focus on the core of the entire medical system. After all, this is a challenge that health needs to address. Nevertheless, whether e-health can gain access to appropriate levels of government funding and overgrow on the role it will play in health care reform (Wickramasingle, Bali, Kirn, & Suomi, 2012).
- A counterpoint to World Economic Statistics. Countries with rapid economic growth, like China or the USA, are a good opportunity to develop e-health. However, there are some financial risks that e-commerce industries will face. The long-term economic risks are real: the economic structure in most countries is unreasonable, and the economic recovery required for countries that have emerged from financial crises requires time. Also, the risks of the pharmaceutical and medical market are accompanied by financial risks. The effectiveness of the medical control system, government medical institutions, corruption, and the unbridled health care market are all factors limiting the development of e-health worldwide (Wickramasingle, Bali, Kirn, & Suomi, 2012).

Finally, electronic prescription systems can be integrated into Electronic Health Record systems. Electronic folders include patient information, like clinical notes, laboratory instructions, and results, as well as Clinical Decision Support Systems (CDSS) (Porterfield, Engelbert, & Coustasse, 2014). The electronic prescription aims mainly at increasing the quality, while at the same time, it contributes to the reduction of the cost of health care. However, the main goal of this system is to offer interested parties the expected cost-effectiveness ratio (Zadeh & Tremblay, 2016).

Sustainability

According to Winston Churchill, "We make a living from what we get. We make a living from what we give." This quote could summarize the concept of sustainability and sustainable development and the economy. This is because sustainable development is not just an international trend; it is how the economics will develop in the future. A philosophy that also affects the investment process and sets specific criteria for responsible investments.

The concept of sustainable development is being put forward for the first time through the World Development and Environment Commission's report, known as the Brundtland Report, and defined as "Development aimed at serving the needs of the present, using the resources available in this way, so that the satisfaction of the needs of future generations is unmortgaged" (Brundtland Report 1987: 15). Sustainability is based on three main pillars:

1. Environmental protection,
2. Economic growth and
3. Social cohesion.

It is, therefore, a dynamic process, where knowledge of the structure and operation of both environmental and human systems, which include the systems of the economy and society, is essential (Kimaro & Nhampossa, 2007). Understanding the relationship and hierarchy between the three pillars of sustainability is equally important. There is a hierarchy between the pillars. The hierarchically superior environment exercises long-term control over the economic and social viability of a system. Besides, the components are not static, but on the contrary, are continually changing and evolving.

Sustainability, otherwise known as sustainability, means maintaining something that already exists over time and often equates to "self-preservation" or "autonomy," implying that no external support is required to continue its existence. In terms of information technology, "sustainability" means the ability to identify and manage risks that threaten the long-term viability of IT (Information Technology) (Kimaro & Nhampossa, 2007). Also, computer sustainability is mainly dependent on the fact that the system can provide reliable and useful information as well as the ability of users at all levels to use it effectively.

In the healthcare sector, health systems have always faced enormous challenges. However, today, their viability is considered a critical condition for their future adaptation. Some of the challenges facing a healthcare system in the medium term, according to Coiera and Hovenga in 2007, are new technology, growing spending to meet the needs of an aging society, education and staffing in the healthcare system and finally quality and the safety of health services (Fischer, 2015). All of these challenges, including others, like culture and the ground, could be described as pillars of the three fundamental components of sustainability.

The assimilation of information systems across the healthcare industry has increased dramatically, especially nowadays, where Information Systems [IS] / Information Technology [IT] can reap enormous medical benefits staff to provide better services. However, the excitement of implementing new computer systems usually worsens dramatically once the system is acquired. This raises an essential issue in the assimilation of new technology that could harm the successful use of the information system.

The assimilation of innovation can be defined as a series of stages that start from the initial evaluation of the organism, for the possible system that will be used, for its official adoption and finally in a well-accepted development of the system to a point where it is a vital part value chain activities. The final stage, where the system becomes part of the value chain activities, is called routine. However, it is worth considering a study that examines the body, the individual, and the technological level, focusing on how technology can be maintained at all stages of innovation simulation and why there is an assimilation gap. Also, how and why people in the body reject or refuse to use innovation after the process of acquiring it by the body, can cause it to stop, thus giving it due importance. Furthermore, there is also a lack of theories developed in a particular type of technology and a specific adoption framework due to the lack of a general theory of technological innovation.

The description of dissemination of innovation is the process in which the knowledge of a change is disseminated to a population. Through the decision of a unit in the organization, the adoption or rejection of the innovation is finally approved. Regarding how to deal with assimilation gaps, it is necessary to examine the elements that affect the process of assimilation of change. Based on the revised bibliography, the Technology – Organization – Environment framework (TOE) provides aspects that companies need to consider when studying the effects on the assimilation of technological innovation (Wickramasingle, Bali, Kirn, & Suomi, 2012). These concepts are grouped into three axes: the technological framework, the organizational framework, and the environment.

Technical framework: Include internal and external technologies related to the business involved in equipment and procedures.

Organizational framework: Include the characteristics and resources of businesses, the structure of management, human resources, the number of loose resources, and the links between employees.

Environmental framework: Include industry structure, business competitors, macroeconomic perception, and regulatory environment (Wickramasingle, Bali, Kirn, & Suomi, 2012).

Sustainable Strategies

The leading emerging e-health solutions can be summarized as follows:

- Distributed electronic health recording systems, interoperable, and interconnected service networks. These solutions create an on-line clinical profile for each patient, where all the data and documents related to the health care of a particular individual can be stored and managed by the stakeholders on a remote basis. As part of a comprehensive online health information network, health records systems can allow for a dramatic simplification of many procedures (like as referrals, prescriptions, reservations, returns and exchange of detailed tests, laboratory results and letters between patients, general and of specialized physicians) related to the complex networked nature of healthcare systems.
- Management and Administration Systems. Support systems, like financial systems and performance control systems, and supply chain management systems, programming systems, charging systems, etc. that health professionals and patients do not usually use clinical support procedures during the provision of health care.
- Clinical Information Systems. Specialized tools for health professionals, like computer-aided diagnostic systems, radiological information systems, surgical training systems, and pharmacy information systems.
- Clinical data analysis systems. Specialized systems for researchers in the field of public health, like statistical programs for the analysis of infectious diseases.

- Social networking and Web 2.0 systems focused on health issues. Web-based systems that connect specific communities (e.g., parents of children suffering from an illness or doctors working in one particular national health system). Also, "traditional" educational health websites are evolving towards the web 2.0 network, that is, from the bottom up and in an interactive way.
- Patient relationship management systems. Customer relationship management systems aimed at understanding the needs, requirements, and satisfaction levels of specific patients, as well as activating customer-oriented problem-solving and organizational innovation processes.
- Telemedicine systems. Personalized disease management services like: teleradiology, telediagnosis, telecounseling, and remote monitoring (Wickramasingle, Bali, Kirn, & Suomi, 2012).

Some of the strategies committed to dominating the healthcare sector in Western developed countries can be described as follows:

1. Improving internet processes. Healthcare systems are networked systems, where they have to manage very complex relationships between the institutions (who take care of them, who decide what kind, who pays for it, who offers it, who evaluates it, etc.). As a result, innovation cannot be limited to individual organizations or factors (e.g., hospital): network processes and network-integrated knowledge are necessary to determine system performance.
2. Development of evaluation processes. Healthcare systems always need evolving measurements and procedures to evaluate system performance, like at least three key aspects: cost control, public health, and well-being performance and customer satisfaction.
3. Development of incentives and deterrents. During the development of health care systems, strategies must be identified to stimulate positive behaviors and prevent harmful ones (like corruption, negligence, or waste).
4. Improving patient safety. Risk management plays a central role in healthcare systems, as well as in all initiatives that may increase patient safety, for example, by improving diagnostic procedures. Again, the role of the knowledge grid will be crucial.
5. Enhancing primary care. A strong primary care network (responsible for all non-specialist care procedures that can be performed before a patient is referred to a hospital, such as family physician services), with easy access for all, is essential for sustainable, effective, and a satisfactory health care system.
6. Strengthen preventive services. Preventive medicine is becoming more and more critical. Proper investment in prevention initiatives (including education, promotion, planned diagnostic testing, guidance, public health data collection) can result in better performance in the health system than could be achieved if some funding for health care "assessment" procedures.
7. Enhancing external care. Outpatient care should become increasingly capable of resolving problems that have previously required in-hospital care. Hospital care should be provided exclusively in specialized services. All solutions that can reduce secondary care without harming clinical performance are likely to prevail. To achieve this goal, hospitals are vital factors: they need to stop behaving like "buildings" and start acting more and more like "territorial networks."
8. Development of specialized structures. Competition in the market will benefit from extensive and specialized structures. Many small hospitals are about to be abandoned and closed, meaning that the development of strategies must provide specialized health care services on a long-term basis effectively.

9. Customer-oriented service development. Healthcare is expected to become more and more customer-oriented: personalization, claims management, transparency, simplicity, human interface, will be increasingly important, and all the innovations that allow patients to save time.
10. Developing the active role of consumers. Consumers are becoming more and more aware, demanding, and willing to take control of their health. Initiatives that can positively develop the part of patients as active and educated members of the health network, selecting and rewarding the best health care providers, are likely to increase the quality and value of the health system.
11. Development of health training services. Long-term "pull" and "push" health training will become increasingly important as a service. It served the patient before (as a preventive activity) and after the onset of a disease, identifying possible threats, monitoring, reminder, encouraging attachment to healthy behaviors, etc.
12. To be able to live independently. Healthcare systems need to focus on solutions that improve the independent living of patients with disabilities and chronic diseases.
13. Protection of privacy and ethics. Healthcare services deal with the most critical data items, namely life and well-being, and the most confidential data. Every innovation in this field is acceptable, must be reliable in terms of privacy protection and ethics.
14. Integration of health, social, and government services. Healthcare is a potential catalyst for building integrated human-centered services (Wickramasingle, Bali, Kirn, & Suomi, 2012).

Sustainable Knowledge

Over the past decade, there has been a scientific consensus on the importance of sustainability, both organizationally and nationally and globally (Murdoch & Clark, 1994). In this context, healthcare organizations are not only expected to be sustainable but also in their interest to invest in sustainability. At the heart of this development is knowledge.

Today, society is experiencing a new era where the proper capitalization of knowledge determines competitiveness and long-term sustainability (Zbuchea, Pinzaru, Busu, Stan, & Bargaoanu, 2018). As a clear indication of this trend, knowledge management has emerged as a key interest for both the academic community and practice, immersed in a variety of sectors and industries. What we once thought to be a modern approach to today's phenomena has quickly become a multi-layered reality: At both macroeconomic and microeconomic levels, the urgent need for knowledge-based processes becomes topical and exciting (Zbuchea, Pinzaru, Busu, Stan, & Bargaoanu, 2018).

At the macroeconomic level, more and more economic development strategies are knowledge-based, thus demonstrating the tendency of decision-makers to collect, store, create, and utilize both existing and new sources of knowledge. The development of new knowledge, as well as the building of the ability to access specific environments, was created as levers for sustainable development. Organically linked to this development, on a tiny scale, a wide range of organizations support strategies and practices that lead to effective knowledge management systems, thus recognizing their impact on business performance and indirectly on sustainable development.

By supporting these perspectives, the new economic and business models give rise to a comprehensive strengthening of the ability of societies, communities, companies, and individuals to use it wisely and to benefit from this knowledge (Zbuchea, Pinzaru, Busu, Stan, & Bargaoanu, 2018). Moreover, they place knowledge at the core of their existence and survival, treating it simultaneously as a strategic resource or as a key business asset.

Nevertheless, attention paid to public organizations, educational institutions, and non-profit organizations allows the examination of a complex frame of reference focusing on internal processes related to knowledge management as well as intellectual capital (Zbuchea, Pinzaru, Busu, Stan, & Bargaoanu, 2018). Therefore, more attention should be paid to inter-organizational processes, including open knowledge transfer within a particular sector or between different sectors, with the aim of sustainable development without exclusions.

Maintaining knowledge is a key aspect for successful healthcare organizations, as knowledge is one of the most valuable assets. However, to be effective, practical knowledge management must be based on a proper set of principles, which leads to increased innovation skills (Zbuchea, Pinzaru, Busu, Stan, & Bargaoanu, 2018). Therefore, in hospital organizations, the effectiveness of overall activity is mediated by knowledge management strategies through enhanced innovation. Nevertheless, knowledge management is part of the equation strategy (Zbuchea, Pinzaru, Busu, Stan, & Bargaoanu, 2018). In a highly technological environment, it complements leadership and strategic management in business performance.

The transfer of knowledge in the healthcare sector between different types of organizations greatly facilitates innovation. Several studies suggest that collaboration between organizations in the value chain is crucial to knowledge creation and innovation. Therefore, the exchange of knowledge, cooperation, and the preservation of knowledge should coexist in a balanced way.

Although many of the world's tasks and resolutions focus on cost and pricing, both in terms of understanding and in terms of the policy, the issues of science and technology have come to the fore in recent years. As a result that the sustainability is increasingly based on innovation skills and the utilization of the necessary knowledge.

In less developed countries, the situation is much more difficult: while GDP (Gross Domestic Product) growth of 6% to 8% per year is necessary to meet all their needs for the next three or four decades, but they must also comply with sustainability requirements while developing the necessary elements and appropriate knowledge, where they are often deprived (Our Common Future - Report of the World Commission on Environment and Development, 1987). The transition to a global, networked economy of knowledge will be one of the most important social and economic changes of the coming decades.

Switzerland, for example, has set up the Swiss Accident Insurance Fund, thus contributing to the management of knowledge through the InWiM (Integrierte Wissensbasen der Medizin, which means "Integrated Knowledge Bases in Medicine") program. It is the most important body for compulsory accident insurance in the country.

The project is part of an ISO 9001 certification program. It includes the definition and documentation of all processes in the field of knowledge management as well as the development of the underlying ITC (International Technology Corporation) infrastructure. The knowledge representation model used to implement Information and Communication Technologies involves the knowledge of a multidimensional network of interconnected information units (Wickramasingle, Bali, Kirn, & Suomi, 2012).

The model allows for comments on a narrative description of the nature of information units (e.g., documents) as well as their interfaces. InWiM manages, in addition to publications, the monitoring of "knowledge gaps," i.e., areas where the correct knowledge is lacking. In particular, it is possible to seek and identify knowledge gaps and solutions that cover the same or a similar issue, thus allowing appropriate convergence and preventing the overlap of work.

Although the majority of e-health initiatives focus on the management of medical records and the interoperability of healthcare IT systems, knowledge management remains one of the cornerstones. Medicine often faces the problem of a variety of views on the events of a case. The coexistence of doctrines leads to insecurity as to the appropriateness of any actions and concerns for physicians and patients. Additional costs can be another negative result. In the late 1970s, a team at McMaster Hamilton University, Ontario, began developing new ideas for clinical decisions. Later, this work emerged in a global movement called Evidence-Based Medicine Working Group (1992) (Wickramasingle, Bali, Kirn, & Suomi, 2012). David Sackett defined medicine based on evidence as "the conscious, explicit and rational use of today's best evidence in making decisions about the care of individual patients" (Sackett et al. 1996) (Wickramasingle, Bali, Kirn, & Suomi, 2012). But before we can make "explicit and prudent use of today's best practices," we need the current best knowledge. Given the oversupply of today's information, a business that wants health professionals to apply Medicine based on evidence needs a proper knowledge management infrastructure.

Medical insurance has always aimed and continues to aim to provide quality health care, but also focuses on the economic efficiency of interventions. This may be one of the reasons why the IT infrastructure for knowledge management has gained ground, especially in the field of insurance medicine. Another reason may be that in health insurance, deviant views not only affect treatment procedures but also carry the risk of different decisions about financial aspects, e.g., whether accident insurance covers a health problem. Patients perceive different decisions in this area as unfair, leading to lawsuits. Also, in terms of ethics, such a discrepancy in judgments is considered unacceptable.

In everyday language, people often do not distinguish the terms "data," "information," and "knowledge," and there are many different terminologies for these terms in the literature (Wickramasingle, Bali, Kirn, & Suomi, 2012). According to the definitions of Rehäuser et al. (1996), which are also adopted by InWiM, the term "data" refers to symbols that represent an event or statement of an event without any relation to other data. The term "information" is used for the data in the box. "Knowledge" is achieved through the process of connecting information into single information based on logical thinking. Schreyögg and Geiger call this "knowledge approach inspired by information theory" (Schreyögg and Geiger 2007). In general, knowledge management systems do not deal with real data. Once the data is subject to knowledge management processes, it has already been set in a context (Wickramasingle, Bali, Kirn, & Suomi, 2012). Therefore, it is more appropriate to use the term "information." In practice, there is no strict framework between information and knowledge. It is useful to consider the path from information to knowledge as continuous. The following example illustrates this: a single scientific publication will provide the reader with information on a particular issue.

A researcher, if he conducts several studies and draws conclusions and conclusions in relation, will gather more and more knowledge about the subject of these studies. Eventually, some authors will write comments about studies, and one day, someone will summarize all the publications, including the primary, secondary, and tertiary literature. In the process, information is transformed into knowledge. Provided that the reviews are done correctly, the knowledge becomes healthier and healthier with each review.

It may be useful in this context to distinguish the term "knowledge" as we understand it here from the concept in everyday language, where knowledge has the meaning of "truth," which distinguishes it from "faith" (Wickramasingle, Bali, Kirn, & Suomi, 2012). However, knowledge in the context of information theory does not imply correctness. Knowledge and information can be wrong or correct. To point out that some knowledge is reliable, one can speak of established knowledge. Evidence-based medicine goes a step further and suggests reliability with evidence-based levels and recommendation levels, for example, if multiple studies of high methodological quality properly support the evidence.

According to Schreyögg and Geiger, "knowledge" is understood as the result of the interconnection of information through brain work (Wickramasingle, Bali, Kirn, & Suomi, 2012). This means knowledge is subjective: Personal knowledge can be seen as a distinct network of interconnected information. It is important to see such a personal knowledge network is multidimensional. This is because information can be interconnected from different points of view. The following example may illustrate this: Suppose we have some studies investigating the effect of various surgical techniques on joint surgery. It will make sense to link all the works that report the same technique to different joints, as well as those that explore different techniques in the same joint. Additional links may indicate judgments about the quality of the study, demographic issues, etc. Unlike hyperlink technology on the World Wide Web, all links are two-way: the target knows the source of the links. Links are therefore called interfaces.

Understanding knowledge "inspired by information theory" (Rehäuser et al., 1996) implies that knowledge must always be examined in a context (Wickramasingle, Bali, Kirn, & Suomi, 2012). A model that claims to be able to visualize knowledge as a network of cross-sectional information units must, therefore, include meta-information not only for the information units (e.g., documents) but also for each cross-link. The meta-information includes not only the author's name, date, edition, etc., but also frees text comments, which allow a narrative description of the point of view based on which a particular cross-link was created.

If we adopt the model that personal knowledge can be considered as a multidimensional information network, corporate knowledge can be seen as the synthesis of all personal knowledge networks in an institution: This means that each user can store their information units (e.g., documents) and these documents are visible to all other users. Each user can then construct as many connections as necessary, which include not only their own but also other users' information units. Comments can be added to both cross-links and information units: the number of comments is unlimited, either per document or per crosses. All comments are visible not only to the creator but to all users.

Economic Cost - Cost Accounting

In the field of health care, there is a growing dynamic in terms of social and economic challenges. In this context, health care institutions are called upon to develop business tools and models for analyzing, calculating, and planning activities and their costs to revenue. This requires a reasonable costing and performance system that supports business and strategic decisions. Surprisingly, to date, it has been observed that there are no appropriate decision support systems to provide automated measurements and related data.

An increasing dynamic in terms of performance and cost structures can be found in the health sector. Demographic developments, limited fiscal resources, the introduction of federal or national DRG (Diagnosis Related Groups), and changes in pricing practices are increasing the pressure and cost of healthcare providers. Besides, national health care spending is steadily rising. This trend is reinforced by structural changes, as evidenced by the increasing importance of corporate relationships between service providers, such as cross-sectoral corporate relationships, mergers, and acquisitions, or medical centers.

The idea behind corporate healthcare relationships is to optimize patient paths in terms of quality and cost and focus on integrated healthcare structures. In this context, the concept of clinical pathways is currently being discussed in science and practice. Clinical avenues represent a standard treatment regimen, which includes all relevant activities. The classic organizational structure of hospitals is the structure of the department that results from the medical disciplines and the dominance of medical matters concerning finance. Combined with the high specialization of these departments and the highly specialized IT and information systems, the flow of information between departments is generally inferior. This situation is increasing in terms of cross-sectional information flow. Usually, the treatment of any disease will combine different departments or different medical fields. Therefore, patients moving from one ward to another or from one sector to another will encounter increased waiting times (which, however, can be avoided) and a doubling of examinations or treatments due to insufficient information flow. The economic dynamics of optimizing the paths that patients follow will substantially contribute to increasing efficiency and medical quality.

Therefore, the goal of integrated care structures is to coordinate the activities and procedures offered by different service providers and the need for specific medical treatments for patients. The ultimate goal of clinical pathways is to increase efficiency and quality. Optimizing the processing process involves selecting the service providers who are most suitable for the best possible exercise of the necessary activity. This shortens the healing process, and therefore the patient can be reintegrated into the professional life faster.

In terms of health insurance, integrated care structures lead to economic benefits by reducing the cost of social security. For patients, double examinations can be avoided with better communication, and a higher level of specialization can ensure high-quality treatments. Therefore, cooperation can help reduce the potential for increased costs in health care and reduce the need for additional financial resources.

Any decision regarding collaborations, human resources, or services offered to patients must be based on a more realistic and accurate knowledge of the underlying cost of the activities. In most Western countries, general hospitals must provide a comprehensive service. Since, in many states, medical treatments are paid at indicative rates per region, based on diagnostic-related DRG groups, the payment unit is not the specific treatment activity.

It is probably all the activities and treatments needed to rehabilitate the patient. This is a substantial financial reason to focus on the cost of clinical pathways. Also, because of the effects of scale, the payments received for specific DRGs, respectively, for all cases of a particular type of disease; do not always cover the total cost of these DRGs. Therefore, to enable the long-term placement of the healthcare provider, the management must at least know which parts of the treatment are causing the corresponding damage.

Since medical services and therapies are not directly comparable to industry services, standardized computational methods developed for industrial purposes are not applicable in the healthcare sector. A model of accounting costs and activity tailored to the specific needs of medical services and clinical pathways is needed.

Only this will provide the necessary support for a rational decision regarding medical services and the classic issue of construction or the market. Despite the long and ongoing debate about costing and clinical methods in practice for science, there is no system available in time that uses the concept of clinical pathways financially and therefore supports operational and strategic decisions.

There is a strong global trend of paying incentives for clinical services, such as Diagnostic Related Groups (DRG). The figure shows the effects associated with this payment plan (Wickramasingle, Bali, Kirn, & Suomi, 2012).

Figure 1.

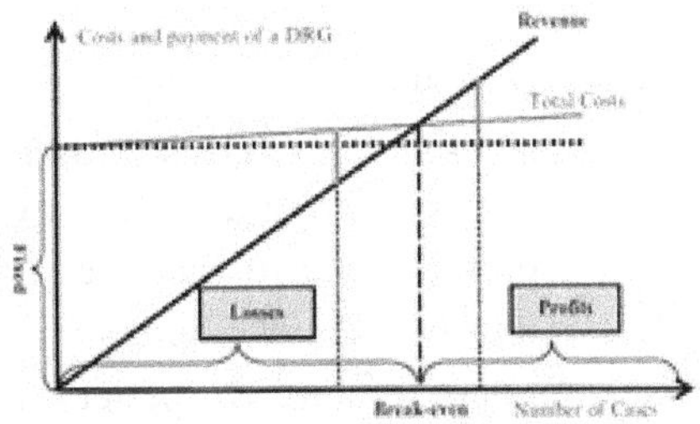

Figure 1: Contribution margin as a function of costs, payment and number of cases

First, the total income generated by a particular DRG is proportional to the number of cases of this DRG; see the bold line of the figure. On the other hand, due to the overwhelming staffing costs, in any hospital, there is a large part of fixed costs and relatively low variable costs, according to the figure with the line "Total Costs." In the number of cases, where both lines intersect, the point of uniform distribution (Break - even) is given. If the number of cases for a particular DRG treated by a hospital is less than the break-even point, losses are created to that specific DRG, and vice versa, if the number of cases is above equilibrium point, profits will result. Clearly, in the short term, general hospitals need to gradually subsidize DRGs that cause losses from the rest, as they are expected to provide all the essential services. But in the long run, management could consider working with other service providers or look for more cost-effective ways to deliver those services.

Second, if the payment is based on full treatment, details on how active the service is willing to become necessary. Therefore, the concept of clinical pathways is fundamental. The term "clinical pathway" is generally understood very differently. The literature uses treatment modalities (clinical), clinical pathways, standardized treatment protocols, integrated treatment pathways, guidelines for treatment, etc. It is essential to take into account that clinical pathways do not represent a binding course of treatment. Activities and flow may vary from patient to patient. Therefore, standardization does not affect the medical aspects but describes an idealized sequence of treatments, which reflects the recommended procedure and activities to a specific diagnosis.

Since a clinical pathway is an abstract shape, it makes no sense to just look at the costs of a route. Instead, we need to look at specific cases of real patients. For them, one can calculate the actual value of the treatments, and also, one can ask how likely it is that this particular treatment will happen and at what actual value. The combination of these individual therapies is the (abstract) clinical pathway. For planning purposes, it is necessary to have an image that expresses the projected cost of a clinical path, which, in simple terms, proves to be the expected cost of possible changes in treatment.

The term "cost of a DRG" needs to be differentiated: Speaking of a real case, always related to an exact DRG, one can collect all the operating costs of this particular case, attributing the actual cost to it. All the expenses of the specific cases related to DRG can be summarized to reach the (real) cost of the period. But by asking the cost of any DRG for programming purposes, one must get the meaning of the expected cost of DRG, that is, the expected value of the cost of all relevant roads.

In short, if the abstract treatment system is being addressed, we need to talk about the planned expenses. These are based on either historical or estimated data. In the case of specific treatment cases, we need to talk about the actual cost of a clinical route. This results from the assignment of the actual services provided by their expenses.

Medical services are affected by several factors. For example, a patient's health status varies from case to case and, therefore, the complications that arise during treatment or not, as well as the precautions to be taken. These and other factors affect treatment, cost, and outcome. Therefore, it is not possible to compare the cases in the process of providing medical services. These events have economic implications for the items to be used as well as for the development of a costing model. Cost items have various definitions in the literature and practice. Usually, the patient is taken as a cost item according to the "case." However, the uniqueness of the cases and their treatments do not allow comparisons. This violates the essential quality of any cost item related to the same issue. To comply with the law of comparability, cost items must be reasonably uniform and standardized. As a result, a new accounting model was developed, the accounting costing model. This model uses patients' "steps - paths" as a cost item.

The accounting costing model is based on the assumption that the routes are similar and so can function as cost items. Therefore, patient paths are considered the optimal form of treatment that has the character of a standard. This makes it possible to match the treatment procedures with this standard and therefore ensures optimal attention. The cost is calculated as the average to take into account the changes associated with each route.

Staff costs are, as is usually the case for service companies, the most significant cost factor in healthcare companies. Mainly, the costs of medical and nursing staff play an essential role and should be considered separately. Well-known cost accounting practices in healthcare often do not differentiate between the different functions and positions of the staff and work with average minute rates for each cost center or business team. However, there is a significant difference in the accuracy of the cost determination if the average cost or a differentiated picture of cost percentages per minute is calculated. A chief physician has a higher salary than an assistant physician resulting in different cost percentages.

As a general rule, each cost center consists of different groups of staff formed by the qualifications and cost percentages. Also, the setting of interest rates should differentiate the time that each person is directly involved in the provision of services (productive time) as long as the person is engaged in other activities (further time), such as time for management activities or meetings. While productive time can be charged directly to cost items, the extra time will be charged as general costs to cost centers, and in the next step will be distributed as indirect costs to the types of activity.

However, flat prices do not allow for different cost and performance analysis. For this reason, in each case, the advantages and disadvantages of a differentiated analysis of activities and the definition of the cost object must be weighed.

Evaluation of Costs and Benefits

The advantages of using smart object technologies in healthcare represent the direct connection of objects or, more precisely, the data generated by them and other systems (state-of-the-art or antiquated systems): Large amounts of data can be shared even without manual intervention, at higher quality and greater distances. Smart objects, perfectly integrated into networks of medical devices, computers, and systems, allow for the provision of active services as well as data intensity services. Another advantage is that RFID (Radio Frequency Identification)-based methods are affordable. This allows for generous use, creating diffuse healthcare systems that monitor the surrounding area in real-time and identify both the context and various potentially critical issues.

Typically, the other value of using smart technologies in business applications can be expressed using the dimensions of time, quality, and cost:

- Time: The data generated by the system are more recent than the data entered manually.
- Quality: With the automation of data production, more and more detailed data are produced. Besides, avoiding manual input reduces the risk of errors.
- Cost: As data are already produced electronically, they can easily communicate between systems and agencies and relieve staff of manual tasks (time) and allow further process optimization (Wickramasingle, Bali, Kirn, & Suomi, 2012).

Especially for the management of assets and reserves, radio technologies (RFID type) have already proven their advantage over linear codes (Wickramasingle, Bali, Kirn, & Suomi, 2012).

The time-consuming search for medical instruments, which can take up to 30 minutes per ward and the shift of work, can be significantly reduced with the help of active monitoring. However, many factors, besides the monetary implications, hinder the assessment of the concept of smart objects. One of them is the number of stakeholders in the healthcare sector and their relationships:

- Users, such as doctors and nurses, expect professional benefits. The simplicity of usage and ease of use are considered to include the most crucial factors in increasing acceptance. However, it is significant to observe there are massive differences in user groups and their participation: nurses, doctors, technicians, management, and others.
- Customers are hospital business partners. The financial benefit and return on investment must be proven. Customers are interested in scalable and reliable solutions as well as in protecting their investments.
- Commercial suppliers, service providers also expect standards that allow mass production. The fragmentation of the healthcare market requires adaptation to each user and customer (Wickramasingle, Bali, Kirn, & Suomi, 2012).

However, there are many other areas involved in providing health services. These may benefit from advanced solutions but do not initially allocate costs. For example, service technicians and technical inspection authorities have faster access to RTLS (Real-Time Locating System) devices. However, as long as service contracts are based on an uninterrupted amount per device, instead of maintenance per call, the result is no cost-sharing. This is why solutions require examining the entire value chain. For example, in a blood transfusion, the appropriate blood should be available and detectable by the donor to the recipient. This finds oneself in need of standardization and interoperability between administrations, organizations, and IT services.

But smart solutions have more to do with financial results. For example, events occurring during a blood transfusion are immediately referred to as an active system. This increases the rate of detection and investigation of errors, helping to improve the quality of the procedure, and patient safety.

CONCLUSION

In conclusion, the vast majority of the population unquestionably recognizes the importance that information and communication technologies can play with enhancing sustainable development. Shortly, health systems need to be integrated and incorporate the necessary knowledge and technology. This is to be able to help primarily and not only developing countries in tackling traditional development problems with innovative solutions and both practical and easily scalable and reproducible approaches. Sustainable technologies can effectively contribute to a country's development goals, especially its goals of reducing income, education, health, the environment, and gender equality. This can be achieved:

- By creating financial opportunities and contributing to poverty reduction.
- By managing the processes of essential services provided (e.g., health care, education) at a lower cost and with more excellent coverage.
- By facilitating access to information and the participation of the principal "shareholders" with greater transparency and support in networking at every stage.
- By enhancing the ability to measure, monitor, and report on progress concerning goals.

Therefore, information and communication technologies in collaboration with the available knowledge are the means or mechanisms and contribute to the significant reduction of financial resources and the increase of quality and effective health care.

REFERENCES

Buntin, M. B., Burke, M. F., Hoaglin, M. C., & Blumenthal, D. (2011). The Benefits Of Health Information Technology: A Review Of The Recent Literature Shows Predominantly Positive Results. *Health Information Technology, 3*, 464-471.

Busse, R., Geissler, A., Quentin, W., & Wiley, M. (2011). *Diagnosis - Related Groups in Europe: Moving Towards Transparency*. Efficiency and Quality in Hospitals.

Carde, S., Hullin, C. M., Chen, R., Schuler, T., Granz, J., Knaup, P., & ... . (2007). Towards Sustainability of Health Information Systems: How Can We Define, Measure and Achieve It? *Studies in Health Technology and Informatics*. PMID:17911901

Coiera, E., & Hovenga, E. S. (2007). *Building a Sustainable Health System*. IMIA and Schattouer GmbH.

Djeflat, A. (2010). Sustainable Knowledge for Sustainable Development: Challenges and opportunities for African development. *World Journal of Science, 7*(2).

Docherty, P., Kira, M., & Shani, A. (2009). *Creating Sustainable Work Systems. USA: Routledge*. Taylor & Francis Group.

Donaldson, C., & Magnussen, J. (1992). *DRGs: The road to hospital efficiency*. Health Policy.

Faggini, M., Cosimato, S., David Nota, F., & Nota, G. (2018). Pursuing Sustainability for Healthcare through Digital Platforms. *Sustainability 2019, 11*(165).

Fineberg, H. V. (2012). A Successful and Sustainable Health System - How to Get There from Here. *The New England Journal of Medicine, 366*(11), 1020–1027. doi:10.1056/NEJMsa1114777 PMID:22417255

Fischer, M. (2015). Fit for the Future? A New Approach in the Debate about What Makes Healthcare Systems Really Sustainable. *Sustainability.*

Kimaro, H. C., & Nhampossa, J. L. (2007). *The challenges of sustainability of health information systems in developing countries: comparative case studies of Mozambique and Tanzania.* JHIDC-Health Informatics in Developing Countries.

Koning, J. (2001). Social Sustainability in a Globalizing World Context, theory and methodology explored. *UNESCO / MOST Meeting.*

Lenz, R. (2009). *Information Systems in Healthcare– State and Steps towards Sustainability.* IMIA and Schattauer GmbH. doi:10.1055-0038-1638640

Mohsin-Shaikh, S., Furniss, D., Blandford, A., McLeod, M., Ma, T., & Beykloo, M. Y. (2019). The impact of electronic prescribing systems on healthcare professionals' working practices in the hospital setting: A systematic review and narrative synthesis. *BMC Health Services Research, 19*(742). PMID:31640689

Murdoch, J., & Clark, J. (1994). Sustainable Knowledge. *Geoforum, 25*(2), 115–132. doi:10.1016/0016-7185(94)90010-8

Nitu, L., & Nitu, L. (2015). *A management system for sustainable development - A new challenge.* ResearchGate.

Ouhbi, S., Fernandez-Aleman, J. L., Toval, A., Pozo, J. R., & Idri, A. (2017). Sustainability requirements for connected health applications. *WILEY Software: Evolution and Process.*

Porterfield, A., Engelbert, K., & Coustasse, A. (2014). Electronic Prescribing: Improving the Efficiency and Accuracy of Prescribing in the Ambulatory Care Setting. *Perspectives in Health Information Management, 11*(1g). PMID:24808808

Ratiu, C., & Beverlee, A. B. (2015). The multiple identities of sustainability. *World Journal of Science, Technology and Sustainable Development, 2042*(5945).

Schiff, G., Mirica, M. M., Dhavle, A. A., Galanter, W. L., Lambert, B., & Wright, A. (2018). A Prescription For Enhancing Electronic Prescribing Safety. *Quality of Care, 11,* 1877–1883. PMID:30395495

United Nations(1987*Our Common Future - Report of the World Commission on Environment and Development.* United Nations.

Wickramasingle, N., Bali, R. K., Kirn, S., & Suomi, R. (2012). *Critical Issues For The Development of Sustainable E-health Solutions.* Springer. doi:10.1007/978-1-4614-1536-7

Zadeh, P. E., & Tremblay, M. C. (2016). A review of the literature and proposed classification on e-prescribing: Functions, assimilation stages, benefits, concerns, and risks. *Research in Social & Administrative Pharmacy, 12*(1), 1–19. doi:10.1016/j.sapharm.2015.03.001 PMID:25847858

Zbuchea, A., Pinzaru, F., Busu, M., Stan, S.-O., & Bargaoanu, A. (2018). Sustainable Knowledge Management and Its Impact on the Performances of Biotechnology Organizations. *Sustainability.*

Chapter 13
Chronic Diseases and Socioeconomic Factors:
Findings From the SHARE 50+ Database

Athanassios Vozikis
Laboratory of Health Economics and Management (LabHEM), University of Piraeus, Greece

Nikoletta Panou
University of Piraeus, Greece

Iris-Panagiota Efthymiou
Laboratory of Health Economics and Management (LabHEM), University of Piraeus, Greece

ABSTRACT

The socioeconomic and behavioral risk factors, as well as advances in technology and medicine, have led to an unprecedented increase in the prevalence of chronic diseases, which have become the leading causes of morbidity and mortality. In this chapter, the authors aim to unveil the relation between socioeconomic and behavioral risk factors and chronic diseases among the Greek population. For this purpose, data from the SHARE 50+ database (specifically from the easySHARE v.6.1.1) were analyzed using the IBM SPSS 26 software. The research sample comprises of 3,584 Greek citizens of age between 50-75 years. The research findings indicate that 60.1% of the research sample have at least one chronic condition, with low education level to associate with a greater prevalence of chronic diseases. Chronic diseases also are correlated to poor self-perceived health and quality of life, especially to those with a low socioeconomic status. The research results are in accordance with the findings in the international literature.

DOI: 10.4018/978-1-7998-5442-5.ch013

INTRODUCTION

Quality of life, prosperity can be defined in multiple ways; and treatment of disease, relief from pain, and protection of health have always been at the centre of human activity and every organised society (Efthymiou I.P, 2016). If there's any confusion around its exact definition, it's because this concept constantly evolves and changes (Kanoura et al., 2003). According to WHO's definition, that concept is defined as "an individual's perception of their position in life in the context of the culture and value systems in which they live and concerning their goals, expectations, standards, and concerns"[1.]

Both health and quality of life constitute multidimensional phenomena, which aligns with the satisfaction of a modern-day individual's needs (Efthymiou and Vozikis, 2017). The term QoL is governed by the principles of both universality and individuality. In other words, it is influenced by the socio-economical conditions of development and prosperity as well as people's physical and mental health and overall wellbeing. In general, the term in question incorporates the interaction of personal and social life's factors and aspects. (Yfantopoulos,2001). Hence, the study of the quality of life is based both on personal experience (subjective consideration) and on estimating external factors that influence life's quality (objective dimension). The correlation between those two considerations reflects the depth and width of the quality of life. (Yfantopoulos and Sarris, 2001). Due to its richness in components, quality of life must be approached with a wide spectrum of estimation questions, so that every aspect is reflected aptly.

Today, chronic diseases are unprecedentedly increased, reaching pandemic levels. Various factors have contributed to that, including the population's ageing, lifestyle changes due to technological development, dietary changes, the evolution of hygiene conditions and science which brought along a spectacular reduction in child mortality and infectious-diseases-related deaths, and the climatic and environmental changes.

The international literature presents a large variation in both the number and the type of diseases that are considered chronic and raises the debate for which should be the period that a condition or disease must last to be considered chronic. There is much divergence concerning the use of the term "chronic disease" even within the medical community. It can be used to only describe diseases such as cardiovascular disease, stroke, cancer, type 2 diabetes, obesity, and arthritis. However, other entities also include Alzheimer's Disease, depression, the human immunodeficiency virus, and chronic respiratory diseases (Bernell et al., 2016). Said differentiations can cause confusion and misinterpretations when one generally talks about the effect, the cost, and the methods of treating and minimising chronic diseases.

According to the World Health Organisation (2014), chronic diseases are responsible for about 68% of deaths worldwide (2012). More than 40% of those were premature, under the age of 70. Approximately the ¾ of total deaths and 82% of people who died prematurely or before the age of 70 occurred in countries of low or middle annual income. More recent WHO data (2016) show an even more solid establishment of chronic diseases as the prevalent cause of death. Out of 57 million deaths worldwide, 41 million (71%) were due to noncommunicable diseases, while chronic diseases were responsible for 75% of premature deaths between the ages of 30 and 69. Most deaths are due to four major diseases: cardiovascular diseases (44%), cancer (42%), chronic respiratory diseases (9%), and diabetes (4%).

Another thing attributed to chronic diseases is about 43% of global disease weight, and it is expected that by the end of 2020, mortality due to chronic diseases will have reached 73% and that they will correspond to 60% of disease weight. It is worth noting that, according to the statistical data concerning the countries of low and middle income for the period 2011-2015, the financial cost that results from the appearance of chronic diseases rises to 7 trillion dollars.

Among the risk factors that are associated with lifestyles, behaviours, and habits and that could have been avoided; we find alcohol consumption, smoking, physical exercise, diet, and obesity. The increasing number of chronic-disease cases leads to increased levels of morbidity and mortality. The prevalence of such diseases is expected to increase because of the population's ageing and the increase of risk factors such as obesity. It has been estimated that 80% of CVD, stroke and diabetes cases, as well as one third of cancer cases, could be avoided.

Literature shows differences between the possibility of developing a chronic disease both among countries and among the different socioeconomic strata within a country. In Europe, 20% of health-related costs are due to socioeconomic inequalities. The resulting health inequalities are expected to grow wider in the context of the current financial crisis and constitute a challenge for health systems (Divajeva et al., 2014).

Health Risk Factors

Health risk factors can be classified as genetic, demographic, environmental, behavioral or physiological. The genetic factors are those associated with genetic predispositions; the demographics are the ones concerning all individuals, namely gender, age, religion, income; the environment are the ones being developed within a broader social, economic and political context – this category includes air pollution, the access to clean water and hygiene conditions of a population; the behavioral are those associated with the actions of an individual or a society, their lifestyle habits, such as dietary habits, smoking, alcohol consumption, exercise, etc.; the physiological are the ones resulting in obesity and/or high levels of cholesterol, glucose, and blood pressure, that is, the factors stemming from a mixture of genetic and behavioral risk factors (WHO, 2009).

Health risk factors can also be divided into the next two categories: modifiable and non-modifiable. The first category includes factors, such as gender, age, and ethnicity. However, our gene-expression profile constitutes an additional factor that seems to cause disputes between health professionals concerning which category it should be conflated in. Obviously, this issue stems from the fact that genes function within a wider physical and social environment where they must cope with external conditions as well as with behaviors and changing emotions. (Fredrickson, 2013).

The second category includes habits which are intrinsic in people's lifestyle. They are acquired in the course of people's life due to their social, economic and cultural characteristics. Among those habitual behaviors are the following: smoking, overconsumption of alcohol, low intake of fruit and vegetables, lack of exercise, obesity, high levels of cholesterol and blood pressure, drug use and unsafe sex. According to Greece's Organization for Economic Cooperation and Development (OECD), the first seven have been established as the main health risk factors in the international literature (Petrelis and Domeyer, 2016).

The factors mentioned above, either individually or combined, underly and lead to a quality of life with a minus sign and today's high percentages of early mortality on a global scale. They are directly linked to chronic noncommunicable diseases such as cardiovascular diseases and cancers, that is, the most frequent causes of death in Europe (World Health Organization – Europe, 2016). It has been estimated that by 2030, noncommunicable diseases will represent the 75% of global mortality, while cardiovascular diseases will be responsible for more deaths in comparison to noncommunicable diseases, such as tuberculosis, malaria or even diseases due to HIV infection, in low-income countries (Bruneau, 2008).

Chronic Diseases in Greece

The increasing incidence and prevalence of chronic non-communicable diseases (CNCD) represents a severe public health problem. CNCDs are associated with complications involving high costs with treatments, as well as compromising people's productivity and quality of life (Bahia et al., 2011).

Greece's National Statistical Service conducts research concerning health in Greece every five years. The research includes data about the health status of the population of over 15 years and how that is associated with the demographic data (e.g., gender, age) and socioeconomic factors (education, professional status) as well as health risk factors (diet, exercise, smoking, alcohol consumption, support from their social environment). The latest research took place in 2014. Among the main findings is that 49.7% of the population suffers from chronic diseases (an increase of 25% since 2009), a percentage that reaches 65% among people of 55 years and older and 90% among 75 and older. Also, almost 90% of deaths are due to chronic diseases, while about half of the deaths are due to cardiovascular diseases. Hypertension, asthma, diabetes, high cholesterol, and strokes appear more frequently in women.

Moreover, more than half the Greeks are overweight or obese, 1 in 3 has a sedentary lifestyle, about half consume fruit and vegetables daily, and 1 in 3 smoke. As results from the research, socioeconomic factors such as income, education, and unemployment are related to morbidity; a lower socioeconomic status means more difficult access to health services. Lastly, the financial crisis seems to increase the inequalities due to a decrease in income and health expenses and lead to dangerous behaviours. It is worth noting that, according to WHO, almost 90% of deaths in Greece is due to chronic diseases.

As many of the risk factors for chronic diseases are related with lifestyle choices, technology can facilitate health care with simple messages and alerts that aid in adherence to treatment (Boulos 2015; Reiners, 2019); for a systematic review of mobile health applications for chronic diseases the interested reader is referred to Debon (2019).

The Survey of Health, Ageing, and Retirement in Europe database

The Survey of Health, Ageing, and Retirement in Europe (SHARE) is a multidisciplinary and cross-national panel database of microdata, available to the entire research community, on health, socio-economic status and social and family networks of about 140,000 individuals aged 50 or older (around 380,000 interviews). SHARE covers 27 European countries and Israel.

The Share 50+ database has been used extensively by the international research community and for a variety of reasons, (e.g., for exploring how the advantaged socioeconomic conditions in childhood are associated with higher cognitive functioning but stronger cognitive decline in older age [1], to study the depression and social involvement among elders (Acciai et al., 2017), to provide current data regarding the prevalence and trends of overweight and obesity of adults from 2005 to 2013 in European countries (Acciai et al., 2018)).

This research has used the spatial data sets of the easySHARE v. 6.1.1, which is a simplified version of the database, where the variables have been condensed and collected in an Excel sheet and concern the country of Greece.

METHODOLOGY

The goal of this research is the epidemiological study of certain chronic diseases within a sample of the Greek population, between 50 and 75 years of age, and to unveil the correlation of those diseases' occurrence concerning a various individual or broader factors concerning the way of life and the socio-economic stratum.

The following key questions have been examined:

(1) What is the frequency of chronic diseases' occurrence?
(2) What is the percentage of the occurrence of one disease, two diseases, three, etc.?
(3) How recently did the studied patients go to a hospital?
(4) Does the occurrence of a chronic disease affect early retirement?
(5) Are socioeconomic factors linked to the occurrence of chronic diseases?
(6) How do the sufferers of chronic diseases self-evaluate their health, and how do socioeconomic factors influence their health self-evaluation?
(7) How is the quality of life evaluated for chronic-disease patients, and how do socioeconomic factors influence the evaluation of their quality of life?
(8) Does gender have a differentiating effect on the occurrence of chronic diseases and the overall health status and quality of life?
(9) Does marital status have a differentiating effect on the occurrence of chronic diseases, on the self-evaluation of health and the quality of life?
(10) Are chronic diseases linked to habit-related risk factors (body mass index, smoking in the present, smoking in the past, physical exercise)?
(11) Is there a correlation between socioeconomic factors (gender, income, education) and risk factors (BMI, smoking, exercise)?

The research used data from the European database SHARE 50+, which includes health data for individuals 50 years and older from various European countries as well as other epidemiological and demographic data and socioeconomic factors.

The data have been collected in multiple waves with questionnaires during personal interviews, and the variables have been appropriately codified in the SHARE 50+ database so that they can be processed by a statistical data processing software, while the data come from easySHARE v. 6.1.1. That is, this study has used data concerning Greece, taken from wave 6 (2015). Greece took part from the beginning in the first three waves but did not participate in the next two (4 and 5).

The research processed about 20 variables in total (secondary variables, because if we also include the variables-items used to evaluate the quality of life and health for individuals with chronic diseases, then the total number gets significantly bigger, reaching perhaps multiple hundreds), which are quantitative, categorical, ordinal or nominal.

The research also examined some additional factors related to chronic diseases, such as marital status as well as risk factors, such as BMI, smoking, and the lack of physical exercise. More analytically, the variables processed by the research are the occurrence of chronic disease, gender, place of residence, education, marital status, early retirement, the household's income as a cumulative percentile and as a measurement variable, the ability to make ends meet in the household, self-evaluation of health, quality of life, body mass index, smoking, smoking in the past and physical exercise.

The Methodology of the Statistical Analysis

The data for this research were collected by the use of questionnaires with the answers written down. After the collection, the data were appropriately codified in the database so that they can be registered in the statistical process software. For the statistical analysis of the data, we used SPSS, created cross-tabulations and did a descriptive statistical analysis and the appropriate statistical checks on a statistical importance level were established at 95%. Specifically, the statistical analysis includes codification and nomenclature of the categorical variables' categories, check that there are enough samples for each category of the categorical variables, construction of total scores for the scales of quality of life and health, drawing of main descriptive statistical measures and development of graphs. Then, we calculated the Pearson factors among continuous variables, did the t-test, ANOVA and logistic regression checks to calculate the simultaneous effect of multiple quantitative and categorical factors on the possibility that a chronic disease occurs.

RESEARCH FINDINGS

The Research Sample Characteristics

The sample for this research comprises 3,584 individuals. Of those, 1,572, a percentage of 44% are men and 2,012, a percentage of 56% are women, as presented in the *Figure 1* below.

As for their origins and place of residence, 23.6% live in urban centers, 7.8% in smaller cities, while 16.5% in rural areas or small villages, as presented in the table below and the corresponding bar chart (relative frequencies) (See *Figure 2*).

As for their level of education, we followed the ISCD classification, which divides the participants into six levels. The results presented in the *Figure 3* below as relative frequencies, show that 30.6% got level 1 (elementary school), 10.7% got level 2 (middle school), 28.5% got level 3 (high school), 2.9% got level 4 (professional specialization faculties), 23.4% got level 5 (bachelor-level university studies), and 0.6% got level 6 (master-level university studies and higher degrees, such as doctorates and post-doctorates).

Concerning the marital status of the sample, it must be noted that 77.8% are married and live together with their spouse, 0.9% live together with their partner, 1.6% are separated, and 4.9% are not married. An additional 6.1% have been divorced, and 8.5% are widows or widowers. The following *Figure 4* presents the absolute and relative frequencies of the marital status of the frequencies, the following bar chart represents the relative frequencies.

The *Tables 1 & 2* below present the results for questions 1 and 2 of the survey. For both the relative and absolute frequency has been calculated as well as the cumulative percent.

Figure 1. Gender Distribution

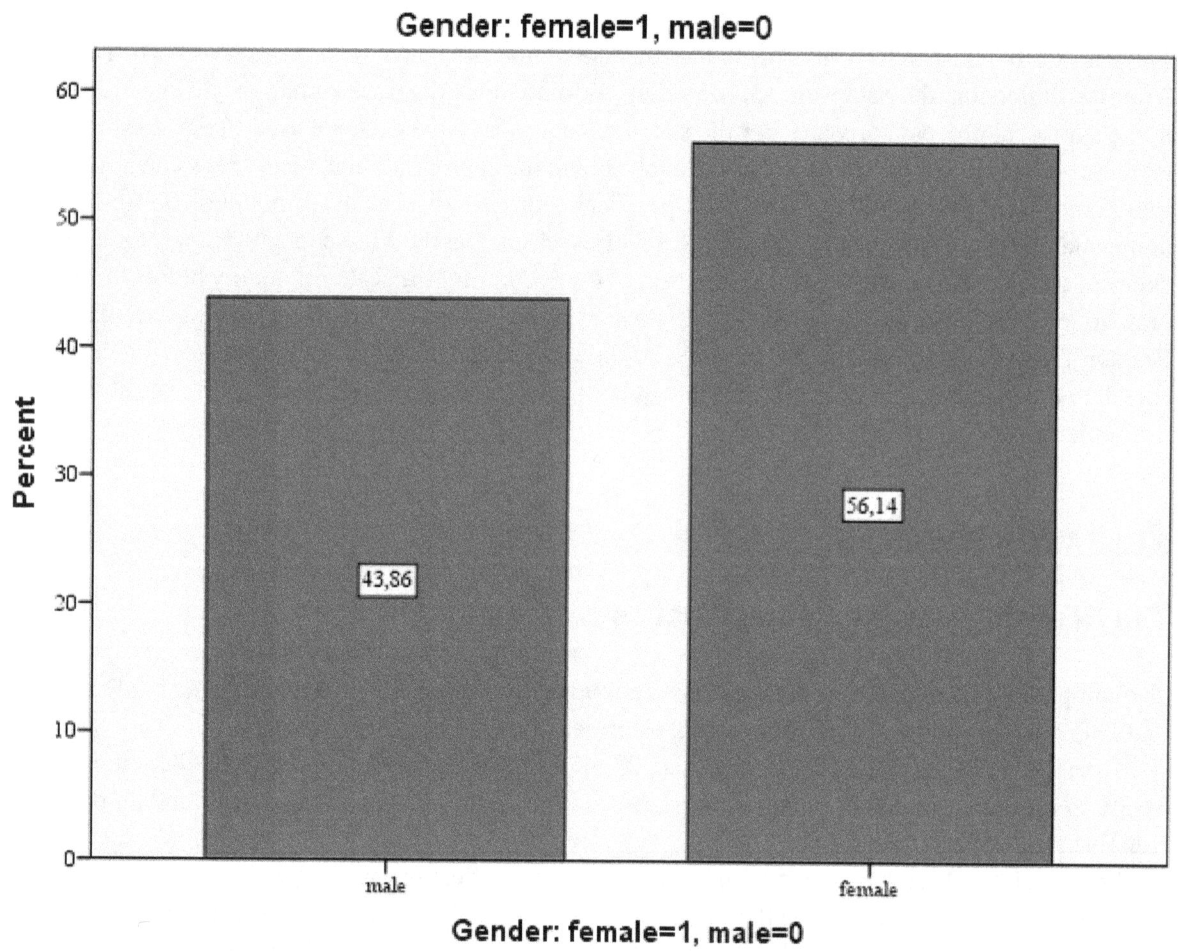

Question 1

The variable "chronic diseases" in this study include heart attack, hypertension, hypercholesterolemia, stroke, diabetes, chronic lung disease, cancer, peptic ulcer, cataracts, Parkinson's disease, and hip fracture., as this is how it is set up in the easySHARE v base. 6.1.1. As shown in the *Table 1* below, the distribution of the occurrence of chronic diseases in the sample collected shows that 60.1% of the sample shows one of the above mentioned chronic diseases, while 39.9% do not show any of them. above chronic diseases. It is noted that the case of multiple chronic diseases is not recorded here, only the existence of at least one chronic disease is recorded.

Question 2

The most common condition of patients with chronic disease is to have a single chronic disease, a condition that occurs in 31.3% of the population (ie about one in three Greeks in the age sample), 50-75 years old suffer from a chronic disease. Multiples of chronic diseases are found as in *Table 2*: two chronic diseases at a rate of 18%, three at a rate of 7.3%, four at a rate of 2.7%, and five and more chronic diseases at a rate of 0.8%. Almost one in five Greeks aged 50-75 develops at least two chronic diseases.

Figure 2. Place of Residence

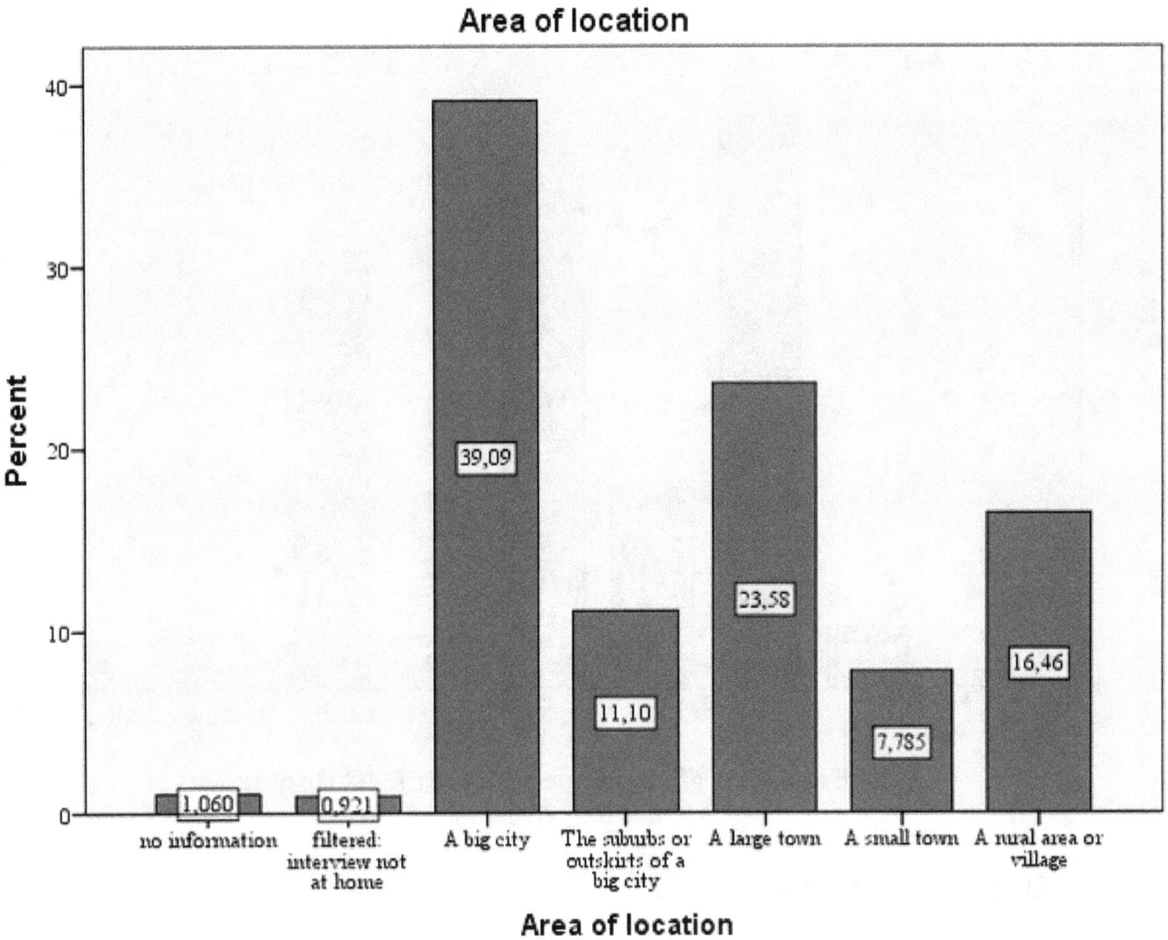

Correlations

Chronic Disease: Hospital Length of Stay

For the total number of people (both patients and healthy) who were hospitalized for one night over the last 12 months, 2.6% accounts of people without chronic disease, compared to 8% of patients were treated. The differentiation is considered statistically significant.

Figure 3. Education Level

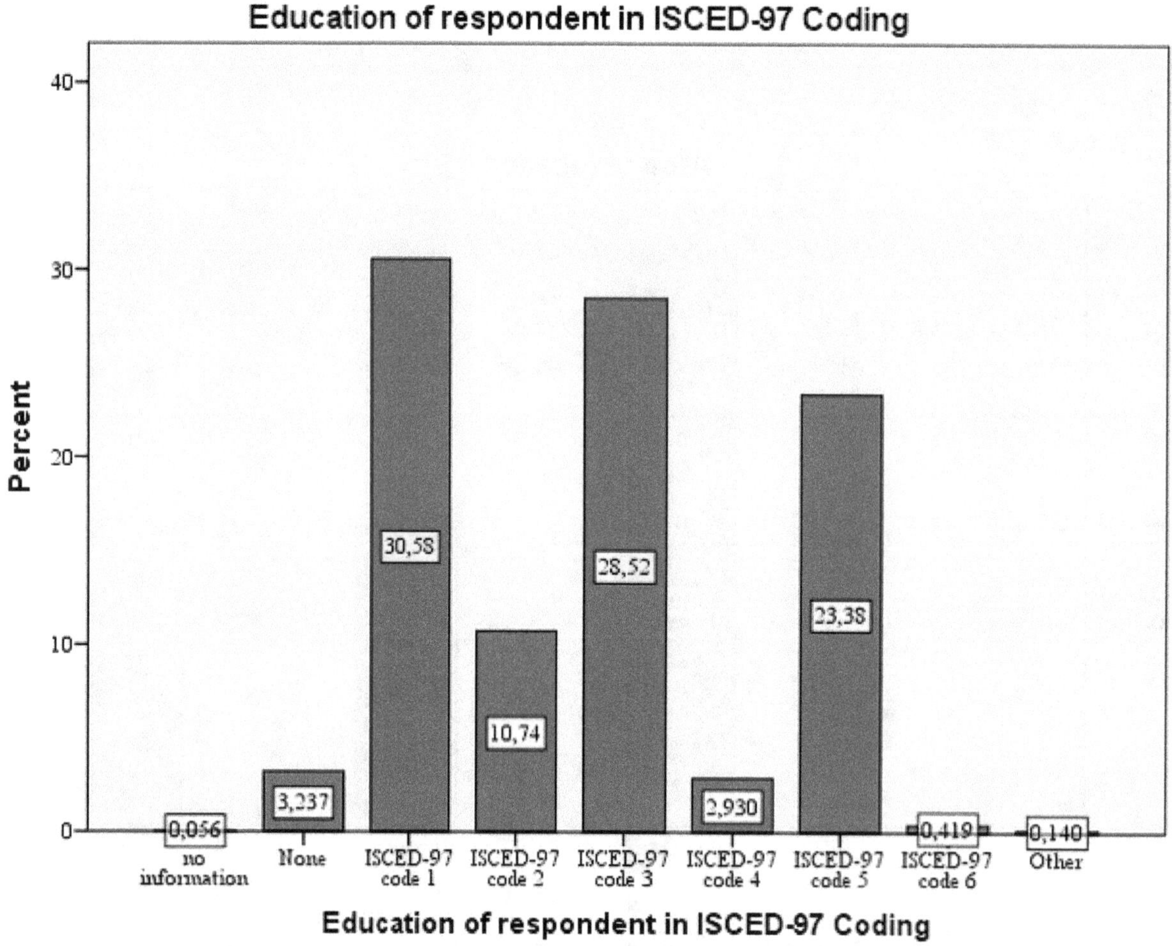

Chronic Disease: Income Level

The grouped box plot (*Figure 5*) shows the correlation between chronic illness and income when family income is expressed in percentages. Average prices: 5.79 (no) and 5.66 (yes), while the *Table 3* shows the descriptive statistical measures to check the correlation between the occurrence of chronic disease and family income as a numerical variable.

In both cases the t-test showed p = 0.2 approximately> 0.05. Correlation tests with the household's ability to meet its needs have yielded similar results.

Figure 4. Marital Status

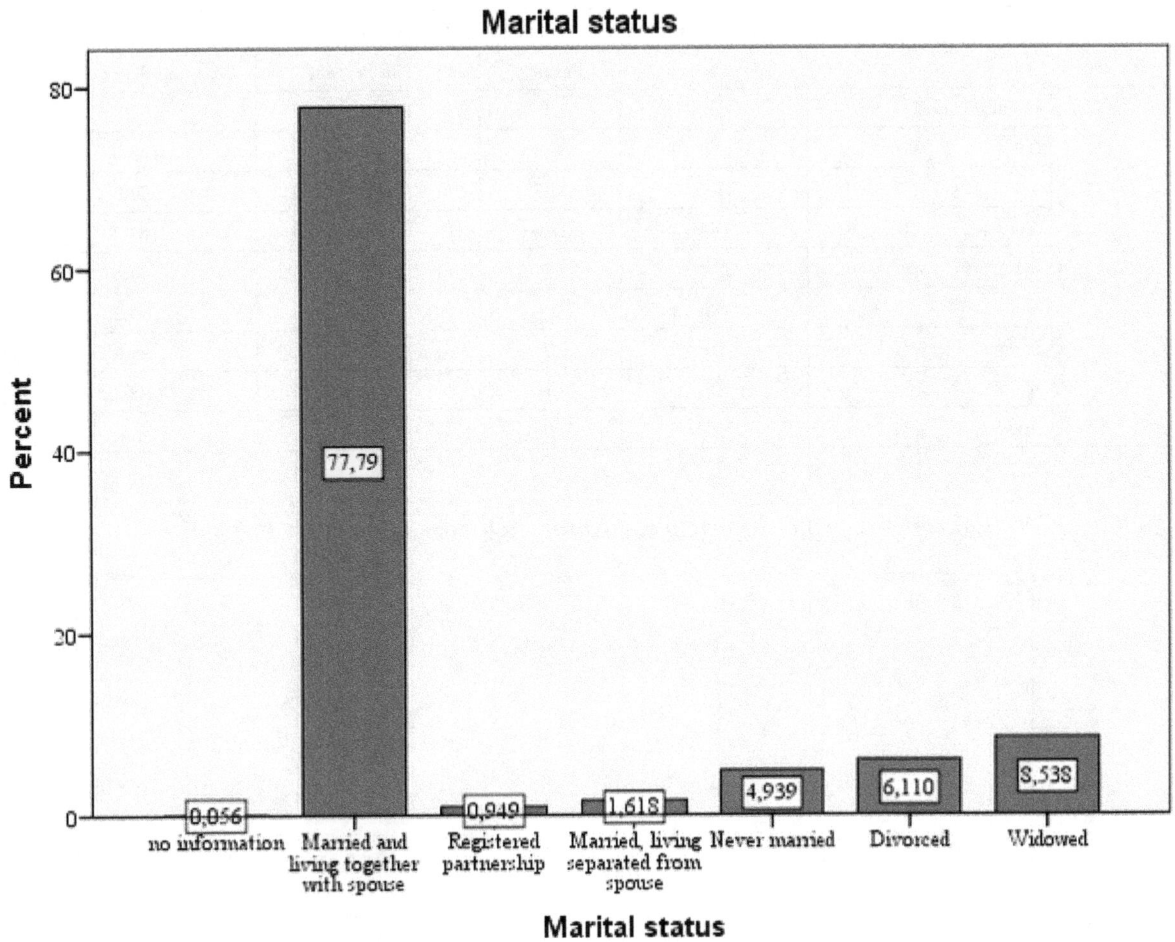

Chronic Disease: Level of Education

The level of education is expressed by the ISCED-97 index which takes values 1-6. In the following *Figure 6*, the grouped box plot from the control of correlation of chronic diseases and t-test education level is presented. The incidence of the chronic disease seems to depend on the level of education at a confidence level of 99%.

Table 1. Frequency of Chronic Diseases' Occurrence

Chronic Disease Existence					
		Frequency	Percent	Valid Percent	Cumulative Percent
Valid	No	1429	39,9	39,9	39,9
	Yes	2155	60,1	60,1	100,0
	Total	3584	100,0	100,0	

Table 2. Number of Chronic Diseases' Occurrence (Comorbidity)

Number of Chronic Diseases (easySHARE version)					
		Frequency	Percent	Valid Percent	Cumulative Percent
Valid	no information	3	,1	,1	,1
	0	1429	39,9	39,9	40,0
	1	1122	31,3	31,3	71,3
	2	644	18,0	18,0	89,2
	3	262	7,3	7,3	96,5
	4	95	2,7	2,7	99,2
	5	26	,7	,7	99,9
	6	3	,1	,1	100,0
	Total	3584	100,0	100,0	

Figure 5. Correlation between the Occurrence of Chronic Disease and Family Income

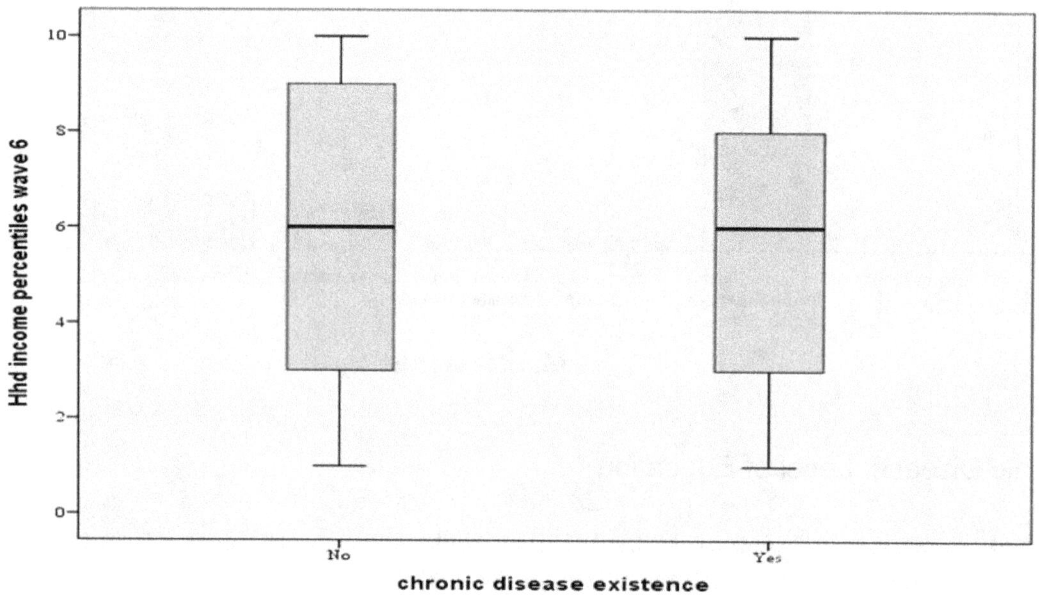

Table 3. Correlation between the Occurrence of Chronic Disease and Family Income

Group Statistics

	chronic disease existence	N	Mean	Std. Deviation	Std. Error Mean
Household net income, imputed	No	1429	15167,61	14571,983	385,481
	Yes	2152	14612,64	15377,227	331,480

Figure 6. Correlation between the Occurrence of Chronic Disease and Education Level

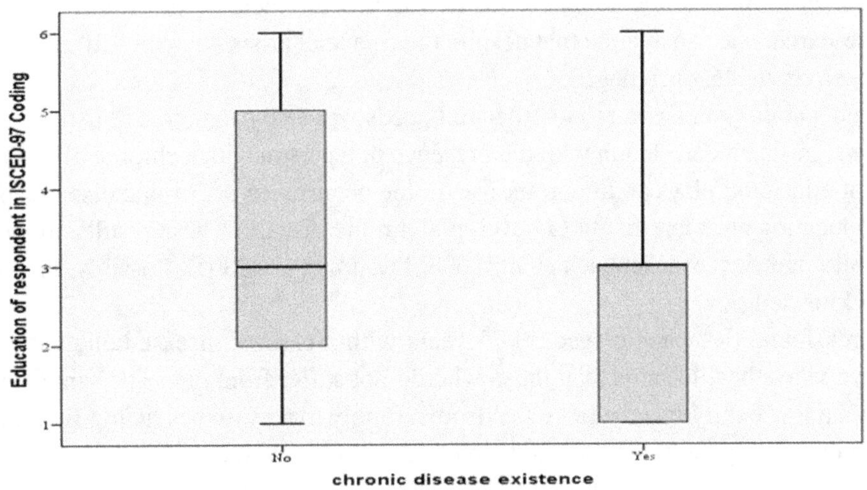

Chronic Disease: Gender

988 men of the sample (62.9%) from 1571 show chronic disease compared to 1164 women (in total of the sample (57.9%)) for the year 2010. The chi -square thes of independence returned a pvalue=0.03.

Chronic Disease: The Quality of Life

Quality of life is calculated with the CASP-12 index, which is based on a questionnaire about how often respondents feel specific emotions or how often they are in specific situations (control ability, autonomy, pleasure, and self-awareness are assessed).

For the correlation between the existence of chronic disease and the quality of life of the individual, the T-test control showed 99% significance level. A positive correlation between quality of life and income was also observed in people with chronic disease, but the Pearson correlation coefficient gave a value of 0.094 which is very low.

DISCUSSION

The main conclusions of the statistical analysis unveiled that 60.1% (that's the sample's majority), suffer from a chronic disease (1 in 3 Greeks between the ages of 50 and 75 of the sample suffer from a single chronic disease; 2 out of 5 suffer from two chronic diseases; less than 1 in 10 suffer from more than two diseases). The frequency of hospitalisation in the last year for chronic disease sufferers is about four times higher, which can have serious implications for health costs and the insurance system. The occurrence of chronic disease does not play a role in early retirement.

Concerning the influence of income in the occurrence of a chronic disease, a slightly larger occurrence of chronic diseases was observed in lower-income individuals along with a respective lower occurrence in higher incomes, but this differentiation is not statistically important, contrary to surveys in other countries (Kim et al., 2016, Campbell et al., 2017), as well as a previous study in Greece (Zavras et al., 2013) which claim that low income leads to a significant increase in chronic diseases. From the result of this research, one can assume that despite the financial crisis, there is still a strong social state, which largely covers the less privileged.

A household's ability to respond to its financial needs was also examined, but still, there was found no statistically significant correlation with the presence of a person with a chronic disease in the family.

The level of education plays an important role in the occurrence of chronic diseases. People with a low level of education are more likely to develop a chronic disease. These results are consistent with surveys in other countries (Mackenbach et al., 2008; Dalstra et al., 2005). Possibly, this has to do with health-related knowledge.

This research found that people aged 50-75 years with a chronic disease believe to a large extent that they have a worse health status than those who do not suffer from one. The same is true when the patient receives a low family income or comes from a family that estimates facing financial difficulties in coping with daily obligations. These results on health self-evaluation go together with other similar surveys (Zavras et al., 2013; HeadetAl., 2018) as well as with logic. Of course, the results show that most patients believe that they generally have good health, which is consistent with previous Greek Statistical Authority data (Health Research, 2014), according to which, ¾ of the population stated that they feel healthy, while half the population suffered from a chronic disease. One explanation for this is perhaps the fact that conditions such as hypertension or hypercholesterolemia are not considered such serious health problems by the sufferers, as long as a normal part of everyday life is treated with medication.

The quality of life on the health of people aged 50-75 years with chronic diseases was studied based on the CASP-12 index which is less subjective and is considered good for ages over 50 years (Borrat-Besson et al., 2015). It was found that people with a chronic disease have a lower level of quality of life than non-sufferers and that the socioeconomic substrate has an effect on the results. The worse the socioeconomic situation, the more negatively the health-related quality of life is affected. A lower-income slightly contributes to a further reduction in the level of quality of life for people suffering from a chronic disease. Similar results have been shown by other such surveys (Kim et al., 2015; Mielck et al., 2014).

Gender proved to be a factor that plays a differentiated role in both the occurrence of chronic diseases and the patients' self-evaluation of health and quality of life. Men aged 50-75 years in this research had a higher percentage of chronic disease occurrence than women but consider that their state of health is better compared to women's respective answers, as shown by other similar researches as well (Zavras Et al., 2013). Those found that men estimate that they have better health. It also seems that they have a better quality of life than women. Specifically, 988 out of 1,571 men of the sample (62.9%) have a chronic disease compared to 1,164 out of 2,010 women of the sample (57.9%). Previous researches show that gender does not always play a role in chronic diseases (Head et al., 2018), but there are data that women are in an unfavourable position, though this depends on the disease. For example, according to data (ELSTAT, 2016), women have a higher likelihood of developing asthma, hypertension, stroke, diabetes, and high blood cholesterol. In terms of quality of life in men and women, the combined influence of other factors can further differentiate the results (Mielck et al., 2014). That could be the subject of further study.

It has been observed that the social environment can affect through various mechanisms the occurrence of chronic diseases and health in general (Cockerham et al., 2017). Depending on the case, the effect may be positive or negative; that is, the social environment can offer support but also lead to dangerous habits (Wilkinson and Marmot, 2003). In this study, a single parameter of the social environment was examined – the marital status – which seemed to be slightly associated with the state of health as married people and those living together with their partner have less chance of suffering from a chronic disease. This probably has to do with the support that they can offer, both mental and physical, or it motivates people to be more attentive to their health since they have to think of their partner as well.

Various risk factors related to behaviours and habits, namely smoking, obesity, and lack of exercise, were also examined. The results confirmed that they could indeed cause the occurrence of chronic disease, as suggested by the international literature as well.

Regarding the body mass index (BMI), it is inversely proportional to the level of education, also mentioned in the literature (Joshua, 2017). No statistically significant correlation was found with gender and income, contrary to the existing literature (Joshua, 2017), arguing that a lower income means an increase in obesity rates.

The exercise was also not found to be linked to gender to a statistically significant degree and is inversely proportional to the level of education but a very small extent. Compared to the sample, other people with a higher income do more physical activity as previous surveys have shown (Shuval et al., 2017).

Contrary to the above, the most important differentiating factor that affects smoking is gender. Men smoke more than women, as evidenced by previous data in Greece and around the world (ELSTAT and WHO). In this research, there was no correlation between smoking and the level of education, though there are studies suggesting that such an association exists, as is the case with income, that is, a higher income leads to a reduction in smoking (Perreti-Watel et al., 2009; Martinez et al., 2018). Such a conclusion is consistent with the results of this research.

CONCLUSIONS AND POLICY PROPOSALS

Studying the spread of chronic diseases in the various social groups and the risk factors that can lead to them is crucial for their treatment. It enables the identification of the most vulnerable groups and individuals of the population so that we know where special attention should be given and that the competent bodies can be mobilised while appropriate measures are taken by the State. It also helps in better designing the health system at the level of healthcare services and insurance coverage, in order to meet the needs with maximum efficiency and effectiveness and minimise the costs caused by increased morbidity and intensified by social inequalities.

The frequent and systematic collection and registering of demographic and other data for chronic diseases allow their systematic monitoring which is of paramount importance for the definition of goals with the ultimate aim of treating them, like the action plan proposed by the WHO, which includes specific objectives to reduce the spread of not only chronic diseases but also their risk factors and to ensure the availability of the appropriate medication. It is also possible to monitor the implementation of those objectives, to what extent they have been achieved, whether there is a positive outcome or whether the strategy should be redefined.

Lastly, based on the conclusions of this research and other relevant ones, we can define goals and implement the appropriate preventative measures that will benefit the society both financially and in terms of disease weight and its negative effects on productivity and psychology. Finally, we suggest diachronic monitoring and analysis of the results' time, the effect of the financial crisis on the social state and the effect on the care for these people, etc.

Limitations of the Research

An objective limitation besets the sample's status. This research is based on data from 2015 (wave 6 of the Share 50+ database) since the wave 7 data, though collected, have not been rendered accessible at the time of the analysis. Furthermore, the alcohol consumption variable has not been registered in the current wave, though alcohol consumption is also considered a health risk factor according to the relevant literature. Also, there was no variable accurately describing the job position, which is a significant socioeconomic factor for chronic diseases.

Research Funding

This research received no specific grant from any funding agency in the public, commercial, or not-for-profit sectors.

REFERENCES

Aartsen, M., Cheval, B., Sieber, S., van der Linden, B., Gabriel, R., Courvoisier, D., Guessous, I., Burton-Jeangros, C., Blane, D., Ihle, A., Kliegel, M., & Cullati, S. (2019). Advantaged Socioeconomic Conditions in Childhood are Associated with Higher Cognitive Functioning but Stronger Cognitive Decline in Older Age. *Proceedings of the National Academy of Sciences of the United States of America, 116*(12), 5478–5486. doi:10.1073/pnas.1807679116 PMID:30804194

Acciai, F., & Hardy, M. (2017). Depression in Later Life: A Closer Look at the Gender Gap. *Social Science Research, 68,* 163–175. Advance online publication. doi:10.1016/j.ssresearch.2017.08.003 PMID:29108595

Acciai, M., Martins, J., Lipert, A., Ramos, M., & Peralta, M. (2018). Prevalence and Trends of Overweight and Obesity in Older Adults from 10 European Countries from 2005 to 2013. *Scandinavian Journal of Public Health.* Advance online publication. doi:10.1177/1403494818764810 PMID:29569525

Bahia, L. R., Araujo, D. V., Schaan, B. D., Dib, S. A., Negrato, C. A., Leão, M. P. S., & … . (2011). *The Costs of Type 2 Diabetes Mellitus Outpatient Care in the Brazilian Public Health System.* Value Heal. doi:10.1016/j.jval.2011.05.009

Borrat-Besson, C., Ryser, V. A., & Gonçalves, J. (2015). *An evaluation of the CASP-12 scale used in the Survey of Ageing and Retirement in Europe (SHARE) to measure Quality of Life among people aged 50+* (No. 4). FORS Working Paper Series.

Boulos, M. N. K., Brewer, A. C., Karimkhani, C., Buller, D. B., & Dellavalle, R. P. (2014). Mobile medical and health apps: State of the art, concerns, regulatory control, and certification. *Online Journal of Public Health Informatics*, *5*(3), 229. PMID:24683442

Bruneau, B.G., (2008). The developmental genetics of congenital heart disease. *Nature, 451*(7181), 943-8. DOI: doi:10.1038/nature06801

Campbell, D. J., Manns, B. J., Weaver, R. G., Hemmelgarn, B. R., King-Shier, K. M., & Sanmartin, C. (2017). Financial barriers and adverse clinical outcomes among patients with cardiovascular-related chronic diseases: A cohort study. *BMC Medicine, 15*(1), 33. doi:10.118612916-017-0788-6 PMID:28196524

Cockerham, W. C., Hamby, B. W., & Oates, G. R. (2017). *The social determinants of chronic disease*. Academic Press.

Dalstra, J. A., Kunst, A. E., Borrell, C., Breeze, E., Cambois, E., Costa, G., ... Regidor, E. (2005). Socio-economic differences in the prevalence of common chronic diseases: An overview of eight European countries. *International Journal of Epidemiology, 34*(2), 316–326. doi:10.1093/ije/dyh386 PMID:15737978

Debon, R., Coleone, J. D., Bellei, E. A., & March, A. C. B. D. (2019). Mobile health applications for chronic diseases: A systematic review of features for lifestyle improvement. *Diabetes & Metabolic Syndrome*.

Divajeva, D., Marsh, T., Logstrup, S., Kestens, M., Vemer, P., Kriaucioniene, V., Peresson, S., O'Kelly, S., Rito, A., & Webber, L. (2014). Economics of chronic diseases protocol: Cost-effectiveness modelling and the future burden of non-communicable disease in Europe. *BMC Public Health, 14*(1), 456. doi:10.1186/1471-2458-14-456 PMID:24886110

Efthymiou, I. P. (2016). *Trends in Healthcare: A Global Challenge*. Xlibris Publishing.

Efthymiou, I. P., & Vozikis, A. (2017). Medicine, A Pan-Historical, Pan-Cultural Science. *Journal of Cardiology & Cardiovascular Therapy*.

Fredrickson, B. L. (2013), Positive emotions broaden and build. Advances in Experimental Social Psychology. Academic Press.

Head, J., Chungkham, H. S., Hyde, M., Zaninotto, P., Alexanderson, K., Stenholm, S., ... Vahtera, J. (2018). Socioeconomic differences in healthy and disease-free life expectancy between ages 50 and 75: A multi-cohort study. *European Journal of Public Health, 29*(2), 267–272. doi:10.1093/eurpub/cky215 PMID:30307554

Joshua, J. (2017a). *The economics of addictive behaviour vol I: the private and social costs of smoking and their remedies. Academic Press.*

Joshua, J. (2017b). *The economics of addictive behaviour vol II: the private and social costs of addictive behaviours and their remedies. Academic Press.*

Joshua, J. (2017c). *The economics of addictive behaviour vol IV: the private and social costs of overeating and their remedies. Academic Press.*

Kavoura, M., Kyriopoulos, G., Geitona, M., & Vandorou, C. (2003). *Poiotita Zois*. JANSSEN-CILAG.

Kim, J. H., & Park, E. C. (2015). Impact of socioeconomic status and subjective social class on overall and health-related quality of life. *BMC Public Health, 15*(1), 783. doi:10.118612889-015-2014-9 PMID:26275823

Kim, S., Lee, B., Park, M., Oh, S., Chin, H. J., & Koo, H. (2016). Prevalence of chronic disease and its controlled status according to income level. *Medicine, 95*(44), e5286. doi:10.1097/MD.0000000000005286 PMID:27858900

Kyriazis, I., Lappa, Th., Koufopoulou, K., Sidiropoulos, S., & Efthymiou, I. P. (n.d.). Impact of Austerity on Dietary Habits and Obesity in Greek Children, Adolescents and Adult. In The Impact and Implications of Crisis: A Comprehensive Approach Combining Elements of Health and Society. Academic Press.

Mackenbach, J. P., Stirbu, I., Roskam, A. J. R., Schaap, M. M., Menvielle, G., Leinsalu, M., & Kunst, A. E. (2008). Socioeconomic inequalities in health in 22 European countries. *The New England Journal of Medicine, 358*(23), 2468–2481. doi:10.1056/NEJMsa0707519 PMID:18525043

Martinez, S. A., Beebe, L. A., Thompson, D. M., Wagener, T. L., Terrell, D. R., & Campbell, J. E. (2018). A structural equation modeling approach to understanding pathways that connect socioeconomic status and smoking. *PLoS One, 13*(2), e0192451. doi:10.1371/journal.pone.0192451 PMID:29408939

Mielck, A., Vogelmann, M., & Leidl, R. (2014). Health-related quality of life and socioeconomic status: Inequalities among adults with a chronic disease. *Health and Quality of Life Outcomes, 12*(1), 58. doi:10.1186/1477-7525-12-58 PMID:24761773

Peretti-Watel, P., Constance, J., Seror, V., & Beck, F. (2009). Cigarettes and social differentiation in France: Is tobacco use increasingly concentrated among the poor? *Addiction (Abingdon, England), 104*(10), 1718–1728. doi:10.1111/j.1360-0443.2009.02682.x PMID:19681803

Petrelis, M., Domeyer, P.R., (2016). 'HEALTH 2020' and Greece. The Contribution of Mass Pre-symptomatic Control. *To Vima tou ASKLIPIOU, 15*(3).

Reiners, F., Sturm, J., Bouw, L. J., & Wouters, E. J. (2019). Sociodemographic factors influencing the use of eHealth in people with chronic diseases. *International Journal of Environmental Research and Public Health, 16*(4), 645. doi:10.3390/ijerph16040645 PMID:30795623

Shuval, K., Leonard, T., Drope, J., Katz, D. L., Patel, A. V., Maitin-Shepard, M., ... Grinstein, A. (2017). Physical activity counseling in primary care: Insights from public health and behavioral economics. *CA: a Cancer Journal for Clinicians, 67*(3), 233–244.

Wilkinson, R. G., & Marmot, M. (Eds.). (2003). *Social determinants of health: the solid facts.* World Health Organization.

World Health Organization. (2009). *Global health risks: Mortality and burden of disease attributable to selected major risks.* World Health Organization.

World Health Organization. (2014). Global status report on noncommunicable diseases 2014 (No. WHO/NMH/NVI/15.1). World Health Organization.

World Health Organization. Management of Substance Abuse Unit. (2014). Global status report on alcohol and health, 2014. World Health Organization.

World Health Organization. (2016). *World health statistics 2016: monitoring health for SDGs sustainable development goals.* World Health Organization.

World Health Organization- Europe. (2016). *Greece-Profile of health and well-being*. Author.

Yfantopoulos, J. (2001c). Health-Related Quality of Life. *Archives of Hellenic Medicine*, *19*, 131–146.

Yfantopoulos, J., & Sarris, M. (2001). *Health-Related Quality of Life* – Methodology of measurement. *Archives of Hellenic Medicine*, *18*(3), 218–229.

Zavras, D., Tsiantou, V., Pavi, E., Mylona, K., & Kyriopoulos, J. (2012). Impact of economic crisis and other demographic and socio-economic factors on self-rated health in Greece. *European Journal of Public Health*, *23*(2), 206–210. doi:10.1093/eurpub/cks143 PMID:23093716

ENDNOTE

1 https://www.who.int/healthinfo/survey/whoqol-qualityoflife/en/

Chapter 14
No One Left Behind?
Migrant and Refugee Health in the COVID-19 Crisis in Greece

Athanassios Vozikis
Laboratory of Health Economics and Management (LabHEM), University of Piraeus, Greece

Theodoros Fouskas
University of West Attica, Greece

Symeon Sidiropoulos
Laboratory of Health Economics and Management (LabHEM), University of Piraeus, Greece

ABSTRACT

Asylum seekers, refugees, and migrants, who are living in RICs, are faced with multiple challenges and vulnerabilities that must be taken into consideration when responding to the COVID-19 pandemic. The chapter focuses on the COVID-19 pandemic concerns over increasing cases recorded in the RICs in Greece. The impact of migration on public health is of particular concern to Greek migration policy, as migrants in the country have a completely different epidemiological profile and higher risks to public health due to the poor living conditions in their countries of origin and during their stay. They live mostly in overcrowded reception and identification centers and accommodation centers under deplorable conditions, lack of proper shelter, extremely unhygienic living conditions. The urgent decongestion of the overcrowded RICs and accommodation centers is required to avoid the risk of rapid spread of the infection.

INTRODUCTION

The first COVID-19 case was diagnosed in Greece on February 26th 2020 (National Public Health Organization, 2020e). As subsequent cases in late February and early March were confirmed. Between March 10-18 educational institutions of all levels nationwide and stores suspended the operation of educational institutions of all levels nationwide, as well as, movie theaters, gyms and courtrooms, shop-

DOI: 10.4018/978-1-7998-5442-5.ch014

ping malls, cafés, restaurants, bars, beauty parlors, museums and archaeological sites, organized beaches and ski resorts and all stores nationwide. On March 23rd, with 695 confirmed cases and 17 deaths, a nation-wide restriction of movement is enforced, whereby citizens can leave their house only for specific reasons and with a special permit (National Public Health Organization, 2020e). In the Reception and Identification Centers (RICs), medical doctors, members of NGOs and asylum seekers, refugees and migrants considered that measures against the spread of the coronavirus are lacking as they live in overcrowded spaces with little access to proper healthcare services. The data presented concern cases from the epidemiological surveillance of the disease by the new coronavirus (COVID-19), based on the data that have been declared to National Public Health Organization and recorded until May 6, 2020 show the following (see Figure 1): The new confirmed laboratory cases of the disease are 21. The total number of cases is 26,631 (daily change + 0.8%), of which 55,5% are men. The new deaths of COVID-19 patients are 1, while a total of 147 deaths have been reported since the outbreak began. The average age of patients who have died is 75 years. The number of patients hospitalized intubated are 36 (77,8% men) (National Public Health Organization, 2020f).

Figure 1. Laboratory number of laboratory confirmed COVID-19 cases in Greece by May 6, 2020
Source: National Public Health Organization, 2020f.

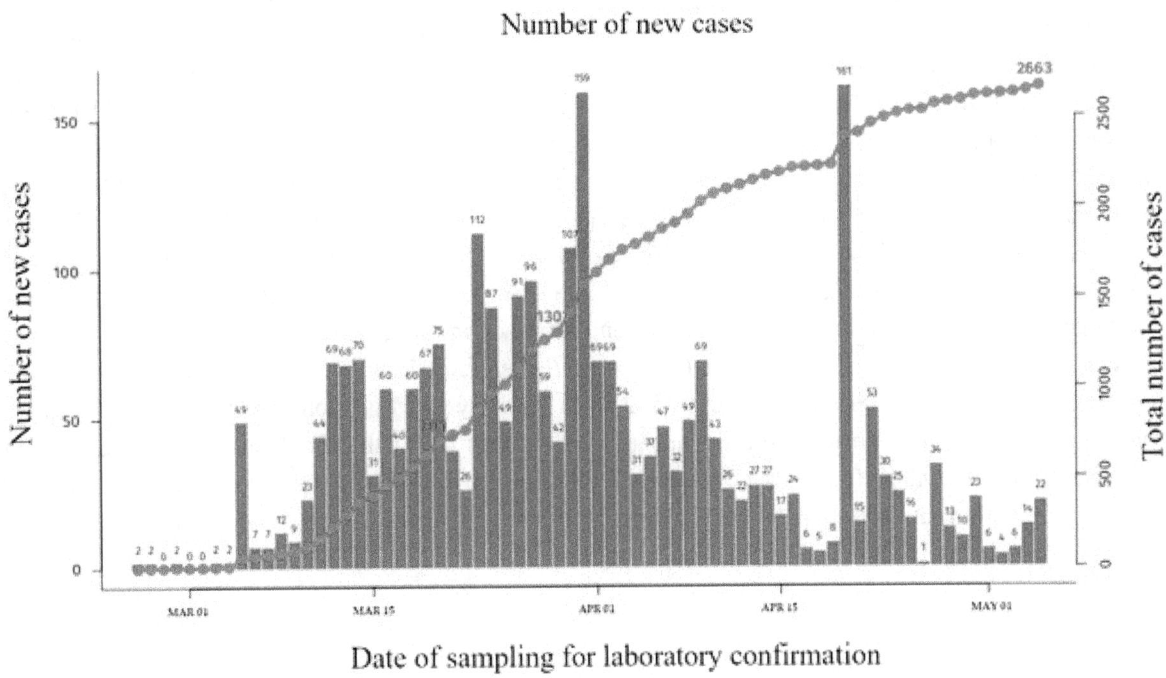

In the Reception and Identification Centers (RICs), medical doctors, members of NGOs and asylum seekers, refugees and migrants considered that measures against the spread of the coronavirus are lacking as they live in overcrowded spaces with little access to proper healthcare services. Greece – due to its geographical position between three continents and because of its extensive coastline comprising a significant part of the external sea borders and part of the land borders of the European Union (EU) and the Schengen area – is the main entry and transit point to the European territory for incoming third-country nationals (TCNs), i.e. asylum seekers, refugees and migrants, who cross the Greek-Turkish land borders in the area of Evros and the sea borders in the Aegean Sea. Since 2015, Greece has faced the largest asylum seeker, refugee and migrant flows of recent years, as a dramatic increase was noted. The country has been the main entry point in Europe for almost a million refugees and migrants seeking safety for themselves and their families. The unprecedented asylum seeker, refugee and migrant flows of 2015 coupled with the movement of the migratory route that comes to Greece from Turkey has tested Greece's already overburdened asylum system and has highlighted the weaknesses and difficulties of handling the dramatic rise of migrants and refugees and meeting their humanitarian needs (Fouskas, 2017, 2019).

Background

Greece set up small Reception and Identification Centers (RICs) on the islands to host asylum seekers but these structures were small, unequipped and unsuitable to host and protect the TCNS. According to the Asylum Service (2020), the number of asylum applications by TCNs within the Greek territory over-multiplied between 2013 and 2019: from 4,814 applications in 2013 (a monthly average of 688 applications), to i.e. a 14.3% rise between 2013 and 2014. In 2014, there were 9,431 applications at a monthly average of 786, a rise of 39.8% between 2014 and 2015. In 2015, there were 13,187 applications at a monthly average of 1,099, an increase by 287, 1% between 2013 and 2014. In 2016, there were 51,053 applications at a monthly average of 4,254, an increase of 14.9% between 2016 and 2017. In 2017, there were 58,637 applications at a monthly average of 4,886, marking an increase of 14.2% between 2017 and 2018. In 2018 there were 66,960 applications at a monthly average of 5,580. In 2019, there were 77,285 applications at a monthly average of 6,440 applications, an increase of 15.4% between 2018 and 2019. The countries of origin with the highest recognition rates were Yemen, Syria, and Palestine, while those with the lowest recognition rates are Georgia, Albania, and India (Asylum Service 2020).

According to the National Center for Social Solidarity (2020), the estimated number of unaccompanied minors amounted to 5,389 (92.7% boys), while 8.8% are under the age of 14. In both the islands and the mainland, the total number of places of long-term accommodation (shelters) is 1,533, while 841 minors are in temporary accommodation (Safe zones/Emergency hotels). As of January 31, 2020, the legally residing TCNs in Greece amount to 542,813 (Ministry of Migration and Asylum 2020). Also, regarding common nationalities of sea arrivals (since 1 January 2020), they are from Afghanistan, Syrian Arab Republic and Others (see Table 1) (UNHCR, 2020). Until 31/01/2020, 42,500 estimated number of refugee and migrant children across Greece. Among them 5,041 are unaccompanied (National Centre for Social Solidarity/UNICEF, 2020).

Table 1. Most common nationalities of sea arrivals (since 1 January 2020)

Country of origin	Data date	Population
Afghanistan	29 Feb 2020	**2,396**
Syrian Arab Rep.	29 Feb 2020	**1,208**
Others	29 Feb 2020	**736**
Dem. Rep. of the Congo	29 Feb 2020	**359**
Iraq	29 Feb 2020	**226**
State of Palestine	29 Feb 2020	**219**
Iran (Islamic Rep. of)	29 Feb 2020	**92**
Stateless	29 Feb 2020	**41**
Algeria	29 Feb 2020	**11**
Pakistan	29 Feb 2020	**5**
Kuwait	29 Feb 2020	**4**

Source: UNHCR, 2020

Asylum seekers, refugees and migrants, who are living in RICs are faced with multiple challenges and vulnerabilities that must be taken into consideration when responding to the COVID-19 pandemic. Multiple stakeholders have been sounding the alarm on the potentially devastating impact on asylum seekers, refugees and migrants due to the COVID-19 pandemic. The chapter focuses on the COVID-19 pandemic concerns over increasing cases recorded in the RICs in Greece.

OVERVIEW OF MIGRANT HEALTH

The Greek society was particularly negatively affected when a study carried out by the Hellenic Center for Disease Control and Prevention (HCDCP/KEELPNO, 2012) was published. According to data from the mobile health units of HCDCP/KEELPNO in Omonia Square, Victoria Square, Attica Square, Larissa Square, Karaiskakis Square, Amerikis Square and Botanicos, there has been an increase in the rates of sexually transmitted diseases, such as genital warts (HPV), syphilis and gonorrhea, as well as outbreaks in infectious diseases such as hepatitis and tuberculosis, and disease recurrence which had been eradicated, like polio as well as multiple cases of infections. These rates were attributed to the increase of male and female prostitution (legal and illegal) and the huge influx of undocumented migrants, who enter and remain without having undergone vaccinations in countries of origin (HCDCP/KEELPNO, 2012). The following data obtained by HCDCP, KEELPNO through recording the medical history of immigrants, was indicative of the prevailing vaccination situation. The impact of migration on public health is of particular concern to Greek migration policy, as the bulk of migrants, asylum seekers and refugees is from Balkan, Asian and African states, have a completely different epidemiological profile from that of the Greeks and other Europeans and higher risks to public health due to the poor living conditions in their countries of origin and during their stay in Greece.

According to national legislation (Law 4368/2016), asylum seekers, persons without social insurance and the vulnerable are entitled free access to necessary health, pharmaceutical and hospital care, including necessary psychiatric care where appropriate (AIDA, 2020b). Nevertheless, actual access to health care services is hindered in practice by significant shortages of resources and capacity for both foreigners and the local population, as a result of the austerity policies followed in Greece, and the lack of adequate cultural mediation in the health services (Fouskas, Sidiropoulos, Vozikis 2019; Fouskas, Gikopoulou, Ioannidi and Koulierakis, 2019). The health challenges that result reflect formal and informal barriers that immigrants, asylum seekers and refugees face in accessing healthcare, such as legal restrictions (stop granting social security numbers to newly arriving migrants) (To Vima, 2020; Kathimerini, 2020a), service provision denial, fear of being denounced to the Authorities and penalization of assistance (for irregular migrants) by health professionals, corruption (under the table payments ("fakelaki" in Greek), language and cultural barriers, as well as unfamiliarity with the way the national health insurance and the health system of the host country function. The situation is further complicated due to the conditions in the Reception and Identification Centers on the Greek islands, where TCNs face existing health problems, live through extreme violence and traumatic events resulting in physical and mental traumas and health issues (AIDA, 2020a), while there is lack of medical staff and cultural health mediators, coupled with persisting overcrowding (AIDA, 2020b). The combination of the restrictive legal framework with the problems that arise in practice exacerbate the inaccessibility of health services by migrants. TCNs in precarious, low-status/low wage jobs seem to develop orientations which question the social importance of heath care services and their significance in their lives (Psimmenos and Kassimati 2003; Psimmenos 2007; Fouskas 2016, 2018).

During the week 8/2020 (17/02 to 23/02), the surveillance system at refugee/migrant care points received data from 32 Accommodation Centers out of a total of 36 Centers participating in the system (89%) (National Public Health Organization, 2020g). During that week, the observed morbidity was within the expected limits, with the exception of the respiratory infection syndrome, for which an increased number of cases was observed on 22/2, at the level of all participating centers. Further investigation did not show any influx of epidemiological interest in individual Accommodation centers. In addition, a case of "meningitis and/or encephalitis" syndrome was reported on 17/2. The case is a 4-month-old baby of Syrian origin, who was confirmed in a laboratory. Instructions have been given for the administration of chemoprophylaxis in close contact. In total, from the end of January to the middle of February, 4 more cases of meningococcal disease from another center (2 serogroup B and 2 serogroup Y) have been reported through the System of Mandatory Disease Statement. They were individuals of Afghan nationality. Massive chemoprophylaxis was given to the population of the center. No cases of the following syndromes/conditions have been reported: Malaria (with positive rapid test), Suspicion of diphtheria, respiratory, or dermal, Jaundice of acute onset, Neurological manifestations of acute onset, Haemorrhagic manifestations with fever, Sepsis or shock (septic, of unknown aetiology), Death of unknown aetiology (see Table 2).

Table 2. Number of cases, proportional morbidity and statistical warning/alert signals by syndrome/ health condition under surveillance, total of reporting centres hosting refugees/migrants, week 8/2020 (from 17/02/2020 to 23/02/2020)

Syndrome	No of visits	Observed proportional morbidity	Expected proportional morbidity	Z-score
i) Respiratory infection with fever	294	4,6	4,0	0,840
ii) Gastroenteritis without blood in the stool	73	1,1	0,8	1,130
iii) Bloody diarrhoea	7	0,1	0,1	0,020
iv) Rash with fever	8	2,6	0,2	-0,396
v) Suspected scabies	165	0,1	3,2	1,269
vi) Suspected pulmonary tuberculosis	4	0,0	0,1	-0,226
vii) Malaria with positive RDT	0	0,0	0,0	-0,261
viii) Suspected diphtheria, respiratory or cutaneous	0	0,0	0,0	0,000
ix) Jaundice of acute onset	0	0,0	0,0	-0,843
x) Neurological manifestations of acute onset	0	0,0	0,0	-0,233
xi) Meningitis and/or encephalitis	1	0,0	0,0	1,544
xii) Haemorrhagic manifestations with fever	0	0,0	0,0	0,000
xiii) Sepsis or shock (septic, of unknown aetiology)	0	0,0	0,0	-0,290
xiv) Death of unknown aetiology	0	0,0	0,0	-0,193

Source: National Public Health Organization, 2020g

CRISIS WITHIN A CRISIS: POOR CONDITIONS IN RECEPTION AND IDENTIFICATION CENTERS ON THE ISLANDS AND ACCOMMODATION CENTERS IN MAINLAND GREECE

Under the "hotspot approach" of the European Commission's Agenda on Migration (2015) as an initial response to the exceptional flows, Reception and Identification Centers (RICs) were established in Greece on the islands of Lesvos, Kos, Chios, Samos, and Leros (Table 3). The competent authority to provide such services was the First Reception Service (FRS), established by Law 3907/2011. First reception procedures included identity and nationality verification, registration, and medical examination. They also provided for any essential care including psychosocial support, provision of information regarding newcomers' obligations and rights, including the conditions under which they can access the asylum procedure.

They also provided for identification of those belonging to vulnerable groups so they could consequently follow the relevant procedure. This was first implemented by the First Reception Centre (FRC) set up in Evros (2013). Joint Ministerial Decision 2969/2015 provided for the establishment of five FRCs in the Eastern Aegean islands. However, under Law 4375/2016 the FRS was succeeded by the Reception and Identification Service (RIS) and was subsumed under what has now been established as the Ministry for Migration Policy.

Figure 2. RIC in Chios
Source: Fouskas, 2019

According to Article 9(1), all TCNs and stateless persons who enter without complying with the legal formalities in the country shall be submitted to reception and identification procedures. Reception and identification procedures include: a) the recording of personal data and fingerprints for those over the age of 14; b) the verification of their identity and nationality; c) their medical screening and provision of any necessary care including psycho-social support; d) provision of information regarding their rights and obligations, in particular the procedure for international protection or the procedure for entering a voluntary return program; e) special attention is being paid to those belonging to vulnerable groups, in order for them to undergo the appropriate procedure and for them to be provided with specialized care and protection; f) referral for initiating the international protection status procedure for those who wish to apply; g) referral to the relevant services for affiliation to read mission or return or expulsion procedures for those who do not submit or for those whose application for international protection has been rejected during their stay at the RIC.

In Greece, thousands of asylum seekers, refugees and migrants are living under inhuman, unsafe and degrading conditions at the camps on the Aegean islands and several others in the mainland Greece (Kathimerini, 2019; Oxfam, 2019) (see Table 3). Camps are severely overcrowded. Multiple deficiencies and lack of medical doctors resulting in numerous health issues, deterioration of health due to weather changes for example there is no insulation or heated accommodation in the winter. Many TCNs feel insecure under these precarious conditions, while many have suffered sexual abuse or traumas.

In the RICs there are frequent outbreaks of violence and fires, for example in Lesvos (March 2020) (Kathimerini, 2020g) and Samos (April 2020) (Kathimerini, 2020h), where the fire spread quickly due to the flammable materials in the camp used for cooking inside the tents. Reasons for these outbreaks are the clashes that break out between migrant groups (for example African and Afghan migrants residing in the camp), poor living conditions, lack of health care and overcrowding as there are multiple different nationalities living in the same area and the psychological breakdown and tensions due to the ongoing limbo/immobility and uncertainty. The RICs have been operating inappropriately. They are mostly set up on old facilities such as abandoned army camps and factories, which are extending beyond their intended borders. Newcomers set up tents wherever they can find available space, in the surrounding area, with no safety and exposure to the weather condition (see Picture 1). TCNs inside the RICs are crammed into small individual tents or separated makeshift spaces (see Picture 2) made by wooden walls under a larger umbrella tent. These spaces offer little or no privacy, use a blanket for the door and mats for the ground, and provide insufficient insulation from weather conditions and temperature changes (summer heat/winter cold). There is no segregation between men, women and children (including unaccompanied minors). At the same time, there are some containers used as living spaces. The RICs, in general, provide poor facilities water and sanitation. In addition, there is insufficient access to electricity and lighting. These conditions have contributed to an environment that is unsafe and full of health risks for TCNs. In the general, a worrying reduction in healthcare provision in all RICs in the islands is clearly observed (MSF, 2017a). TCNs current squalid living conditions in Greece are inadequate (for example in Samos' Vathy camp, in Chios' Vial camp, in Lesvos' Moria camp). The uncertainty about their futures, the threat of deportation, and the lack of access to appropriate healthcare that they emphasise during consultations as negatively affecting their mental wellbeing (MSF, 2017b).

MIGRANTS IN GREECE IN THE MIDST OF THE COVID-19 PANDEMIC

As the World Health Organization (WHO) (2020) states, asylum seekers, refugees and migrants are exposed to increased risks of contracting diseases, such as COVID-19, as they live mostly in overcrowded facilities and conditions without basic public health conditions as they are lacking the relevant means (WHO, 2020). In addition, the lack of sufficient crucial services (Jalbout, 2020), overcrowded accommodation centers, along with deplorable conditions, the lack of proper shelter, the extremely unhygienic living conditions (including lack of water and sanitation) with improper sewage system (Veizis, 2020) deteriorated by insufficient health services (Jalbout, 2020) (including medical doctors, social workers psychologists). The daily routine of asylum seekers, refugees and migrants includes 2-3 hours waiting in queues receiving food and meals, often described of low quality which leads some to abandon the wait or not to eat due to untastiness (Jalbout, 2020). Shortages of medicines and lack of healthcare facilities compromise the capacity of asylum seekers, refugees and migrants to access healthcare services in humanitarian settings (WHO, 2020).

Table 3. National Situational Picture Regarding the Islands at Eastern Aegean Sea (Occupancy/Capacity) on 05/05/2020

PLACE/LOCATION		LESVOS		CHIOS		SAMOS		LEROS		KOS		OTHER ISLANDS		TOTAL	
		OCC.	CAP.	OCC.	CAP.	OCC.	CAP.	OCC.	CAP.	OCC.	CAP.	OCC.	CAP.	OCC.	CAP.
R.I.C.		17940	2757	5025	1014	6842	648	1906	860	2387	816			**34100**	**6095**
OTHER ACCOMMODATION FACILITIES		1067						65	120					**1132**	
HELLENIC POLICE FACILITIES	P.D.C.	187	210							442	474			**629**	**684**
	DETENTION FACILITIES	34		10		26		1		1		34		**106**	
U.N.H.C.R.		680	782	269	297	300	282	116	136	186	213	62	81	**1613**	**1791**
N.C.S.S.		142	148	14	18	17	18							**173**	**184**
OTHER N.G.O.s		68	100											**68**	
MAKESHIFT CAMPS		0		0		0		0		0		0		**0**	
MIGRANTS PRESENT ON THE ISLAND		**20118**		**5318**		**7185**		**2088**		**3016**		**96**		**37821**	
ARRIVALS		0		0		0		0		0		0		**0**	
TRANSPORTS TO THE MAINLAND		18		0		0		0		0		0		**18**	
DEPARTURES (EU-TURKEY STATEMENT)		0		0		0		0		0		0		**0**	
DEPARTURES (1.0.M.)		0		0		0		0		0		0		**0**	
TOTAL DEPARTURES FROM THE ISLAND		**0**		**0**		**0**		**0**		**0**		**0**		**0**	

Source: Hellenic Republic Ministry of Citizen Protection National Coordination Center for Border Control, Immigration and Asylum, 2020

Multiple COVID-19 cases detected in accommodation centers in mainland Greece (see Table 4). During the first case of COVID-19 multiple attempts to enter via the Greek-Turkish land border recorded leading to the prevention of entry and border sealing policy (Mitarakis, 2020g) and the suspension of asylum applications due to the instrumentalization of migrant flows and securitization to limit the spread of COVID-19 respectively (Asylum Service, 2020d) for migrants who arrived in Greece from Turkey. A total of 26,532 people prevented from entering the country illegally and 218 individuals were arrested by March 3, 2020 (Kathimerini, 2020k). Since March 13, 2020 (Asylum Service, 2020a) which extended to April 10, 2020, the suspension of reception and public service by the Asylum Service was extended until Friday, May 15, 2020 in order to limit the spread of COVID-19 virus and protect public health. All administrative actions were suspended (recordings, interviews, filing of an appeal, etc.). The validity of Asylum Seekers' Certificates and Residence Permits expiring before May 15 was extended (Asylum Service, 2020b, 2020c). The Greek Government and the relevant ministries aimed at preventing the dispersion of the COVID-19 and to protect employees at Migration Services and the public (European Migration Network, 2020). Remedying subsequent dysfunctions concerning legal migration procedures and legality of stay for TCNs was necessary (European Migration Network, 2020). The Ministerial Decision regarding the extension of residence titles of TCNs, so as to ensure their legal status of residence, was issued on 30 March 2020 (European Migration Network, 2020; Government Gazette, 2020d).

20 patients are among 63 residents of the Ritsona camp who were tested immediately on March 31, 2020 after a woman was found to be infected at a hospital in Athens after giving birth. Greek health officials have placed a refugee and migrant camp northeast of the capital in quarantine after 20 of its residents tested positive for Covid-19 (Kathimerini, 2020c; NPHO/EODY, 2020a).

Figure 3. RIC in Lesvos
Source: Fouskas, 2019

Table 4. COVID-19 case detected in Accommodation Centers for Migrants and Refugees in Greece

Area/Facility	Date	COVID-19 case detected
Ritsona	March 31, 2020	23
Malakasa, East Attica	April 5, 2020	1
Porto Heli, Argolida	April 19, 2020	1
Kranidi, Argolida	April 20, 2020	150
Megala Therma, Lesvos	May 12, 2020	2
Efthalou beach, Southern Lesvos	May, 15, 2020	2

Moreover, the accommodation centers in Malakasa placed under total sanitary isolation for a 14 days period following the detection of a case of COVID-19 on an individual who is a resident of the structure (NPHO/EODY, 2020b; Kathimerini, 2020n). Confinement measures in his/her family member and tracing procedures of identifying individuals who came in contact with him/her were implemented, including staff and army personnel. However, later within the same month 5 TCNs broke quarantine rules and went missing from the Malakasa center (Kathimerini, 2020e) and arrested later (Kathimerini, 2020o).

In addition, a positive COVID-19 case in the temporary facility outside Porto Heli (IOM Office in Greece, 2020a, 2020b) was discovered on April 19, 2020. A 6-month pregnant woman from Africa who was hosted at the temporary facility for asylum seekers in Porto Heli, Southern Greece, required hospitalization in relation to her pregnancy and was transferred with an ambulance to the hospital of Nafplio (IOM Office in Greece, 2020a, 2020b). According to the measures developed by the Greek Authorities and through relevant health checks and medical examinations, the woman was found positive to COVID-19 (IOM Office in Greece, 2020a, 2020b). A total of 150 migrants in an accommodation center tested positive for the new coronavirus (COVID-19) on April 20, 2020 (Kathimerini, 2020a; NPHO/EODY, 2020c). Greek health officials concerned about outbreaks in the migrant community (Kathimerini, 2020b). The residents of Kranidi expressed alarm and are demanding widespread testing in the community as the migrants regularly visited supermarkets, banks and other venues (To Vima, 2020).

Moreover, two cases of migrants with COVID-19 disease without symptoms detected in a temporary refugee accommodation center at Megala Therma, Lesvos island in a random sampling testing of the National Public Health Organisation (NPHO/EODY) on May 12, 2020 (ANA-MPA, 2020; NPHO/EODY, 2020d). In addition, two more migrants who arrived in Lesvos on May 6, 2020 were found positive positive to COVID-19. Thus 4 out of the 70 migrants who arrived on the island from Turkey between May 11-15 were found positive. The cases were identified as part of a random sampling testing. All cases are not placed in the RIC in Moria, as from March 1, 2020, those irregular migrants who arrive to the Greek islands are not admitted to the RICs, but to special quarantine structures.

Protection measures against COVID-19 in the RICS, in the accommodation centers and in the Asylum Service were implemented in mid-March 2020. The measures included the suspension of all special activities and facilities in the camp like schools, libraries and exercise areas, while new arrivals should be checked for a fever and other symptoms of Covid-19 and confined to quarantine if found to be ailing (Kathimerini, 2020d). Sanitation measures were implemented to ensure that all common-use areas are regularly cleaned and updated and instructions for to the five island camps' residents twice a day in multiple languages including Arabic, Farsi, English, French and Greek were distributed (Kathimerini, 2020d). TCNs in camps were discouraged from strolling around the facilities or going outside the RICs, not even for getting supplies. Moreover, controlled entry and exit in two camps in Malakasa and Sintiki of Serres was announced and expected to operate as the two centers were set up to function as closed centers (Kathimerini, 2020m). In addition, preventive measures to avoid the spread of COVID-19 in RICs and accommodation centers of TCNs was launched, via a National Action Plan of Crisis Management in Refugee Structures entitled: AGNODIKI. It included delivery points by companies and the installatrion of ATMs within the structures (GOVGR, 2020).

An extension of restrictive movement measures in the RICs and accommodation centers throughout the territory of Greece, until May 21, 2020 due to COVID-19 extended via the relevant Joint Ministerial Decision (Government Gazette, 2020c). Before the restrictive movement measures were until April, 21, 2020 and May 10, 2020 (Government Gazette 2020a; 2020b). In all cases the personnel of the Ministry of Migration and Asylum and the NPHO/EODY deployed in the areas and attempt to track down TCNs, state employees, staff of international organization and non-governmental organizations and others who were in contact with infected individuals. At the RICs of Moria (Lesvos) a medical evaluation space was created on the perimeter of the center (Mitarakis, 2020d) and the operation of Outpatient Clinics in RIC of Chios launched, both in early May 2020 (Mitarakis, 2020e).

According to the statements of the Greek Minister of Migration and Asylum in the Parliament, the cases of COVID-19 in accommodation centers correspond to 0.2% of the total migrant population and are limited to 3 of the 93 total structures in Greece, underlying that "Our islands remain 'clean'" (Mitarakis, 2020a). However, a possible COVID-19 outbreak could decimate this vulnerable population.

SOLUTIONS AND RECOMMENDATIONS

The cases of COVID-19 in the mainland are raising concerns for the regarding virus spreads in the RICs and accommodation centers. Promotion of information on health in the RICs and accommodation centers is deemed necessary. No visitors including state employees, staff of international organization and non-governmental organizationsshould be allowed into the RICs both on the islands and mainland Greece to prevent a coronavirus outbreak in any of the facilities (Kathimerini, 2020c). It is extremely difficult to take the necessary precaution measures, for example keeping and maintaining distance between individuals of all ages and to implement required hygiene rules, so that better management can be done when the virus spreads (IOM – Office in Greece, 2020c). The urgent decongestion of the overcrowded structures on the islands is deemed necessary. Any efforts to create new housing in the mainland is crucial. The containment of COVID-19 virus in a humane and dignified way cannot be achieved without the urgent decongestion of the RICs and accommodation centers.

FUTURE RESEARCH DIRECTIONS

Concerning the COVID-19 pandemic, future research should focus the understand how the pandemic is impacting the health, employment, living conditions, human rights and well-being of communities of migrants, asylum seekers and refugees as well as to unravel the effect of policy responses in the reception societies via critical analyses that examine the responses under the prisms of exclusionary migration regimes and sorting of individuals according to access to immunization and the ability to prove it.

CONCLUSION

The RICs on the Greek islands are not adequately prepared for COVID-19 (HRW, 2020) and the situations is like a a tinderbox waiting to ignite as physical distancing is impossible and containment of an outbreak in such settings will be devastating. Asylum seekers, refugees and migrants have to deal with unsanitary conditions while fearing the spread of COVID-19 in the overcrowded RICs and accommodation centers that are without adequate medical staff to take care potential patients. In this precarious context, a possible outbreak of Covid-19 could devastate this crammed vulnerable population. 37,821 individuals on the five islands, including families with children, unaccompanied refugee minors and young children, pregnant women, new mothers, the elderly, others with chronic illnesses and disabilities, are particularly vulnerable under inhumane conditions, close quarters and limited access to sanitation. Providing information in multiple languages to stay healthy via quarantine guidelines is necessary. The Hellenic police used loudspeakers to broadcast messages about related movement restrictions due to COVID-19 in 10 languages including Arabic, Urdu and Farsi. Such measures should be accompanied with information via social media and provision of free bottled water, sanitation and hygiene products (masks, antiseptics, gloves) to vulnerable groups. The transfer of individuals from the RICs and accommodation centers to apartments and hotels will safeguard public health via alternative methods of accommodation, providing solidarity, dignity and safety to asylum seeker, refugees and migrants. The urgent decongestion of the overcrowded RICs and accommodation centers is required to avoid the risk of rapid spread of the infection. Forced lockdown of asylum seekers, refugees and migrants on grounds of public health, as taken on March 17, 2020 (Kathimerini, 2020l), cannot be justified as a supreme measure, as what should be guaranteed is that that no asylum seeker, refugee and migrant is left behind amid the COVID-19 pandemic.

REFERENCES

ANA-MPA. (2020). *Two migrants with Covid-19 in quarantine at Lesvos refugee facility, Migration Ministry confirms.* Retrieved from https://www.amna.gr/en/article/457131/Two-migrants-with-Covid-19-in-quarantine-at-Lesvos-refugee-facility--Migration-Ministry-confirms

Asylum Information Database (AIDA)/European Council on Refugees and Exiles (ECRE). (2020a). *Conditions in Reception Facilities: Greece.* European Programme for Integration and Migration (EPIM). Retrieved from https://www.asylumineurope.org/reports/country/greece/reception-conditions/housing/conditions-reception-facilities

Asylum Information Database (AIDA)/European Council on Refugees and Exiles (ECRE). (2020b). *Health care: Greece.* European Programme for Integration and Migration (EPIM). Retrieved from https://www.asylumineurope.org/reports/country/greece/reception-conditions/health-care

Asylum Service. (2020a). *Statistical Data of the Greek Asylum Service (from 07.06.2013 to 29/02/2020).* Retrieved from http://asylo.gov.gr/en/wp-content/uploads/2020/03/Greek_Asylum_Service_data_February_2020_en.pdf

Asylum Service. (2020b). *Important notice of the Asylum Service- Reception and service of the publicis suspended. (in Greek)* Retrieved from http://asylo.gov.gr/wp-content/uploads/2020/03/ΑΝΑΚΟΙΝΩΣΗ-ΑΝΑΣΤΟΛΗΣ-ΥΠΟΔΟΧΗΣ-ΚΑΙ-ΕΞΥΠΗΡΕΤΗΣΗΣ-ΚΟΙΝΟΥ-ΕΛΛΗΝΙΚΑ-13-3-2020-1.pdf

Asylum Service. (2020c). *The suspension of reception and public service by the Asylum Service is extended until May 15. (in Greek)* Retrieved from http://asylo.gov.gr/wp-content/uploads/2020/04/Greek-Announcement-April-10_MK1.pdf

Asylum Service. (2020d). *Suspension of asylum applications. (in Greek)* Retrieved from http://asylo.gov.gr/wp-content/uploads/2020/03/pnp-anastolh-ths-ypovolis-aithseon-asylou.pdf

Council of Europe. (2019). *Committee for the Prevention of Torture, Report to the Greek Government on the visits to Greece carried out by the European Committee for the Prevention of Torture and Inhuman or Degrading Treatment or Punishment (CPT), 26 September 2017, CPT/Inf (2017) 25.* Retrieved from https://rm.coe.int/1680930c9a

European Commission. (2015). *A European Agenda on Migration.* Retrieved from https://ec.europa.eu/home-affairs/sites/homeaffairs/files/what-we-do/policies/european-agenda-migration/background-information/docs/communication_on_the_european_agenda_on_migration_en.pdf

European Migration Network (EMN)/European Commission. (2018). *Asylum and Migration Glossary 6.0.* Retrieved from https://ec.europa.eu/home-affairs/sites/homeaffairs/files/what-we-do/networks/european_migration_network/docs/interactive_glossary_6.0_final_version.pdf

European Migration Network (EMN)/European Commission. (2020). *Special Annex to the 30th EMN Bulletin EU Member States & Norway: responses to COVID-19 in the migration and asylum area January – March 2020.* Retrieved from: https://ec.europa.eu/home-affairs/sites/homeaffairs/files/00_eu_30_emn_bulletin_annex_covid_19.pdf

Fouskas, T. (2016). Repercussions of Precarious, Low-Status Work Regarding Access to Healthcare Services by Immigrant Workers in Greece: Seasonal Flu or an Incurable Disease? In J. Jaworski (Ed.), *Advances in Sociology Research* (Vol. 18, pp. 13–74). Nova Science Publishers.

Fouskas, T. (2017). EU societies divided in unity? The changing influx of asylum seekers in 2014-2016 in Greece and the EU Member States' societies disunity due to the impact of the migration and refugee crisis. *European Politeia, 5*(1-2), 123–169.

Fouskas, T. (2018). Precarious employment repercussions on migrants' perceptions regarding healthcare. *International Journal of Human Rights in Healthcare, 11*(3), 298–311. doi:10.1108/IJHRH-01-2018-0010

Fouskas, T. (2019). Unravelling solidarity and hostility: Mobilizations concerning migrants, asylum seekers and refugees in anti-migrant times in Greece. In A. Rea, M. Martiniello, P. De Cuyper, B. Meuleman, & A. Mazzola (Eds.), *The refugee reception crisis in anti-immigrant times. Polarization of the public opinion, local mobilizations and reception practices in Europe* (pp. 121–143). ULB Press.

Fouskas, T., Gikopoulou, P., Ioannidi, E., & Koulierakis, G., (2019). Health inequalities and female migrant domestic workers: Accessing healthcare as a human right and barriers due to precarious employment in Greece. *Collectivus, 6*(2), 71-90.

Fouskas, T., Sidiropoulos, S., & Vozikis, A. (2019). Leaving no one out? Public health aspects of migration: Health risks, responses and accessibility by asylum seekers, refugees and migrants in Greece. *International Journal of Health Research and Innovation, 7*(1), 13–28.

Government Gazette. (2011). *Law 3907/2011 on the Establishment of an Asylum Service and a First Reception Service, adaptation of the Greek legislation to the provisions of Directive 1008/115/EC 'with regard to the common rules and procedures in Member States for the return of illegally staying third-country nationals' and other provisions. (Greek).* Retrieved from http://www.yptp.gr/images/stories/2011/law%203907.pdf.pdf

Government Gazette. (2015). *Joint Ministerial Decision 2969/2015: Establishment of First Reception Centers and temporary Hosting Structures for asylum seekers and vulnerable groups of third-country nationals.* https://bit.ly/3bovbh4

Government Gazette. (2016). *Law 4368/2016. Article 33: Health coverage of uninsured and vulnerable social groups.* Retrieved from http://asylo.gov.gr/wp-content/uploads/2020/02/φεκ-ΠΑΑΥΠΑ.pdf

Government Gazette. (2016). *Law 4375/2016 on the On the organization and operation of the Asylum Service, the Appeals Authority, the Reception and Identification Service, the establishment of the General Secretariat for Reception, the transposition into Greek legislation of the provisions of Directive 2013/32/EC "on common procedures for granting and withdrawing the status of international protection (recast) (L 180/29.6.2013), provisions on the employment of beneficiaries of international protection and other provisions.* Retrieved from http://asylo.gov.gr/en/wp-content/uploads/2016/05/GREECE_Law_4375_2016_EN_final.pdf

Government Gazette. (2020a). *Decision D1A/20030 (Government Gazette B – 9 85 – 22.03.2020) Taking measures against the occurrence and spread of cases of COVID-19 in the Reception and Identification Centers, in the total of the Territory, for the period from 21.03.2020 to 21.04.2020 (in Greek).* Retrieved from http://asylo.gov.gr/wp-content/uploads/2020/03/RICS_RESTRICTIONS_MINISTERIAL-DECISION.pdf

Government Gazette. (2020b). *Joint Ministerial Decision D1A/25768/2020 – 1472 – 17-4-2020 (in Greek).* Retrieved from http://www.elinyae.gr/sites/default/files/2020-04/1472b_2020.pdf

Government Gazette. (2020c). *Extension of D1A 20030/21.3.2020 (B 985) of Joint Decision of the Ministers of Civil Protection, Health and Migration and Asylum until 21.5.2020 (in Greek).* Retrieved from http://www.elinyae.gr/sites/default/files/2020-05/1771b_2020.pdf

Government Gazette. (2020d). *B 1086/30.03.2020 (in Greek)* http://www.immigration.gov.gr/documents/20182/31271/YA+12889_2020_ΠΑΡΑΤΑΣΗ+ΙΣΧΥΟΣ+ΤΩΝ+ΑΔΕΙΩΝ+ΔΙΑΜΟΝΗΣ_ΦΕΚ+1086+B.pdf/229a7015-dd96-4d3b-84c6-7c84153ac04b

GOVGR. (2020). *Preventive measures to avoid the spread of coronavirus disease in Reception and Identification Centers and Third Country Nationals' Accommodation centers.* Retrieved from https://covid19.gov.gr/proliptika-metra-gia-tin-apofygi-tis-diasporas-tou-koronoiou-se-kentra-ypodochis-ke-taftopiisis-ke-domes-filoxenias-politon-triton-choron/

Hellenic Center for Disease Control and Prevention (HCDCD/KEELPNO). (2012). *Intervention Project in the Historical Centre of Athens. In: Hellenic Ministry of Health/Office of Parliamentary Control, Answer to the Question: Programmatic Convention of the Hellenic Ministry of Health and the Municipality of Athens/ANS 699/03/08/2012.* Athens: Hellenic Ministry of Health/Office Parliamentary Control: 3-64 (in Greek). Retrieved from https://www.hellenicparliament.gr/UserFiles/67715b2c-ec81-4f0c-ad6a-476a34d732bd/7737577.pdf

Hellenic Republic Ministry of Citizen Protection National Coordination Center for Border Control. Immigration And Asylum (NCCBCIA) (2020). *National Situational Picture Regarding the Islands at Eastern Aegean Sea (5/5/2020).* Retrieved from https://infocrisis.gov.gr/8818/national-situational-picture-regarding-the-islands-at-eastern-aegean-sea-5-5-2020/?lang=en

Human Rights Watch (HRW). (2020). *Greece: Island Camps Not Prepared for Covid-19 Government Neglect Puts Lives at Risk.* Retrieved from https://www.hrw.org/news/2020/04/22/greece-island-camps-not-prepared-covid-19

Kathimerini. (2020a). *150 people test positive for Covid-19 at Kranidi refugee facility.* Retrieved from https://www.ekathimerini.com/251872/article/ekathimerini/news/150-people-test-positive-for-covid-19-at-kranidi-refugee-facility

Kathimerini. (2020b). *Health, migration ministers discuss measures for migrant camps.* Retrieved from https://www.ekathimerini.com/251908/article/ekathimerini/news/health-migration-ministers-discuss-measures-for-migrant-camps

Kathimerini. (2020c). *Ritsona refugee camp quarantined after 20 test positive for Covid-19.* Retrieved from https://www.ekathimerini.com/251258/article/ekathimerini/news/ritsona-refugee-camp-quarantined-after-20-test-positive-for-covid-19

Kathimerini. (2020d). *Migrant camps on islands put on lockdown.* Retrieved from https://www.ekathimerini.com/250739/article/ekathimerini/news/migrant-camps-on-islands-put-on-lockdown

Kathimerini. (2020e). *Concern over outbreaks at migrant camps.* Retrieved from https://www.ekathimerini.com/252057/article/ekathimerini/news/concern-over-outbreaks-at-migrant-camps

Kathimerini. (2020f). *Oxfam report details inhumane conditions at Greek migrant camps.* Retrieved from https://www.ekathimerini.com/236377/article/ekathimerini/news/oxfam-report-details-inhumane-conditions-at-greek-migrant-camps

Kathimerini. (2020g). *Fire destroys warehouse at Lesvos migrant center.* Retrieved from https://www.ekathimerini.com/250368/article/ekathimerini/news/fire-destroys-warehouse-at-lesvos-migrant-center

Kathimerini. (2020h). *Fires erupt amid unrest at Samos migrant camp.* Retrieved from https://www.ekathimerini.com/252108/article/ekathimerini/in-images/fires-erupt-amid-unrest-at-samos-migrant-camp

Kathimerini. (2020i). *Greece slammed over conditions at migrant camps.* Retrieved from https://www.ekathimerini.com/237845/article/ekathimerini/news/greece-slammed-over-conditions-at-migrant-camps

Kathimerini. (2020j). *NGOs urge Greek gov't to restore health coverage for migrants.* Retrieved from http://www.ekathimerini.com/248700/article/ekathimerini/news/ngos-urge-greek-govt-to-restore-health-coverage-for-migrants

Kathimerini. (2020k). *We shield the national borders against provocative actions. Retrieved* from https://www.kathimerini.gr/1067403/article/epikairothta/ellada/panw-apo-26000-oi-apotrapeises-prospa8eies-eisodoy-sth-xwra

Kathimerini. (2020l). *Prevention measures against coronavirus by the Ministry of Immigration and Asylum (in Greek).* Retrieved from https://www.kathimerini.gr/1069591/article/epikairothta/ellada/metra-prolhyhs-kata-toy-korwnoioy-apo-to-ypoyrgeio-metanasteyshs-kai-asyloy

Kathimerini. (2020m). *Entrance, exit with cards in closed structures (in Greek).* Retrieved from https://www.kathimerini.gr/1073748/article/epikairothta/ellada/eisodos-e3odos-me-kartes-se-kleistes-domes

Kathimerini. (2020n). *EODY unit in the accommodation center in Malakasa after a coronavirus case (in Greek).* Retrieved from https://www.kathimerini.gr/1072471/article/epikairothta/ellada/klimakio-toy-eody-sthn-domh-filo3enias-sth-malakasa-meta-apo-kroysma-korwnoioy

Kathimerini. (2020o). *All five third-country nationals who "broke" the quarantine in the Malakassa structure were arrested (in Greek).* Retrieved from https://www.kathimerini.gr/1075324/article/epikairothta/ellada/synelhf8hsan-kai-oi-pente-allodapoi-poy-espasan-thn-karantina-sth-domh-malakasas

Medicines sans Frontiers (MSF). (2017a). *A Dramatic Deterioration for Asylum Seekers on Lesbos.* Retrieved from https://msf.gr/sites/default/files/msfpublications//msf_report_vulnerable_lesvos_en.pdf

Medicines sans Frontiers (MSF). (2017b). *Confronting the mental health emergency on Samos and Lesvos: Why the containment of asylum seekers on the Greek islands must end.* Retrieved from https://msf.gr/sites/default/files/msfpublications//2017_10_mental_health_greece_report_lowres_spreads.pdf

Ministry of Migration and Asylum. (2020). *Monthly Statistics.* Retrieved from http://www.immigration.gov.gr/documents/20182/45177/StatsMetan_20200131.zip/adddf678-d929-4f44-a429-f0f9517eef0d

Mitarakis, N. (2020a). *Speech by the Minister of Migration and Asylum, Member of Parliament for Chios, Notis Mitarakis, during the discussion of Current Queries in the Hellenic Parliament.* Retrieved from https://www.mitarakis.gr/gov/migration/6115-dt-yma-vouli-epikairi-erwtisi

Mitarakis, N. (2020b). *JMD Restriction of Movement in the Accommodation centers of Asylum Seekers.* Retrieved from https://www.mitarakis.gr/gov/migration/1962-dt-yma-kya-periorismou

Mitarakis, N. (2020c). *New 6-measures bundle of the Ministry of Migration and Asylum for the protection of the coronavirus in the RICs on the Islands and other host structures.* Retrieved from https://www.mitarakis.gr/gov/migration/1970-dt-yma-nea-desmi-metrwn

Mitarakis, N. (2020d). *Visit of the Minister of Immigration and Asylum in Lesvos.* Retrieved from https://www.mitarakis.gr/gov/migration/6121-dt-yma-programma-lesvou

Mitarakis, N. (2020e). *Operation of Outpatient Clinics in KYT Chios.* Retrieved from https://www.mitarakis.gr/gov/migration/1958-dt-yma-kyt-vial-iatreia

Mitarakis, N. (2020f). *Measures to protect against coronavirus in the Reception and Identification Centers, in the accommodation centers and in the Asylum Service.* Retrieved from https://www.mitarakis.gr/gov/migration/1956-μέτρα-προστασίας-από-τον-κορωνοϊό-στα-κέντρα-υποδοχής-και-ταυτοποίησης,-στις-δομές-φιλοξενίας-και-στην-υπηρεσία-ασύλου

Mitarakis, N. (2020g). *We shielded the national borders against provocative actions (in Greek).* Retrieved from https://www.mitarakis.gr/gov/migration/1931-dt-yma-ant1-prwinoi-typoi

National Center for Social Solidarity. (2020). *Situation Update: Unaccompanied Children (UAC) in Greece 15 January 2020.* Retrieved from https://www.e-pronoia.gr/sites/default/files/ekka_dashboard_15-1-2020.pdf

National Centre for Social Solidarity/UNICEF. (2020). *Refugee and migrant children in Greece (As of 31 January 2020).* Retrieved from https://www.unicef.org/eca/media/10861/file

National Public Health Organisation (NPHO/EODY). (2020a). *Update of accredited health journalists by the Crisis Management and Civil Protection Minister Nikos Hardalias and the representative of the Ministry of Health for the new coronavirus, Professor Sotiris Tsiodras April 1, 2020.* Retrieved from: https://eody.gov.gr/dienergeia-elegchoy-gia-sars-cov-2-sto-kfpm-ritsonas-kai-ichnilatisi-epafon-apo-to-prosopiko-toy-eody/ and https://eody.gov.gr/enimerosi-diapisteymenon-syntakton-ygeias-apo-ton-yfy-poyrgo-politikis-prostasias-kai-diacheirisis-kriseon-niko-chardalia-kai-ton-ekprosopo-toy-ypoyrgeioy-ygeias-gia-to-neo-koronoio-kathigiti-sotir-5/

National Public Health Organisation (NPHO/EODY). (2020b). *Update of accredited health journalists by the Crisis Management and Civil Protection Minister Nikos Hardalias and the representative of the Ministry of Health for the new coronavirus, Professor Sotiris Tsiodras April 5, 2020.* Retrieved from: https://eody.gov.gr/klimakio-toy-eody-sto-kfpm-malakasas-gia-tin-ygeionomiki-prostasia-prosopikoy-kai-filoxenoymenon/

National Public Health Organisation (NPHO/EODY). (2020c). *Update of accredited health journalists by the Crisis Management and Civil Protection Minister Nikos Hardalias and the representative of the Ministry of Health for the new coronavirus, Professor Sotiris Tsiodras April 21, 2020.* Retrieved from: https://eody.gov.gr/0421_briefing_covid19/

National Public Health Organisation (NPHO/EODY). (2020d). *Update of accredited health journalists by the Crisis Management and Civil Protection Minister Nikos Hardalias and the representative of the Ministry of Health for the new coronavirus, Professor Sotiris Tsiodras May 14, 2020.* Retrieved from: https://eody.gov.gr/0514_briefing_covid19/

National Public Health Organization (NPHO/EODY). (2020e). *Current state of Covid-19 outbreak in Greece and timeline of key containment events. News, Press Releases. 04.03.2020.* Retrieved from https://eody.gov.gr/en/current-state-of-covid-19-outbreak-in-greece-and-timeline-of-key-containment-events/

National Public Health Organization (NPHO/EODY). (2020f). *Daily epidemiological surveillance report new coronavirus infection (COVID-19) Data until May 6, 2020, at 15:00.* Retrieved from https://eody.gov.gr/wp-content/uploads/2020/05/covid-gr-daily-report-20200506.pdf

National Public Health Organization (NPHO/EODY). (2020g). *Weekly Epidemiological Surveillance Report in points of healthcare of refugees/migrants Week 8/2020 (17 to 23 February 2020).* Retrieved from https://eody.gov.gr/wp-content/uploads/2020/03/Surveillance-refugees-weekly-GR-2020-08.pdf

Office, I. O. M. (2020a). *IOM announcement on positive COVID 19 case in the temporary facility in Porto Heli.* Retrieved from https://greece.iom.int/en/news/iom-announcement-positive-covid-19-case-temporary-facility-porto-heli

Office, I. O. M. (2020b). *IOM Responding to New COVID-19 Outbreak in Greece.* Retrieved from https://www.iom.int/news/iom-responding-new-covid-19-outbreak-greece

Office, I. O. M. (2020c). *IOM Raises Concern Over Increasing COVID-19 Cases Recorded in Greece Mainland Refugee and Migrant Camp.* Retrieved from https://greece.iom.int/en/news/iom-raises-concern-over-increasing-covid-19-cases-recorded-greece-mainland-refugee-and-migrant

Oxfam. (2019). *Vulnerable and abandoned.* Retrieved from https://oi-files-d8-prod.s3.eu-west-2.amazonaws.com/s3fs-public/file_attachments/2019-01_greece_media_briefing_final.pdf

Psimmenos, I. (2007). Work Culture and Migrant Women's Welfare Marginalization. *Epitheorese Koinonikon Ereunon, 124*(3), 9–33. doi:10.12681/grsr.112

Psimmenos, I., & Kassimati, K. (2003). Immigration Control Pathways: Organizational Culture and Work Values of Greek Welfare Officers. *Journal of Ethnic and Migration Studies, 29*(2), 337–371. doi:10.1080/1369183032000079639

To Vima. (2019). *Vroutsis blocks simplified AMKA issuance for non-EU nationals.* Retrieved from https://www.tovima.gr/2019/07/12/international/vroutsis-blocks-simplified-amka-issuance-for-non-eu-nationals/

To Vima. (2020). *Civil protection authority: Kranidi quarantined after 150 migrants test positive for COVID-19.* Retrieved from https://www.tovima.gr/2020/04/21/english-edition/civil-protection-under-minister-kranidi-quarantined-after-150-migrants-test-positive-for-covid-19/

UNHCR. (2020). *Most common nationalities of sea arrivals (since 1 January 2020).* Retrieved from https://data2.unhcr.org/en/situations/mediterranean/location/5179

Veizis, A. (2020). "Leave No One Behind" and Access to Protection in the Greek Islands in the COVID-19 Era. *International Migration (Geneva, Switzerland), 58*(3), 264–266. Advance online publication. doi:10.1111/imig.12721 PMID:32518422

ADDITIONAL READING

Brandenberger, J., Baauw, A., Kruse, A., & Ritz, N. (2020). The global COVID-19 response must include refugees and migrants. *Swiss Medical Weekly, 150*, w20263. PMID:32343357

Hargreaves, S., Kumar, B., McKee, M., Jones, L., & Veizis, A. (2020). Europe's migrant containment policies threaten the response to covid-19. *BMJ (Clinical Research Ed.), 368*, m12. PMID:32217531

Kluge, H., Jakab, Z., Bartovic, J., D'Anna, V., & Severoni, S. (2020). Refugee and migrant health in the COVID-19 response. *Lancet, 395*(10232), 1237–1239. doi:10.1016/S0140-6736(20)30791-1 PMID:32243777

World Health Organization (WHO). (2020). Coronavirus disease 2019 (COVID-19) Situation Report – 72. Retrieved from https://apps.who.int/iris/bitstream/handle/10665/331685/nCoVsitrep01Apr2020-eng.pdf

KEY TERMS AND DEFINITIONS

Access to Healthcare: Healthcare entitlements enjoyed by third-country nationals (migrants, applicants for international protection and refugees) in EU Member States and in their countries of origin.

Accommodation Centre: A place used for the collective housing of applicants for international protection.

Asylum Seeker: In the global context, a person who seeks protection from persecution or serious harm in a country other than their own and awaits a decision on the application for refugee status under relevant international and national instruments. In the EU context, a third-country national or stateless person who has made an application for protection under the Geneva Refugee Convention and Protocol in respect of which a final decision has not yet been taken.

Migrant: In the global context, a person who is outside the territory of the State of which they are nationals or citizens and who has resided in a foreign country for more than one year irrespective of the causes, voluntary or involuntary, and the means, regular or irregular, used to migrate.

Refugee: In the global context, either a person who, owing to a well-founded fear of persecution for reasons of race, religion, nationality, political opinion or membership of a particular social group, is outside the country of nationality and is unable or, owing to such fear, is unwilling to avail themselves of the protection of that country, or a stateless person, who, being outside of the country of former habitual residence for the same reasons as mentioned before, is unable or, owing to such fear, unwilling to return to it. In the EU context, either a third-country national who, owing to a well-founded fear of persecution for reasons of race, religion, nationality, political opinion or membership of a particular social group, is outside the country of nationality and is unable or, owing to such fear, is unwilling to avail themselves of the protection of that country, or a stateless person, who, being outside of the country of former habitual residence for the same reasons as mentioned above, is unable or, owing to such fear, unwilling to return to it, and to whom Art. 12 (Exclusion) of Directive 2011/95/EU (Recast Qualification Directive) does not apply.

Third-Country National: Any person who is not a citizen of the European Union within the meaning of Art. 20(1) of TFEU and who is not a person enjoying the European Union right to free movement, as defined in Art. 2(5) of the Regulation (EU) 2016/399 (Schengen Borders Code).

Chapter 15
Agro-Tourism Customer Satisfaction Analysis Based on the Theory of Attractive Quality

Maria Lachnidaki
Technical University of Crete, Greece

Evangelos Grigoroudis
https://orcid.org/0000-0001-8613-9350
Technical University of Crete, Greece

Constantin Zopounidis
Technical University of Crete, Greece & Audencia Business School, Greece

ABSTRACT

Customer satisfaction measurement is an important and critical issue of all businesses in the tourism sector, since it gives a better understanding of the needs and requirements of tourists. The Kano model is one of the most widely used approaches in analyzing customer satisfaction, aiming to categorize the quality attributes of products/services in different quality dimensions. The main aim of this chapter is to analyze customer satisfaction from an agro-tourism farm and prioritize potential improvement actions. The preliminary analysis has identified a large set of customer satisfaction criteria that refer to the location, the personnel, the activities, the cost, and the "green" aspects of the agro-tourism farm. The MUSA method has been applied in the context of Kano's theory of attractive quality. The main results show which satisfaction criteria mostly affect satisfaction and/or dissatisfaction, identifying the strong and the weak points of the farm. Finally, combining the previous results, it is possible to prioritize improvement actions.

DOI: 10.4018/978-1-7998-5442-5.ch015

INTRODUCTION

Tourism in rural areas is directly linked to food, health, and accommodation. In particular, agro-tourism is defined as the act of visiting a working farm or any agricultural, horticultural, or agribusiness operation for the purpose of enjoyment, education or active involvement in the activities of the farm operation (Lobo, 2012).

Customers of all kinds of tourism, including agro-tourism, are the most important factor for tourist businesses. According to Kozark and Rimmington (2000), customer satisfaction is very important for tourism marketing, as it can influence the selection of the destination, service/product consumption and the decision to return. Gursoy et al. (2007) also argue that customer's satisfaction is one the areas most researched in many tourism studies, due to its importance in determining the success and the continued existence of the tourism business.

Customer satisfaction mainly concerns the final outcome of actions taken in the process of purchasing or consuming a service/product. However, in the case of tourism, customer satisfaction is the visitors' state of emotion after their experience with a tour (Baker and Crompton, 2000; Sanchez et al., 2006). Regarding the quality of services in the tourism industry, the main aim is not only to satisfy tourists, but also to improve the image of the destination, differentiate the destination from others, and try to persuade tourists to return to the destination and/or to make a positive oral advertising. Finally, according to the survey of Tian-Cole and Crompton (2003), both the overall quality of services and overall satisfaction of tourists directly affect the destination selection and are confirmed as different constructs.

Customer satisfaction measurement and analysis are important in the tourism industry, given the contribution of the tourism sector in local economies, as well as the intense competition among alternative tourism destinations. As noted by Tsitsiloni et al. (2012), the tourism sector is heavily influenced by significant external factors from the global economic environment, and it is thus necessary to improve the quality of the services offered in order to gain competitive advantages and increase tourist loyalty.

In this context, the importance of tourist satisfaction from a destination is emphasized in several studies, although the concept of "destination satisfaction" is rather general and ambiguous, given the complexity of its elements and characteristics. This is the main reason why several scholars focus on particular components of experiences that affect tourist satisfaction within different tourism and hospitality contexts (e.g., guest satisfaction with hotels and restaurant services, satisfaction with destination services, satisfaction with recreational services, satisfaction with tours or cruise travel) (Yuksel, 2001). In most cases, a confirmation/disconfirmation approach is applied in order to analyze tourist satisfaction (Bowen and Clarke, 2002). In the expectancy disconfirmation theory, satisfaction may be defined as a pleasant past-purchasing experience from a product or service, given the ante-purchasing expectancy of the customer (Oliver, 1977, 1980, 1997). In this context, tourist satisfaction is the result of interaction between tourist's experience at the destination areas and the expectations she/he had about that destination (Pizam et al., 1978). The HOLSAT model is a widely adopted approach for destination satisfaction measurement that relies on the aforementioned disconfirmatory paradigm (Tribe and Snaith, 1998). It is based on the SERVQUAL model (Parasuraman et al., 1988, 1991) in order to estimate the difference between "expectation" and "experience" scores for each attribute, which gives a quantitative measure of the level of satisfaction shown by the vacationers (Truong and Foster, 2006).

On the other hand, agro-tourism, as noted above, has distinguished characteristics, although it is not clear if emphasis is given in the "agro" or in the "tourism" dimension. In any case, it is important to note that in order for a rural area to be considered suitable for tourism development, agro-tourism should be (OECD, 1994):

- Located in rural areas;
- Based on elements of the "rural world" (i.e., small outdoor businesses, contact with nature, culture);
- Sustainable, in the sense that its development should contribute to its preservation.

Furthermore, agro-tourism should have a traditional character, representing the rural environment, economy, and history.

The concept of agro-tourism is also recognized as an alternative business that could use agricultural resources in several ways. According to Maetzold (2004), agro-tourism is an alternative business associated with added value or with non-traditional agricultural products and marketing. Thus, farmers, as entrepreneurs, may acquire a larger share of income with differentiated activities.

The main aim of this chapter is to analyze and measure agro-tourists' expectations and satisfaction. For this purpose, the Kano model of attractive quality is adopted, in order to categorize the quality attributes of an agro-tourism farm in Crete. According to this model, quality attributes have not the same impact on customer satisfaction/dissatisfaction, since some of them may only cause satisfaction, some may cause dissatisfaction, and other may cause both satisfaction and dissatisfaction. In order to categorize the customer satisfaction criteria, the MUSA method has been applied in the context of the Kano model. The MUSA method is a multicriteria approach for measuring and analyzing customer satisfaction that follows the principles of ordinal regression analysis, and, taking into account the qualitative form of customer data, may provide quantitative satisfaction measures. The main results of the study focus on the identification of attributes, which mostly affect satisfaction and/or dissatisfaction and the prioritization of potential improvement actions.

The chapter is organized in three more sections. Section 2 gives a brief overview of the applied methodological framework, including the presentation of the Kano's theory of attractive quality, the presentation of the MUSA method, and the derived importance assessment through the MUSA method and the better-worse diagram. The case study of an agro-tourism farm is presented in section 3, while section 4 summarizes some concluding remarks and gives potential future research directions.

BACKGROUND

MUSA Method

The MUSA method infers an additive collective value function Y^* and a set of partial satisfaction functions X_i^*, given customer's global satisfaction Y and partial satisfaction X_i according to the i-th criterion. The main objective of the method is to achieve the maximum consistency between the value function Y^* and the customers' judgements Y. Based on the modeling of preference disaggregation approach, the ordinal regression equation becomes as follows (Grigoroudis and Siskos, 2002):

$$\tilde{Y}^* = \sum_{i=1}^{n} b_i X_i^* - \sigma^+ + \sigma^- \text{ with } \sum_{i=1}^{n} b_i = 1$$

where $\tilde{Y}^*$ is the estimation of Y^*, b_i is the weight of the i-th criterion, n is the number of criteria, and σ^+ and σ^- are the overestimation and the underestimation errors, respectively (see Figure 1).

The customer satisfaction evaluation problem may be formulated as a linear program (LP) that minimizes the sum of errors under the constraints: ordinal regression equation for each customer, normalization constraints for Y^* and X_i^* in the interval [0, 100], and monotonicity constraints for Y^* and X_i^*. In order to reduce the size of the mathematical program, removing the monotonicity constraints for Y^* and X_i^*, the following transformation equations are used:

$$\begin{cases} z_m = y^{*m+1} - y^{*m} \text{ for } m = 1,2,\ldots,\alpha - 1 \\ w_{ik} = x_i^{*k+1} - x_i^{*k} \text{ for } k = 1,2,\ldots,\alpha_i - 1 \text{ and } i = 1,2,\ldots,n \end{cases}$$

where y^{*m} is the value of the y^m satisfaction level, x_i^{*k} is the value of the x_i^k satisfaction level, and α and α_i are the number of the global and partial satisfaction levels, respectively.

Figure 1. Error variables for the j-th customer (Grigoroudis and Siskos, 2010)

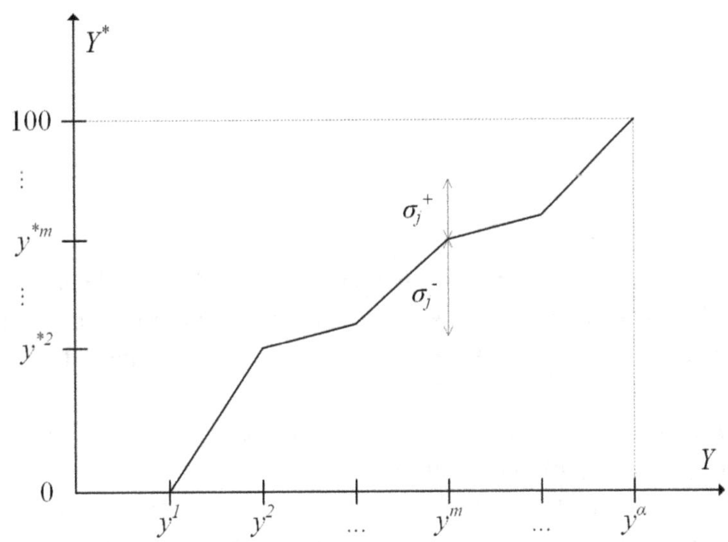

According to the aforementioned definitions and assumptions, the basic estimation model can be written in a LP formulation, as follows

$$\begin{cases} [\min]F = \sum_{j=1}^{M} \sigma_j^+ + \sigma_j^- \\ \text{subject to} \\ \sum_{i=1}^{n}\sum_{k=1}^{x_i^j-1} w_{ik} - \sum_{m-1}^{y^j-1} z_m - \sigma_j^+ + \sigma_j^- = 0 \ \text{ for } j = 1,2,\ldots,M \\ \sum_{m=1}^{\alpha-1} z_m = 100 \\ \sum_{i=1}^{n}\sum_{k=1}^{\alpha_i-1} w_{ik} = 100 \\ z_m, w_{ik}, \sigma_j^+, \sigma_j^- \geq 0 \ \ \forall i,j,k,m \end{cases}$$

where M is the size of the customer sample and y^j, x_i^j are the j-th level on which variables Y and X_i, respectively, are estimated (i.e., global and partial satisfaction judgments of the j-th customer).

The MUSA method includes also a post optimality analysis stage in order to overcome the problem of model stability. The final solution is obtained by exploring the polyhedron of multiple or near optimal solutions, which is generated by the constraints of the previous LP. This solution is calculated by n LPs (equal to the number of criteria) of the following form:

$$\begin{cases} [\max]F' = \sum_{k=1}^{\alpha_i-1} w_{ik} \\ \text{subject to} \\ F \leq F^* + \varepsilon \\ \text{all the constraints of LP(3)} \end{cases}$$

where F^* is the optimal value of the objective function of LP and ε is a small percentage of F^*. The average of the optimal solutions given by the n LPs may be considered as the final solution of the problem.

The MUSA method provides several quantitative results, including the following:

- **Value functions**: they show the real value (in a normalized interval 0-100) that customers give for each level of the global or partial ordinal satisfaction scale.
- **Criteria weights**: they represent the relative importance of the assessed satisfaction criteria.
- **Average satisfaction indices**: they show, in a range of 0 to 100%, the level of customers' satisfaction.
- **Average demanding indices**: they show, in a range of -100% to 100%, the demanding level of customers.
- **Average improvement indices**: they show, in the interval [0, 1], the improvement efforts, and they depend on the importance of satisfaction criteria and their contribution to dissatisfaction as well.

Combining criteria weights and average satisfaction indices, a series of action diagrams can be developed (Figure 2). These diagrams indicate the strong and the weak points of customer satisfaction, and define the required improvement efforts. The action diagrams are also mentioned as decision, strategic, perceptual, and performance-importance maps (Dutka, 1995; Motorola, 1995; Naumann and Giel, 1995), or gap analysis (Hill, 1996; Woodruff and Gardial, 1996; Vavra, 1997), and they are similar to SWOT (Strengths-Weaknesses-Opportunities-Threats) analysis. Each of these diagrams is divided into quadrants, according to performance (high/low) and importance (high/low) that may be used to classify actions:

- **Status quo** (low performance and low importance): Generally, no action is required, given that these satisfaction criteria are not considered as important by the customers.
- **Leverage opportunity** (high performance/high importance): This area can be used as advantage against competition. In several cases, these satisfaction criteria are the most important reasons why customers have purchased the product/service under study.
- **Transfer resources** (high performance/low importance): Regarding the particular satisfaction criteria, company's resources may be better used elsewhere (i.e. improve satisfaction criteria located in the action opportunity quadrant).
- **Action opportunity** (low performance/high importance): These are the criteria that need attention; improvement efforts should be focused on these, in order to increase the global customer satisfaction level.

Figure 2. Action diagram (Grigoroudis and Siskos, 2010)

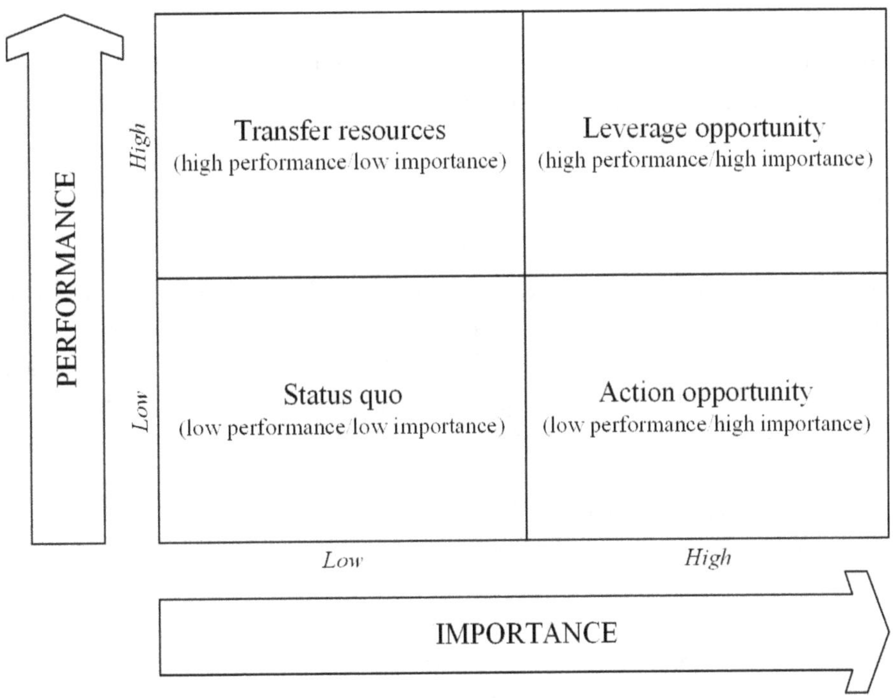

This grid can be also used in order to identify priorities for improvement. The bottom right quadrant is obviously the first priority, for the attributes are important to customers but company's performance is rated moderately low. The second priority may be given to the satisfaction criteria in the top right quadrant, especially if there is a significant improvement margin. The third priority issues are indicated in the bottom left quadrant; although these issues are not terribly pertinent at the time of the analysis, they may become more important in the future, and company's performance is certainly not good. Finally, last priority for improvement should be given to the criteria in the top left quadrant, because this category is the least important and company's performance is relatively good. Apparently, priorities for improvement may vary among different companies, depending on the potential capabilities of improving the particular category.

The action diagrams can indicate which satisfaction dimensions should be improved, but they cannot determine the output or the extent of improvement efforts. For this reason, combining the average improvement and demanding indices, a series of improvement diagrams can be developed. As shown in Figure 3, each of these diagrams is divided into quadrants according to demanding (high/low), and effectiveness (high/low), that may be used to rank improvement priorities:

- **First priority**: this area indicates direct improvement actions, since these dimensions are highly effective and customers are not demanding.
- **Second priority**: it includes satisfaction dimensions that have either a low demanding index or a high improvement index.
- **Third priority**: it refers to satisfaction dimensions that have small improvement margin and need substantial effort.

The principles and the main methodological framework of the MUSA method have been developed by Siskos et al. (1998) and Grigoroudis and Siskos (2002), while a discussion and a more detailed presentation of the method may also be found in Grigoroudis and Siskos (2010).

Kano Model

The Kano model allows the measurement and evaluation of the level of customer satisfaction by providing a practical approach for categorizing the attributes of the products/services, based on how well they can satisfy customer's needs (Kano et al., 1984). In fact, this categorization is based on the fact that each customer's opinion is different, so different quality levels are formed to determine how important an attribute is for each customer. This enables firms to focus on the specific attributes, which are expected to lead to high satisfaction (Gustafsson et al., 1999).

According to Dominici and Palumbo (2013), the main differences between the Kano model and other satisfaction models (e.g., SERVQUAL) are the following:

- The Kano model can be applied both before consumer experience (to design the product/service ideal for customer) and beyond (to measure satisfaction). According to Chauda et al. (2011), the Kano model is a useful tool for identifying customers' needs, as these requirements determine the design and engineering specifications and production details of a product/service.
- The Kano model does not presume that there is a linear relationship between product/service performance and customer satisfaction. For example, according to the model, customer satisfaction is

Figure 3. Improvement diagram (Grigoroudis and Siskos, 2010)

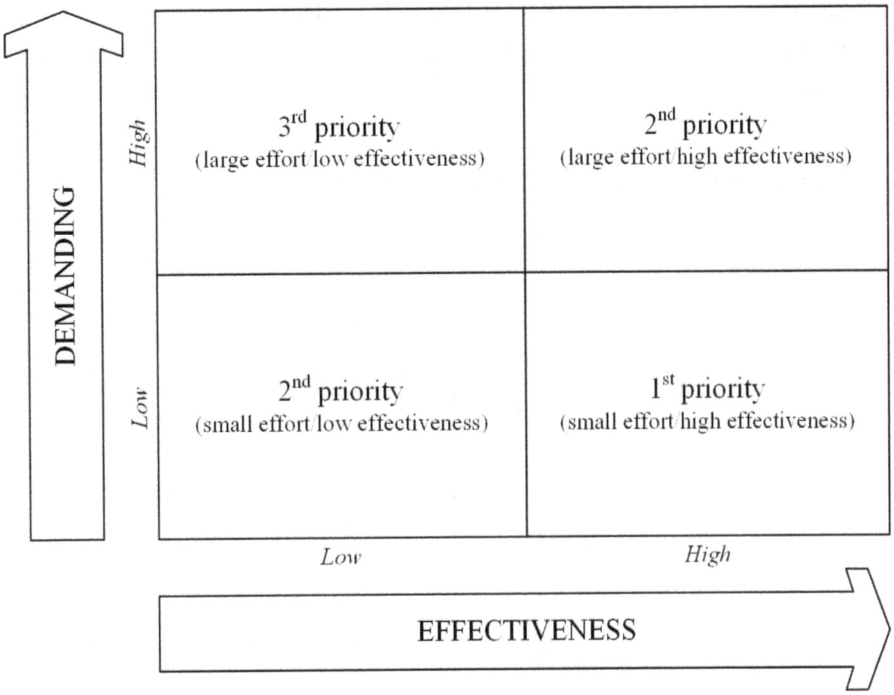

not always proportionate to the performance of a product/service, implying that quality does not necessarily lead to higher satisfaction.

- The Kano model is a rather simple and practical approach that may categorize quality attributes and define different customer satisfaction strategies.

The theory of attractive quality is shown in Figure 4, and it is inspired by Herzberg's motivation theory (dual factor theory) (Hertzberg et al., 1959). According to this theory, the Kano model is based on two models: one-dimensional and two-dimensional model. It also includes three basic dimensions of the perceived quality: the attractive, desired, and expected quality. The theory of attractive quality was developed to help separate the factors that lead to satisfaction from those that lead to dissatisfaction, providing a better understanding of how customers assess a product/service and helping companies to focus on the most important attributes to improve (Gustafsson et. al., 1999). The basic dimensions that are mentioned include the following:

- **Expected quality**: The expected quality includes must-be requirements, which are considered standard by customers and do not need to be required. Essentially, they are the characteristics of a product/service which "must" be of good quality without any excuse (DeLayne, 2016). The lack of these attributes leads to customer's dissatisfaction, but their existence does not lead to greater satisfaction.
- **Desired quality**: This dimension includes one-dimensional requirements, which are related to the customer's basic needs and desires. Increasing the positive performance of one-dimensional requirements leads to a higher level of satisfaction and vice versa.

- **Attractive quality**: Attractive quality includes attractive requirements, which usually have the greatest influence on customer satisfaction. Customers do not expect the availability of these attributes, thus their existence creates high satisfaction, while, at the same time, their absence does not cause dissatisfaction.

Figure 4. The theory of attractive quality (Kano et. al., 1984)

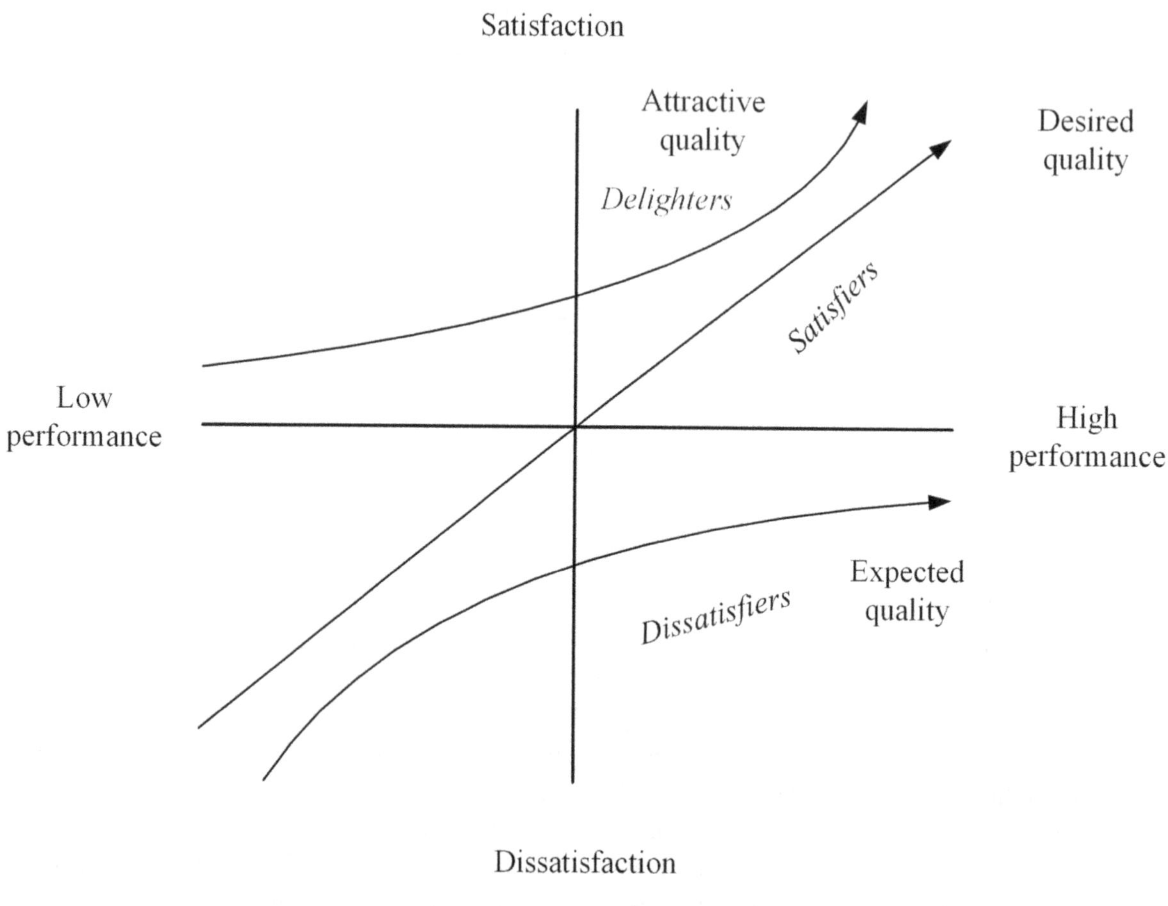

In simple words, customers do not expect or express the attractive requirements, but they express the one-dimensional requirements and consider as standards the Must-be requirements. Therefore, it is understandable that each category of requirements has different consequences on customer satisfaction and dissatisfaction. For this reason, the Kano model is one of the most appropriate methods for measuring and analyzing customer satisfaction, as it can identify the attributes of a product/service that can achieve the highest level of customer satisfaction.

Categorizing Satisfaction Criteria

The Kano model is quite flexible and can be incorporated in several customer satisfaction methods and tools. Krassadaki and Grigoroudis (2018) propose a methodological framework that categorizes quality attributes combining the MUSA method and the Kano model. The basic idea of this methodological framework is to carry out a comparative examination of the relationship between the derived importance of satisfied and dissatisfied customers.

The main steps of this approach are shown in Figure 5 and include the following:

1. The first step concerns the collection of customer satisfaction data. Customers are asked to express their satisfaction level on a set of product/service quality attributes (usually these questions have the following form "How satisfied are you with the criterion …?").
2. In the next step, satisfied customers are separated from dissatisfied customers. Having n quality attributes, a total of $2 \bullet n$ different datasets are developed.
3. In the next step, the MUSA method is applied in the previous $2 \bullet n$ datasets. As a result, for each quality attribute i, two different weights are estimated: b_i^S, which is the weight of satisfied customers and b_i^D, which is the weight of dissatisfied customers.
4. The fourth step concerns the normalization of the previous weights in order to avoid potential comparability problems. These normalized relative weights compare the importance of each satisfaction criterion to the importance of the other criteria, and they are calculated using the following formula:

$$b_i'^S = \frac{b_i^S - \bar{b}_i^S}{\sum_i (b_i^S - \bar{b}_i^S)^2} \text{ and } b_i'^D = \frac{b_i^D - \bar{b}_i^D}{\sum_i (b_i^D - \bar{b}_i^D)^2}$$

where $b_i'^S$ and $b_i'^D$ are the relative weights of b_i^S and b_i^D, respectively, and $\bar{b}_i^S$ and $\bar{b}_i^D$ are the averages of b_i^S and b_i^D, respectively.

5. The last step concerns the development of a better-worse diagram, which is a representation of the relative weights of the two customer groups (see Figure 6). The main purpose of this diagram is to categorize quality attributes according to the Kano model by benchmarking the relationship between the derived importance of satisfied and dissatisfied customers.

As shown in this diagram, the first and third quadrants include attributes which are either of high or of low importance for both customer groups, respectively. Based on the Kano model, the dimension of the desired quality is related to the fact that the attributes of a low-performance product/service cause dissatisfaction, while those with high performance generate satisfaction. The first and third quadrants therefore include one-dimensional requirements.

The second quadrant contains features that are of high importance to dissatisfied customers and of low importance to satisfied customers. In fact, this means that this discontent has arisen because of the low performance of these attributes. Thus, the second quadrant contains must-be requirements.

The fourth quadrant contains features that are of low importance to dissatisfied customers and of high importance to satisfied customers. This means that the level of customer satisfaction is not affected by the low performance of the product/service attribute, and improving the quality of the product would create an unexpected increase in satisfaction. That means that the fourth quadrant includes the attractive requirements.

Figure 5. Combining the Kano model and the MUSA method (Krassadaki and Grigoroudis, 2018)

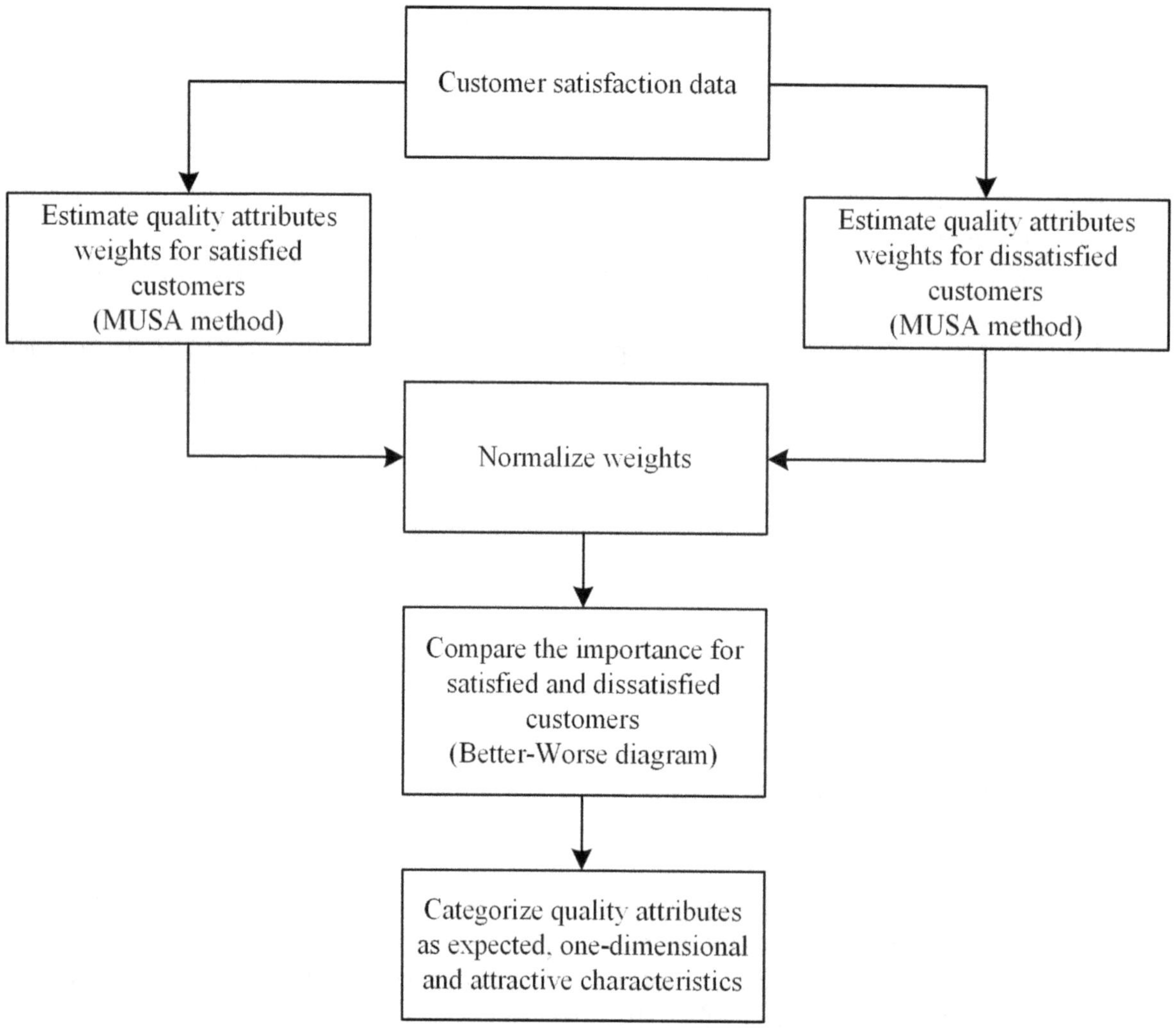

APPLICATION IN AN AGRO-TOURISM FARM GROUND

Satisfaction Criteria and Survey Details

The case study presented in this chapter aims to analyze customer satisfaction of an agro-tourism farm. The farm is located in the southeastern part of Crete, Greece. The facilities are in the middle of olive fields surrounded by mountains with fragrant pine trees. The farm also produces a variety of local products, including olive oil, goat cheese, eggs, organic wine, and thyme honey.

In a general context, satisfaction criteria are defined as the product/service characteristics that directly affect either customer satisfaction or dissatisfaction. The assessment of satisfaction criteria is a first important step in a customer survey. In the presented application, based on the results of an extended literature review and a preliminary analysis with agro-tourism farm managers, the following criteria were defined:

Figure 6. Better-worse diagram (Krassadaki and Grigoroudis, 2018)

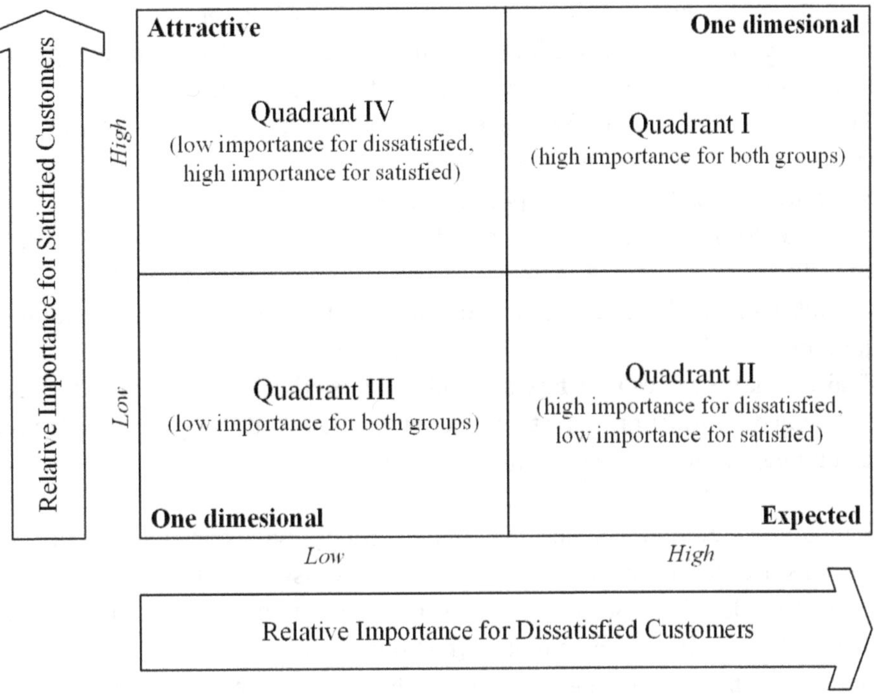

1. **Location attractiveness**: Previous studies have shown that site attractiveness plays a key role in agro-tourism satisfaction. This is justified by the nature-based character of agro-tourism.
2. **Location access**: Easy access to agro-tourism location is an important aspect of an agro-tourism facility (Srikatanyoo and Campiranon, 2010).
3. **"Green" accommodation**: Agro-tourists expect a "green" or ecological environment, as they believe that this represents the main concept of agro-tourism and differentiates it from other forms of tourism.
4. **Friendliness and courtesy of personnel**: The linkages between employee-friendly behavior and customer satisfaction are strong in agro-tourism services (Lemmink and Mattsson, 1998).
5. **Communication with personnel**: Based on the relevant literature, communication with personnel, particularly regarding agro-tourism activities, is an expected feature that affects agro-tourist satisfaction.
6. **Personnel adequacy**: Similarly, studies show that the sufficient number of personnel and their availability can affect the agro-tourist satisfaction.
7. **Quality of food** (from local and agricultural products): This is an important satisfaction criterion, since agro-tourists expect their food to come from local agricultural products.
8. **Variety of farm activities**: Almost all agro-tourists expect to participate in agricultural activities (Wicks and Merrett 2003, Carpio 2006).
9. **Variety of educational/recreational activities**: Agro-tourism facilities also provide non-agricultural educational and recreational activities (Jolly and Reynolds, 2005).
10. **Shopping opportunities**: Agro-tourists often wish to buy fruits, vegetables, herbs and other agricultural products from the farm (Maetzold, 2002, Langworthy et al., 2006).

11. **Quiet and peaceful environment**: Based on secondary data, agro-tourists expect to find a quiet and peaceful environment, thus having the opportunity to gain the experience of rural life.
12. **Contact with nature**: This is one of the most important factors that can affect agro-tourist satisfaction, since it is related to the core characteristics of agro-tourism.
13. **Contact with local community/farmers**: Agro-tourists consider agro-tourism as an opportunity to contact with the local community and local culture.
14. **Accommodation cost**: Cost is an important factor that affects customer satisfaction. It is related with the "value for money", which is considered as "a standard feature of satisfaction" (Yu and Goulden, (2006).
15. **Cost of food/drinks**: This refers to the cost of food and drinks offered within the agro-tourism farm facilities.
16. **Cost of additional services/purchases**: Similarly, this criterion is related to the cost of additional services (e.g., educational or recreational activities) or additional purchases from the store with local agricultural products located in the farm.

The questionnaire of the survey was developed based on the previous satisfaction criteria. Agro-tourists expressed their satisfaction using a 5point ordinal scale (Very dissatisfied, Dissatisfied, Neither satisfied/Nor Dissatisfied, Satisfied, Very satisfied). The survey was conducted within the period September-December 2019 and the final sample consisted of 94 questionnaires.

Figure 7 presents some profile characteristics of the final customer sample. As shown, the sample has been quite balanced with regard to sex and age.

Figure 7. Customer sample profile (sex and age)

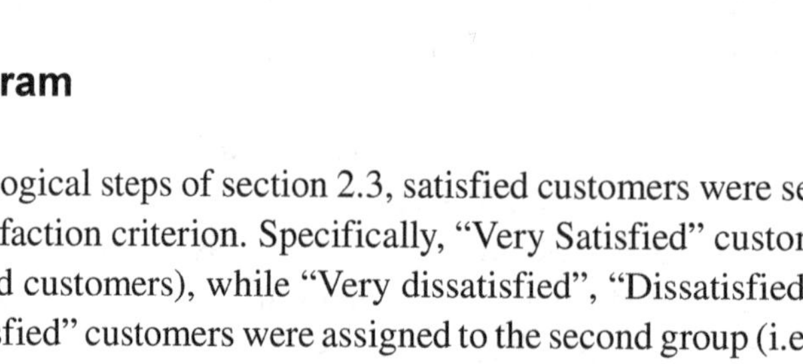

Better-Worse Diagram

Following the methodological steps of section 2.3, satisfied customers were separated from dissatisfied customers for each satisfaction criterion. Specifically, "Very Satisfied" customers were assigned to the first group (i.e., satisfied customers), while "Very dissatisfied", "Dissatisfied", "Neither Satisfied/Nor Dissatisfied", and "Satisfied" customers were assigned to the second group (i.e., dissatisfied customers). This grouping was justified by the highly skewed distribution of customer satisfaction. Having a total of 16 satisfaction criteria, 32 different datasets have been developed.

The MUSA method was then applied to the previous datasets, leading to the estimated criteria weights, which are analytically shown in Table 1. The second and third column of the table shows the weights of dissatisfied and satisfied customers, b_i^D and b_i^S, respectively, while the last two columns correspond to the relative weights $b_i'^D$ and $b_i'^S$, respectively.

Table 1. Estimated criteria weights for satisfied and dissatisfied customers

Criteria	b_i^D	b_i^S	$b_i'^D$	$b_i'^S$
Location attractiveness	0.063	0.068	-0.019	0.670
Location access	0.060	0.054	-0.073	-0.216
"Green" accommodation	0.063	0.055	-0.018	-0.204
Friendliness and courtesy of personnel	0.058	0.054	-0.142	-0.218
Communication with personnel	0.059	0.053	-0.095	-0.324
Personnel adequacy	0.063	0.056	-0.001	-0.143
Quality of food	0.098	0.058	0.864	-0.002
Variety of farm activities	0.077	0.057	0.332	-0.034
Variety of educational/recreational activities	0.061	0.059	-0.057	0.103
Shopping opportunities	0.061	0.056	-0.067	-0.127
Quiet and peaceful environment	0.060	0.055	-0.073	-0.181
Contact with nature	0.061	0.055	-0.063	-0.181
Contact with local community/farmers	0.055	0.061	-0.196	0.209
Accommodation cost	0.059	0.060	-0.112	0.172
Cost of food/drinks	0.058	0.063	-0.134	0.309
Cost of additional services/purchases	0.057	0.060	-0.149	0.166

Based on the estimated weights, the better-worse diagram for the examined agro-tourism farm was further developed, as shown in Figure 8. This diagram may classifies the satisfaction criteria according to the Kano's quality dimensions.

It is important to note that according to the diagram, some satisfaction criteria appear to be located quite close to the axes. Such criteria include: location attractiveness, personnel adequacy, "green" accommodation, and quality of food. In practical terms, this means that that their categorization is less "certain" compared to the other satisfaction criteria. The aforementioned criteria may be categorized in two alternative Kano's quality dimensions. These results may be justified by the small sample or the existence of non-homogenous customer references and expectations.

Table 2 summarizes the assignment of satisfaction criteria to the three Kano's dimensions: attractive quality, expected quality, and desired (one-dimensional quality). Based on the previous comments, a parentheses is used when the assignment is not certain and a criterion may be assigned to two alternative quality dimensions.

Specifically, the results of Table 2 show that the desired or one-dimensional requirements include the following criteria: location access, friendliness and courtesy of personnel, communication with personnel, shopping opportunities, quiet and peaceful environment, and contact with nature. Also, the attributes of "green" accommodation, personnel adequacy, and quality of food may also belong to the desired quality category. The low performance of these satisfaction criteria causes dissatisfaction and their high performance causes satisfaction. Thus, customer satisfaction is proportional to the level of fulfillment. Desired requirements are usually explicitly demanded by the customer.

Figure 8. Better-worse diagram for the agro-tourism farm

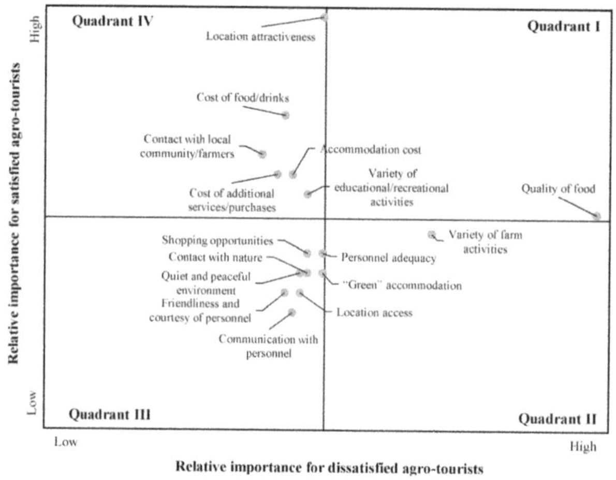

Table 2. Classification of satisfaction criteria

Criteria	Attractive quality	Expected quality	Desired quality
Location attractiveness	(X)	(X)	
Location access			X
"Green" accommodation		(X)	(X)
Friendliness and courtesy of personnel			X
Communication with personnel			X
Personnel adequacy		(X)	(X)
Quality of food		(X)	(X)
Variety of farm activities		X	
Variety of educational/recreational activities	X		
Shopping opportunities			X
Quiet and peaceful environment			X
Contact with nature			X
Contact with local community/farmers	X		
Accommodation cost	X		
Cost of food/drinks	X		
Cost of additional services/purchases	X		

Accordingly, the expected or must-be requirements include the variety of farm activities. The low performance of this criterion causes high dissatisfaction but their high performance does not entail higher satisfaction. Thus, agro-tourists consider that such farm activities should exist during their stay on the farm. The criteria of location attractiveness, "green" accommodation, personnel adequacy, and quality of food may be also categorized in the set of expected quality attributes. These attributes may be considered as a basic need or desire for the visitors of the agro-tourism farm. They can also be considered as "unspoken motivators", since these requirements are often so elementary that customers would not state them unless the company fails to perform them. Agro-tourists consider the previous expected requirements as prerequisites, he/she take them for granted and therefore does not explicitly demand them.

Finally, attractive requirements include the following attributes: variety of educational/recreational activities, contact with local community/farmers, accommodation cost, cost of food/drinks, and cost of additional services/purchases. When the performance of these criteria is low/moderate, it does not cause dissatisfaction but their unexpected improvement may result in high satisfaction. It is important to note that all cost-related criteria are considered as attractive requirements. This implies that agro-tourists are willing to accept higher costs, without this causing dissatisfaction. In addition, agro-tourists do not expect a large variety of educational/recreational activities and opportunities to contact with the local community. However, if this happens, it is likely to generate high satisfaction. The criterion of location attractiveness may also be categorized as attractive requirement. It is important to note that attractive requirements have the greatest influence on how satisfied a customer will be with a given product/service. They are neither explicitly expressed nor expected by the customer.

Action and Improvement Diagrams

In addition to the results obtained from the MUSA-Kano methodological framework, the MUSA method has been separately applied in the customer satisfaction dataset. Thus, additional results regarding the analysis of customer satisfaction in the agro-tourism farm may be considered, in order to identify potential improvement actions. The MUSA results show that the overall customer satisfaction is very high, given that the average global satisfaction index is 96.51%. At the same time, customer appear to have a neutral demanding level (average global demanding index -27.44).

The action diagram that combines satisfaction criteria weights and average satisfaction indices may identify the strengths and weaknesses of the agro-tourism farm. As shown in Figure 9, there are several satisfaction criteria located in the action opportunity quadrant: variety of farm activities, variety of educational/recreational activities, shopping opportunities and marginally personnel adequacy. These criteria appear to be have relatively low satisfaction indices, while they are considered important by agro-tourists. The specific criteria need to be improved in order to improve customer satisfaction level. For this reason, the farm should focus on the criteria that belong to this quadrant, and improving them should be a top priority.

On the other hand, location attractiveness and quality of food appear as the most important strengths of the farm. These criteria may be considered as the major competitive advantages of the agro-tourism farm, and their relatively high performance should be retained.

Similarly, the transfer resources quadrant contains the following satisfaction criteria: contact with nature, friendliness and courtesy of personnel, communication of personnel, quiet and peaceful environment, and marginally "green" environment. These criteria appear to have relatively higher satisfaction performance, although they are not considered important by agro-tourists. Usually in such cases, company's resources may be better used elsewhere (i.e., improvement of satisfaction criteria located in the action opportunity quadrant).

Finally, the cost-related criteria (accommodation cost, cost of additional services/purchases, and cost of foods/drinks) and the contacts with local community/farmers are located in the status quo quadrant. Although no action is generally required for these satisfaction criteria, given that they are not considered very important by agro-tourists, their relatively low satisfaction performance shows that these are potential threats. Therefore, the farm should monitor its performance on these criteria.

The improvement diagram shown in Figure 10 combines the average improvement and demanding indices, as given by the MUSA method. This diagram takes into account the demanding level of agro-tourists for each satisfaction criterion and may prioritize improvement actions.

Figure 9. Action diagram for the agro-tourism farm

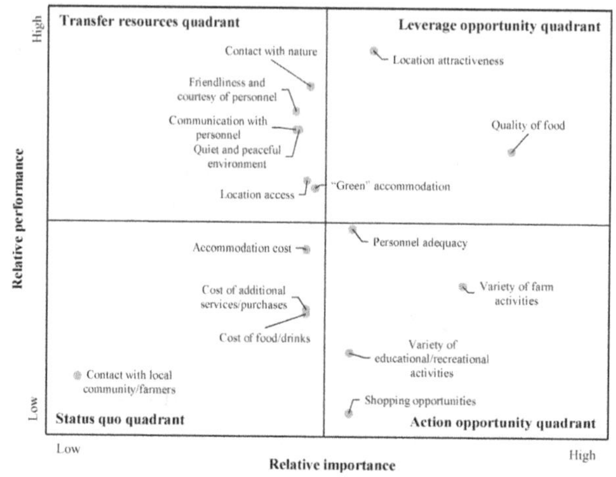

Figure 10. Improvement diagram for the agro-tourism farm

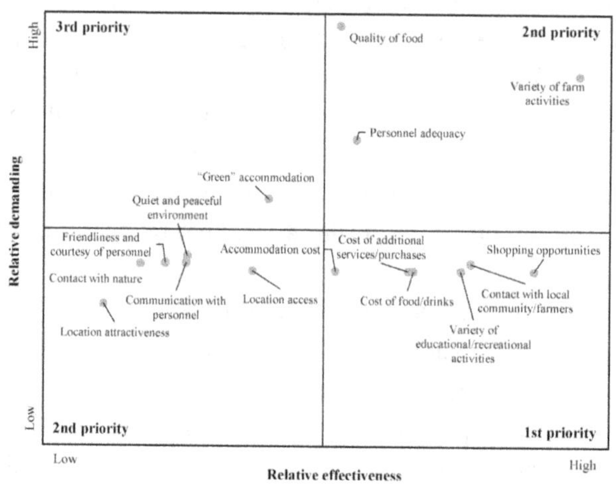

As shown in Figure 10, the following criteria may be the top priority for improvement by the owners of the farm: accommodation cost, cost for foods/drinks, cost for additional services/purchases, variety of educational/recreational activities, contact with local community/farmers, and shopping opportunities. Since agro-tourists appear non-demanding, the farm should firstly focus on these particular attributes mainly because a small improvement effort may be quite effective overall. On the other hand, "green" accommodation appears as the last improvement priority due to the high customer demanding level and its low effectiveness. All the other satisfaction criteria are located in the second priority quadrants since they have either a low average demanding index or a high improvement index.

CONCLUSION AND FUTURE RESEARCH DIRECTIONS

The aim of this chapter has been to analyze the satisfaction of agro-tourists by categorizing the attributes of the services offered by an agro-tourism farm in Crete (Greece), and examining how well these attributes can satisfy customers' needs. The presented analysis has been based on the MUSA method, while the categorization of the examined satisfaction criteria has been based on the Kano model (theory of attractive quality).

According to the classification of criteria at the three quality levels of the Kano model, and in relation to perceived needs/experiences as expressed by the farm's customers, the following should be noted:

- **Attractive quality**: the attributes of low/moderate performance that do not cause dissatisfaction while their unexpected improvement will create high satisfaction include the following: variety of educational/recreational activities, contact with local community/farmers, accommodation cost, cost for foods/drinks, cost for additional services/purchases, and potentially location attractiveness.
- **Expected quality**: the attributes whose low performance generated high dissatisfaction while high performance does not imply high satisfaction include the following: variety of farm activities and potentially location attractiveness, "green" environment, personnel adequacy, and quality of food (i.e., food from local agricultural products).
- **Desired quality**: the attributes whose low performance creates dissatisfaction while high performance causes satisfaction include the following: location access, friendliness and courtesy of personnel, communication with personnel, shopping opportunities, quiet and peaceful environment, contact with the natural environment, and potentially "green" accommodation, personnel adequacy, and quality of food.

Furthermore, combining the results of the better-worse, the action, and the improvement diagrams, it is possible to develop a prioritization focus on which characteristics need to be improved:

The first improvement priority should concern the variety of farm activities, since according to the better-worse diagram is a must-be requirement, while it is located in the action opportunity quadrant in the action diagram, and it has been identified as a second priority in the improvement diagram.

In a similar way, the second improvement priority refers to shopping opportunities. This attribute is a one-dimensional requirement according to the better-worse diagram, it is located in the in the action opportunity quadrant in the action diagram, and it is located in the first priority quadrant in the improvement diagram.

Finally, the personnel adequacy should be the third improvement priority, given that it is a one-dimensional requirement based on the better-worse diagram, it is located in the in the action opportunity quadrant according to the action diagram, and it has been identified as a second priority in the improvement diagram.

However, given that the categorization of some attributes has been found to be less certain, it is important to extend the current study in larger customer samples. Other future research directions include the segmentation of customer sample and the analysis of different customer groups, based on their overall profile, which may have different expectations and preferences. Finally, the presented methodological framework may be applied in additional agro-tourism accommodations/farms, in order to acquire a more integrated view of agro-tourists' needs and expectations.

REFERENCES

Baker, D. A., & Crompton, J. L. (2000). Quality, satisfaction, and behavioral intentions. *Annals of Tourism Research, 27*(3), 785–804. doi:10.1016/S0160-7383(99)00108-5

Bowen, D., & Clarke, J. (2002). Reflections on tourist satisfaction research: Past, present and future. *Journal of Vacation Marketing, 8*(4), 297–308. doi:10.1177/135676670200800401

Carpio, C. (2006). *Two-constraints models of consumer demand: An application to the demand for agritourism in the United States* (PhD Thesis). North Carolina State University, Raleigh.

Chaudha, A., Jain, R., Singh, A. R., & Mishra, P. K. (2011). Integration of Kano's model into quality function deployment. *International Journal of Advanced Manufacturing Technology, 53*(5-8), 689–698. doi:10.100700170-010-2867-0

DeLayne, S. J. (2016). *The Kano analysis: Customer needs are ever changing, iSixSigma*. Retrieved from: https://www.isixsigma.com/tools-templates/kano-analysis/kano-analysis-customer-needs-are-ever-changing

Dominici, G., & Palumbo, F. (2013). The drivers of customer satisfaction in the hospitality industry: Applying the Kano model to Sicilian hotels. *International Journal of Leisure and Tourism Marketing, 3*(3), 215–236. doi:10.1504/IJLTM.2013.052623

Dutka, A. (1995). *AMA handbook of customer satisfaction: A guide to research, planning, and implementation*. NTC Publishing Group.

Grigoroudis, E., & Siskos, Y. (2002). Preference disaggregation for measuring and analyzing customer satisfaction: The MUSA method. *European Journal of Operational Research, 143*(1), 148–170. doi:10.1016/S0377-2217(01)00332-0

Grigoroudis, E., & Siskos, Y. (2010). *Customer satisfaction evaluation: Methods for measuring and implementing service quality*. Springer. doi:10.1007/978-1-4419-1640-2

Gursoy, D., McCleary, K. W., & Lepsito, L. R. (2007). Propensity to complain: Effects of personality and behavioral factors. *Journal of Hospitality & Tourism Research (Washington, D.C.), 31*(3), 358–386. doi:10.1177/1096348007299923

Gustafsson, A., Ekdahl, F., & Edvardsson, B. (1999). Customer focused service development in practice: A case study at Scandinavian Airlines System (SAS). *International Journal of Service Management, 10*(4), 344–358.

Herzberg, F., Bernard, M., & Snyderman, B. B. (1959). *The motivation to work*. John Wiley and Sons.

Hill, N. (1996). *Handbook of customer satisfaction measurement*. Gower Publishing.

Jolly, D. A., & Reynolds, K. A. (2005). *Consumer demand for agricultural and on-farm nature tourism, UC Small Farm Centre Research Brief*. Retrieved from: http://www.ucanr.org/sites/sfp/files/143466.pdf

Kano, N., Seraku, N., Takahashi, F., & Tsjui, S. (1984). Attractive quality and must-be quality. *Hinshitsu, 14*(2), 147–156.

Kozark, M., & Rimmington, M. (2000). Tourist satisfaction with Mallorca, Spain, as an off-season holiday destination. *Journal of Travel Research, 38*(3), 260–269. doi:10.1177/004728750003800308

Krassadaki, E., & Grigoroudis, E. (2018). Analyzing perceived quality of health care services: A multicriteria decision analysis approach based on the theory of attractive quality. In N. Matsatsinis & E. Grigoroudis (Eds.), *Preference disaggregation in multiple criteria decision analysis: Essays in honor of Yannis Siskos* (pp. 211–235). Springer. doi:10.1007/978-3-319-90599-0_10

Langworthy, A., Howard, J., & Mawson, F. (2006). Building the relationship between agriculture and tourism-models and benefits of co-operation. Report Prepared for the Centre for Agriculture and Business, Swinburne University of Technology, Melbourne.

Lemmink, J., & Mattsson, J. (1998). Warmth during non-productive retail encounters: The hidden side of productivity. *International Journal of Research in Marketing, 15*(5), 505–517. doi:10.1016/S0167-8116(98)00016-0

Lobo, R. (2012). *Agritourism definition, UC Small Farm Program, University of California Corporative Extension*. Retrieved from: http://sfp.ucdavis.edu/agritourism/Definitions/

Maetzold, J. A. (2002). *Nature-based tourism and agritourism trends: Unlimited opportunities*. Retrieved from: https://www.agmrc.org/media/cms/agritourism_E6794269B3FF6.pdf

Maetzold, J. A. (2004). *Nature-based tourism and agritourism trends: Unlimited opportunities, National Alternative Enterprises and Agritourism Leader*. USDA/NRCS.

Motorola. (1995). *Customer satisfaction assessment guide*. Motorola University Press.

Naumann, E., & Giel, K. (1995). *Customer satisfaction measurement and management: Using the voice of the customer*. Thomson Executive Press.

OECD. (1994). *Tourism strategies and rural development*. OECD Publication.

Oliver, R. L. (1977). Effect of expectation and disconfirmation on postexposure product evaluations: An alternative interpretation. *The Journal of Applied Psychology, 62*(4), 480–486. doi:10.1037/0021-9010.62.4.480

Oliver, R. L. (1980). A cognitive model of the antecedents and consequences of satisfaction decisions. *JMR, Journal of Marketing Research*, *17*(11), 460–469. doi:10.1177/002224378001700405

Oliver, R. L. (1997). *Satisfaction: A behavioral perspective on the customer*. McGraw-Hill.

Parasuraman, A., Zeithaml, V. A., & Berry, L. L. (1988). SERVQUAL: A multiple item scale for measuring consumer perceptions of service quality. *Journal of Retailing*, *64*(1), 14–40.

Parasuraman, A., Zeithaml, V. A., & Berry, L. L. (1991). Refinement and reassessment of the SERVQUAL scale. *Journal of Retailing*, *67*(4), 420–450.

Pizam, A., Neumann, Y., & Reichel, A. (1978). Dimensions of tourist satisfaction with a destination area. *Annals of Tourism Research*, *5*(3), 314–322. doi:10.1016/0160-7383(78)90115-9

Sanchez, J., Callarisa, L., Rodriguez, R., & Moliner, M. (2006). Perceived value of the purchase of a tourism product. *Tourism Management*, *27*(3), 394–409. doi:10.1016/j.tourman.2004.11.007

Siskos, Y., Grigoroudis, E., Zopounidis, C., & Saurais, O. (1998). Measuring customer satisfaction using a collective preference disaggregation model. *Journal of Global Optimization*, *12*(2), 175–195. doi:10.1023/A:1008262411587

Srikatanyoo, N., & Campiranon, K. (2010). Agritourist needs and motivations: The Chiang Mai case. *Journal of Travel & Tourism Marketing*, *27*(2), 166–178. doi:10.1080/10548400903579795

Tian-Cole, S., & Crompton, J. (2002). A conceptualization of the relationships between service quality and visitor satisfaction, and their links to destination selection. *Leisure Studies*, *22*(1), 65–80. doi:10.1080/02614360306572

Tribe, J., & Snaith, T. (1998). From SERVQUAL to HOLSAT: Holiday satisfaction in Varadero, Cuba. *Tourism Management*, *19*(1), 25–34. doi:10.1016/S0261-5177(97)00094-0

Truong, T. H., & Foster, D. (2006). Using HOLSAT to evaluate tourist satisfaction at destinations: The case of Australian holidaymakers in Vietnam. *Tourism Management*, *27*(5), 842–855. doi:10.1016/j.tourman.2005.05.008

Tsitsiloni, M., Grigoroudis, E., & Zopounidis, C. (2012). A service quality evaluation method in the tourism industry: A multicriteria analysis. *Journal of Computational Optimization in Economics and Finance*, *3*(3), 185–202.

Vavra, T. G. (1997). *Improving your measurement of customer satisfaction: A guide to creating, conducting, analyzing, and reporting customer satisfaction measurement programs*. ASQC Quality Press.

Wicks, B. E., & Merrett, C. D. (2003). Agritourism: An economic opportunity for Illinois. *Rural Research Report*, *14*(9), 1–8.

Woodruff, R. B., & Gardial, S. F. (1996). *Know your customer: New approaches to understanding customer value and satisfaction*. Blackwell Publishers.

Yu, L., & Goulden, M. (2006). A comparative analysis of international tourists' satisfaction in Mongolia. *Tourism Management*, *27*(6), 1331–1342. doi:10.1016/j.tourman.2005.06.003

Yuksel, A. (2001). Managing customer satisfaction and retention: A case of tourist destinations, Turkey. *Journal of Vacation Marketing*, *7*(2), 153–168. doi:10.1177/135676670100700205

ADDITIONAL READING

Arabatzis, G., & Grigoroudis, E. (2010). Visitors' satisfaction, perceptions and gap analysis: The case of Dadia-Lefkimi-Souflion National Park. *Forest Policy and Economics*, *12*(3), 163–172. doi:10.1016/j.forpol.2009.09.008

Bowen, D., & Clarke, J. (2002). Reflections on tourist satisfaction research: Past, present and future. *Journal of Vacation Marketing*, *8*(4), 297–308. doi:10.1177/135676670200800401

Hsu, T.-K., Tsai, Y.-F., & Wu, H.-H. (2009). The preference analysis for tourist choice of destination: A case study of Taiwan. *Tourism Management*, *30*(2), 288–297. doi:10.1016/j.tourman.2008.07.011

Karakitsiou, A., Mavrommati, A., Migdalas, A., & Tsiakali, K. (2007). Customer satisfaction evaluation in the tourism industry: The case of Chania. *Foundations of Computing and Decision Sciences*, *32*(2), 111–124.

Kozak, M. (2001). Comparative assessment of tourist satisfaction with destinations across two nationalities. *Tourism Management*, *22*(4), 391–401. doi:10.1016/S0261-5177(00)00064-9

Oh, H. (2001). Revisiting importance-performance analysis. *Tourism Management*, *22*(6), 617–627. doi:10.1016/S0261-5177(01)00036-X

Pawitra, T. A., & Tan, K. C. (2003). Tourist satisfaction in Singapore: A perspective from Indonesian tourists. *Managing Service Quality*, *13*(5), 399–411. doi:10.1108/09604520310495868

Pizam, A., Neumann, Y., & Reichel, A. (1978). Dimensions of tourist satisfaction with a destination area. *Annals of Tourism Research*, *5*(3), 314–322. doi:10.1016/0160-7383(78)90115-9

Rozman, C., Potočnik, M., Pažek, K., Borec, A., Majkovič, D., & Bohanec, M. (2009). A multi-criteria assessment of tourist farm service quality. *Tourism Management*, *30*(5), 629–637. doi:10.1016/j.tourman.2008.11.008

Ryan, C., & Cessford, G. (2003). Developing a visitor satisfaction monitoring methodology: Quality gaps, crowding and some results. *Current Issues in Tourism*, *6*(6), 457–507. doi:10.1080/13683500308667966

Tonge, J., & Moore, S. A. (2007). Importance-satisfaction analysis for marine park hinterlands: A western Australian case study. *Tourism Management*, *28*(1), 768–776. doi:10.1016/j.tourman.2006.05.007

Wade, D. J., & Eagles, P. (2003). The use of importance-performance analysis and market segmentation for tourism management in parks and protected areas: An application to Tasmania's National Parks. *Journal of Ecotourism*, *2*(3), 196–212. doi:10.1080/14724040308668144

KEY TERMS AND DEFINITIONS

Agro-Tourism: A form of tourism in which tourists stay in farms or rural villages and take part in farm or village activities.

Attractive Quality: Excitement product/service attributes that are usually unforeseen by customers and may cause high satisfaction.

Customer Satisfaction: The consumer's response to the evaluation of the perceived discrepancy between prior expectations and the perceived performance of the product service.

Desired Quality: Performance product/service attributes that are usually explicitly demanded by the customer and may cause both satisfaction and dissatisfaction.

Expected Quality: Basic product/service attributes that the product must have in order to meet customer basic requirements.

Kano Model: A theory of customer satisfactions used to classify customer needs and prioritize product/service attributes.

MUSA Method: A multicriteria decision analysis method for measuring and analyzing customer satisfaction based on the principles of ordinal regression.

SERVQUAL: A multi-dimensional research instrument developed to measure the gap between consumer expectations and perceptions of a service along a predefined set of five quality dimensions.

SWOT Analysis: A strategic planning technique used to identify strengths, weaknesses, opportunities, and threats related to business competition.

Chapter 16
Assessing the Impact of Cigarette Taxation on the Supply Chain Stakeholders' Revenue Shares in Greece

Athanassios Vozikis
Laboratory of Health Economics and Management (LabHEM), University of Piraeus, Greece

Yannis A. Pollalis
University of Piraeus, Greece

Archontoula Armoutaki
University of Piraeus, Greece

ABSTRACT

The chapter aims to analyze the impact of cigarette taxes evolution over the period 1992-2017 on the revenue share of the main tobacco supply chain stakeholders in Greece. This empirical analysis uses a pooled time series from 1992 through 2017 including a data set of retail prices, three tax groups levied on cigarettes (specific tax, ad valorem tax, and V.A.T.), and revenue shares for three categories of stakeholders. The results indicate that the revenue share of the stakeholders is decreasing over the whole period, and specifically, their shares drop by half in the last 15 years. The regression results show that the revenue shares of tobacco companies are most affected by both excise taxes in a negative way, and similarly, both price and the excises affect significantly the revenue share of retailers, whereas the findings regarding distributors' revenues are insignificant. This knowledge is likely to be useful for policymakers in the development of effective tobacco control policies.

DOI: 10.4018/978-1-7998-5442-5.ch016

INTRODUCTION AND LITERATURE REVIEW

Tobacco consumption is the leading preventable cause of mortality and one of the largest avoidable health risks in the European Union (EU) (see e.g. WHO, 2008; Gruer et al., 2009; Eriksen et al., 2015). Especially in Greece every year, more than 22,700 of its people are killed by tobacco-caused disease (see e.g., Drope and Schluger, 2018). Still, more than 12,000 children, aged 10-14 years and 256,000 adults (15+ years old) continue to use tobacco each day. Besides, the economic cost of smoking in Greece amounts to 4,663 million euro, including direct costs related to healthcare expenditures and indirect costs related to lost productivity due to early mortality and morbidity. Tobacco taxation is an efficient tobacco control tool for reducing the smoking prevalence and its harmful consequences. On the flip side, though, it is an important source of revenue, making it an interesting topic for branches of economics concerned with this issue (see e.g., Goodchild et al., 2016). Tobacco supply chain involves many types of activities employing a substantial number of employees.

Our paper aims to analyze the impact of cigarette taxation and price evolution over the period 1992-2017 on the revenue share of the main tobacco supply chain stakeholders in Greece; Tobacco Companies, including producers and importers, Wholesalers – Distributors and Retailers. In particular, we indent to show whether and how the changes in cigarette prices, and every type of tax levied on cigarettes are related to the changes in the revenue shares of each stakeholder.

The tobacco market in Greece is highly competitive and dominated by four large multinational companies: Philip Morris International (PMI) (including Papastratos), British American Tobacco (BAT), Japan Tobacco International (JTI), Imperial Tobacco (IT) and three Greek tobacco companies; Karelia Tobacco Company, SEKAP SA and Georgiades Th. D. Cigarettes Industry S.A (see e.g. Executive Agency for Health and Consumers, 2013). The four multinational companies control almost the 79% of the cigarette market. Despite the strong presence of multinationals, the domestic manufacturer Karelia Tobacco Company manages to rank second in 2015 with an 18% retail volume share (see e.g. Euromonitor International, 2016). This increasingly competitive tobacco environment is characterized by continuous mergers and acquisitions, causing a significant loss of volume share for Greek manufacturers during the crisis period of 2009 and afterwards (see e.g. Harvard School of Public Health, 2011).

The distribution of tobacco products remained fairly stable in Greece in 2018, as the dominant channel of newsagent-tobacconists/kiosks continued to gradually lose retail volume share to convenience stores and the number of kiosks continued to decline due to the recession (see e.g. Harvard School of Public Health, 2011). Regarding the retail of tobacco products, legislative changes mid-review period contributed to the growth in the sales of tobacco through convenience stores, as it allowed the sale of tobacco on the same premises as fresh groceries, which means that all convenience stores can sell tobacco products, whereas previously only a limited number of chains were qualified to do so.

Smoking prevalence throughout EU of varies greatly between the 28 Member States, while the highest shares of current smokers for 2014 were recorded in Bulgaria (34.7%) and Greece (32.6%) (see e.g. Eurostat, 2016). The heterogeneity of smoking prevalence between EU Member States reflects, at least in part, a failure by governments to prioritize public health over the tobacco industry (Bogdanovica et al., 2011). Although globally, states intervene through taxes or other tobacco control policies in order to reduce tobacco consumption (see e.g., Chow, 2017), a growing evidence highlight tobacco industry's active role in opposing tobacco control legislation by biasing both scientific and public opinion (see e.g., Dearlove et al., 2002; Muggli et al., 2004; Diethelm et al., 2005; Gruning et al., 2006). Evidence also reveals that Transactional Tobacco Companies lobby governments for keeping cigarettes affordable in

Finland (see e.g., Hiilamo, 2003), Hungary (Szilágyi and Chapman, 2004), and other countries (see e.g. Gilmore et al., 2007; Krasovsky, 2010).

The taxation policy has long been supported by the World Health Organization (WHO). The EU is party to the WHO Framework Convention on Tobacco Control (FCTC). The article 6 of the WHO FCTC, "Price and Tax Measures to Reduce the Demand for Tobacco recognizes the importance of this policy and calls on governments to implement tax and price policies to contribute to their national health objectives. The Directive 2011/64/EU, which is broadly coherent with the principles of WHO, regulates the minimum rates of excise duties levied on cigarettes. Cigarettes manufactured in the EU and those imported from third countries are subject to two types of Excise Duty, a specific excise duty per unit of the product and an ad valorem excise duty calculated on the basis of the maximum retail selling price. In addition to the excise duty, all tobacco products are charged with Value Added Tax (VAT), which is calculated on the basis of the retail selling price of each product (see e.g. European Commission, 2018). In this context, the Member States are free to apply excise rates above the minimum levels of taxation according to their own national needs. The *Figure 1* depicts the different level of prices and excises levied on cigarettes across E.U. on 1/7/2017.

As depicted in the Graph 1, there are different excise tax structures across E.U. with some MS relying more on specific excise duty and other MS favoring ad valorem taxation (see e.g. Delipalla, 1998). There is theoretical and empirical evidence (see e.g. WHO, 2010) on alternative approaches to the choice of specific and ad valorem taxes and their effects on price, consumption, quality and variety of tobacco products, government revenue and tax administration. The choice between specific and ad valorem taxes is a long-standing issue in tax policy. Different excises may have a different effect on government's revenues, manufacturer's revenue, wholesaler's revenue, retailer's revenue, consumer's price, product's "quality1" and variety, and ability to administer taxes (see e.g. Delipalla and Keen, 1992; Schröder, 2004). Theoretical work has shown that the incidence of ad valorem and specific taxes may differ and each may be over or under-shifted onto consumers in the presence of imperfect competition, (see e.g. Skeath and Trandel, 1994; Myles, 1996). Empirical evidence of undershifting of both taxes indicates that specific tax have a significantly greater impact on price (see e.g., Delipalla and O'Donnell, 1999). Specific taxes may also provide incentives for more appealing and higher-priced products, as well as greater variety (upgrading effect), whereas the ad valorem component of excise duty may encourage more "trading down" in favor of cheaper cigarettes reducing health benefit (see e.g. Keen, 1998). Given that ad valorem taxation has a "multiplier effect", that is, to increase producer price by 1, consumer price has to increase by 1/(1 tv) > 1 (tv is the ad valorem tax rate). Thus, when the ad valorem tax increases, it is likely to lead to a reduction in quality and, consequently, a price rise lower than the amount of the tax increase (see e.g., Anderson et al., 1997).

On the other hand, ad valorem taxes have the benefit of being indexed for inflation, but the impact of the tax can be mitigated by low cost production (see e.g. Chaloupka, 2012). Finally, the ad valorem tax requires stronger tax administration with technical capacity and this can lead to higher administrative burden (see e.g., Yurekli & Beyer, 2011). The differences in the preferred structure of cigarette taxation among countries in the EU have impeded agreement on harmonization so as to ensure the proper functioning of internal market.

This paper aims to present the evolution of cigarettes tax rates in Greece over the period 1992-2017 and examine how these tax fluctuations have influenced the revenue shares of three main stakeholders of tobacco supply chain. To succeed this, our analysis is based on econometrics offering an estimate of the price and taxation impact on revenue shares.

Figure 1. Tax structure and level of prices (W.A.P. per 1 pack) in EU on 1/7/2017
**TIRSP: Retail Selling Price, all Taxes Included*
Source: European Commission, Taxation and Customs Union & our calculations

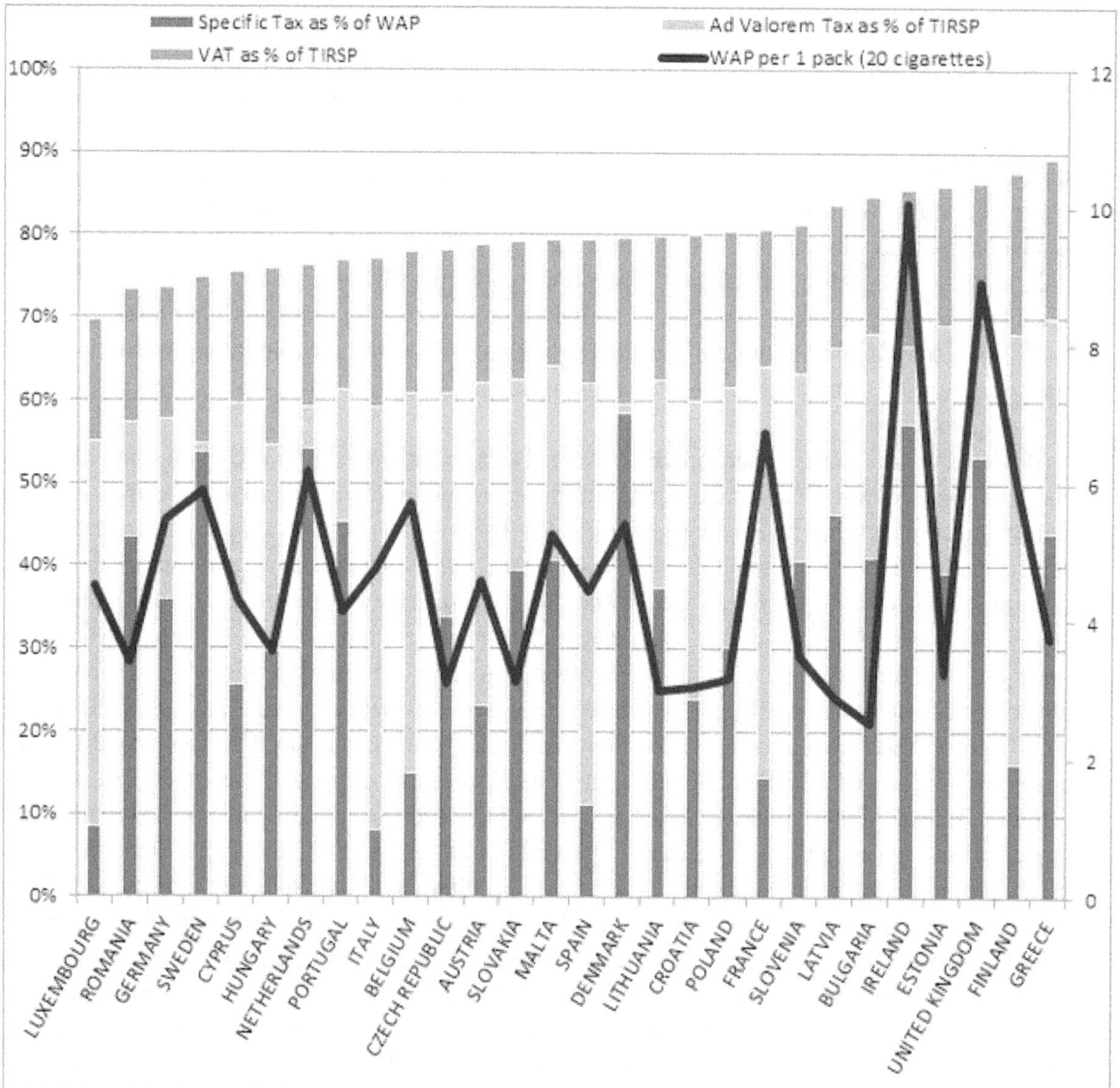

The reminder of the paper is structured as follows: section 2 describes the theoretical background and data; section 3 introduces the methodology; section 4 presents the results; and section 5 discusses the findings and concludes.

Theoretical Background: Data

Tobacco taxation in Greece is firstly determined by EU legislation requiring Member States to levy a minimum rate of excise duties on cigarettes. According to the Directive 2011/64/EU, the specific component should be of between 7.5% and 76.5% of the total tax burden. Second, the overall excise duty on cigarettes should represent at least 60% of the weighted average retail selling price of cigarettes released for consumption. Finally, the excise duty should not be less than 90 euros per 1,000 cigarettes irrespective of the weighted average retail selling price. Member States that apply excise duty of 115 euro or more, however, do not need to comply with the 60% criterion above.

As presented in the Graph 1, Greece is classified at the highest position regarding the cigarette tax burden across the EU for 2017, followed by Finland and United Kingdom. Yet, the level of cigarette prices remains among the lowest in EU. The preferred excise tax structure levied on cigarettes in Greece is set by the National Customs Code of Greece (L.2960/2001), requiring the total amount of excise duty not to be lower than 117.50 euros per 1,000. To explain the way that total taxation is levied on cigarettes, we construct the following *Table 1* presenting the Greek tax structure of cigarettes for the year 2016.

Table 1. Cigarette pricing and taxation in Greece for the year 2016

TAX	TAX RATE	TAX BASE	VALUE per 1000 cigs) *	VALUE FOR 1 PACK (20 cigs)
SPECIFIC TAX (1)	82.5 €	* 1,000 cigs	82.50 €	1.65 €
AD VALOREM TAX (2)	20%	** 185.5 €	37.10 €	0.74 €
VAT (3)	18.70%	** 185.5 €	34.69 €	0.69 €
TOTAL TAX (4)=(1)+(2)+(3)			154.29 €	3.09 €
VALUE EXCLUDING TAXES (5)=(6)-(4)	185.5 – 154.29		31.21 €	0.62 €
W.A.P. (6) = (4)+(5)			185.5 €	3.71 €
* 50 packs * 20 cigarettes = 1,000 cigarettes				
** W.A.P.: 50 *3.71 € = 185.5 €				

Source: Ministry of Economics of Greece, Directorate General of Customs and Excise Duties (G.D.T. & E.F.K.)

The above Weighted Average Retail Price (W.A.P.) is defined as the weighted average price for cigarettes calculated by reference to the total value of all cigarettes released for consumption, based on the retail selling price including all taxes, divided by the total quantity of cigarettes released for consumption during the previous year (2015). Until January 2011, excise taxes were calculated on the basis of the retail selling price of highest selling category, defined by price (i.e. the most popular price category - MPPC), in each country (see e.g., European Commission, 2008). Directive 2010/12/EU, which entered into effect on 1 January 2011, changed the reference point for calculating taxes from the MPPC to the W.A.P.

Data on excise tax rates (specific excise and ad valorem excise) applicable to cigarettes, the weighted average price (WAP) and tax revenues accrued from cigarette consumption are derived from the Excise Duties Tables series compiled by Directorate-General for Taxation and the Customs Union (DG TAXUD) and published in the European Commission data base, the official site of European Union.

Note that prices are referred to one pack which consists of twenty cigarettes and they are expressed in euro. Until 2001, prices were expressed in drachma, the ex national currency. Yet, all price data were converted into euro in favor of comparison and statistical analysis. Data on revenue shares of three elements of tobacco industry were obtained from the Trade Attica Union of Retail Business, Kiosk owners & Tobacco Sellers.

The following *Table 2* presents the changes in the cigarette tax structure, the prices and the revenue shares over the period 1992-2017.

It should be noted that that there are two data points for 2005, as there had been a V.A.T increase during this year leading to differential revenue shares among three components of tobacco industry. Moreover, three different data points are noticed for 2010, albeit price and tax incidence remained the same. The reason is that tobacco industries and retailers changed their tax policy three times during 2010, causing correspondingly a gradual revenue share reduction of these two stakeholders.

As might be anticipated, the tax burden on cigarettes increases gradually during the period 1992-2017, while the mixture of tax incidence changes. The two types of excise experience a smooth increase, with the exception of the last seven years. The reason is that the government initiated a major shift away from ad valorem taxation towards specific taxation. The ad valorem component of excise on cigarettes decreased by more than 50% in 2013 and has been compensated for with an abrupt increase of the specific excise duty component. On the contrary, the increase rate of VAT remained unchanged over the whole period. The price and taxation evolution occurred in Greece over the period 1992-2017 is illustrated in the *Figure 2*.

According to the *Table 2*, the revenue shares of the stakeholders follow a stable course until 2009-2010, where their earnings almost reach their highest point. It is obvious that all stakeholders' shares declined by half during the last four years. Moreover, the proportion of the revenue shares changed essentially and specifically for wholesalers and distributors, whose share seems to have been absorbed mostly by the retailers. After all, tobacco companies remain the prominent profitable category.

Methodology

Following the empirical literature, we construct an empirical model that investigates the non-linear effect of the changes of excise taxes and prices on revenue share of each stakeholder involved in the cigarette supply chain, i.e., as below:

$$Ri = \alpha i + \alpha 1 \text{ pricei} + \alpha 2 \text{ tax1i} + \alpha 3 \text{ tax2i} + \alpha 4 \text{ tax3i}$$

Where

i denotes the different type of stakeholder; tobacco companies – either manufacturers or importers – distributers and retailers,

Ri is the revenue share of each stakeholder,
pricei is the retail selling price of one pack containing 20 cigarettes,
tax1i is the specific component of excise duty levied on cigarettes as a percentage of price,
tax2i is the ad-valorem component of excise duty levied on cigarettes and
tax3i is the Value Added Tax (VAT).

The first analysis of stakeholders' revenue shares uses a pooled time series from 1992 through 2017 and includes a data set of retail prices, three tax groups levied on cigarettes (specific tax, ad valorem tax and VAT) and revenue shares of the stakeholders.

Table 2. Prices, Residual Revenue Value and Revenue Share of the three main stakeholders over the period 1992-2017

YEAR	MMPC - W.A.P.*	RESIDUAL REVENUE VALUE	TAX AS % OF MMPC OR WAP			REVENUE SHARE OF THE STAKEHOLDERS					
						Tobacco companies (producers and importers)		Wholesalers - Distributors		Retailers	
			SPECIFIC TAX	AD VALOREM TAX	VAT	As % of MMPC or W.A.P.	As % of Total Revenue	As % of MMPC or W.A.P.	As % of Total Revenue	As % of MMPC or W.A.P.	As % of Total Revenue
1992	0.59	0.17	3.58%	41.45%	26.47%	16.76%	60.40%	2.58%	9.30%	8.41%	30.31%
1993	0.99	0.46	3.23%	53.39%	15.25%	12.48%	60.40%	1.92%	9.30%	6.26%	30.30%
1994	1.64	0.69	3.23%	53.39%	15.25%	11.80%	60.40%	1.86%	9.49%	5.88%	30.11%
1995	1.54	0.58	3.88%	53.39%	15.25%	15.10%	60.40%	2.37%	9.50%	7.53%	30.11%
1996	1.47	0.50	4.04%	53.86%	15.25%	16.42%	60.40%	2.58%	9.50%	8.18%	30.10%
1997	1.81	0.64	3.64%	53.86%	15.25%	14.64%	60.40%	2.30%	9.49%	7.30%	30.11%
1998	1.93	0.65	3.64%	53.86%	15.25%	15.01%	60.40%	2.36%	9.49%	7.48%	30.11%
1999	1.77	0.49	3.64%	53.86%	15.25%	17.70%	60.40%	2.78%	9.50%	8.82%	30.11%
2000	1.98	0.59	3.64%	53.86%	15.25%	15.87%	60.40%	2.50%	9.50%	7.91%	30.11%
2001	2.06	0.57	3.64%	53.86%	15.25%	16.39%	60.40%	2.58%	9.49%	8.17%	30.11%
2002	2.34	0.64	3.64%	53.86%	15.25%	16.67%	60.94%	2.56%	9.38%	8.12%	29.69%
2003	2.50	0.68	3.64%	53.86%	15.25%	17.60%	60.27%	2.80%	9.59%	8.80%	30.14%
2004	2.50	0.68	3.64%	53.86%	15.25%	17.60%	60.27%	2.80%	9.59%	8.80%	30.14%
2005 (a)	2.70	0.74	3.67%	53.83%	15.25%	16.30%	60.27%	2.59%	9.59%	8.15%	30.14%
2005 (b)	2.70	0.72	3.67%	53.83%	15.97%	15.93%	60.56%	2.59%	9.86%	7.78%	29.58%
2006	2.80	0.74	3.67%	53.83%	15.97%	15.36%	60.56%	2.50%	9.86%	7.50%	29.58%
2007	2.80	0.74	3.67%	53.83%	15.97%	15.36%	60.56%	2.50%	9.86%	7.50%	29.58%
2008	3.00	0.80	3.67%	53.83%	15.97%	16%	60%	2.67%	10%	8.00%	30%
2009	3.00	0.80	3.67%	53.83%	15.97%	17%	61.45%	2.67%	9.64%	8.00%	28.92%
Jan-10	3.20	0.46	8.57%	58.43%	18.70%	12.81%	60.29%	2.81%	13.24%	5.63%	26.47%
Mar-10	3.20	0.46	8.57%	58.43%	18.70%	10.63%	59.65%	2.81%	15.79%	4.38%	24.56%
May-10	3.20	0.46	8.57%	58.43%	18.70%	8.75%	60.87%	2.81%	19.57%	2.81%	19.57%
2011	3.13	0.51	12.55%	52.45%	18.70%	10.22%	60.38%	2.88%	16.98%	3.83%	22.64%
2012	3.25	0.53	12.56%	52.45%	18.70%	10.48%	59.65%	2.77%	15.79%	4.31%	24.56%
2013	3.28	0.41	48.75%	20%	18.70%	8.53%	63.64%	1.83%	13.64%	3.05%	22.73%
2014	3.50	0.50	47.10%	20%	18.70%	9.42%	62.26%	1.14%	7.55%	4.57%	30.19%
2015	3.64	0.58	45.38%	20%	18.70%	10.31%	62.50%	1.38%	8.33%	4.81%	29.17%
2016	3.71	0.59	44.50%	20%	18.70%	11.33%	61.76%	1.08%	5.88%	5.93%	32.35%
2017	3.80	0.43	44.05%	26%	19.35%	7.11%	61.78%	0.26%	2.29%	4.13%	35.93%

* *W.A.P. since 1/1/2011 (for 1 pack of 20 cigarettes)*

Source: European Commission, Trade Attica Union of Retail Business, Kiosk owners & Tobacco Sellers & authors' calculations

Figure 2. Prices, deflated prices and tax structure of cigarettes in Greece over the period 1992-2017
Source: Hellenic Statistical Authority, (https://www.statistics.gr/el/statistics/-/publication/DKT87/2015-M10)

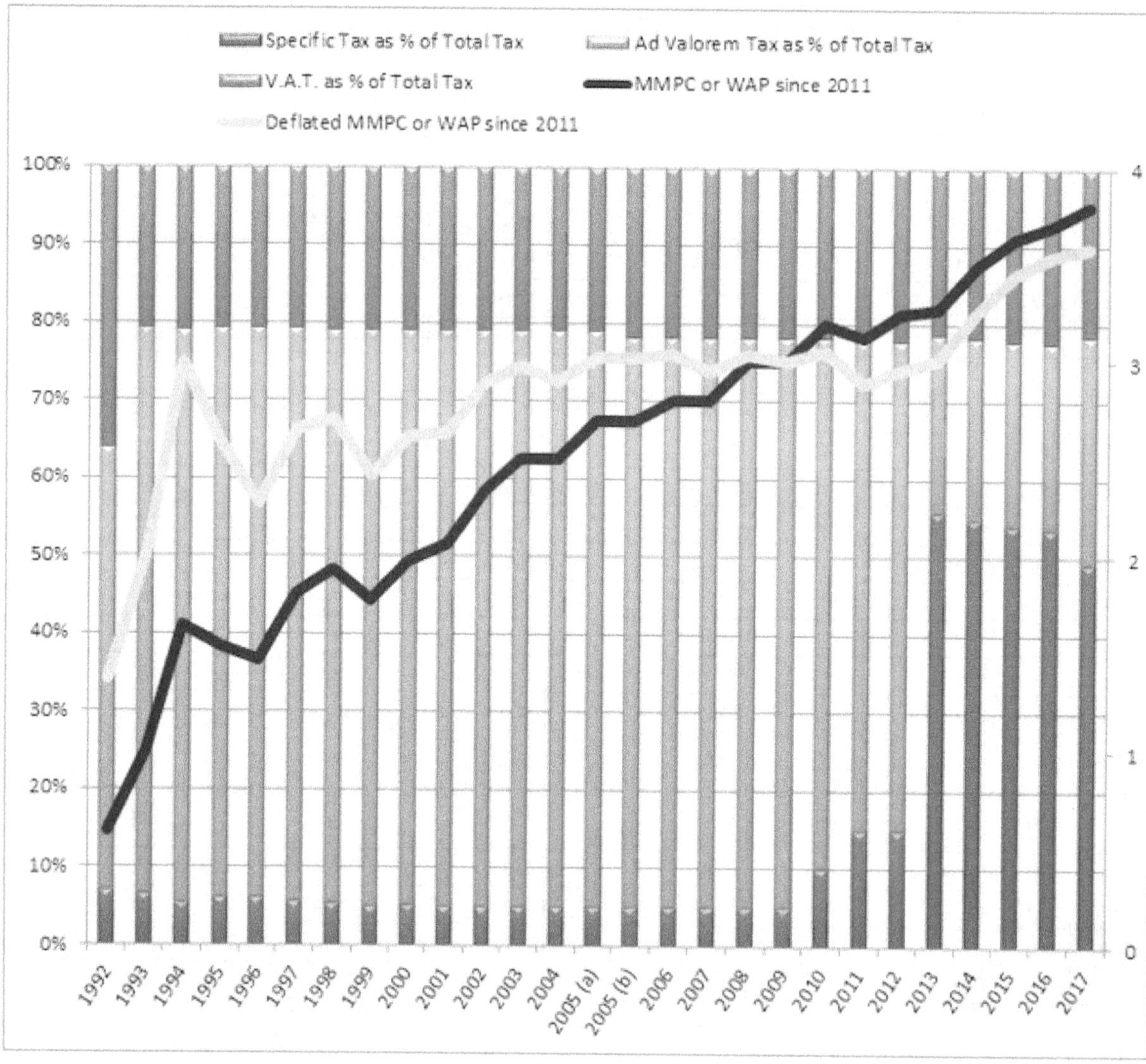

First, we applied the Kolmogorov–Smirnov test in order to test that our data follow the normality of the distribution. The *Table 3* presents the results confirming the normal distribution of our sample (sig>0.1), as P-value exaggerate the confidence interval of 10%. As a result, we can apply regression analyses so as to evaluate the effect of prices' and taxes' changes on the evolution of revenue shares.

Results

First, we apply linear regression to all categories of variables (all types of revenue shares and all types of taxation) and we extract the below table of variation analysis.

Table 3. Tests of Normality

	Kolmogorov-Smirnov[a]			Shapiro-Wilk		
	Statistic	df	Sig.	Statistic	df	Sig.
Price	.119	29	.200*	.955	29	.245
Total tax	.112	29	.200*	.956	29	.268
Specific tax	.343	29	.000	.584	29	.000
Ad valorem tax	.125	29	.200*	.942	29	.114
VAT	.128	29	.200*	.950	29	.179
Revenue share of Tobacco companies (RTCs)	.107	29	.200*	.958	29	.289
Revenue share of Wholesalers/ Distributors RWDs	.149	29	.099	.944	29	.130
Revenue share of Retailers (RRs)	.141	29	.144	.959	29	.309
Total Revenue	.103	29	.200*	.962	29	.361

*. *This is a lower bound of the true significance.*

a. Lilliefors Significance Correction

Box 1. ANOVA[a]

Model		Sum of Squares	df	Mean Square	F	Sig.
1	Regression	.702	2	.351	122.823	.000[b]
	Residual	.074	26	.003		
	Total	,776	28			

a. Dependent Variable: Total revenue

b. Predictors: (Constant), total tax, price

According to the P-value (column sig.), the results show that the association between these variables is statistically significant. Additionally, the estimators show a high fitting that reaches 90.4% (R2 = 0.904), as presented below.

Box 2. Model Summary

Model	R	R Square	Adjusted R Square	Std. Error of the Estimate
1	.951[a]	.904	.897	.0534408501

a. Predictors: (Constant), total tax, price

Based on the above regression, we obtain the following equation;

Total revenue = 0.029 + 0.895 *– 0.875 * total tax.

Table 4. Predictors associated to the revenue shares of each stakeholder

Coefficients - Dependent Variable: Revenue share of Tobacco companies						
Dependent Variable: Revenue share of Tobacco companies		**Unstandardized Coefficients**		**Standardized Coefficients**	**t**	**Sig.**
		B	**Std. Error**	**Beta**		
1	(Constant)	.013	.020		.659	.516
	Price	.476	.042	4.481	11.391	.000
	Specific tax	-.390	.099	-2.482	-3.957	.001
	Ad valorem tax	-.359	.102	-1.670	-3.527	.002
	VAT	-.719	.266	-1.354	-2.699	.013
	R Square	.932				
	Adjusted R Square	.921				
	Std. Error of the Estimate	.0283789993				
ANOVA		Sum of Squares	df	Mean Square	F	Sig.
	Regression	.265	4	.066	82.127	.000[b]
	Residual	.019	24	.001		
	Total	.284	28			
Coefficients - Dependent Variable: Revenue share of Wholesalers - Distributors						
Dependent Variable: Revenue share of Wholesalers/Distributors		**Unstandardized Coefficients**		**Standardized Coefficients**	**t.**	**Sig**
		B	**Std. Error**	**Beta**		
2	(Constant)	.004	.004		.922	.366
	Price	-.008	.009	-.355	-.957	.348
	Specific tax	.024	.021	.678	1.146	.263
	Ad valorem tax	.069	.021	1.434	3.212	.004
	VAT	-.031	.056	-.264	-.558	.582
	R Square	.939				
	Adjusted R Square	.929				
	Std. Error of the Estimate	.0059521383				
ANOVA		Sum of Squares	df	Mean Square	F	Sig.
	Regression	.013	4	.003	93.062	.000[b]
	Residual	.001	24	.000		
	Total	.014	28			
Coefficients - Dependent Variable: Revenue share of Retailers						
Dependent Variable: Revenue share of Retailers		**Unstandardized Coefficients**		**Standardized Coefficients**	**t**	**Sig.**
		B	**Std. Error**	**Beta**		
3	(Constant)	.004	.014		.297	.769
	Price	.312	.029	5.546	10.723	.000
	Specific tax	-.304	.069	-3.654	-4.429	.000
	Ad valorem tax	-.308	.071	-2.710	-4.353	.000
	VAT	-.364	.185	-1.296	-1.965	.061
	R Square	. 882				
	Adjusted R Square	.863				
	Std. Error of the Estimate	.0197509750				
ANOVA		Sum of Squares	df	Mean Square	F	Sig.
	Regression	.070	4	.018	44.975	.000[b]
	Residual	.009	24	.000		
	Total	.080	28			

Note: Three stars (***) indicate statistical significance at 1% level, two stars (**) at 5% level, and one star (*) at 10% level.

After estimating the above relationship between all the categories of estimators, we can apply three separate equations, which evaluate the effect of the price and each tax (specific tax, ad valorem tax, VAT) on each stakeholder's revenue share. The results are presented in the below *Table 4*.

According to the above Table, we obtain the following equations with the estimated coefficients and p-values (in parenthesis) as below:

(1) RTCs = 0.013 + 0.476 * price – 0.39 * specific tax – 0.359 * ad valorem tax – 0.719 * VAT
(0.000) (0.001) (0.002) (0.013)
(2) RWDs = 0.004 - 0.008 * price + 0.024 * specific tax + 0.069 * ad valorem tax – 0.031 * VAT
(0.348) (0.263) (0.004) (0.582)
(3) RRs = 0.004 + 0.312 * price - 0.304 * specific tax - 0.308 * ad valorem tax – 0.364 * VAT
(0.000) (0.000) (0.000) (0.061)

The above-mentioned equations provide a better view of the dynamics of the relations between revenues, taxes and prices. They explain how any change in the price, the excise tax or the VAT may affect the revenues of each category of stakeholder. As expected, prices and taxes are proved to affect statistically significantly the revenue share of TCs and Rs. Specifically, the revenues of TCs are most affected by both excise taxes in a negative way and at the 1% significance level. This means that PTCs are expected to decrease by almost €0.4 corresponding to each 1€ increase in each excise element (specific or ad valorem). The effect of VAT changes is also significant but at the 10% significance level.

Regarding retailers, their revenue share is expected to reduce by almost 0.3 for each €1 increase in every excise component and to increase by the same amount after a €1 increase in retail selling price of cigarettes. Both price and the excises affect significantly the revenue share of retailers at the 1% level of confidence and the association with VAT is also significant, but at the significance level of 10%. On the contrary, the revenue share of wholesalers and distributors is related significantly only with the ad valorem tax but in an unexpectedly positive way. Therefore, this finding may not contribute to our research.

Given the afore-mentioned regressions, we create four ratios:

1. The first calculates the total tax as a percentage of the cigarette price (total tax/price).
2. The second estimates the total tax as a percentage of the revenue share of tobacco companies (total tax/ RTCs)
3. The third calculates the total tax as a percentage of the revenue share of wholesalers and distributors (total tax/ RWDs)
4. The fourth estimates the total tax as a percentage of the revenue share of retailers (total tax/ RRs).

Based on the *Table 5* above and using the Spearman test, we find four statistically significant combinations of associations. As we can notice, the strongest correlation is between the total tax as a percentage of the revenue share of tobacco companies (total tax/ RTCs) and the total tax as a percentage of the revenue share of wholesalers and distributors (total tax/ RWDs). Specifically, this relationship is positive and equal to 0.570. Similarly, the rest correlations among the stakeholders have positive signs but they are less related (0.471, 0.504, 0.559).

Table 5. Correlations across the above four ratios

		total tax/price	total tax/ RTCs	total tax/ RWDs	total tax/ RRs
total tax/price	Pearson Correlation	1	.471*	.504**	-.280
	Sig. (2-tailed)		.010	.005	.142
	N	29	29	29	29
total tax/ RTCs	Pearson Correlation	.471*	1	.570***	.559**
	Sig. (2-tailed)	.010		.001	.002
	N	29	29	29	29
total tax/ RWDs	Pearson Correlation	.504**	.570***	1	-.185
	Sig. (2-tailed)	.005	.001		.338
	N	29	29	29	29
total tax/ RRs	Pearson Correlation	-.280	.559**	-.185	1
	Sig. (2-tailed)	.142	.002	.338	
	N	29	29	29	29

DISCUSSION AND CONCLUSION

Literature review suggests that tobacco taxation is unarguably the most effective tool for reducing the tobacco consumption and hence the health care costs associated with treating smoking-caused diseases (see e.g. Knuchel-Takano et al., 2017). Besides the reduction of smoking prevalence, tobacco taxes contribute to government revenue accruable from tobacco manufacturers, wholesalers and retailers. Our research examined firstly the changes occurred at the revenue shares of the main stakeholders of tobacco supply chain in Greece resulting from the cigarette tax increases over the period 1992-2017.

Based on a pooled time series from 1992 through 2017, the findings of this research model the effects of the price, the excise tax (specific and ad valorem) and the VAT on the revenues of each category of stakeholder. In the tobacco taxation literature, there is a gap regarding such revenue share data. Our analysis shows that revenues of tobacco industries remain constant in contrast with the other two categories of tobacco supply chain (distributors and retailers), whose revenue shares seem to be affected mostly by special legislation or institutional changes (i.e. progressive liberalization procedures) rather than tax increases.

As pointed out in the paper, the level of cigarette prices in Greece remains among the lowest in EU, despite its highest cigarette tax burden recorded across the EU for 2017. Specifically, Greece is among the EU-28 countries with the highest specific excise and with the lowest ad valorem excises. Delipalla and Keen (1998) claim that the specific tax has a significantly greater effect on price than the ad valorem one. This is consistent with other studies' findings demonstrating that the specific excise tax on cigarettes, compared to the ad valorem tax, is a more efficient policy device to achieve fiscal policy, as well as public health objectives (see e.g. Chaloupka, et al., 2010). Specific duties reduce relative price differences and minimize the variability of prices. Consequently, specific duties have an advantage from a health point of view. Specific duties also have a number of other benefits. A scenario that sets all duties to specific duties would entail a narrowing of price differences at EU level by 50%. Therefore, greater reliance on specific duties can lead to further approximation of retail selling prices within the EU (see e.g. Commission of the European Communities, 2008).

It should be highlighted that in 2012, in Greece, an important change in the structure of excise taxes came to effect. The specific excise tax per 1000 cigarettes increased by 300%, the W.A.P. per 1000 cigarettes increased by 8.5% while the ad valorem excise decreased by 61.9%. This change resulted in a significant increase in the price of low-price cigarettes and a decrease of the gap between the low and high-priced cigarettes (see e.g. Georgikopoulos and Pinedo, 2016).

One limitation of our research is worth noting. The regression models are expressed in a simplified way excluding other control variables in order to highlight how the revenues of three main stakeholders' categories have been fluctuated in average as a result of price and tax policies. But, this constrain is likely to create a bias in our findings.

Though, the novelty of this paper lies in the fact that for the first time in literature our research documents the impact of cigarette tax structure and price fluctuation on the cigarette supply chain revenue shares in Greece. The evidence that this paper provides is likely to be useful for policy-makers in development of effective tobacco control policies. Our analysis aspires to help authorities with setting up optimal taxation scheme from a tax revenue perspective as well as consumers' health perspective.

Declarations

The authors declare that;

- There is no support from any organisation for the submitted work, no financial relationships with any organisations that might have an interest in the submitted work in the previous three years, no other relationships or activities that could appear to have influenced the submitted work.
- They have no competing interests.
- The manuscript is an honest, accurate and transparent account of the study being reported; that no important aspects of the study have been omitted; and that any discrepancies from the study as planned have been explained.

REFERENCES

Anderson, S. P., & Kreider, B. (1997). *Tax incidence in differentiated product oligopoly, mimeo*. University of Virginia.

Bogdanovica, I., McNeill, A., Murray, R., & Britton, J. (2011). What Factors Influence Smoking Prevalence and Smoke Free Policy Enactment across the European Union Member States. *PMC PLoS One, Vol, 6*(8), e23889. doi:10.1371/journal.pone.0023889 PMID:21909375

Chaloupka, F. J., & Peck, J. A. (2010). *Cigarette Excise Taxation: The Impact Of Tax Structure On Prices, Revenues, And Cigarette Smoking*. NBER Working Paper No. 16287. JEL, No. H2, I18.

Chaloupka, F. J., Yurekli, A., & Fong, G. T. (2012). Tobacco Taxes As A Tobacco Control Strategy. *Tobacco Control, Vol, 21*(3), 172–180. doi:10.1136/tobaccocontrol-2011-050417 PMID:22345242

Chow C. K., Corsi D. J., & Gilmore A. B. (2017) Tobacco control environment: cross-sectional survey of policy implementation, social unacceptability, knowledge of tobacco health harms and relationship to quit ratio in 17 low-income, middle-income and high-income countries. *BMJ*, *7*(3).

Commission of the European Communities (2008). *Report from the Commission to the European Parliament and the Council On the structure and rates of excise' duty applied on cigarettes and other manufactured tobacco products.* Author.

Dearlove, J., Bialous, S., & Glantz, S. A. (2002). Tobacco industry manipulation of the hospitality industry to maintain smoking in public places. *Tobacco Control, Vol, 11*(2), 94–104. doi:10.1136/tc.11.2.94 PMID:12034999

Delipalla, S., & Keen, M. J. (1992). The comparison between ad valorem and specific taxation under imperfect competition. *Journal of Public Economics, Vol, 49*(3), 351–167. doi:10.1016/0047-2727(92)90073-O

Delipalla, S., & O'Donnell, O. (1998). *The Comparison Between Ad Valorem and Specific Taxation under Imperfect Competition: Evidence from the European Cigarette Industry*. Department of Economics Discussion Paper. No. 9802, University of Kent, Department of Economics, Canterbury.

Delipalla, S., & O'Donnell, O. (1999). *Estimating Tax Incidence, Market Power and Market Conduct: The European Cigarette Industry*. Academic Press.

Diethelm, P. A., Rielle, J.-C., & McKee, M. (2005). The whole truth and nothing but the truth? The research that Philip Morris did not want you to see. *Lancet, 366*(9479), 86–92. doi:10.1016/S0140-6736(05)66474-4 PMID:15993237

Drope, J., & Schluger, N. W. (2018). *The Tobacco atlas*. American Cancer Society, Inc.

Eriksen, M., Mackay, J., Schluger, N., Gomeshtapeh, F. I., & Drope, J. (2015) The Tobacco Atlas. American Cancer Society. World Lung Foundation.

Euromonitor International. (2016). *Country report*. Tobacco in Greece.

European Commission. (2018). *Taxation and Customs Union. Excise Duties; Cigarettes*. https://ec.europa.eu/taxation_customs/business/excise-duties-alcohol-tobacco-energy/excise-duties-tobacco/excise-duties-cigarettes_en

European Union (2011) Council Directive 2011/64/EU on the structure and rates of excise duty applied to manufactured tobacco. Official Journal of the European Union. Chapter 3, "Provisions Applicable To Cigarettes

European Union (2011) Council Directive 2011/64/EU on the structure and rates of excise duty applied to manufactured tobacco. Official Journal of the European Union. Chapter 3, "Provisions Applicable To Cigarettes". Articles 8-10.

Eurostat. (2016). Tobacco Consumption., news-release. *European Health Interview Survey, 245.*

Gilmore, A., Collin, J., & Townsend, J. (2007). Transnational Tobacco Company Influence on Tax Policy During Privatization of a State Monopoly: British American Tobacco and Uzbekistan. *Am J Public Health, Vol, 97*(11), 2001–2009. doi:10.2105/AJPH.2005.078378 PMID:17138915

Goodchild, M., Perucica, A., & Nargisb, N. (2016). Modelling the impact of raising tobacco taxes on public health and finance. *Bull World Health Organ, Vol, 94*(4), 250–257. doi:10.2471/BLT.15.164707 PMID:27034518

Government Gazette. (2001). *National Customs Code of Greece 2960/2001, Government Gazette Issue No A'265/22.11.01. Article 96*. Tax Determination.

Gruer, L., Hart, C. L., Gordon, D. S., & Watt, C. M. (2009). Effect of tobacco smoking on survival of men and women by social position: a 28 year cohort study. *BMJ, 338*, b480.

Gruning, T., Gilmore, A. B., & McKee, M. (2006). Tobacco industry influence on science and scientists in Germany. *American Journal of Public Health, 96*(1), 20–32. doi:10.2105/AJPH.2004.061507 PMID:16317203

Hiilamo, H. (2003). Tobacco industry strategy to undermine tobacco control in Finland. *Tobacco Control, Vol, 12*(4), 414–423. doi:10.1136/tc.12.4.414 PMID:14660780

Keen, M. (1998). The Balance between Specific and Ad Valorem Taxation. *Fiscal Studies, Vol, 19*(1), 1–37. doi:10.1111/j.1475-5890.1998.tb00274.x

Knuchel-Takano, A., Hunt, D., Jaccard, A., Bhimjiyani, A., Brown, M., Retat, L., Brown, K., Hinde, S., Selvarajah, C., Bauld, L., & Webber, L. (2018). Modelling the implications of reducing smoking prevalence: The benefits of increasing the UK tobacco duty escalator to public health and economic outcomes. *Tobacco Control, 27*(e2), e124–e129. doi:10.1136/tobaccocontrol-2017-053860 PMID:29212863

Krasovsky, K. S. (2010). The lobbying strategy is to keep excise as low as possible" - tobacco industry excise taxation policy in Ukraine. *Tobacco Induced Diseases, 8*(1), 10. doi:10.1186/1617-9625-8-10 PMID:20807434

Muggli, M. E., Hurt, R. D., & Becker, L. B. (2004). *Turning free speech into corporate speech: Philip Morris' efforts to influence U.S. and European journalists regarding the U.S. EPA report on secondhand smoke*. Academic Press.

Myles, G. (1996). Imperfect competition and the optimal combination of ad valorem and specific taxation. *International Tax and Public Finance, 3*(1), 29–44. doi:10.1007/BF00400145

Schröder, P. J. H. (2004). The Comparison between Ad Valorem and Unit Taxes under Monopolistic Competition. *Journal of Economics, 83*(3), 281–292. doi:10.100700712-004-0096-y

Skeath, S. E., & Trandel, G. E. (1994). A Pareto comparison of ad valorem and specific taxation in noncompetitive environments. *Journal of Public Economics, 53*(1), 53–71. doi:10.1016/0047-2727(94)90013-2

Szilágyi, T., & Chapman, S. (2004). Tobacco industry efforts to keep cigarettes affordable: A case study from Hungary. *Central European Journal of Public Health, 11*(4), 223–228. PMID:14768787

Vardavas, C. L., & Kafatos, A. G. (2006). Smoking policy and prevalence in Greece: An overview. *European Journal of Public Health, 17*(2), 211–213. doi:10.1093/eurpub/ckl094 PMID:17065175

World Health Organization. (2008). *WHO Report on the global tobacco epidemic. The MPOWER package*. WHO.

World Health Organization. (2012). *Global Progress Report on the Implementation of the WHO Framework Convention on Tobacco Control*. WHO.

World Health Organization (WHO). (2010). *WHO Technical Manual On Tobacco Tax Administration*. WHO.

Yurekli, A., & Beyer, J. (2011). *Design and Administer Tobacco Taxes. World Bank Economics of Tobacco Toolkit. Tool 4*. World Health Organization.

ENDNOTE

1 Quality here does not refer in any way to the health impact of the product. It may be evaluated based on the packaging or the blend used for the cigarette, or anything that makes the product more appealing to consumers. In that sense, cigarettes might be of "higher or lower quality" but they are equally harmful.

Chapter 17
An Overview of the Cruise Industry and Its Economic Impacts:
Lessons From Europe

Georgia Papadopoulou
https://orcid.org/0000-0002-9170-4278
Al Ghurair University, UAE

ABSTRACT

The cruise industry is one of the dynamic markets with high investment in port facilities worldwide and in building luxury cruise ships as well. It incorporates elements of the product market, educational activities, and has developed into a spectacular destination. Cruising includes a great part in the tourism chain, including transport, catering, tourism, leisure, and travel. The aim of the chapter is to examine the economic and statistical data of the cruise industry and cruise tourism through a European approach, based on secondary data and research. It further explores the economic impact of the European cruise industry on the European economies. Moreover, the chapter tries to compare the economic impact of the European cruise among the previous years and to ascertain whether these impacts benefit the European economies or not. The evolution of the direct economic impacts are captured and a discussion about the total economic impacts of the European cruise sector is given. This study would offer facilitation in directing future research relevant to cruise tourism.

1. INTRODUCTION

Tourism is among the world's largest and most rapidly growing industries, has a key vital role in job generation, export revenues, and the added national value and contributes directly to the country's GDP. The cruise industry, in particular, the cruise industry has been regarded as the fastest-growing tourism sector. According to the World Tourism Organization (2008), over the last 20 years, global demand for cruises has grown at a cumulative annual rate of 8 percent. In its turn, CLIA (2015) marks out that over

DOI: 10.4018/978-1-7998-5442-5.ch017

the last 10 years, the global demand for cruises, despite the economic crisis, has risen by 84 percent from 13.1 million passengers in 2004 to 22.0 million in 2004. The main objective of the activity of maritime tourism is to provide the cruise tourism product. As Wild and Dearing (2000) say, the main sectors of the cruise as a tourism chain are transport, tourism, entertainment, and the trip.

Over the past two decades, the cruise line industry has become one of the fastest-growing segments in the travel industry. The cruise industry has already developed into a mass-market using large vessels in accordance with the cruise industry of the 1960s (Rodrigue and Notteboom, 2012). According to Chin (2008), Weaver (2005) and Wood (2000), the cruise industry has become a symbol of globalization of the tourism industry in terms of its market coverage, its practices and the mobility of its assets. MacNeill and Wozniak (2018) emphasis the importance of the cruise industry by considering it one of the major growth areas of international tourism.

With a contribution of 10.4% of the Gross World Product, and serving one in every 10 jobs in 2017, the tourism sector has an important role to play in economies of the world (WTTC, 2018). In comparison with other tourism products, cruise tourism is a specific type of tourism which has experienced significant growth around the world throughout the last decade. Cruise tourism demand is increasing in particular in Europe, which is growing from 4.05 million in 2007 to 6.96 million in 2017 for the number of people who buy a cruise holiday (UNWTO, 2018: CLIA, 2018).

From its origins in the Caribbean (Wood 2000), the geography of the cruising has increased to cover the most part of the world. As emphasized by Rodrigue & Notteboom (2013,p. 13) that cruising 'has become the salient symbol of the globalization of the tourist industry in terms of its coverage, its practices and the mobility of its assets.' Today, the major cruise markets are the Caribbean, the Mediterranean, Europe, Alaska, Trans-Canal, West USA, Hawaii, and South America. Europe is an important destination for cruisers growing its market share on the world map.

According to new figures published by Cruise Lines International Association (CLIA) in the European Economic Contribution Report, the cruise industry made a significant contribution with € 47.86 billion to the European Economy in the year 2017. The cruise industry is still making significant contributions to the European Economy. The positive economic advantages are apparent, as cruise ships continue to play a major part in the European economy through smart sustainable growth. This represents an increase of 16.9% compared with the previous figure in 2015.

This is thanks to more European cruise passengers choosing a cruise holiday, more cruise passengers traveling in Europe, and the construction of more cruise ships in the EU shipyards. All of this represents significant economic advantages for the entire region. In 2015 and 2017, over 43,000 new jobs were generated by the cruise industry throughout Europe, with 40,621 currently working for cruise and shipping companies. Salaries and other incentives to European workers totaled € 12,77 billion, while the direct expenditure derived by the cruise industry reached € 19,70 billion last year, up from 16,9 billion € in 2015.

The European cruise industry, in general, is quite developed, with a total operating income of about €32,2 billion and approximately €14,2 billion in direct costs, resulting from passengers and cruise companies. Specifically the Mediterranean, which along with the Caribbean comprises the two main cruise destinations with over the 60% of cruise passengers worldwide. In this respect Greece is placed in a developed market, facing direct competition from countries in the Eastern Mediterranean and the Black Sea.

Europe remains a vibrant hub for cruising, as the international cruise industry continues to develop and expand to new attractions. Three important factors support this trend:

- Europe secures its position as the second most popular cruise destination in the world. The study shows that in the year 2017, 6.50 million passengers embarked on their cruises from the European ports which are 6.1 percent higher than in 2015.
- Europe is the second-largest passenger source market globally –in 2017 only, 6.96 million Europeans took a cruise vacation which shows a growth of 7.8% as compared to 2015.
- European shipyards are the center of the global cruise shipbuilding industry and continue to build the most inventive and largest ships in the world.

In 2017, cruise lines expended € 5.6 billion on European shipyards, a 22.4 percent increase from 2015. 66 cruise ships currently operate on European shipyard order books for delivery by 2021, totaling over € 29.4 billion in the economic value of European shipyards. This contribution stems from the spectacular growth that the cruise sector experienced in 2017, with 26.7 million passengers at ocean cruises worldwide. There are a number of cruise ports in Europe in different regions. The top ten leading cruise ports in the Mediterranean are in Balcelona, Civitavecchia, Venice, Palma Majorca, Marseille, Naples, Piraeus, Savona, Genoa and Dubrovnik (CLIA, 2015).

According to the European Cruise Council (2017) and Cruise Lines International Association (CLIA, 2018), 40 cruise lines were operated by 137 cruise ships operating in Europe with a capacity of approximately 164,000 lower berths in 2017. In Europe, 23 non-European lines have deployed another 75 vessels, with a capacity of around 95,000 lower berths. In the Mediterranean region, Baltic and other European regions, the vast majority of these cruises visited ports generating 34.10 million passenger visits, and in a total of about 260 European port cities which shows 9.4 percent increase from 2015. Moreover, an estimated 16.8 million crew is recorded at European ports.

Cruise tourism has grown considerably in economic significance in recent years, however, still a lack of clarity of real economic value. Though there are some studies which tried to evaluate the impact of cruise tourism, a scarcity of descriptive research is visible to examine the economic impact of the cruise industry on European economies. Hence, this study employed an investigative and comparative approach to analyze the cruise industry and cruise tourism in Europe and the way it economically impacts the European economies.

This chapter includes an analysis of the cruise industry and more specifically of the European Cruise Market. The study employed a descriptive approach based on the extensive review of the literature. Descriptive research is defined as research which describes the situation, subject, behavior or phenomenon. used to study an unclear problem. It leads to a better understanding of the present problem but not make accurate predictions, and they do not determine cause and effect. This study relies on secondary data such as survey reports, documentaries, government census reports and published articles. Section 2 refers to the background of the cruise tourism, section 3 talks about the economic impact of the European cruise sector and the section 4 discusses the global economic impact of cruise tourism. Continuously, section 5 analyzes the economic impact of cruise industry on European economies, section 6 describes the five-year growth trend of European cruise industry and section 7 concludes.

2. BACKGROUND

Cruise tourism indicates a vacation trip by cruise ships, known as floating resorts addressed to leisure (Petrick & Durko, 2016). According to the Research Centre for Coastal Tourism (2012, p. 3), cruise tourism is a "luxurious form of travelling, involving an all-inclusive holiday on a cruise ship of at least 48 hours, according to a specific itinerary, in which the cruise ship calls at several ports or cities". Thus, cruise tourism brings together many holiday package options to choose between a minimum of two days and maximum to several months (Marti, 2004). Tourism with a cruise ship is a trip which lasts a minimum of 60 hours by a ship mainly for pleasure including all meals and accommodation (Kizielewicz, 2013). Cruise tourism is the fastest grwoing sector in tourism experiencing a stable gwroth globally during the last three decades with North America and Europe are becoming leaders the market. (Petrick & Durko, 2016; Masiero, 2016).

Tourism can lead to improvements in economic, social, cultural and environmental aspects of community development according to the United Nations World Tourism Organization (2016, 2017). Many social, economic and environmental issues are raised in relation to the growth of the cruise tourism sector, while cruise tourism earns an admirable international demand (Weeden et al. (2011); Vayá et al. (2016); Van et al. (2017). Many concerns are raised from the economic angle on diverse issues such as port user fees, and the construction and maintenance of terminal facilities (Klein, 2011). Since the arrival of mega cruise ships with a capacity of over 5,000 passengers, the sociao-economic and environmental sustainability of the cruise tourism becomes an important concern in destinations like Venice and Barcelona (De Cantis, 2016).

Tourism deveelopment can be measured not only through the economic impacts but also through the evaluation of multidimensional community assets, according to some theorists (Alkire, 2002). In order to measure the impacts on tourism, some studies explore passenger surveys tin an attempt to estimate the expenses of the cruisers compared to terrestrial tourists (Brida & Zapata, 2010a; Brida et al., 2012; Larsen et al., 2013). Other researchers exploti data from the national income accounts and other industry reports by implementing different models to estimate the multiplier effects of tourists' expenditures on the national income (Brida & Zapata, 2010b; Dwyer, Ngaire, & Zelco, 2004; Marques & Cruse, 2015; McKee & Chase, 2003). The above mentioned studeis find that economic multipliers are low with cruise tourism because the passenfers are eating, drinking, sleeping and spenfind money on board a cruise ship with credit cards and most cruise ships are registered to tax-haven ports.

3. ECONOMIC IMPACT OF THE EUROPEAN CRUISE SECTOR

There is widespread recognition of the economic importance of tourism. The tourism industry is highly dynamic despite its maturity, as shown in recent years by the European exponential boom in cruise tourism. The European cruise industry is defined as those cruise-related activities that take place within Europe including cruise itineraries that visit European ports and destinations and also directly impact businesses and individuals located in Europe. It is broadly defined to include cruise lines and their employees; the direct suppliers to the cruise lines, such as wholesale distributors, stevedoring firms, and financial and business service providers, such as insurers and consultants; shipyards; and cruise passengers.

Nowadays, cruising is completely different from the original activity back at the 1970s (Bull, 1996). Douglas and Douglas (2004) and MacNeill and Wozniak (2018) mention that cruise is an interesting case of a complex globalized industry in an environment of international competition, capital mobility and labor migration. Bull (1996) indicates that the structure of the cruise market can be considered either as an oligopoly or as a monopolistic competition (Chang, Park, Liu & Roh, 2016). Ashcroft (2005) divides the cruise market into three groups of companies: Carnival Group (43%), Royal Caribbean International (20%) and Star Cruises (8%).

According to the European Cruise Council (2010), a unique characteristic of the cruise industry is that its impacts, whether they are positive or negative, affect not only one destination but the whole cruise society. Cruise industry faces huge income flows for hosting countries yearly, of which €14.5 billion have as the final recipient the European economies. Edgell (2016) highlighted that the effect on ports growth owing to the elevated financial impact is resulting from the expenditure of cruisers and crews, shore-side employees (tour operations), port duties and maintenance.

4. GLOBAL ECONOMIC IMPACT OF CRUISE TOURISM

The boom in cruise tourism, being a highly dynamic sector, has led many countries to consider cruise transport as a key product of their tourism development in the recent decade. It has a significant global economic impact worldwide. For instance, as reported by CLIA (2017b), the total worldwide output from the tourism industry is $134 billion in the year 2017. Besides, it provides jobs to 1,108,676 people all around the world. Moreover, it offered $45.6 billion in wages and salaries and entertains $26.7 million passengers. The USA is the main contributor of the cruise tourism industry as 11.9 million passengers come from USA. China comes second with 2.4 million passengers and Germany third with 2.19 million passengers (CLIA, 2017b).

The economic impact of the cruise industry, many researchers suggest a traditional methodology in impact studies through the quantification of three types of effects such as direct impact, indirect impact and induced impact (Brida, & Zapata Aguirre, 2009; Murillo et al., 2013). The economic impact of cruise tourism is divided into two segments such as direct economic impact and total economic impact.

4.1 Direct Economic Impact

The direct economic impact involves the production, jobs, and compensation for employees derived by European companies providing goods and services for cruise lines and to the passengers and crew. Besides, the direct impact contains the compensation remunerated to European cruise lines employees. According to Vayá et al. (2016), the direct impact is the initial expenditure by three agents engaged in cruising: carriers, cruise passengers and crew. The initial direct costs of shipping companies comprise all of the goods and services required for cruise ships to dock at a port.

Expenses included in this category are: services provided by shipping agents, cruise terminal services (baggage, safety, handling, chickening, etc.); services rendered by the Port Authority of Barcelona (including taxes and port charges); marine pilotage and ships mooring and unmooring— collection and treatment of waste technical services; fuel services; food, drink and other services. Initial direct cruise passenger expense includes travel and other cultural activities, accommodation (hotels, hostels and resort accommodation), cost (restaurants and cafes), various shopping activities (remembrances, clothing and footwear etc.) ; in-house city transports (including airport / train to Port transfers) ; shopping activities (souvenirs, clothing and footwear etc.). Lastly, direct expenses of the city's crew include: cost (restaurants and cafés), various shopping (souvenirs, footwear, and clothes, etc.), and intra-city transport. The direct impacts on businesses affected by the cruise industry have an immediate impact. An increase in cruise ships would raise port dues paid in the port of call by cruise lines. Also, when some cruise ships land in a destination, their expenses have a direct impact on business sales (Styliadis 2012).

4.2 Indirect Economic Impact

The indirect impact regarded as the effect on other sectors of the economy created by the goods and services that are required for companies receiving direct spendings, such as the purchase of a set of goods (such as textiles, food, etc.) and services (cleaning, transport, etc.) for hotel passenger accommodation. Likewise, mooring enterprises and pilot boats require a range of goods and services to operate at the port based on the cruise enterprises, which require goods and services to develop their activities and so on for the "second-order" provider. The purchases of direct supplies result in indirect effects (Brida & Zapata, 2009).

4.3 Induced Economic Impact

The induced impact is that of consumer spending on jobs created directly and indirectly (in cruise activities), which are owed to cruise shipping in the Port of Barcelona by people occupying such jobs directly or indirectly. These people receive salary income allocated in part to consuming goods and services in their place of residence and employment (after tax deduction, contributions and saving) which therefore reactivates a chain of intersectoral relations that results in an increase in turnover in various economic sectors. Induced effects arise from the expenditures of direct and indirect recipients produced from their increased incomes (Brida & Zapata, 2009).

The economic impact of the cruise is multiple-characters (Stefanidaki & Lekakou, 2012). The total cruising industry economic impact can be calculated by summing up the direct and secondary (indirect and induced) effects. The economic impact (EI) from cruise passengers' spending is calculated as follows (Stefanidaki and Lekakou, 2012):

$$EI = N*AS*M$$

Where:

- EI is the economic impact
- N is the number of cruise passengers;
- AS is the average spending;
- M is the appropriate multiplier

5. ECONOMIC IMPACT OF CRUISE INDUSTRY ON EUROPEAN ECONOMIES

In the year 2017, direct investment in cruise ship construction and rehabilitation of existing ships with European yards was € 19.70 billion including €5,63 billion, while expenses for European companies for goods and services supported cruiser operations by cruise lines amounted to € 8.17 billion (CLIA, 2018a). The European admin staff and crew of the cruise lines were paid € 4.23 billion of cruise passengers ' expenditure and €1.67 billion of wages and salaries as well as the other benefit (CLIA, 2018a).

The cruise industry's contribution to the national economy depends on the level of expenditure realized by the producers and consumers of the cruise product (Wilkinson, 1999). MacNeill and Wozniak (2018) argue that the economic impact of cruise tourism is spread to the whole economy via an income multiplier effect. According to Gouveia and Eusébio (2018), Someone ought to distinguish whether the port is a home port or a stopover port in order to identify economic impact flows.

Chen (2016) states that someone must consider whether a cruise company is of a foreign interest or whether national enterprises are developing their business. This segregation is of huge consideration due to the categories of expenditure and the pertinent financial consequences can vary (Stefanidaki & Lekakou, 2012). As highlighted by Dwyer, Forsyth and Spurr, (2004), mainland ports and small island ports might be the major categorization.

On the other hand, Dwyer, Pham, Forsyth and Spurr, (2014) argues that in response to a given change in marketing costs, the extent to which tourism numbers increase depends on marketing elasticity of demand. They describe further that marketing return measures can inform private and public sector organizations on the effective allocation of marketing resources and on the kinds of tourism development that best meet the targets of the operator and the target manager. In the same vein, Lundberg (1990) shed light on the six phases of tourism development such as rapid growth, awareness of problems, short-run success, tourism recession, difficulties, and reflection.

These expenses have created jobs and compensation for workers in a variety of industries and practically any country in which passengers or cruise ships have received calls. 195,241 direct jobs were created with direct expenses of € 19.70 billion and employee compensation amounted to € 6,023 billion (CLIA, 2018b).

Out of 100%, 79% of the direct economic impacts of the European cruise industry accounted for the following three economic sectors:

Manufacturing industry caused 49% of direct spending in the cruise business, 27% of direct jobs and 34% of direct compensation to staff, all of which shows an increasing trend from 2015 as a result of increased growth in the shipbuilding industry.

European cruise lines ' employees made up of 35% and 28% of the compensation for direct jobs created by cruise companies, which were slightly lower than in 2015.

The Transportation and Utility sector, including tour operators and travel agents among others, accounted for 22 percent of the direct expenditures, 14 percent of the direct jobs and 16 percent of the compensation impacts.

The total economic impacts represent the total of direct, indirect, and induced impact. The direct impact is due to the expenditure on the goods and services required for the cruise sector by directly linked companies. The induced impacts result in spending by the impacted employees for the goods and services belongs to the household. The indirect impacts therefore primarily affect business-to-business companies, whilst the induced impact mainly affects consumer firms.

There have been widespread economic impacts on cruise tourism across Europe. However, the greater part of these impacts are largely belonged to five countries, representing approximately 79 percent of the impact of the cruise industry across Europe. Among them, 63% of the cruise industry's direct spending was done by Italy, the United Kingdom, and Germany. The cumulative increases in direct expenditure for these three countries were 16 percent from 2015. The direct expenditures by the top five countries for 2017 were €15,613 with a growth of 17.2% compared with 2015 (CLIA, 2018b).

6. FIVE-YEAR GROWTH TREND OF EUROPEAN CRUISE INDUSTRY

European passengers grew by 13 percent since 2012 from 6,14 million passengers to 6,96 million passengers in 2017, an average increase of about 2,5 percent a year over the last five years. Embarkations at European ports have increased by almost 13% in the 5-year period, from 5.77 million in 2012 to 6.5 million in 2017. European ports have increased by 6.1% in 2017 and have reached a new high. In the period of 2012-17, port call visits have increased by 19% from 28.69 million passengers to 34.15 million and port call visitors increased by 9.6% compared with 2015 (CLIA, 2018b).

In terms of expenditures, the average annual growth over the five-year period was 4.9 percent, up by 27%, from € 15.5 billion in 2012 to $19.7 billion in 2017 (CLIA 2018b)

Over the five year period, the general direct expenditure in the cruise industry steadily increased and the growth in category spending varied, with a total increase of 24 percent over the last five years, averaging an average of 4.4 percent per year. The total output of the industries affected by the direct, indirect and induced impacts of the European cruise industry. The overall output in industries affected by the impact of the European cruise industry directly, indirectly and induced has risen by 26 percent, up from € 37.86 billion in 2012, to € 47.86 billion in 2017. y has risen by 26 percent from €37.86 billion in 2012 to €47.86 billion in 2017. This increase in total output mirrors the growth in direct expenditures reported above.

In terms of employment trend, 2012 – 2017, total employment related to the above-mentioned total output of cruise industry in Europe has increased by 23 percent, for instance, from 326,9000 jobs in 2012 to 403,600 in 2017; since 2012, it grew annually to an average of 4,3 percent annually.

The cruise industry in Europe is a dynamic source of economic activity providing economic benefits to virtually all industries and countries throughout Europe. Over 6 million European residents booked cruises in 2011, a 9.0% increase over 2010. In 2011, Europeans represented nearly 30% of all cruise passengers worldwide, compared with 22% ten years earlier. The vast majority visited ports in the Mediterranean, the Baltic and other European regions and generated 27.8 million passenger visits during 2011, a 9.7% increase over 2010 (European Cruise Council, 2011).

Over the five-year period from 2012 to 2016, 24 cruise vessels have been scheduled for delivery for worldwide trading with capacity for 67,325 passengers. Out of the total, 13 ships with 30,375 berths (45.1%) are primarily for the European source market, representing investment of €9.5 billion; two of these ships with capacity for 6,500 passengers are to be built in Japan. Many of the others will visit European destinations (European Cruise Council, 2011).

An expenditure breakdown can conveniently be divided into passenger expenses and operator expenses. Passenger expenditures include the cost of the cruise, but this is an amount paid to a foreign company, and it is not included. Cruise related expenditures include the add-on expenditure (accommodation, meals, shopping, excursions), the port expenditures (meals, excursions and travel, shopping), the operator expenditures, crew expenditures, ship maintenance (Dwyer & Forsyth, 1998).

According to Lekakou et al. (2009) the Mediterranean market reached 3.8 million passengers in 2008 and is divided into three main areas: Western, Eastern and Adriatic. Bartolome et al. (2009) say that the Western Mediterranean section attains most of the Mediterranean cruise demand of which the majority is of European origin. The ports in the Mediterranean involved with cruise are located in different countries, such as Spain, France, Italy, and Greece.

Cruise ports are generally classified into three categories based on passenger embarkation and disembarkation volume shares: transit ports, homeports and hybrid ports (Lekakou et al., 2010). Bagis and

Dooms (2012) say that homeports are the crucial link of the itinerary system as the continuity of the system depends on the performance of the homeport.

In 2012 a total of 168 cruise ships were active in Mediterranean waters with a capacity of 221,214 lower berths with an average of 1,317 berths per ship. Collectively these ships carried a potential 3.78 million passengers on 2,650 cruises, offering a total capacity of 30.74 million pax-nights, giving an average cruise length of 8.1 nights. A further 480,000 potential passengers cruised the Atlantic Isles. In 2012, North American operators deployed 59 ships with 83,229 lower berths in the Mediterranean, including some ships targeted at European markets. In comparison, European domiciled lines operated 107 vessels, which offered 135,336 lower berths (Cruise Lines International Association, 2013).

The Mediterranean Sea is an area with great success concerning the cruise industry and it emerged to be the world's second most popular cruising area behind the Caribbean. In 2012 the Mediterranean accounted for almost 20% of the global market. Each year of the past decade (with the single exception of 2002) the Mediterranean area has faced a positive variation on the previous one reaching 24.2 million (Risposte Turismo elaborations based on CLIA Market Overview (various editions) (2013) in MedCruise Yearbook 2013/14).

The passenger movement in the Mediterranean reaches its peak in August with more than 3.4 million passengers, while for cruise calls, October did not pass the mark of 2,000 cruise calls in 2012 (MedCruise Yearbook 2013/14).

Many of the leading ports are regarded as "must see" or "marquee" destinations that destination planners will wish to include in their itineraries. Other ports, some of which are also marquee ports in their own right, have advantages of strategic position, access to major hub airports and suitable bed-stock, enabling them to feature prominently as Home Ports. The leading Mediterranean ports of 2012 were Civitavecchia, Barcelona, Venice, Piraeus, Maples Livorno, Palma Majorca, Dubrovnik, Santorini and Savona (CLIA, 2013).

The market share of the Mediterranean increased into 18.25% in 2009 comparatively to 2000 (11.65%) (Torbianelli, 2010). The Mediterranean cruise market holds more than 21% of the total cruise market, being the second largest cruise market worldwide after the Caribbean. Pastena (2011) says that the Mediterranean cruise market presents a huge potential for growth, which it could led to serious reclassifications in the cruise geography. In 2011 the Mediterranean attracted almost 5.6 million passengers comparatively to 2000 (2.06 million).

The Mediterranean is the world's second most popular cruising arena, behind the Caribbean. In 2010 it accounted for 18% of the global cruise market representing 18,538 bed days. In Northern Europe In 2011 a total of 102 cruise ships were active in Northern European waters with a capacity of 111,188 lower berths with an average of 1,090 berths per ship. These carried a potential of 1.26 million passengers on 1,051 cruises, offering a total capacity of 11.18 million pax-nights, giving an average cruise length of 8.9 nights.

According to the European Cruise Council (2011) the Northern European market grew by around 15% in 2011 and is expected to expand further in 2012 to around 13.2 million pax-nights and to continue to grow in 2013. In 2011, North American operators deployed 29 ships, with 38,153 lower berths in Northern Europe. European domiciled cruise lines operated 59 vessels with 71,216 lower berths. The balance was largely made up of niche market ships visiting the polar-regions.

The Baltic is the largest segment in the Northern Europe market, generating about 3.6 million passenger port visits in 2011 and around 4 million in 2012. Many of the leading European ports are regarded as "must see" or "marquee" destinations that destination planners will wish to include in their itineraries. Other ports, some of which are also marquee ports in their own right, have advantages of strategic position, access to major hub airports and suitable bed-stock, enabling them to feature prominently as Home Ports.

7. CONCLUSION

This research developed a European Cruise Industry assessment framework. This framework applies primarily to the European cruise market, the cruise industry's significance and effect. This article seeks to determine a cruise's financial effect on the European strategy. Regarding financial spending, passenger preferences and also cruise ship orders, the European cruise industry is being examined.

Analyzing the economic impacts, it was found that there was an increase in overall jobs in the European cruise sector. (from € 187.252 billion in 2005 to € 315.5 billion in 2011) derived from an increase in overall economic benefit (from € 37.86 billion in 2012, to € 47.86 billion in 2017). Furthermore, it is confirmed that financial development in the top five European nations (Italy, UK, Germany, Spain and France) was 3.3% compared to the prior year. Moreover, fresh vessels are in order for 2014, showing Europe's cruise industry development.

In recent years, the European cruise industry has maintained a strong economic trend with a further performance, increasing its participation towards the economies of Europe by 3 percent. Cruise tourists are likely to have an impact on a country economy that is similar to that of other foreign visitors in many respects (Vayá et al., 2018).

The foreign exchange income and the economic impact on the host economy will rely on visitors expenditure, directly and indirectly, in relation to goods and services originating in the region or from elsewhere (ECC 2011).

The number of direct and indirect jobs generated by the cruise sector was over 403621 compared with some 315000 five years ago. The overwhelming majority of the world's cruise ships are built by European shipyards (up to 2021 they are scheduled to deliver 7 new major vessels with a combined capacity of 23000 passengers). The number of people who chose a cruise holiday in Europe has more than doubled in the past decade up to 26.7 Million (CLIA, 2018).

Generally, cruise is making inroads in the tourism sector although just 2% of the total tourist market took a cruise in 2011. There seems a hopeful future for cruise in Europe, as there are more jobs offered and more and more tourists prefer to experience the cruise comparatively to the previous years. This study gives an overview of the situation in the global and, in particular, European cruise industry in the contemporary globalized era and suggests that further studies can assess and compare the impact of each cruise line's initiatives and country's specific role with regards to cruise industry. Emperical studies in this context would be helpful to understand the phenoman.

REFERENCES

Alkire, S. (2002). Dimensions of Human Development. *World Development*, *30*(2), 181–205. doi:10.1016/S0305-750X(01)00109-7

Ashcroft, C. (2005). Overview of the global cruise industry. In European Cruise Council (Eds.), The ECC 2004/2005 annual review. Ashcroft & Associates Ltd.

Bagis, O., & Dooms, M. (Eds.). (2013). *Turkey's potential on becoming a cruise hub for the East Mediterranean Region: The case of Istanbul: Proceedings of the International Association of Maritime Economists (IAME) Conference*. Academic Press.

Bartolome, A., McAleer, M., Ramos, V., & Rey-Maquieira, J. (2009). *Cruising is risky business.* CIRJE Discussion paper. Retrieved from http://www.cirje.e.u-tokyo.ac.jp/research/dp/2009/2009cf664.pdf

Briassoulis, H. (1991). Methodological issues. Tourism input–output analysis. *Annals of Tourism Research*, *18*(3), 485–495. doi:10.1016/0160-7383(91)90054-F

Brida, J., Bukstein, D., Carrido, N., & Tealde, E. (2012). Cruise Passengers' Expenditure in the Caribbean Port of Call of Cartagena de Indias: A Cross-Section Data Analysis. *Tourism Economics*, *18*(2), 431–447. doi:10.5367/te.2012.0115

Brida, J., & Zapata, S. (2008). *The Impacts of the Cruise Industry on Tourism Destinations*. Research Unit on Sustainable Development. University of Milano Bicocca University of Bolzano.

Brida, J., & Zapata, S. (2009). Cruise Tourism: Economic, Socio-Cultural and Environmental Impacts. *International Journal of Leisure and Tourism Marketing*, *1*(3), 205–226. doi:10.1504/IJLTM.2010.029585

Brida, J., & Zapata, S. (2010a). Economic Impacts of Cruise Tourism: The Case of Costa Rica. *Anatolia.*, *21*(2), 322–338. doi:10.1080/13032917.2010.9687106

Brida, J., & Zapata, S. (2010b). Cruise tourism: Economic, socio-cultural and environmental impacts. *International Journal of Leisure and Tourism Marketing*, *1*(3), 205–226. doi:10.1504/IJLTM.2010.029585

Bull, O. A. (1996). The economics of cruising: An application to the short ocean cruise market. *Journal of Tourism Studies*, *7*(2), 28–35.

Chin, C. B. N. (2008). *Cruising in the global economy: Profits, pleasure and work at sea.* Ashgate.

Cruise Lines International Association (CLIA). (2011). *Cruise Market Profile Study.* Retrieved from https://www.slideshare.net/cruiseship/cruise-market-profile-clia-2011

De Cantis, S., Ferrante, M., Kahani, A., & Shoval, N. (2016). Cruise passengers' behavior at the destination: Investigation using GPS technology. *Tourism Management*, *52*, 133–150. doi:10.1016/j.tourman.2015.06.018

Diakomichalis, M. (2007). Greek Maritime Tourism: Evolution, Structures and Prospects. *Research in Transportation Economics*, *21*, 419–455. doi:10.1016/S0739-8859(07)21013-3

Douglas, N., & Douglas, N. (2004). Cruise ship passenger spending patterns in Pacific island ports. *International Journal of Tourism Research*, *6*(4), 251–261. doi:10.1002/jtr.486

Dwyer, L., Douglas, N., & Livaic, Z. (2004). Estimating the economic contribution of a cruise ship visit. *Tourism in Marine Environments, 1*(1), 5–16. doi:10.3727/154427304774865841

Dwyer, L., & Forsyth, P. (1998). Economic Significance of Cruise Tourism. *Annals of Tourism Research, 25*(2), 393–415. doi:10.1016/S0160-7383(97)00098-4

Dwyer, L., Forsyth, P., & Spurr, R. (2004). Evaluating tourism's effects: New and old approaches. *Tourism Management, 25*(3), 307–317. doi:10.1016/S0261-5177(03)00131-6

Ellis, C., & Kriwoken, L. K. (2006). Off the beaten track: A case study of expedition cruise ships in south-west Tasmania, Australia. In R. K. Dowling (Ed.), *Cruise Ship Tourism* (pp. 251–258). CABI. doi:10.1079/9781845930486.0251

European Cruise Council. (2010). *European Cruise Council 2010/2011 Report*. European Cruise Council. Retrieved from http://www.ashcroftandassociates.com/images/ECC-LR.pdf

European Cruise Council (ECC). (2011). *The Cruise Industry: A Leader in Europe's Economic Recovery*. Retrieved from https://www.cruise-norway.no/viewfile.aspx?id=3186

Hobson, J. S. P. (1993). Analysis of the US Cruise Line Industry. *Tourism Management, 13*(6), 453–462. doi:10.1016/0261-5177(93)90098-6

Kizielewicz, J. (2013). Cruise ship tourism – a case study in Poland. *Zesz. Nauk., 35*(107), 65–75.

Klein, R. A. (2011). Responsible cruise tourism: Issues of cruise tourism and sustainability. *Journal of Hospitality and Tourism Management, 18*(1), 107–116. doi:10.1375/jhtm.18.1.107

Kwak, S. J., Yoo, S. H., & Chang, J. I. (2005). The role of the maritime industry in the Korean national economy: An input–output analysis. *Marine Policy, 29*(4), 371–383. doi:10.1016/j.marpol.2004.06.004

Larsen, S., Wolff, K., Marnburg, E., & Øgaard, T. (2013). Belly full, purse closed: Cruise line passengers' expenditures. *Tourism Management Perspectives, 6*, 142–148. doi:10.1016/j.tmp.2013.02.002

Lekakou, M. B., Pallis, A. A., & Vaggelas, G. K. (2009). *The potential of Piraeus as a major Mediterranean cruise home-port: A selection criteria analysis*. 4th International Conference Planning for the Future – Learning from the Past: Contemporary Developments in Tourism, Travel & Hospitality, Rhodes, Greece.

Lekakou, M. B., Pallis, A. A., & Vaggelas, G. K. (2010). Which homeport in Europe: The cruise industry's selection criteria. *Tourismos: An International Multidisciplinary Journal of Tourism, 4*(4), 215–240.

Livas, P. C. (1994). *Input output analysis*. Stamoulis.

Lundberg, D. E. (1990). *The tourist business* (6th ed.). Van Nostrand-Reinhold.

Marques, B., & Cruse, R. (2015). Cruise tourism in Dominica: Benefits and beneficiaries. *ARA Journal of Tourism Research., 5*(2), 7–19.

Marti, B. E. (2004). Trends in world and extended-length cruising (1985–2002). *Marine Policy, 28*(3), 199–211. doi:10.1016/j.marpol.2003.09.004

Masiero, L. Transportation. In J. Jafari & H. Xiao (Eds.), *Encyclopedia of Tourism* (pp. 965–967). Springer International Publishing.

McKee, D. L., & Chase, G. L. (2003). The economic impact of cruise tourism on Jamaica. *Journal of Tourism Studies*, *14*(2), 16–22.

MedCruise Yearbook 2013/14. (2014). *A Directory of Cruise Ports and Professionals in the Mediterranean and Adjoining Seas*. Academic Press.

Pastena, L. (2011). *Modernizing the Med*. World Cruise Industry Review.

Peisley, T. (1992). *World Cruise Ship Industry in the 1990's*. Economist Intelligence Unit.

Petrick, J. F., & Durko, A. (2016). Cruise tourism. In J. Jafari & H. Xiao (Eds.), *Encyclopedia of Tourism* (pp. 206–285). Springer International Publishing. doi:10.1007/978-3-319-01384-8_42

Raguž, I. V., Perucic, D., & Pavlic, I. (2012). Organization and Implementation of Integrated Management System Processes - Cruise Port Dubrovnik. *International Review of Management and Marketing*, *2*(4), 199–209.

Research Centre for Coastal Tourism. (2012). *Cruise Tourism: from a Broad Perspective to a Focus on Zeeland*. Retrieved from http://www.kenniscentrumtoerisme.nl/l/library/download/13920

Rodrigue, J. P., & Notteboom, T. (2012). The geography of cruises: Itineraries, not destinations. *Applied Geograghy*, *38*, 31–42. doi:10.1016/j.apgeog.2012.11.011

Seidl, A., Guiliano, F., & Pratt, L. (2006). Cruise Tourism and Economic Developmet in Central America and the Caribbean: The case of Costa Rica. *Revista de Turismo y Patrimonio Cultural*, *4*(2), 213–224. doi:10.25145/j.pasos.2006.04.015

Sinclair, T. M., & Stabler, M. (1997). *The economics of tourism*. Routledge. doi:10.4324/9780203195437

Stefanidaki, E., & Lekakou, M. (2012). Generated Economic Impact on Cruise Destinations: The Piraeus Case. In *Cruise Tourism and Society* (pp. 69–84). Springer-Verlag.

Styliadis, T. (2012). *Greek Sea is not in Crisis. An institutional approach of the economic impact cruise ships have on local societies: the case of two Ionian Islands* (MSc Thesis). Erasmus University, Rotterdam.

Torbianelli, J. (2010). *Mediterranean Cruises as Drivers of Investment and Relations between Ports and Cities*. University of Trieste.

UNWTO. (2016). *Yearbook of tourism Statistics*. United Nations World Tourism Organization.

UNWTO. (2017). *UNWTO panel on indigenous Tourism: Promoting equitable partnerships*. Berlin: United Nations World Tourism Organization. Retrieved from: http://ethics.unwto.org/event/unwto-panel-indigenous-tourism-promotingequitable-partnerships

Vaggelas, G. (2011). Cruise Tourism: Economic Benefits, Sustainable Development and Port City. Plan Bleu Regional Seminar – Tourism and sustainable development in the Mediterranean, Genoa, Italy.

Van Bets, L. K., Lamers, M. A., & Van Tatenhove, J. P. (2017). Collective self-governance in a marine community: Expedition cruise tourism at Svalbard. *Journal of Sustainable Tourism, 25*(11), 1583–1599. doi:10.1080/09669582.2017.1291653

Vayá, E., Garcia-Sanchís, J. R., Murillo, J., Romaní, J., & Suriñach, J. (2016). *Economic Impact of Cruise Activity: the Port of Barcelona*. Retrieved from https://poseidon01.ssrn.com/delivery.php?ID=859095117066064115106126113118101112023069025084090022029080122123075072073 02506508504504 300701702504302 607010511306807101900 304600202703500009600411300608 01221160 97068050008064 10700700208700 210200212 5102087 0950 96 119064107016081071112111082008082125120&EXT=pdf

Vina, L., & Ford, J. (1998). Economic impact of proposed cruise ship business. *Annals of Tourism Research, 25*, 205–221.

Vogel, M. (2009). Onboard revenue: The secret of the cruise industry's success? In A. Papathanassis (Ed.), *Cruise sector growth: Managing emerging markets, human resources, processes and systems* (pp. 3–15). Gabler. doi:10.1007/978-3-8349-8346-6_1

Weaver, A. (2005). The McDonaldization thesis and cruise shipping. *Annals of Tourism Research, 32*(2), 346-366.

Weeden, C., Lester, J. A., & Thyne, M. (2011). Cruise tourism: Emerging issues and implications for a maturing industry. *Journal of Hospitality and Tourism Management, 18*(1), 26–29. doi:10.1375/jhtm.18.1.26

Wild, G. P. (International) Ltd., B. (2012). The Cruise Industry: Contribution of Cruise Tourism to the Economies of Europe. European Cruise Council.

Wild, P., & Dearing, J. (2000). Development of and prospects for cruising in Europe. *Maritime Policy & Management, 27*(4), 315–337. doi:10.1080/030888300416522

Wilkinson, P. (1999). Caribbean cruise tourism: Delusion? Illusion? *Tourism Geographies, 1*(3), 261–282. doi:10.1080/14616689908721321

Wood, R. E. (2000). Caribbean cruise tourism: Globalization at sea. *Annals of Tourism Research, 27*(2), 345–370. doi:10.1016/S0160-7383(99)00073-0

Zhou, D., Yanagida, F., Chakravorty, U., & Leung, P. (1997). Estimating the economic impacts from tourism. *Annals of Tourism Research, 24*(1), 76–89. doi:10.1016/S0160-7383(96)00035-7

Compilation of References

Aartsen, M., Cheval, B., Sieber, S., van der Linden, B., Gabriel, R., Courvoisier, D., Guessous, I., Burton-Jeangros, C., Blane, D., Ihle, A., Kliegel, M., & Cullati, S. (2019). Advantaged Socioeconomic Conditions in Childhood are Associated with Higher Cognitive Functioning but Stronger Cognitive Decline in Older Age. *Proceedings of the National Academy of Sciences of the United States of America*, *116*(12), 5478–5486. doi:10.1073/pnas.1807679116 PMID:30804194

Abduelkarem, A. R., Sackville, M. A., Morgan, R. M., Sackville, M. P., & Hildreth, A. J. (2003). An assessment of the level of type 2 diabetes patients' satisfaction with community pharmacists' services. *International Journal of Pharmacy Practice*, *11*(3), 161–168. doi:10.1211/0022357021990

Abdullah, L., & Najib, L. (2014). A new type-2 fuzzy set of linguistic variables for the fuzzy analytic hierarchy process. *Expert Systems with Applications*, *41*(7), 3297–3305. doi:10.1016/j.eswa.2013.11.028

Abdullah, L., & Zulkifli, N. (2015). Integration of fuzzy AHP and interval type-2 fuzzy DEMATEL: An application to human resource management. *Expert Systems with Applications*, *42*(9), 4397–4409. doi:10.1016/j.eswa.2015.01.021

Abolhosseini, S., Heshmati, A., & Altmann, J. (2020). *A Review of Renewable Energy Supply and Energy Efficiency Technologies*. IZA Institute of Labor Economics: https://www.iza.org/

Acciai, F., & Hardy, M. (2017). Depression in Later Life: A Closer Look at the Gender Gap. *Social Science Research*, *68*, 163–175. Advance online publication. doi:10.1016/j.ssresearch.2017.08.003 PMID:29108595

Acciai, M., Martins, J., Lipert, A., Ramos, M., & Peralta, M. (2018). Prevalence and Trends of Overweight and Obesity in Older Adults from 10 European Countries from 2005 to 2013. *Scandinavian Journal of Public Health*. Advance online publication. doi:10.1177/1403494818764810 PMID:29569525

Akash, B. A., Mamlook, R., & Mohsen, M. S. (1999). Multi-criteria selection of electric power plants using analytical hierarchy process. *Electric Power Systems Research*, *52*(1), 29–35.

Akay, D., & Atak, M. (2007). Grey prediction with rolling mechanism for electricity demand forecasting of Turkey. *Energy*, *32*(9), 1670–1675. doi:10.1016/j.energy.2006.11.014

Akbaba, A. (2006). Measuring service quality in the hotel industry: A study in a business hotel in Turkey. *International Journal of Hospitality Management*, *25*(2), 170–192. doi:10.1016/j.ijhm.2005.08.006

Akyildiz, B., Kadaifci, C., & Topcu, I. (2015). A decision framework proposal for customer order prioritization: A case study for a structural steel company. *International Journal of Production Economics*, *169*, 21–30. doi:10.1016/j.ijpe.2015.07.004

Alam, M. S., & Mondal, M. (2019). Assessment of sanitation service quality in urban slums of Khulna city based on SERVQUAL and AHP model: A case study of railway slum, Khulna, Bangladesh. *Journal of Urban Management*, *8*(1), 20–27. doi:10.1016/j.jum.2018.08.002

Al-Arifi, M. N. (2012). Patients' perception, views and satisfaction with pharmacists' role as health care provider in community pharmacy setting at Riyadh, Saudi Arabia. *Saudi Pharmaceutical Journal*, *20*(4), 323–330. doi:10.1016/j.jsps.2012.05.007 PMID:23960807

Alexander, G. J., Sharpe, W. F., & Bailey, J. V. (2003). *Fundamentals of investments/fundamentos de inversiones*. Pearson Educación.

Alhassan, R. K., Spieker, N., van Ostenberg, P., Ogink, A., Nketiah-Amponsah, E., & de Wit, T. F. R. (2013). Association between health worker motivation and healthcare quality efforts in Ghana. *Human Resources for Health*, *11*(1), 1–11. doi:10.1186/1478-4491-11-37 PMID:23945073

Alkire, S. (2002). Dimensions of Human Development. *World Development*, *30*(2), 181–205. doi:10.1016/S0305-750X(01)00109-7

Amjady, N. (2006). Day-ahead price forecasting of electricity markets by a new fuzzy neural. *IEEE Transactions on Power Systems*, *21*(2), 887–896. doi:10.1109/TPWRS.2006.873409

Amjady, N., & Hemmati, M. (2006). Energy price forecasting - problems and proposals for such predictions. *IEEE Power & Energy Magazine*, *4*(2), 20–29. doi:10.1109/MPAE.2006.1597990

ANA-MPA. (2020). *Two migrants with Covid-19 in quarantine at Lesvos refugee facility, Migration Ministry confirms*. Retrieved from https://www.amna.gr/en/article/457131/Two-migrants-with-Covid-19-in-quarantine-at-Lesvos-refugee-facility--Migration-Ministry-confirms

Anderson, S. P., & Kreider, B. (1997). *Tax incidence in differentiated product oligopoly, mimeo*. University of Virginia.

Antunes, L. O., Gomes, J. J., & Cavaco, A. M. (2015). How pharmacist–patient communication determines pharmacy loyalty? Modeling relevant factors. *Research in Social & Administrative Pharmacy*, *11*(4), 560–570. doi:10.1016/j.sapharm.2014.11.003 PMID:25577547

Aouam, T., Geryl, K., Kumar, K., & Brahimi, N. (2018). Production planning with order acceptance and demand uncertainty. *Computers & Operations Research*, *91*, 145–159. doi:10.1016/j.cor.2017.11.013

Apergis, N., & Danuletiu, D. C. (2014). Renewable Energy and Economic Growth: Evidence from the Sign of Panel Long-Run Causality. *International Journal of Energy Economics and Policy*, *4*(4), 578–587.

Apergis, N., & Payne, J. E. (2012). The Electricity Consumption-Growth Nexus: Renewable Versus Non-Renewable Electricity in Central America. *Energy Sources. Part B, Economics, Planning, and Policy*, *7*(4), 423–431. doi:10.1080/15567249.2011.639336

Apergis, N., & Payne, J. E. (2014). Renewable Energy, Output, CO2 Emissions, and Fossil Fuel Prices in Central America: Evidence from a Nonlinear Panel Smooth Transition Vector Error Correction Model. *Energy Economics*, *42*, 226–232. doi:10.1016/j.eneco.2014.01.003

Ariff, M. S. M., Yun, L. O., Zakuan, N., & Jusoh, A. (2012). Examining dimensions of electronic service quality for internet banking services. *Procedia: Social and Behavioral Sciences*, *65*(3), 854–859. doi:10.1016/j.sbspro.2012.11.210

Arnau, R. C., & Thompson, B. (2000). Second Order Confirmatory Factor Analysis of the WAIS-III. *Assesssment*, *7*(3), 237–246. doi:10.1177/107319110000700304 PMID:11037391

Arneson, D., Jacobs, E. W., Scott, D. M., & Murray, W. J. (2011). Patronage Motives of Community Pharmacy Patrons. *Journal of Pharmaceutical Marketing & Management*, *4*(2), 3–22. doi:10.3109/J058v04n02_02

Arredondo, F., & Martinez, E. (2010). Learning and adaptation of a policy for dynamic order acceptance in make-to-order manufacturing. *Computers & Industrial Engineering*, *58*(1), 70–83. doi:10.1016/j.cie.2009.08.005

Arvizu, D., & Balaya, P. (2012). *Direct Solar Energy. In IPCC Special Report on Renewable Energy Sources and Climate Change Mitigation Special Report of the Intergovernmental Panel on Climate Change.* Cambridge University Press.

Ashcroft, C. (2005). Overview of the global cruise industry. In European Cruise Council (Eds.), The ECC 2004/2005 annual review. Ashcroft & Associates Ltd.

Asylum Information Database (AIDA)/European Council on Refugees and Exiles (ECRE). (2020a). *Conditions in Reception Facilities: Greece.* European Programme for Integration and Migration (EPIM). Retrieved from https://www.asylumineurope.org/reports/country/greece/reception-conditions/housing/conditions-reception-facilities

Asylum Information Database (AIDA)/European Council on Refugees and Exiles (ECRE). (2020b). *Health care: Greece.* European Programme for Integration and Migration (EPIM). Retrieved from https://www.asylumineurope.org/reports/country/greece/reception-conditions/health-care

Asylum Service. (2020a). *Statistical Data of the Greek Asylum Service (from 07.06.2013 to 29/02/2020).* Retrieved from http://asylo.gov.gr/en/wp-content/uploads/2020/03/Greek_Asylum_Service_data_February_2020_en.pdf

Asylum Service. (2020b). *Important notice of the Asylum Service- Reception and service of the publicis suspended. (in Greek)* Retrieved from http://asylo.gov.gr/wp-content/uploads/2020/03/ΑΝΑΚΟΙΝΩΣΗ-ΑΝΑΣΤΟΛΗΣ-ΥΠΟΔΟΧΗΣ-ΚΑΙ-ΕΞΥΠΗΡΕΤΗΣΗΣ-ΚΟΙΝΟΥ-ΕΛΛΗΝΙΚΑ-13-3-2020-1.pdf

Asylum Service. (2020c). *The suspension of reception and public service by the Asylum Service is extended until May 15. (in Greek)* Retrieved from http://asylo.gov.gr/wp-content/uploads/2020/04/Greek-Announcement-April-10_MK1.pdf

Asylum Service. (2020d). *Suspension of asylum applications. (in Greek)* Retrieved from http://asylo.gov.gr/wp-content/uploads/2020/03/pnp-anastolh-ths-ypovolis-aithseon-asylou.pdf

Athavale, A. S., Banaham, B. F., Bentley, J. P., & West-Strum, D. S. (2015). Antecedents and consequences of pharmacy loyalty behavior. *International Journal of Pharmaceutical and Healthcare Marketing*, *9*(1), 36–55. doi:10.1108/IJPHM-02-2013-0003

Awasthi, A., Chauhan, S. S., Omrani, H., & Panahi, A. (2011). A hybrid approach based on SERVQUAL and fuzzy TOPSIS for evaluating transportation service quality. *Computers & Industrial Engineering*, *61*(3), 637–646. doi:10.1016/j.cie.2011.04.019

Aydemir, E., Fevzi, B., & Gültekin, Ö. (2013) Gri sistem teorisi ve uygulamaları: bilimsel yazın taraması, *Süleyman Demirel Üniversitesi İktisadi ve İdari Bilimler Fakültesi Dergisi*, *18*(3), 188.

Aydın, İ. (2013). Balıkesir'de Rüzgar Enerjisi. *Eastern Geographical Review*, *18*(29), 29–50.

Ayvaz, B., & Kusakci, A. O. (2017). Electricity consumption forecasting for turkey with nonhomogeneous discrete grey model. *Energy Sources. Part B, Economics, Planning, and Policy*, *12*(3), 260–267. doi:10.1080/15567249.2015.1089337

Bader, J. D. (2009). Challenges in quality assessment of dental care. *The Journal of the American Dental Association*, *140*(12), 1456–1464. doi:10.14219/jada.archive.2009.0084 PMID:19955057

Bagis, O., & Dooms, M. (Eds.). (2013). *Turkey's potential on becoming a cruise hub for the East Mediterranean Region: The case of Istanbul: Proceedings of the International Association of Maritime Economists (IAME) Conference.* Academic Press.

Bahari, M. B., & Ling, Y. P. (2009). Factors contributing to customer satisfaction with community pharmacies in Malaysia. *Journal of Public Health, 18*(1), 35–41. doi:10.100710389-009-0280-7

Bahia, L. R., Araujo, D. V., Schaan, B. D., Dib, S. A., Negrato, C. A., Leão, M. P. S., & ... . (2011). *The Costs of Type 2 Diabetes Mellitus Outpatient Care in the Brazilian Public Health System*. Value Heal. doi:10.1016/j.jval.2011.05.009

Bahrami, S., Hooshmand, R. A., & Parastegari, M. (2014). Short term electric load forecasting by wavelet transform and grey model ımproved by pso (particle swarm optimization) algorithm. *Energy, 72*, 434–442. doi:10.1016/j.energy.2014.05.065

Baker, D. A., & Crompton, J. L. (2000). Quality, satisfaction, and behavioral intentions. *Annals of Tourism Research, 27*(3), 785–804. doi:10.1016/S0160-7383(99)00108-5

Balackandran, K., & Schaefer, M. E. (1981). Optimal acceptance of job orders. *International Journal of Production Research, 19*(2), 195–200. doi:10.1080/00207548108956641

Balcerzak, A. P., & Pietrzak, M. B. (2016). Application of TOPSIS method for analysis of sustainable development in European Union countries. *Chapters, 1*, 82–92.

Bal, H., Örkcü, H. H., & Çelebioğlu, S. (2010). Improving the Discrimination Power and Weights Dispersion in The Data Envelopment Analysis. *Computers & Operations Research, 37*(1), 99–107. doi:10.1016/j.cor.2009.03.028

Barroso, J. M. (2013). *Letter of President Barroso to the Members of the European Council*. MEMO/13/416. Retrieved January 3, 2020, https://europa.eu/rapid/press-release_MEMO-13-416_en.htm

Bartolome, A., McAleer, M., Ramos, V., & Rey-Maquieira, J. (2009). *Cruising is risky business*. CIRJE Discussion paper. Retrieved from http://www.cirje.e.u-tokyo.ac.jp/research/dp/2009/2009cf664.pdf

Barut, M., & Sridharan, V. (2005). Revenue management in order-driven production systems. *Decision Sciences, 36*(2), 287–316. doi:10.1111/j.1540-5414.2005.00074.x

Basfirinci, C., & Mitra, A. (2015). A cross cultural investigation of airlines service quality through integration of Servqual and the Kano model. *Journal of Air Transport Management, 42*, 239–248. doi:10.1016/j.jairtraman.2014.11.005

Basto-Pereira, M., Furtado, S. I. F., Silva, R. J. P., Fachado González, F., Vara Fernandes, T. M., Correia de Sousa, J., & Yaphe, J. (2015). Performance indicators for clinical practice management in primary care in Portugal: Consensus from a Delphi study. *The European Journal of General Practice, 21*(1), 52–57. doi:10.3109/13814788.2014.907268 PMID:24845034

Behzadian, M., Otaghsara, S. K., Yazdani, M., & Ignatius, J. (2012). A state-of the-art survey of TOPSIS applications. *Expert Systems with Applications, 39*(17), 13051–13069. doi:10.1016/j.eswa.2012.05.056

Belwal, R., & Amireh, M. (2018). Service quality and attitudinal loyalty: Consumers' perception of two major telecommunication companies in Oman. *The Arab Economics and Business Journal, 13*(2), 197–208. doi:10.1016/j.aebj.2018.11.003

Bennet, J. A. (1997). A case for theory triangulation. *Nursing Science Quarterly, 10*(2), 97–106. doi:10.1177/089431849701000210 PMID:9197723

Benzécri, J.-P. (1973). L'Analyse des Données.: Vol. II. *L'Analyse des Correspondances*. Dunod.

Bhattacharyya, R., Kar, S., & Majumder, D. D. (2011). Fuzzy mean–variance–skewness portfolio selection models by interval analysis. *Computers & Mathematics with Applications (Oxford, England), 61*(1), 126–137. doi:10.1016/j.camwa.2010.10.039

Bianco, V., Manca, O., Nardini, S., & Minea, A. A. (2010). Analysis and forecasting of nonresidential electricity consumption in Romania. *Applied Energy, 87*(11), 3584–3590. doi:10.1016/j.apenergy.2010.05.018

Bigerna, S., Bollino, C. A., & Micheli, S. (2015). *The Sustainability of Renewable Energy in Europe. İsviçre*. Springer International Publishing.

Bilbao-Terol, A., Pérez-Gladish, B., Arenas-Parra, M., & Rodríguez-Uría, M. V. (2006). Fuzzy compromise programming for portfolio selection. *Applied Mathematics and Computation, 173*(1), 251–264. doi:10.1016/j.amc.2005.04.003

Bock, F. (1958). An algorithm for solving travelling-salesman and related network optimization problems. *Operations Research, 6*(6), 897–897.

Boender, C. G. E., De Graan, J. G., & Lootsma, F. A. (1989). Multi-criteria decision analysis with fuzzy pairwise comparisons. *Fuzzy Sets and Systems, 29*(2), 133–143. doi:10.1016/0165-0114(89)90187-5

Bogdanovica, I., McNeill, A., Murray, R., & Britton, J. (2011). What Factors Influence Smoking Prevalence and Smoke Free Policy Enactment across the European Union Member States. *PMC PLoS One, Vol, 6*(8), e23889. doi:10.1371/journal.pone.0023889 PMID:21909375

Borrat-Besson, C., Ryser, V. A., & Gonçalves, J. (2015). *An evaluation of the CASP-12 scale used in the Survey of Ageing and Retirement in Europe (SHARE) to measure Quality of Life among people aged 50+* (No. 4). FORS Working Paper Series.

Boström, K. (2011). *Consumer behaviour of pharmacy customers. Choice of pharmacy and over-the-counter medicines* (Degree Thesis). Degree Program in International Business, Helsinki.

Boulos, M. N. K., Brewer, A. C., Karimkhani, C., Buller, D. B., & Dellavalle, R. P. (2014). Mobile medical and health apps: State of the art, concerns, regulatory control, and certification. *Online Journal of Public Health Informatics, 5*(3), 229. PMID:24683442

Bowen, D., & Clarke, J. (2002). Reflections on tourist satisfaction research: Past, present and future. *Journal of Vacation Marketing, 8*(4), 297–308. doi:10.1177/135676670200800401

Bozdağ, C. E., Kahraman, C., & Ruan, D. (2003). Fuzzy group decision making for selection among computer integrated manufacturing systems. *Computers in Industry, 51*(1), 13–29. doi:10.1016/S0166-3615(03)00029-0

BP. (2019). *BP Statistical Review of World Energy 2019*. London: BP.

Briassoulis, H. (1991). Methodological issues. Tourism input–output analysis. *Annals of Tourism Research, 18*(3), 485–495. doi:10.1016/0160-7383(91)90054-F

Brida, J., Bukstein, D., Carrido, N., & Tealde, E. (2012). Cruise Passengers' Expenditure in the Caribbean Port of Call of Cartagena de Indias: A Cross-Section Data Analysis. *Tourism Economics, 18*(2), 431–447. doi:10.5367/te.2012.0115

Brida, J., & Zapata, S. (2008). *The Impacts of the Cruise Industry on Tourism Destinations*. Research Unit on Sustainable Development. University of Milano Bicocca University of Bolzano.

Brida, J., & Zapata, S. (2009). Cruise Tourism: Economic, Socio-Cultural and Environmental Impacts. *International Journal of Leisure and Tourism Marketing, 1*(3), 205–226. doi:10.1504/IJLTM.2010.029585

Brida, J., & Zapata, S. (2010a). Economic Impacts of Cruise Tourism: The Case of Costa Rica. *Anatolia., 21*(2), 322–338. doi:10.1080/13032917.2010.9687106

Briesacher, B., & Corey, R. (1997). Patient satisfaction with pharmaceutical services at independent and chain pharmacies. *American Journal of Health-System Pharmacy, 54*(5), 531–536. doi:10.1093/ajhp/54.5.531 PMID:9066860

Bruneau, B.G., (2008). The developmental genetics of congenital heart disease. *Nature, 451*(7181), 943-8. Doi:10.1038/nature06801

Buckley, J. J. (1985). Fuzzy hierarchical analysis. *Fuzzy Sets and Systems*, *17*(3), 233–247. doi:10.1016/0165-0114(85)90090-9

Buckley, J. J., Feuring, T., & Hayashi, Y. (2001). Fuzzy hierarchical analysis revisited. *European Journal of Operational Research*, *129*(1), 48–64. doi:10.1016/S0377-2217(99)00405-1

Bukhari, O. M. (2019). (in press). Dental Caries Experience and Oral Health Related Quality of Life in Working Adults. *The Saudi Dental Journal*. Advance online publication. doi:10.1016/j.sdentj.2019.11.003

Bull, O. A. (1996). The economics of cruising: An application to the short ocean cruise market. *Journal of Tourism Studies*, *7*(2), 28–35.

Buntin, M. B., Burke, M. F., Hoaglin, M. C., & Blumenthal, D. (2011). The Benefits Of Health Information Technology: A Review Of The Recent Literature Shows Predominantly Positive Results. *Health Information Technology, 3*, 464-471.

Burden, F., & Winkler, D. (2008). Bayesian regularization of neural networks. *Methods in Molecular Biology (Clifton, N.J.)*, *458*, 25–44. doi:10.1007/978-1-60327-101-1_3 PMID:19065804

Bush, J., Langley, C. A., & Wilson, K. A. (2009). The corporization of community pharmacy: Implications for service provision, the public health function, and pharmacy's claims to professional status in the United Kingdom. *Research in Social & Administrative Pharmacy*, *5*(4), 305–331. doi:10.1016/j.sapharm.2009.01.003 PMID:19962674

Busse, R., Geissler, A., Quentin, W., & Wiley, M. (2011). *Diagnosis - Related Groups in Europe: Moving Towards Transparency*. Efficiency and Quality in Hospitals.

Buyukozkan, G., Kahraman, C., & Ruan, D. (2004). A fuzzy multi-criteria decision approach for software development strategy selection. *International Journal of General Systems*, *33*(2-3), 259–280. doi:10.1080/03081070310001633581

Byrne, B. M. (2005). Factor Analytic Models: Viewing the Structure of an Assessment Instrument From Three Perspectives. *Journal of Personality Assessment*, *85*(1), 17–32. doi:10.120715327752jpa8501_02 PMID:16083381

Byrne, M. J., Tickle, M., Glenny, A. M., Campbell, S., Goodwin, T., & O'Malley, L. (2019). A systematic review of quality measures used in primary care dentistry. *International Dental Journal*, *69*(4), 252–264. doi:10.1111/idj.12453 PMID:30430563

Byrne, M., & Tickle, M. (2019). Conceptualising a framework for improving quality in primary dental care. *British Dental Journal*, *227*(10), 865–868. doi:10.103841415-019-0900-8 PMID:31758120

Calosso, T., Cantamessa, M., Vu, D., & Villa, A. (2003). Production planning and order acceptance in business to business electronic commerce. *International Journal of Production Economics*, *85*(2), 233–249. doi:10.1016/S0925-5273(03)00112-9

Calthrop, E., & Proost, S. (1998). Road Transport Externalities. *Environmental and Resource Economics*, *11*(3/4), 335–348. doi:10.1023/A:1008267917001

Campbell, D. J., Manns, B. J., Weaver, R. G., Hemmelgarn, B. R., King-Shier, K. M., & Sanmartin, C. (2017). Financial barriers and adverse clinical outcomes among patients with cardiovascular-related chronic diseases: A cohort study. *BMC Medicine*, *15*(1), 33. doi:10.118612916-017-0788-6 PMID:28196524

Campbell, S. M., Roland, M. O., & Buetow, S. A. (2000). Defining quality of care. *Social Science & Medicine*, *51*(11), 1611–1625. doi:10.1016/S0277-9536(00)00057-5 PMID:11072882

Campbell, S., & Tickle, M. (2013). How do we improve quality in primary dental care? *British Dental Journal*, *215*(5), 239–243. doi:10.1038j.bdj.2013.831 PMID:24029991

Campbell, S., & Tickle, M. (2013). What is quality primary dental care? *British Dental Journal*, *215*(3), 135–139. doi:10.1038j.bdj.2013.740 PMID:23928610

Can, G. F., & Delice, E. K. (2020). An advanced human error assessment approach: HEART and AV-DEMATEL. *Human Factors and Ergonomics in Manufacturing & Service Industries*, *30*(1), 29–49. doi:10.1002/hfm.20819

Cansın, Y., & Sohtaoğlu, N. H. (2009). OECD/IEA Ülkelerinin Ar-Ge Haracamalarındaki Eğilimler Kapsamında Yenilenebilir Enerji Teknolojilerindeki Gelişmelerin İncelenmesi. In *Yenilenebilir Enerji Kaynakları Sempozyumu Bildiriler Kitabı* (pp. 29-36). Diyarbakır: Elektrik Mühendisleri Odası. http://www.emo.org.tr/ekler/0e3a6c01095e487_ek.pdf

Carde, S., Hullin, C. M., Chen, R., Schuler, T., Granz, J., Knaup, P., & .... (2007). Towards Sustainability of Health Information Systems: How Can We Define, Measure and Achieve It? *Studies in Health Technology and Informatics*. PMID:17911901

Carpio, C. (2006). *Two-constraints models of consumer demand: An application to the demand for agritourism in the United States* (PhD Thesis). North Carolina State University, Raleigh.

Carrion, M., Conejo, A., & Arroyo, J. (2007). Forward contracting and selling price determination for retailer. *IEEE Transactions on Power Systems*, *22*(4), 2105–2114. doi:10.1109/TPWRS.2007.907397

Castaldo, S., Grosso, M., Mallarini, E., & Rindone, M. (2016). The missing path to gain customers loyalty in pharmacy retail: The role of the store in developing satisfaction and trust. *Research in Social & Administrative Pharmacy*, *12*(5), 699–712. doi:10.1016/j.sapharm.2015.10.001 PMID:27555108

Catalao, J., Mariano, S., Mendes, V., & Ferreira, L. (2007). Short-term electricity prices forecasting in a competitive market: A neural network approach. *Electric Power Systems Research*, *77*(10), 1297–1304. doi:10.1016/j.epsr.2006.09.022

Celik, E., & Akyuz, E. (2018). An interval type-2 fuzzy AHP and TOPSIS methods for decision-making problems in maritime transportation engineering: The case of ship loader. *Ocean Engineering*, *155*, 371–381. doi:10.1016/j.oceaneng.2018.01.039

Celik, E., Gumus, A. T., & Alegoz, M. (2014). A trapezoidal type-2 fuzzy MCDM method to identify and evaluate critical success factors for humanitarian relief logistics management. *Journal of Intelligent & Fuzzy Systems*, *27*(6), 2847–2855. doi:10.3233/IFS-141246

Cerjan, M., Krželj, I., Vidak, M., & Delimar, M. (2013). A literature review with statistical analysis of electricity price forecasting methods. In I. Kuzle, T. Capuder, & H. Pandžić (Eds.), *Eurocon 2013* (pp. 756–763). IEEE. doi:10.1109/EUROCON.2013.6625068

Cerjan, M., Petricic, A., & Delimar, M. (2019). HIRA model for short-term electricity. *Energies*, *12*(3), 568–600. doi:10.3390/en12030568

Cesaret, B., Oğuz, C., & Salman, F. S. (2012). A tabu search algorithm for order acceptance and scheduling. *Computers & Operations Research*, *39*(6), 1197–1205. doi:10.1016/j.cor.2010.09.018

Cevik Onar, S., Oztaysi, B., & Kahraman, C. (2014). Strategic decision selection using hesitant fuzzy TOPSIS and interval type-2 fuzzy AHP: A case study. *International Journal of Computational Intelligence Systems*, *7*(5), 1002–1021. doi:10.1080/18756891.2014.964011

Chaloupka, F. J., & Peck, J. A. (2010). *Cigarette Excise Taxation: The Impact Of Tax Structure On Prices, Revenues, And Cigarette Smoking*. NBER Working Paper No. 16287. JEL, No. H2, I18.

Chaloupka, F. J., Yurekli, A., & Fong, G. T. (2012). Tobacco Taxes As A Tobacco Control Strategy. *Tobacco Control, Vol*, *21*(3), 172–180. doi:10.1136/tobaccocontrol-2011-050417 PMID:22345242

Chamodrakas, I., Alexopoulou, N., & Martakos, D. (2009). Customer evaluation for order acceptance using a novel class of fuzzy methods based on TOPSIS. *Expert Systems with Applications*, *36*(4), 7409–7415. doi:10.1016/j.eswa.2008.09.050

Chan, F. S. (2003). Interactive selection model for supplier selection process: An analytical hierarchy process approach. *International Journal of Production Research*, *41*(15), 3549–3579.

Chang, C. H., Lin, J. J., Lin, J. H., & Chiang, M. C. (2010). Domestic open-end equity mutual fund performance evaluation using extended TOPSIS method with different distance approaches. *Expert Systems with Applications*, *37*(6), 4642–4649. doi:10.1016/j.eswa.2009.12.044

Chang, D. Y. (1996). Applications of the extent analysis method on fuzzy AHP. *European Journal of Operational Research*, *95*(3), 649–655. doi:10.1016/0377-2217(95)00300-2

Charnes, A., Cooper, W. W., & Rhodes, E. (1978). Measuring the Efficiency of Decision Making Units. *European Journal of Operational Research*, *2*(6), 429–444. doi:10.1016/0377-2217(78)90138-8

Chatzoglou, P., Chatzoudes, D., Vraimaki, E., & Leivaditou, E. (2014). Measuring citizen satisfaction using the SERVQUAL approach: The case of the 'Hellenic post'. *Procedia Economics and Finance*, *9*, 349–360. doi:10.1016/S2212-5671(14)00036-7

Chaudha, A., Jain, R., Singh, A. R., & Mishra, P. K. (2011). Integration of Kano's model into quality function deployment. *International Journal of Advanced Manufacturing Technology*, *53*(5-8), 689–698. doi:10.100700170-010-2867-0

Che, J., Wang, J., & Wang, G. (2012). An adaptive fuzzy combination model based on self-organizing map and support vector regression for electric load forecasting. *Energy*, *37*(1), 657–664. doi:10.1016/j.energy.2011.10.034

Chen, C. F. (2006). Applying the analytical hierarchy process (AHP) approach to convention site selection. *Journal of Travel Research*, *45*(2), 167–174. doi:10.1177/0047287506291593

Cheng, C. H. (1997). Evaluating naval tactical missile systems by fuzzy AHP based on the grade value of membership function. *European Journal of Operational Research*, *96*(2), 343–350. doi:10.1016/S0377-2217(96)00026-4

Cheng, C. H., & Mon, D. L. (1994). Evaluating weapon system by analytical hierarchy process based on fuzzy scales. *Fuzzy Sets and Systems*, *63*(1), 1–10. doi:10.1016/0165-0114(94)90140-6

Cheng, C. H., Yang, K. L., & Hwang, C. L. (1999). Evaluating attack helicopters by AHP based on linguistic variable weight. *European Journal of Operational Research*, *116*(2), 423–435. doi:10.1016/S0377-2217(98)00156-8

Chen, K. K., Chang, C. T., & Lai, C. S. (2009). Service quality gaps of business customers in the shipping industry. *Transportation Research Part E, Logistics and Transportation Review*, *45*(1), 222–237. doi:10.1016/j.tre.2008.02.005

Chen, M. F., Tzeng, G. H., & Ding, C. G. (2008). Combining fuzzy AHP with MDS in identifying the preference similarity of alternatives. *Applied Soft Computing*, *8*(1), 110–117. doi:10.1016/j.asoc.2006.11.007

Chen, S. J., & Hwang, C. L. (1992). Fuzzy multiple attribute decision making methods. In *Fuzzy multiple attribute decision making* (pp. 289–486). Springer. doi:10.1007/978-3-642-46768-4_5

Chen, S. M. (1996). Evaluating weapon systems using fuzzy arithmetic operations. *Fuzzy Sets and Systems*, *77*(3), 265–276. doi:10.1016/0165-0114(95)00096-8

Chen, X., Zhou, Y., Tang, Z., & Luo, Q. (2017). A hybrid algorithm combining glowworm swarm optimization and complete 2-opt algorithm for spherical travelling salesman problems. *Applied Soft Computing*, *58*, 104–114. doi:10.1016/j.asoc.2017.04.057

Chen, Z., Lu, M., Ming, X., Zhang, X., & Zhou, T. (2020). Explore and evaluate innovative value propositions for smart product service system: A novel graphics-based rough-fuzzy DEMATEL method. *Journal of Cleaner Production*, *243*, 118672. doi:10.1016/j.jclepro.2019.118672

Chien, T., & Hu, J. L. (2007). Renewable energy and macroeconomic efficiency of OECD and non-OECD economies. *Energy Policy*, *37*(7), 3606–3615. doi:10.1016/j.enpol.2006.12.033

Chin, C. B. N. (2008). *Cruising in the global economy: Profits, pleasure and work at sea.* Ashgate.

Chou, Y. C., Yen, H. Y., Dang, V. T., & Sun, C. C. (2019). Assessing the Human Resource in Science and Technology for Asian Countries: Application of Fuzzy AHP and Fuzzy TOPSIS. *Symmetry*, *11*(2), 251. doi:10.3390ym11020251

Chow C. K., Corsi D. J., & Gilmore A. B. (2017) Tobacco control environment: cross-sectional survey of policy implementation, social unacceptability, knowledge of tobacco health harms and relationship to quit ratio in 17 low-income, middle-income and high-income countries. *BMJ*, *7*(3).

Chowdhury, A., Ghosh, A., Sinha, S., Das, S., & Ghosh, A. (2013). A novel genetic algorithm to solve travelling salesman problem and blocking flow shop scheduling problem. *International Journal of Bio-inspired Computation*, *5*(5), 303–314. doi:10.1504/IJBIC.2013.057172

Chu, C. H., Yeung, C. Y. Y. J., & Lo, E. C. M. (2001). Monitoring patient satisfaction with university dental services under two fee-paying systems. *Community Dentistry and Oral Epidemiology*, *29*(5), 390–398. doi:10.1034/j.1600-0528.2001.290509.x PMID:11553112

Chu, S. C., Roddick, J. F., & Pan, J. S. (2004). Ant colony system with communication strategies. *Information Sciences*, *167*(1-4), 63–76. doi:10.1016/j.ins.2003.10.013

Clare, D. A., & Sanford, D. G. (1984). Cooperation and conflict between industrial sales and production. *Industrial Marketing Management*, *13*(3), 163–169. doi:10.1016/0019-8501(84)90029-4

Cockerham, W. C., Hamby, B. W., & Oates, G. R. (2017). *The social determinants of chronic disease.* Academic Press.

Coiera, E., & Hovenga, E. S. (2007). *Building a Sustainable Health System.* IMIA and Schattouer GmbH.

Coleman, B., Martin, C., Barber, N., & Patterson, D. (2004). An evaluation of the safety and acceptability of an anticoagulation clinic in a community pharmacy setting—A pilot study. *The Pharmaceutical Journal*, *273*(7328), 822–824.

Commission of the European Communities (2008). *Report from the Commission to the European Parliament and the Council On the structure and rates of excise' duty applied on cigarettes and other manufactured tobacco products.* Author.

Conejo, A., Plazas, M., Espínola, R., & Molina, A. (2005). Day-ahead electricity price forecasting using the wavelet transform and ARIMA models. *IEEE Transactions on Power Systems*, *20*(2), 1035–1042. doi:10.1109/TPWRS.2005.846054

Contreras, J., Espínola, R., Nogales, F., & Conejo, A. (2003). ARIMA models to predict next-day electricity prices. *IEEE Transactions on Power Systems*, *18*(3), 1014–1020. doi:10.1109/TPWRS.2002.804943

Contreras-Ramírez, R., Davis-Toledo, G., & Nuñez-Franz, L. (2018). Patient satisfaction and quality of emergency dental care in Chilean public health services. *Journal of Oral Research*, *7*(6), 236–243. doi:10.17126/joralres.2018.057

Cormen, T., Leiserson, C., Rivest, R., & Stein, C. (2001). *Introduction to Algorithms.* MIT Press and McGraw-Hill.

Council of Europe. (2019). *Committee for the Prevention of Torture, Report to the Greek Government on the visits to Greece carried out by the European Committee for the Prevention of Torture and Inhuman or Degrading Treatment or Punishment (CPT), 26 September 2017, CPT/Inf (2017) 25.* Retrieved from https://rm.coe.int/1680930c9a

Crittenden, V. L. (1992). Close the marketing/manufacturing gap. *MIT Sloan Management Review*, *33*(3), 41.

Croes, G. A. (1958). A method for solving traveling-salesman problems. *Operations Research*, *6*(6), 791–812. doi:10.1287/opre.6.6.791

Cruise Lines International Association (CLIA). (2011). *Cruise Market Profile Study*. Retrieved from https://www.slideshare.net/cruiseship/cruise-market-profile-clia-2011

Csutora, R., & Buckley, J. J. (2001). Fuzzy hierarchical analysis: The Lambda-Max method. *Fuzzy Sets and Systems*, *120*(2), 181–195. doi:10.1016/S0165-0114(99)00155-4

Cucchiella, F., & Gastaldi, M. (2014). Data Envelopment Analysis to Compare Renewable Energy Efficiency in The Italian Regions. *Advanced Materials Research, 912-914*, 1607-1611. www.scientific.net/AMR.912-914.1607

Dağdeviren, M., Yavuz, S., & Kılınç, N. (2009). Weapon selection using the AHP and TOPSIS methods under fuzzy environment. *Expert Systems with Applications*, *36*(4), 8143–8151. doi:10.1016/j.eswa.2008.10.016

Dalstra, J. A., Kunst, A. E., Borrell, C., Breeze, E., Cambois, E., Costa, G., ... Regidor, E. (2005). Socioeconomic differences in the prevalence of common chronic diseases: An overview of eight European countries. *International Journal of Epidemiology*, *34*(2), 316–326. doi:10.1093/ije/dyh386 PMID:15737978

Dantzig, G., Fulkerson, R., & Johnson, S. (1954). Solution of a large-scale traveling-salesman problem. *Journal of the Operations Research Society of America*, *2*(4), 393–410. doi:10.1287/opre.2.4.393

Daskalou, K. (2019). *Forecasting of day ahead electricity prices in coupled markets. Implementation in the Italian day ahead market* (MBA Dissertation). Hellenic Open University, Patra, Greece.

De Cantis, S., Ferrante, M., Kahani, A., & Shoval, N. (2016). Cruise passengers' behavior at the destination: Investigation using GPS technology. *Tourism Management*, *52*, 133–150. doi:10.1016/j.tourman.2015.06.018

De Clercq, E., Van Casteren, V., Bossuyt, N., Goderis, G., & Moreels, S. (2014). EPR-based, quality-related process parameters: A nationwide assessment. *Studies in Health Technology and Informatics*, 121–125. PMID:25160158

De Man, J. C., Andersen, B. S., & Strandhagen, J. O. (2017). Addressing the performance of order acceptance. Procedia CIRP, 63, 301–306. doi:10.1016/j.procir.2017.03.101

Dearlove, J., Bialous, S., & Glantz, S. A. (2002). Tobacco industry manipulation of the hospitality industry to maintain smoking in public places. *Tobacco Control, Vol*, *11*(2), 94–104. doi:10.1136/tc.11.2.94 PMID:12034999

Debon, R., Coleone, J. D., Bellei, E. A., & March, A. C. B. D. (2019). Mobile health applications for chronic diseases: A systematic review of features for lifestyle improvement. *Diabetes & Metabolic Syndrome*.

DeLayne, S. J. (2016). *The Kano analysis: Customer needs are ever changing, iSixSigma*. Retrieved from: https://www.isixsigma.com/tools-templates/kano-analysis/kano-analysis-customer-needs-are-ever-changing

Delias, P., Doumpos, M., Grigoroudis, E., Manolitzas, P., & Matsatsinis, N. (2015). Supporting healthcare management decisions via robust clustering of event logs. *Knowledge-Based Systems*, *84*, 203–213. doi:10.1016/j.knosys.2015.04.012

Delipalla, S., & O'Donnell, O. (1998). *The Comparison Between Ad Valorem and Specific Taxation under Imperfect Competition: Evidence from the European Cigarette Industry*. Department of Economics Discussion Paper. No. 9802, University of Kent, Department of Economics, Canterbury.

Delipalla, S., & O'Donnell, O. (1999). *Estimating Tax Incidence, Market Power and Market Conduct: The European Cigarette Industry*. Academic Press.

Delipalla, S., & Keen, M. J. (1992). The comparison between ad valorem and specific taxation under imperfect competition. *Journal of Public Economics, Vol, 49*(3), 351–167. doi:10.1016/0047-2727(92)90073-O

Deng, J. L. (1982). Control problems of grey systems. *Systems & Control Letters, 5*, 288–294.

Deng, J. L. (1982). *Grey System Fundamental Method.* Huazhong University of Science and Technology Wuhan.

Dental Quality Alliance. (2019). *Quality Measurement in Dentistry. A Guidebook.* Retrieved from https://www.ada.org/~/media/ADA/DQA/2019_Guidebook.pdf?la=en

De, P., Ghosh, J. B., & Wells, C. E. (1993). Job selection and sequencing on a single machine in a random environment. *European Journal of Operational Research, 70*(3), 425–431. doi:10.1016/0377-2217(93)90252-I

Diakomichalis, M. (2007). Greek Maritime Tourism: Evolution, Structures and Prospects. *Research in Transportation Economics, 21*, 419–455. doi:10.1016/S0739-8859(07)21013-3

Diethelm, P. A., Rielle, J.-C., & McKee, M. (2005). The whole truth and nothing but the truth? The research that Philip Morris did not want you to see. *Lancet, 366*(9479), 86–92. doi:10.1016/S0140-6736(05)66474-4 PMID:15993237

Dijkstra, E. (1959). A note on two problems in connexion with graphs. *Numerische Mathematik, 1*(1), 269–271. doi:10.1007/BF01386390

Dinçer, H., Yüksel, S., & Martínez, L. (2019). Analysis of balanced scorecard-based SERVQUAL criteria based on hesitant decision-making approaches. *Computers & Industrial Engineering, 131*, 1–12. doi:10.1016/j.cie.2019.03.026

Dinçer, H., Yüksel, S., & Pınarbaşı, F. (2020). Kano-Based Measurement of Customer Expectations in Retail Service Industry Using IT2 DEMATEL-QUALIFLEX. In *Handbook of Research on Positive Organizational Behavior for Improved Workplace Performance* (pp. 349–370). IGI Global. doi:10.4018/978-1-7998-0058-3.ch020

Ding, S., Hipel, K. W., & Dang, Y. G. (2018). Forecasting china's electricity consumption using a new grey prediction model. *Energy, 149*, 314–328. doi:10.1016/j.energy.2018.01.169

Divajeva, D., Marsh, T., Logstrup, S., Kestens, M., Vemer, P., Kriaucioniene, V., Peresson, S., O'Kelly, S., Rito, A., & Webber, L. (2014). Economics of chronic diseases protocol: Cost-effectiveness modelling and the future burden of non-communicable disease in Europe. *BMC Public Health, 14*(1), 456. doi:10.1186/1471-2458-14-456 PMID:24886110

Dixon, C.I., & Leach, B. (1978). Sampling methods for geographical research. *CATMOG 17.*

Djeflat, A. (2010). Sustainable Knowledge for Sustainable Development: Challenges and opportunities for African development. *World Journal of Science, 7*(2).

Docherty, P., Kira, M., & Shani, A. (2009). *Creating Sustainable Work Systems. USA: Routledge.* Taylor & Francis Group.

Dogan, T. (2005). *An analysis of the Pan-European transport network* (MSc thesis). School of Social Sciences, Middle East Technical University, Turkey.

Dominici, G., & Palumbo, F. (2013). The drivers of customer satisfaction in the hospitality industry: Applying the Kano model to Sicilian hotels. *International Journal of Leisure and Tourism Marketing, 3*(3), 215–236. doi:10.1504/IJLTM.2013.052623

Donaldson, C., & Magnussen, J. (1992). *DRGs: The road to hospital efficiency.* Health Policy.

Dorigo, M., Birattari, M., & Stutzle, T. (2006). Artificial ants as a computational intelligence technique. *IEEE Computational Intelligence Magazine, 1*, 28–39. doi:10.1109/MCI.2006.329691

Dorigo, M., & Gambardella, L. M. (1997). Ant colonies for the travelling salesman problem. *Bio Systems*, *43*(2), 73–81. doi:10.1016/S0303-2647(97)01708-5 PMID:9231906

Douglas, N., & Douglas, N. (2004). Cruise ship passenger spending patterns in Pacific island ports. *International Journal of Tourism Research*, *6*(4), 251–261. doi:10.1002/jtr.486

Drope, J., & Schluger, N. W. (2018). *The Tobacco atlas*. American Cancer Society, Inc.

Dutka, A. (1995). *AMA handbook of customer satisfaction: A guide to research, planning, and implementation*. NTC Publishing Group.

Dwyer, L., Douglas, N., & Livaic, Z. (2004). Estimating the economic contribution of a cruise ship visit. *Tourism in Marine Environments*, *1*(1), 5–16. doi:10.3727/154427304774865841

Dwyer, L., & Forsyth, P. (1998). Economic Significance of Cruise Tourism. *Annals of Tourism Research*, *25*(2), 393–415. doi:10.1016/S0160-7383(97)00098-4

Dwyer, L., Forsyth, P., & Spurr, R. (2004). Evaluating tourism's effects: New and old approaches. *Tourism Management*, *25*(3), 307–317. doi:10.1016/S0261-5177(03)00131-6

Ebben, M. J., Hans, E. W., & Weghuis, F. O. (2005). Workload based order acceptance in job shop environments. *OR-Spektrum*, *27*(1), 107–122. doi:10.100700291-004-0171-9

ECMT. (2005). Strengthening inland waterway transport, Pan-European co-operation for progress. Proceedings of Pan-European Co-operation towards Strong Inland Waterway Transport: On the Move.

Economou, C., Kaitelidou, D., Karanikolos, M., & Maresso, A. (2017). Greece: Health system review. *Health Systems in Transition*, *19*(5), 1–192. PMID:29972131

Efthymiou, I. P., & Vozikis, A. (2017). Medicine, A Pan-Historical, Pan-Cultural Science. *Journal of Cardiology & Cardiovascular Therapy*.

Efthymiou, I. P. (2016). *Trends in Healthcare: A Global Challenge*. Xlibris Publishing.

El Hajj, M. S., Salem, S., & Mansoor, H. (2011). Public attitudes toward community pharmacy in Qatar: A pilot study. *Patient Preference and Adherence*, *5*, 405–422. doi:10.2147/PPA.S22117 PMID:21949604

Elani, H. W., Simon, L., Ticku, S., Bain, P. A., Barrow, J., & Riedy, C. A. (2018). Does providing dental services reduce overall health care costs?: A systematic review of the literature. *The Journal of the American Dental Association*, *149*(8), 696–703. doi:10.1016/j.adaj.2018.03.023 PMID:29866364

El-Fouly, T. H. M., El-Saadany, E. F., Salama, M. M. A., El-Fouly, T. H. M., El-Saadany, E. F., & Salama, M. M. A. (2006). Grey predictor for wind energy conversion systems output power prediction. *IEEE Transactions on Power Systems*, *21*(3), 1450–1452. doi:10.1109/TPWRS.2006.879246

Ellabban, O., Abu-Rub, H., & Blaabjerg, F. (2014). Renewable Energy Resources: Current Status, Future Prospects and Their Enabling Technology. *Renewable & Sustainable Energy Reviews*, *39*, 748–764. doi:10.1016/j.rser.2014.07.113

Ellis, C., & Kriwoken, L. K. (2006). Off the beaten track: A case study of expedition cruise ships in south-west Tasmania, Australia. In R. K. Dowling (Ed.), *Cruise Ship Tourism* (pp. 251–258). CABI. doi:10.1079/9781845930486.0251

Enea, M., & Piazza, T. (2004). Project selection by constrained fuzzy AHP. *Fuzzy Optimization and Decision Making*, *3*(1), 39–62. doi:10.1023/B:FODM.0000013071.63614.3d

Energy Exchange Group. (n.d.). *Price coupling of regions (PCR)*. Retrieved January 3, 2020, from http://www.enexgroup.gr/agores/agores-monteloy-stochoy/syzeyxi-perifereiakon-agoron/

EREC. (2010). *Renewable Energy in Europe: Markets, Trends and Technologies. Londra*. European Renewable Energy Council.

Erensal, Y. C., Öncan, T., & Demircan, M. L. (2006). Determining key capabilities in technology management using fuzzy analytic hierarchy process: A case study of Turkey. *Information Sciences*, *176*(18), 2755–2770. doi:10.1016/j.ins.2005.11.004

Eriksen, M., Mackay, J., Schluger, N., Gomeshtapeh, F. I., & Drope, J. (2015) The Tobacco Atlas. American Cancer Society. World Lung Foundation.

Esmaeilbeigi, R., Charkhgard, P., & Charkhgard, H. (2016). Order acceptance and scheduling problems in two-machine flow shops: New mixed integer programming formulations. *European Journal of Operational Research*, *251*(2), 419–431. doi:10.1016/j.ejor.2015.11.036

Euromonitor International. (2016). *Country report*. Tobacco in Greece.

European Commission (EC). (1998). Status of the Pan-European Transport Corridors and Transport Areas. TINA office, Vienna, Austria: EC.

European Commission Statistical Pocketbook. (2019). *Mobility and Transportation. Publication Office of the European Union.*

European Commission. (2015). *A European Agenda on Migration*. Retrieved from https://ec.europa.eu/home-affairs/sites/homeaffairs/files/what-we-do/policies/european-agenda-migration/background-information/docs/communication_on_the_european_agenda_on_migration_en.pdf

European Commission. (2018). *Taxation and Customs Union. Excise Duties; Cigarettes*. https://ec.europa.eu/taxation_customs/business/excise-duties-alcohol-tobacco-energy/excise-duties-tobacco/excise-duties-cigarettes_en

European Cruise Council (ECC). (2011). *The Cruise Industry: A Leader in Europe's Economic Recovery*. Retrieved from https://www.cruise-norway.no/viewfile.aspx?id=3186

European Cruise Council. (2010). *European Cruise Council 2010/2011 Report*. European Cruise Council. Retrieved from http://www.ashcroftandassociates.com/images/ECC-LR.pdf

European Migration Network (EMN)/European Commission. (2018). *Asylum and Migration Glossary 6.0*. Retrieved from https://ec.europa.eu/home-affairs/sites/homeaffairs/files/what-we-do/networks/european_migration_network/docs/interactive_glossary_6.0_final_version.pdf

European Migration Network (EMN)/European Commission. (2020). *Special Annex to the 30th EMN Bulletin EU Member States & Norway: responses to COVID-19 in the migration and asylum area January – March 2020*. Retrieved from: https://ec.europa.eu/home-affairs/sites/homeaffairs/files/00_eu_30_emn_bulletin_annex_covid_19.pdf

European Union (2011) Council Directive 2011/64/EU on the structure and rates of excise duty applied to manufactured tobacco. Official Journal of the European Union. Chapter 3, "Provisions Applicable To Cigarettes

European Union (2011) Council Directive 2011/64/EU on the structure and rates of excise duty applied to manufactured tobacco. Official Journal of the European Union. Chapter 3, "Provisions Applicable To Cigarettes". Articles 8-10.

European Union (EU). (2001). Commission underlines the importance of Pan-European Corridors. IP/01/522, Brussels, Belgium: EU.

Eurostat. (2016). Tobacco Consumption., news-release. *European Health Interview Survey, 245.*

Eurostat. (2020). *Energy from Renewable Sources.* https://ec.europa.eu/: https://ec.europa.eu/eurostat/web/energy/data/shares

Evans, D. B., Edejer, T. T. T., Lauer, J., Frenk, J., & Murray, C. J. (2001). Measuring quality: From the system to the provider. *International Journal for Quality in Health Care, 13*(6), 439–446. doi:10.1093/intqhc/13.6.439 PMID:11769745

Faggini, M., Cosimato, S., David Nota, F., & Nota, G. (2018). Pursuing Sustainability for Healthcare through Digital Platforms. *Sustainability 2019, 11*(165).

Feng, C. M., & Wang, R. T. (2000). Performance evaluation for airlines ıncluding the consideration financial ratios. *Journal of Air Transport Management, 6*(3), 6. doi:10.1016/S0969-6997(00)00003-X

Feng, G. C., & Ma, R. (2020). Identification of the factors that influence service innovation in manufacturing enterprises by using the fuzzy DEMATEL method. *Journal of Cleaner Production, 253*, 120002. doi:10.1016/j.jclepro.2020.120002

Feng, S. J., Ma, Y. D., Song, Z. L., & Ying, J. (2012). Forecasting the energy consumption of china by the grey prediction model. *Energy Sources. Part B, Economics, Planning, and Policy, 7*(4), 376–389. doi:10.1080/15567240903330426

Fenlon, M. R., & Sherriff, M. (2004). Investigation of new complete denture quality and patients' satisfaction with and use of dentures after two years. *Journal of Dentistry, 32*(4), 327–333. doi:10.1016/j.jdent.2004.01.007 PMID:15053917

Fiechter, C. N. (1994). A parallel tabu search algorithm for large traveling salesman problems. *Discrete Applied Mathematics, 51*(3), 243–267. doi:10.1016/0166-218X(92)00033-I

Fineberg, H. V. (2012). A Successful and Sustainable Health System - How to Get There from Here. *The New England Journal of Medicine, 366*(11), 1020–1027. doi:10.1056/NEJMsa1114777 PMID:22417255

Fischer, M. (2015). Fit for the Future? A New Approach in the Debate about What Makes Healthcare Systems Really Sustainable. *Sustainability.*

Flores-Mir, C., Palmer, N. G., Northcott, H. C., Khurshed, F., & Major, P. W. (2006). Perceptions and attitudes of Canadian dentists toward digital and electronic technologies. *Journal of the Canadian Dental Association, 72*(3), 243.

Fornell, C., & Larcker, D. F. (1981). Evaluating structural equation models with unobservable variables and measurement error. *JMR, Journal of Marketing Research, 18*(1), 39–50. doi:10.1177/002224378101800104

Fouskas, T., Gikopoulou, P., Ioannidi, E., & Koulierakis, G., (2019). Health inequalities and female migrant domestic workers: Accessing healthcare as a human right and barriers due to precarious employment in Greece. *Collectivus, 6*(2), 71-90.

Fouskas, T. (2016). Repercussions of Precarious, Low-Status Work Regarding Access to Healthcare Services by Immigrant Workers in Greece: Seasonal Flu or an Incurable Disease? In J. Jaworski (Ed.), *Advances in Sociology Research* (Vol. 18, pp. 13–74). Nova Science Publishers.

Fouskas, T. (2017). EU societies divided in unity? The changing influx of asylum seekers in 2014-2016 in Greece and the EU Member States' societies disunity due to the impact of the migration and refugee crisis. *European Politeia, 5*(1-2), 123–169.

Fouskas, T. (2018). Precarious employment repercussions on migrants' perceptions regarding healthcare. *International Journal of Human Rights in Healthcare, 11*(3), 298–311. doi:10.1108/IJHRH-01-2018-0010

Fouskas, T. (2019). Unravelling solidarity and hostility: Mobilizations concerning migrants, asylum seekers and refugees in anti-migrant times in Greece. In A. Rea, M. Martiniello, P. De Cuyper, B. Meuleman, & A. Mazzola (Eds.), *The refugee reception crisis in anti-immigrant times. Polarization of the public opinion, local mobilizations and reception practices in Europe* (pp. 121–143). ULB Press.

Fouskas, T., Sidiropoulos, S., & Vozikis, A. (2019). Leaving no one out? Public health aspects of migration: Health risks, responses and accessibility by asylum seekers, refugees and migrants in Greece. *International Journal of Health Research and Innovation*, *7*(1), 13–28.

Fredrickson, B. L. (2013), Positive emotions broaden and build. Advances in Experimental Social Psychology. Academic Press.

Freisleben, B., & Merz, P. (1996, May). A genetic local search algorithm for solving symmetric and asymmetric traveling salesman problems. In *Proceedings of IEEE international conference on evolutionary computation* (pp. 616-621). IEEE. 10.1109/ICEC.1996.542671

Freris, L., & Infield, D. (2008). *Renewable Energy in Power Systems*. John Wiley & Sons, Ltd.

Fridleifsson, I. B. (2001). Geothermal Energy for The Benefit of The People. *Renewable & Sustainable Energy Reviews*, *5*(3), 299–312. doi:10.1016/S1364-0321(01)00002-8

Garcia, R., Contreras, J., Akkeren, M., & Garcia, J. (2003). A GARCH forecasting model to predict day-ahead electricity prices. *IEEE Transactions on Power Systems*, *2*, 867–874.

Gastelurrutia, M. A., de San Vicente, O. G., Erauncetamurgil, O., Odriozola, I., & Fernandez-Llimos, F. (2006). Customers' expectations and satisfaction with a pharmacy not providing advanced cognitive services. *Pharmacy World & Science*, *8*(6), 374–376. doi:10.100711096-006-9049-z PMID:17120133

Gavilan, D., Avello, M., & Abril, C. (2014). Shopper marketing: A new challenge for Spanish community pharmacies. *Research in Social & Administrative Pharmacy*, *10*(6), e125–e136. doi:10.1016/j.sapharm.2014.02.008 PMID:24690551

Gen, M., & Cheng, R. (1997). *Genetic algorithms and engineering design*. John Wiley&Sons.

Gestore Mercati Energetici. (n.d.). *Zones*. Retrieved January 3, 2020, from https://www.mercatoelettrico.org/En/Mercati/MercatoElettrico/Zone.aspx

Gharehgozli, A. H., Rabbani, M., Zaerpour, N., & Razmi, J. (2008). A comprehensive decision-making structure for acceptance/rejection of incoming orders in make-to-order environments. *International Journal of Advanced Manufacturing Technology*, *39*(9-10), 1016–1032. doi:10.100700170-007-1275-6

Ghasemi, M. R., Ignatius, J., & Emrouznejad, A. (2014). A bi-Objective Weighted Model for Improving the Discrimination Power in MCDEA. *European Journal of Operational Research*, *233*(3), 640–650. doi:10.1016/j.ejor.2013.08.041

Ghosh, J. B. (1997). Job selection in a heavily loaded shop. *Computers & Operations Research*, *24*(2), 141–145. doi:10.1016/S0305-0548(96)00045-7

GIFT. (2013). *GIFT Green Freight Transport Observatory: System design and platform manual. Deliverable 5.2.1, Green Intermodal Freight Transport project*. SEE Programme.

Gilmore, A., Collin, J., & Townsend, J. (2007). Transnational Tobacco Company Influence on Tax Policy During Privatization of a State Monopoly: British American Tobacco and Uzbekistan. *Am J Public Health, Vol*, *97*(11), 2001–2009. doi:10.2105/AJPH.2005.078378 PMID:17138915

Giove, S., Funari, S., & Nardelli, C. (2006). An interval portfolio selection problem based on regret function. *European Journal of Operational Research*, *170*(1), 253–264. doi:10.1016/j.ejor.2004.05.030

Glurich, I., Nycz, G., & Acharya, A. (2017). Status update on translation of integrated primary dental-medical care delivery for management of diabetic patients. *Clinical Medicine & Research*, *15*(1-2), 21–32. doi:10.3121/cmr.2017.1348 PMID:28373288

Goel, S. (2015). *Financial ratios*. Business Expert Press.

Gogus, O., & Boucher, T. O. (1997). A consistency test for rational weights in multi-criterion decision analysis with fuzzy pairwise comparisons. *Fuzzy Sets and Systems*, *86*(2), 129–138. doi:10.1016/0165-0114(95)00410-6

Gogus, O., & Boucher, T. O. (1998). Strong transitivity, rationality and weak monotonicity in fuzzy pairwise comparisons. *Fuzzy Sets and Systems*, *94*(1), 133–144. doi:10.1016/S0165-0114(96)00184-4

Goldbarg, E. F., Goldbarg, M. C., & de Souza, G. R. (2008). Particle swarm optimization algorithm for the traveling salesman problem. In *Traveling Salesman Problem*. IntechOpen. doi:10.5772/5580

Goldberg, D. E. (1989). *Genetic algorithms in search optimization and machine learning*. Addison-Wesley Publishing Company.

Göleç, A., & Taşkın, H. (2007). Novel methodologies and a comparative study for manufacturing systems performance evaluations. *Information Sciences*, *177*(23), 5253–5274. doi:10.1016/j.ins.2007.06.024

González, G. Z., Klazinga, N., ten Asbroek, G., & Delnoij, D. M. (2006). Performance indicators used to assess the quality of primary dental care. *Community Dental Health*, *23*(4), 228–235. PMID:17194070

Goodchild, M., Perucica, A., & Nargisb, N. (2016). Modelling the impact of raising tobacco taxes on public health and finance. *Bull World Health Organ, Vol*, *94*(4), 250–257. doi:10.2471/BLT.15.164707 PMID:27034518

Goodrick, E., & Reay, T. (2011). Constellations of institutional logics: Changes in the professional work of pharmacists. *Work and Occupations*, *38*(3), 372–416. doi:10.1177/0730888411406824

Government Gazette. (2001). *National Customs Code of Greece 2960/2001, Government Gazette Issue No A'265/22.11.01. Article 96*. Tax Determination.

Government Gazette. (2011). *Law 3907/2011 on the Establishment of an Asylum Service and a First Reception Service, adaptation of the Greek legislation to the provisions of Directive 1008/115/EC 'with regard to the common rules and procedures in Member States for the return of illegally staying third-country nationals' and other provisions. (Greek)*. Retrieved from http://www.yptp.gr/images/stories/2011/law%203907.pdf.pdf

Government Gazette. (2015). *Joint Ministerial Decision 2969/2015: Establishment of First Reception Centers and temporary Hosting Structures for asylum seekers and vulnerable groups of third-country nationals*. https://bit.ly/3bovbh4

Government Gazette. (2016). *Law 4368/2016. Article 33: Health coverage of uninsured and vulnerable social groups*. Retrieved from http://asylo.gov.gr/wp-content/uploads/2020/02/φεκ-ΠΑΑΥΠΑ.pdf

Government Gazette. (2016). *Law 4375/2016 on the On the organization and operation of the Asylum Service, the Appeals Authority, the Reception and Identification Service, the establishment of the General Secretariat for Reception, the transposition into Greek legislation of the provisions of Directive 2013/32/EC "on common procedures for granting and withdrawing the status of international protection (recast) (L 180/29.6.2013), provisions on the employment of beneficiaries of international protection and other provisions*. Retrieved from http://asylo.gov.gr/en/wp-content/uploads/2016/05/GREECE_Law_4375_2016_EN_final.pdf

Government Gazette. (2020a). *Decision D1A/20030 (Government Gazette B – 9 85 – 22.03.2020) Taking measures against the occurrence and spread of cases of COVID-19 in the Reception and Identification Centers, in the total of the Territory, for the period from 21.03.2020 to 21.04.2020 (in Greek).* Retrieved from http://asylo.gov.gr/wp-content/uploads/2020/03/RICS_RESTRICTIONS_MINISTERIAL-DECISION.pdf

Government Gazette. (2020b). *Joint Ministerial Decision D1A/25768/2020 – 1472 – 17-4-2020 (in Greek).* Retrieved from http://www.elinyae.gr/sites/default/files/2020-04/1472b_2020.pdf

Government Gazette. (2020c). *Extension of D1A 20030/21.3.2020 (B 985) of Joint Decision of the Ministers of Civil Protection, Health and Migration and Asylum until 21.5.2020 (in Greek).* Retrieved from http://www.elinyae.gr/sites/default/files/2020-05/1771b_2020.pdf

Government Gazette. (2020d). *B 1086/30.03.2020 (in Greek)* http://www.immigration.gov.gr/documents/20182/31271/ΥΑ+12889_2020_ΠΑΡΑΤΑΣΗ+ΙΣΧΥΟΣ+ΤΩΝ+ΑΔΕΙΩΝ+ΔΙΑΜΟΝΗΣ_ΦΕΚ+1086+Β.pdf/229a7015-dd96-4d3b-84c6-7c84153ac04b

GOVGR. (2020). *Preventive measures to avoid the spread of coronavirus disease in Reception and Identification Centers and Third Country Nationals' Accommodation centers.* Retrieved from https://covid19.gov.gr/proliptika-metra-gia-tin-apofygi-tis-diasporas-tou-koronoiou-se-kentra-ypodochis-ke-taftopiisis-ke-domes-filoxenias-politon-triton-choron/

Goyal, K. K., Jain, P. K., & Jain, M. (2012). Optimal configuration selection for reconfigurable manufacturing system using NSGA II and TOPSIS. *International Journal of Production Research, 50*(15), 4175–4191. doi:10.1080/00207543.2011.599345

Greenacre, M. (2007). *Correspondence Analysis in Practice* (2nd ed.). Chapman & Hall/CRC. doi:10.1201/9781420011234

Grigoroudis, E., & Siskos, Y. (2002). Preference disaggregation for measuring and analyzing customer satisfaction: The MUSA method. *European Journal of Operational Research, 143*(1), 148–170. doi:10.1016/S0377-2217(01)00332-0

Grigoroudis, E., & Siskos, Y. (2010). *Customer satisfaction evaluation: Methods for measuring and implementing service quality*. Springer. doi:10.1007/978-1-4419-1640-2

Gruer, L., Hart, C. L., Gordon, D. S., & Watt, C. M. (2009). Effect of tobacco smoking on survival of men and women by social position: a 28 year cohort study. *BMJ, 338*, b480.

Gruning, T., Gilmore, A. B., & McKee, M. (2006). Tobacco industry influence on science and scientists in Germany. *American Journal of Public Health, 96*(1), 20–32. doi:10.2105/AJPH.2004.061507 PMID:16317203

Guerrero, H. H., & Kern, G. M. (1988). How to more effectively accept and refuse orders. *Production and Inventory Management Journal, 29*(4), 59–62.

Gülcü, Ş., Mahi, M., Baykan, Ö. K., & Kodaz, H. (2018). A parallel cooperative hybrid method based on ant colony optimization and 3-Opt algorithm for solving traveling salesman problem. *Soft Computing, 22*(5), 1669–1685. doi:10.100700500-016-2432-3

Güler, Ö. (2005). Dünyada ve Türkiye'de Rüzgar Enerjisi. In *TMMOB Türkiye V. Energy Conference Proceeding Book* (s. 209-2015). Ankara: UCTEA The Chamber of Electrical Engineers.

Gul, M., Celik, E., Gumus, A. T., & Guneri, A. F. (2016). Emergency department performance evaluation by an integrated simulation and interval type-2 fuzzy MCDM-based scenario analysis. *European Journal of Industrial Engineering, 10*(2), 196–223. doi:10.1504/EJIE.2016.075846

Gültekin, Ö. (2009). *Bursa İli Orta Dönem Elektrik Talep Tahmini.* Yüksek Lisans Tezi, Dumlupınar Üniversitesi, Fen Bilimleri Enstitüsü, Kütayha.

Güngör, Z., Serhadlıoğlu, G., & Kesen, S. E. (2009). A fuzzy AHP approach to personnel selection problem. *Applied Soft Computing*, *9*(2), 641–646. doi:10.1016/j.asoc.2008.09.003

Gursoy, D., McCleary, K. W., & Lepsito, L. R. (2007). Propensity to complain: Effects of personality and behavioral factors. *Journal of Hospitality & Tourism Research (Washington, D.C.)*, *31*(3), 358–386. doi:10.1177/1096348007299923

Gustafsson, A., Ekdahl, F., & Edvardsson, B. (1999). Customer focused service development in practice: A case study at Scandinavian Airlines System (SAS). *International Journal of Service Management*, *10*(4), 344–358.

Gyaneshwari, K. (2015). A study on consumer behavior towards OTC products. *International Journal of Management and Social Science Research. RE:view*, *1*(13), 134–137.

Hagan, M. T., Demuth, H. B., Beale, M. H., & De Jesús, O. (n.d.). *Neural network design* (2nd ed.). Retrieved January 3, 2020, from http://hagan.okstate.edu/NNDesign.pdf

Hagan, M., & Menhaj, M. (1994). Training feedforward networks with the Marquardt algorithm. *IEEE Transactions on Neural Networks*, *5*(6), 989–993. doi:10.1109/72.329697 PMID:18267874

Hair, J. F., Black, W. C., Babin, B. J., & Anderson, R. E. (2010). *Multivariate Data Analysis*. Prentice Hall.

Hambali, A., Sapuan, S. M., Ismail, N., & Nukman, Y. (2009). Application of analytical hierarchy process in the design concept selection of automotive composite bumper beam during the conceptual design stage. *Scientific Research and Essays*, *4*(4), 198–211.

Hamzacebi, C., & Es, H. A. (2014). Forecasting the annual electricity consumption of turkey using an optimized grey model. *Energy*, *70*, 165–171. doi:10.1016/j.energy.2014.03.105

Harris, F. H. D., & Pinder, J. P. (1995). A revenue management approach to demand management and order booking in assemble-to-order manufacturing. *Journal of Operations Management*, *13*(4), 299–309. doi:10.1016/0272-6963(95)00029-1

Harris, R., Bridgman, C., Ahmad, M., Bowes, L., Haley, R., Saleem, S., Singh, R., & Taylor, S. (2011). Introducing care pathway commissioning to primary dental care: Measuring performance. *British Dental Journal*, *211*(11), 22–29. doi:10.1038j.bdj.2011.1005 PMID:22158196

Harris, S. B., Green, M. E., Brown, J. B., Roberts, S., Russell, G., Fournie, M., Webster-Bogaert, S., Paquette-Warren, J., Kotecha, J., Han, H., Thind, A., Stewart, M., Reichert, S., Tompkins, J. W., & Birtwhistle, R. (2015). Impact of a quality improvement program on primary healthcare in Canada: A mixed-method evaluation. *Health Policy (Amsterdam)*, *119*(4), 405–416. doi:10.1016/j.healthpol.2014.10.019 PMID:25476553

Harteloh, P. P. (2003). Quality systems in health care: A sociotechnical approach. *Health Policy (Amsterdam)*, *64*(3), 391–398. doi:10.1016/S0168-8510(02)00183-5 PMID:12745175

Harteloh, P. P., & Verheggen, F. W. (1994). Quality assurance in health care. From a traditional towards a modern approach. *Health Policy (Amsterdam)*, *27*(3), 261–270. doi:10.1016/0168-8510(94)90120-1 PMID:10134583

Hatami-Marbini, A., & Toloo, M. (2017). An Extended Multiple Criteria Data Envelopment Analysis Model. *Expert Systems with Applications*, *73*, 201–219. doi:10.1016/j.eswa.2016.12.030

Head, J., Chungkham, H. S., Hyde, M., Zaninotto, P., Alexanderson, K., Stenholm, S., ... Vahtera, J. (2018). Socioeconomic differences in healthy and disease-free life expectancy between ages 50 and 75: A multi-cohort study. *European Journal of Public Health*, *29*(2), 267–272. doi:10.1093/eurpub/cky215 PMID:30307554

Heale, R., Wenghofer, E., James, S., & Garceau, M. L. (2018). Quality of care for patients with diabetes and mulitmorbidity registered at nurse practitioner-led clinics. *Canadian Journal of Nursing Research*, *50*(1), 20–27. doi:10.1177/0844562117744137 PMID:29216748

Heard, A., Nath, D. K., & Loevinsohn, B. (2013). Contracting urban primary healthcare services in B angladesh–effect on use, efficiency, equity and quality of care. *Tropical Medicine & International Health*, *18*(7), 861–870. doi:10.1111/tmi.12113 PMID:23611146

Hecht-Nielsen, R. (1990). *Neurocomputing*. Addison-Wesley.

Hellenic Center for Disease Control and Prevention (HCDCD/KEELPNO). (2012). *Intervention Project in the Historical Centre of Athens. In: Hellenic Ministry of Health/Office of Parliamentary Control, Answer to the Question: Programmatic Convention of the Hellenic Ministry of Health and the Municipality of Athens/ANS 699/03/08/2012*. Athens: Hellenic Ministry of Health/Office Parliamentary Control: 3-64 (in Greek). Retrieved from https://www.hellenicparliament.gr/UserFiles/67715b2c-ec81-4f0c-ad6a-476a34d732bd/7737577.pdf

Hellenic Republic Ministry of Citizen Protection National Coordination Center for Border Control. Immigration And Asylum (NCCBCIA) (2020). *National Situational Picture Regarding the Islands at Eastern Aegean Sea (5/5/2020)*. Retrieved from https://infocrisis.gov.gr/8818/national-situational-picture-regarding-the-islands-at-eastern-aegean-sea-5-5-2020/?lang=en

Herzberg, F., Bernard, M., & Snyderman, B. B. (1959). *The motivation to work*. John Wiley and Sons.

Hiilamo, H. (2003). Tobacco industry strategy to undermine tobacco control in Finland. *Tobacco Control, Vol*, *12*(4), 414–423. doi:10.1136/tc.12.4.414 PMID:14660780

Hill, N. (1996). *Handbook of customer satisfaction measurement*. Gower Publishing.

Hirschfeld, H. O. (1935). A connection between correlation and contingency. *Proceedings of the Cambridge Philosophical Society*, *31*(4), 520–524. doi:10.1017/S0305004100013517

Hobson, J. S. P. (1993). Analysis of the US Cruise Line Industry. *Tourism Management*, *13*(6), 453–462. doi:10.1016/0261-5177(93)90098-6

Holland, J. H. (1975). *Adaptation in natural and artificial systems: an introductory analysis with applications to biology, control, and artificial intelligence*. University of Michigan Press.

Homburg, C., Droll, M., & Totzek, D. (2008). Customer prioritization: Does it pay off, and how should it be implemented? *Journal of Marketing*, *72*(5), 110–130. doi:10.1509/jmkg.72.5.110

Horvat, N., & Kos, M. (2011). Slovenian pharmacy performance: A patient –centered approach to patient satisfaction survey content development. *International Journal of Pharmacy Practice*, *33*(6), 985–996. PMID:21984227

Hsu, C. C., & Chen, C. Y. (2003). Applications of improved grey prediction model for power demand forecasting. *Energy Conversion and Management*, *44*(14), 2241–2249. doi:10.1016/S0196-8904(02)00248-0

Huang, J. (2008, September). Combining entropy weight and TOPSIS method for information system selection. In *2008 IEEE Conference on Cybernetics and Intelligent Systems* (pp. 1281-1284). IEEE. doi:10.1109/ICCIS.2008.4670971

Huang, C. Y. (2012). To model, or not to model: Forecasting for customer prioritization. *International Journal of Forecasting*, *28*(2), 497–506. doi:10.1016/j.ijforecast.2011.04.004

Huang, J. J., Tzeng, G. H., & Ong, C. S. (2006). A novel algorithm for uncertain portfolio selection. *Applied Mathematics and Computation*, *173*(1), 350–359. doi:10.1016/j.amc.2005.04.074

Hui, G., & Bifeng, S. (2009). Study on effectiveness evaluation of weapon systems based on grey relational analysis and TOPSIS. *Journal of Systems Engineering and Electronics*, *20*(1), 106–111.

Human Rights Watch (HRW). (2020). *Greece: Island Camps Not Prepared for Covid-19 Government Neglect Puts Lives at Risk*. Retrieved from https://www.hrw.org/news/2020/04/22/greece-island-camps-not-prepared-covid-19

Hwang, C. L., & Yoon, K. (1981). Multiple attribute decision making: Methods and Applications (*Vol. 104*). Springer-Verlag. doi:10.1007/978-3-642-48318-9

IEA. (2019). *World Energy Outlook 2019*. International Energy Agency.

Inuiguchi, M., & Tanino, T. (2000). Portfolio selection under independent possibilistic information. *Fuzzy Sets and Systems*, *115*(1), 83–92. doi:10.1016/S0165-0114(99)00026-3

Jafari-Marandi, R., & Smith, B. K. (2017). Fluid genetic algorithm (FGA). *Journal of Computational Design and Engineering*, *4*(2), 158–167. doi:10.1016/j.jcde.2017.03.001

Jain, A., Mao, J., & Mohiuddin, K. (1996). Artificial neural networks: A tutorial. *IEEE Computer*, *29*(3), 31–44. doi:10.1109/2.485891

Janic, M. (2007). Modeling the full costs of an intermodal and road freight transport network. *Transportation Research Part D, Transport and Environment*, *12*(1), 33–44. doi:10.1016/j.trd.2006.10.004

Jati, G. K. (2011, September). Evolutionary discrete firefly algorithm for travelling salesman problem. In *International Conference on Adaptive and Intelligent Systems* (pp. 393-403). Springer. 10.1007/978-3-642-23857-4_38

Jayakrishna, K., Vimal, K. E. K., Vibha, M., Jain, S., & Kulatunga, A. K. (2020). Fostering systematic eco-innovation in an industrial symbiosis environment using DEMATEL. *International Journal of Industrial and Systems Engineering*, *34*(1), 20–42. doi:10.1504/IJISE.2020.104314

Jeeradist, T., Thawesaengskulthai, N., & Sangsuwan, T. (2016). Using TRIZ to enhance passengers' perceptions of an airline's image through service quality and safety. *Journal of Air Transport Management*, *53*, 131–139. doi:10.1016/j.jairtraman.2016.02.011

Jiang, P., Zhou, Q., Jiang, H., & Dong, Y. (2014). An optimized forecasting approach based on grey theory and cuckoo search algorithm: a case study for electricity consumption in new south wales. In *Abstract And Applied Analysis*. Hindawi.

Jiang, S., Shi, H., Lin, W., & Liu, H. C. (2020). A large group linguistic Z-DEMATEL approach for identifying key performance indicators in hospital performance management. *Applied Soft Computing*, *86*, 105900. doi:10.1016/j.asoc.2019.105900

Jiao, J., Wei, M., Yuan, Y., & Zhao, T. (2020). Risk Quantification and Analysis of Coupled Factors Based on the DEMATEL Model and a Bayesian Network. *Applied Sciences (Basel, Switzerland)*, *10*(1), 317. doi:10.3390/app10010317

John, J., Yatim, F. M., & Mani, S. A. (2011). Measuring service quality of public dental health care facilities in Kelantan, Malaysia. *Asia-Pacific Journal of Public Health*, *23*(5), 742–753. doi:10.1177/1010539509357341 PMID:20460281

Jolly, D. A., & Reynolds, K. A. (2005). *Consumer demand for agricultural and on-farm nature tourism, UC Small Farm Centre Research Brief*. Retrieved from: http://www.ucanr.org/sites/sfp/files/143466.pdf

Joshua, J. (2017a). *The economics of addictive behaviour vol I: the private and social costs of smoking and their remedies. Academic Press.*

Joshua, J. (2017b). *The economics of addictive behaviour vol II: the private and social costs of addictive behaviours and their remedies. Academic Press.*

Joshua, J. (2017c). *The economics of addictive behaviour vol IV: the private and social costs of overeating and their remedies. Academic Press.*

Junior, F. R. L., Osiro, L., & Carpinetti, L. C. R. (2014). A comparison between Fuzzy AHP and Fuzzy TOPSIS methods to supplier selection. *Applied Soft Computing, 21*, 194–209. doi:10.1016/j.asoc.2014.03.014

Kagel, A., Bates, D., & Gawell, K. (2007, April 1). *A Guide to Geothermal Energy and the Environment.* http://www.charleswmoore.org: http://www.charleswmoore.org/pdf/Environmental%20Guide.pdf

Kahar, P., Chapman, C., & Gupta, J. (2019). Assessment of the Effect of Oral Health on Quality of Life and Oral-Health Indicators among ESRD Patients in Southwest Florida: A Pilot Study. *International Journal of Dentistry, 2019*, 1–10. doi:10.1155/2019/1608329 PMID:31662758

Kahraman, C., Cebeci, U., & Ruan, D. (2004). Multi-attribute comparison of catering service companies using fuzzy AHP: The case of Turkey. *International Journal of Production Economics, 87*(2), 171–184. doi:10.1016/S0925-5273(03)00099-9

Kahraman, C., Öztayşi, B., Sarı, İ. U., & Turanoğlu, E. (2014). Fuzzy analytic hierarchy process with interval type-2 fuzzy sets. *Knowledge-Based Systems, 59*, 48–57. doi:10.1016/j.knosys.2014.02.001

Kahraman, C., Ruan, D., & Doğan, I. (2003). Fuzzy group decision-making for facility location selection. *Information Sciences, 157*, 135–153. doi:10.1016/S0020-0255(03)00183-X

Kalaja, R., Myshketa, R., & Scalera, F. (2016). Service quality assessment in health care sector: The case of Durres public hospital. *Procedia: Social and Behavioral Sciences, 235*, 557–565. doi:10.1016/j.sbspro.2016.11.082

Kalantari, M., Rabbani, M., & Ebadian, M. (2011). A decision support system for order acceptance/rejection in hybrid MTS/MTO production systems. *Applied Mathematical Modelling, 35*(3), 1363–1377. doi:10.1016/j.apm.2010.09.015

Kaldellis, J. K., & Zafirakis, D. (2011). The Wind Energy (R)evolution: A Short Review of a Long History. *Renewable Energy, 36*(7), 1887–1901. doi:10.1016/j.renene.2011.01.002

Kamariotou, M., & Kitsios, F. (2017). Information Systems Phases and Firm Performance: A conceptual Framework. In A. Kavoura, D. Sakas, & P. Tomaras (Eds.), *Strategic Innovative Marketing* (pp. 553–560). Springer. doi:10.1007/978-3-319-33865-1_67

Kandananond, K. (2011). Forecasting electricity demand in Thailand with an artificial neural network approach. *Energies, 4*(8), 1246–1257. doi:10.3390/en4081246

Kannan, G., Haq, A. N., & Sasikumar, P. (2008). An application of the analytical hierarchy process and fuzzy analytical hierarchy process in the selection of collecting centre location for the reverse logistics multicriteria decision-making supply chain model. *International Journal of Management and Decision Making, 9*(4), 350–365. doi:10.1504/IJMDM.2008.019360

Kano, N., Seraku, N., Takahashi, F., & Tsjui, S. (1984). Attractive quality and must-be quality. *Hinshitsu, 14*(2), 147–156.

Karagul, K., Aydemir, E., & Tokat, S. (2016). Using 2-Opt based evolution strategy for travelling salesman problem. *An International Journal of Optimization and Control: Theories & Applications, 6*(2), 103–113. doi:10.11121/ijocta.01.2016.00268

Karayılmazlar, S., Saraçoğlu, N., Çabuk, Y., & Kurt, R. (2011). Biyokütlenin Türkiye'de Enerji Üretiminde Değerlendirilmesi. *Bartin Orman Fakültesi Dergisi, 13*(19), 63–75.

Karp, R. M. (1972). Reducibility among Combinatorial Problems. In R. E. Miller, J. W. Thatcher, & J. D. Bohlinger (Eds.), *Complexity of Computer Computation* (pp. 85–103). Springer. doi:10.1007/978-1-4684-2001-2_9

Karydis, A., Komboli-Kodovazeniti, M., Hatzigeorgiou, D., & Panis, V. (2001). Expectations and perceptions of Greek patients regarding the quality of dental health care. *International Journal for Quality in Health Care*, *13*(5), 409–416. doi:10.1093/intqhc/13.5.409 PMID:11669569

Kathimerini. (2020a). *150 people test positive for Covid-19 at Kranidi refugee facility*. Retrieved from https://www.ekathimerini.com/251872/article/ekathimerini/news/150-people-test-positive-for-covid-19-at-kranidi-refugee-facility

Kathimerini. (2020b). *Health, migration ministers discuss measures for migrant camps*. Retrieved from https://www.ekathimerini.com/251908/article/ekathimerini/news/health-migration-ministers-discuss-measures-for-migrant-camps

Kathimerini. (2020c). *Ritsona refugee camp quarantined after 20 test positive for Covid-19*. Retrieved from https://www.ekathimerini.com/251258/article/ekathimerini/news/ritsona-refugee-camp-quarantined-after-20-test-positive-for-covid-19

Kathimerini. (2020d). *Migrant camps on islands put on lockdown*. Retrieved from https://www.ekathimerini.com/250739/article/ekathimerini/news/migrant-camps-on-islands-put-on-lockdown

Kathimerini. (2020e). *Concern over outbreaks at migrant camps*. Retrieved from https://www.ekathimerini.com/252057/article/ekathimerini/news/concern-over-outbreaks-at-migrant-camps

Kathimerini. (2020f). *Oxfam report details inhumane conditions at Greek migrant camps*. Retrieved from https://www.ekathimerini.com/236377/article/ekathimerini/news/oxfam-report-details-inhumane-conditions-at-greek-migrant-camps

Kathimerini. (2020g). *Fire destroys warehouse at Lesvos migrant center*. Retrieved from https://www.ekathimerini.com/250368/article/ekathimerini/news/fire-destroys-warehouse-at-lesvos-migrant-center

Kathimerini. (2020h). *Fires erupt amid unrest at Samos migrant camp*. Retrieved from https://www.ekathimerini.com/252108/article/ekathimerini/in-images/fires-erupt-amid-unrest-at-samos-migrant-camp

Kathimerini. (2020i). *Greece slammed over conditions at migrant camps*. Retrieved from https://www.ekathimerini.com/237845/article/ekathimerini/news/greece-slammed-over-conditions-at-migrant-camps

Kathimerini. (2020j). *NGOs urge Greek gov't to restore health coverage for migrants*. Retrieved from http://www.ekathimerini.com/248700/article/ekathimerini/news/ngos-urge-greek-govt-to-restore-health-coverage-for-migrants

Kathimerini. (2020k). *We shield the national borders against provocative actions. Retrieved* from https://www.kathimerini.gr/1067403/article/epikairothta/ellada/panw-apo-26000-oi-apotrapeises-prospa8eies-eisodoy-sth-xwra

Kathimerini. (2020l). *Prevention measures against coronavirus by the Ministry of Immigration and Asylum (in Greek)*. Retrieved from https://www.kathimerini.gr/1069591/article/epikairothta/ellada/metra-prolhyhs-kata-toy-korwnoioy-apo-to-ypoyrgeio-metanasteyshs-kai-asyloy

Kathimerini. (2020m). *Entrance, exit with cards in closed structures (in Greek)*. Retrieved from https://www.kathimerini.gr/1073748/article/epikairothta/ellada/eisodos-e3odos-me-kartes-se-kleistes-domes

Kathimerini. (2020n). *EODY unit in the accommodation center in Malakasa after a coronavirus case (in Greek)*. Retrieved from https://www.kathimerini.gr/1072471/article/epikairothta/ellada/klimakio-toy-eody-sthn-domh-filo3enias-sth-malakasa-meta-apo-kroysma-korwnoioy

Kathimerini. (2020o). *All five third-country nationals who "broke" the quarantine in the Malakassa structure were arrested (in Greek)*. Retrieved from https://www.kathimerini.gr/1075324/article/epikairothta/ellada/synelhf8hsan-kai-oi-pente-allodapoi-poy-espasan-thn-karantina-sth-domh-malakasas

Kavoura, M., Kyriopoulos, G., Geitona, M., & Vandorou, C. (2003). *Poiotita Zois*. JANSSEN-CILAG.

Kaytez, F., Taplamacioglu, M. C., Cam, E., & Hardalac, F. (2015). Forecasting electricity consumption: A comparison of regression analysis, neural networks, and least squares support vector machines. *International Journal of Electrical Power & Energy Systems*, *67*, 431–438. doi:10.1016/j.ijepes.2014.12.036

Keay, M. (2013). *The EU "Target Model" for electricity markets: fit for purpose?* Oxford Energy Comment, The Oxford Institute for Energy Studies. Retrieved from https://www.oxfordenergy.org/publications/the-eu-target-model-for-electricity-markets-fit-for-purpose/?v=f214a7d42e0d

Keen, M. (1998). The Balance between Specific and Ad Valorem Taxation. *Fiscal Studies, Vol*, *19*(1), 1–37. doi:10.1111/j.1475-5890.1998.tb00274.x

Kevrekidis, D. P., Minarikova, D., Markos, A., Ivona Malovecká, I., & Minárik, P. (2018). Community pharmacy customer segmentation based on factors influencing their selection of pharmacy and over-the-counter medicines. *Saudi Pharmaceutical Journal*, *26*(1), 33–43. doi:10.1016/j.jsps.2017.11.002 PMID:29379331

Keyhani, A., Marwali, M. N., & Dai, M. (2010). *Integration of Green and Renewable Energy in Electric Power Systems*. John Wiley & Sons.

Kilbourne, W. E., Duffy, J. A., Duffy, M., & Giarchi, G. (2004). The applicability of SERVQUAL in cross-national measurements of health-care quality. *Journal of Services Marketing*, *18*(7), 524–533. doi:10.1108/08876040410561857

Kilincci, O., & Onal, S. A. (2011). Fuzzy AHP approach for supplier selection in a washing machine company. *Expert Systems with Applications*, *38*(8), 9656–9664. doi:10.1016/j.eswa.2011.01.159

Kimaro, H. C., & Nhampossa, J. L. (2007). *The challenges of sustainability of health information systems in developing countries: comparative case studies of Mozambique and Tanzania*. JHIDC-Health Informatics in Developing Countries.

Kim, J. H., & Park, E. C. (2015). Impact of socioeconomic status and subjective social class on overall and health-related quality of life. *BMC Public Health*, *15*(1), 783. doi:10.118612889-015-2014-9 PMID:26275823

Kim, S., Lee, B., Park, M., Oh, S., Chin, H. J., & Koo, H. (2016). Prevalence of chronic disease and its controlled status according to income level. *Medicine*, *95*(44), e5286. doi:10.1097/MD.0000000000005286 PMID:27858900

Kitsios, F., & Kamariotou, M. (2016b). Critical success factors in service innovation strategies: An annotated bibliography on NSD. In *Proceedings of British Academy of Management (BAM) Conference 2016* (pp. 1-28). Academic Press.

Kitsios, F., & Kamariotou, M. (2016a). Decision Support Systems and Business Strategy: A conceptual framework for Strategic Information Systems Planning. In *Proceedings of the 6th IEEE International Conference on IT Convergence and Security (ICITCS2016)* (pp. 149-153). Prague, Czech Republic: IEEE 10.1109/ICITCS.2016.7740323

Kitsios, F., & Kamariotou, M. (2019a). Business Strategy Modelling based on Enterprise Architecture: A State of the Art Review. *Business Process Management Journal*, *25*(4), 606–624. doi:10.1108/BPMJ-05-2017-0122

Kitsios, F., & Kamariotou, M. (2019b). Mapping New Service Development: A Review and Synthesis of Literature. *Service Industries Journal*.

Kitsios, F., & Kamariotou, M. (2019c). (in press). Service innovation process digitization: Areas for exploitation and exploration. *Journal of Hospitality and Tourism Technology*. Advance online publication. doi:10.1108/JHTT-02-2019-0041

Kitsios, F., Kamariotou, M., & Manthou, V. (2019a). Hospital Information Systems Planning: Strategic IT Alignment in Healthcare. In D. P. Sakas & D. K. Nasiopoulos (Eds.), *Strategic Innovative Marketing* (pp. 203–209). Springer. doi:10.1007/978-3-030-16099-9_25

Kitsios, F., Kamariotou, M., & Talias, M. (2020). Corporate Sustainability Strategies and Decision Support Methods: A Bibliometric Analysis. *Sustainability, 12*(521), 1–21.

Kitsios, F., Stefanakakis, S., Kamariotou, M., & Dermentzoglou, L. (2019b). E-service Evaluation: User Satisfaction Measurement and Implications in Health Sector. *Computer Standards & Interfaces Journal, 63*, 16–26. doi:10.1016/j.csi.2018.11.002

Kizielewicz, J. (2013). Cruise ship tourism – a case study in Poland. *Zesz. Nauk., 35*(107), 65–75.

Klein, R. A. (2011). Responsible cruise tourism: Issues of cruise tourism and sustainability. *Journal of Hospitality and Tourism Management, 18*(1), 107–116. doi:10.1375/jhtm.18.1.107

Knuchel-Takano, A., Hunt, D., Jaccard, A., Bhimjiyani, A., Brown, M., Retat, L., Brown, K., Hinde, S., Selvarajah, C., Bauld, L., & Webber, L. (2018). Modelling the implications of reducing smoking prevalence: The benefits of increasing the UK tobacco duty escalator to public health and economic outcomes. *Tobacco Control, 27*(e2), e124–e129. doi:10.1136/tobaccocontrol-2017-053860 PMID:29212863

Konijnendijk, P. A. (1994). Coordinating marketing and manufacturing in ETO companies. *International Journal of Production Economics, 37*(1), 19–26. doi:10.1016/0925-5273(94)90004-3

Koning, J. (2001). Social Sustainability in a Globalizing World Context, theory and methodology explored. *UNESCO / MOST Meeting.*

Korpela, J., Kyläheiko, K., Lehmusvaara, A., & Tuominen, M. (2002). An analytic approach to production capacity allocation and supply chain design. *International Journal of Production Economics, 78*(2), 187–195. doi:10.1016/S0925-5273(01)00101-3

Köse, E., Erol, S., & Temiz, İ. (2010). grey system approach for eqq models. *Journal of Engineering and Natural Sciences, 28*, 298–309.

Kovács, N., Pálinkás, A., Sipos, V., Nagy, A., Harsha, N., Kőrösi, L., Papp, M., Ádány, R., Varga, O., & Sándor, J. (2019). Factors Associated with Practice-Level Performance Indicators in Primary Health Care in Hungary: A Nationwide Cross-Sectional Study. *International Journal of Environmental Research and Public Health, 16*(17), 1–15. doi:10.3390/ijerph16173153 PMID:31470573

Kozark, M., & Rimmington, M. (2000). Tourist satisfaction with Mallorca, Spain, as an off-season holiday destination. *Journal of Travel Research, 38*(3), 260–269. doi:10.1177/004728750003800308

Kradjan, W. A., Schulz, R., Christensen, D. B., Stergachis, A., Sullivan, S., & Fullerton, D. S. (1999). Patients' perceived benefit from and satisfaction with asthma-related pharmacy services. *Journal of the American Pharmaceutical Association, 39*(5), 658–666. doi:10.1016/S1086-5802(15)30350-8 PMID:10533347

Krasovsky, K. S. (2010). The lobbying strategy is to keep excise as low as possible" - tobacco industry excise taxation policy in Ukraine. *Tobacco Induced Diseases, 8*(1), 10. doi:10.1186/1617-9625-8-10 PMID:20807434

Krassadaki, E., & Grigoroudis, E. (2018). Analyzing perceived quality of health care services: A multicriteria decision analysis approach based on the theory of attractive quality. In N. Matsatsinis & E. Grigoroudis (Eds.), *Preference disaggregation in multiple criteria decision analysis: Essays in honor of Yannis Siskos* (pp. 211–235). Springer. doi:10.1007/978-3-319-90599-0_10

Kulak, O., & Kahraman, C. (2005). Fuzzy multi-attribute selection among transportation companies using axiomatic design and analytic hierarchy process. *Information Sciences, 170*(2-4), 191–210. doi:10.1016/j.ins.2004.02.021

Kumar, S., & Chacko, R. K. (2018). Assessing perceptions of oral health related quality of life in dental implant patients. Experience of a tertiary care center in India. *Journal of Oral Biology and Craniofacial Research, 8*(2), 74-77.

Kumar, A., & Schei, T. (2012). *Hydropower. In IPCC Special Report on Renewable Energy Sources and Climate Change Mitigation Special Report of the Intergovernmental Panel on Climate Change*. Cambridge University Press.

Kutlu, A. C., & Ekmekçioğlu, M. (2012). Fuzzy failure modes and effects analysis by using fuzzy TOPSIS-based fuzzy AHP. *Expert Systems with Applications, 39*(1), 61–67. doi:10.1016/j.eswa.2011.06.044

Kwak, S. J., Yoo, S. H., & Chang, J. I. (2005). The role of the maritime industry in the Korean national economy: An input–output analysis. *Marine Policy, 29*(4), 371–383. doi:10.1016/j.marpol.2004.06.004

Kwiesielewicz, M. (1998). A note on the fuzzy extension of Saaty's priority theory. *Fuzzy Sets and Systems, 95*(2), 161–172. doi:10.1016/S0165-0114(96)00329-6

Kyriazis, I., Lappa, Th., Koufopoulou, K., Sidiropoulos, S., & Efthymiou, I. P. (n.d.). Impact of Austerity on Dietary Habits and Obesity in Greek Children, Adolescents and Adult. In The Impact and Implications of Crisis: A Comprehensive Approach Combining Elements of Health and Society. Academic Press.

Lacagnina, V., & Pecorella, A. (2006). A stochastic soft constraints fuzzy model for a portfolio selection problem. *Fuzzy Sets and Systems, 157*(10), 1317–1327. doi:10.1016/j.fss.2005.10.002

Lago, J., Ridder, F., & Schutter, B. (2018a). Forecasting spot electricity prices: Deep learning approaches and empirical comparison of traditional algorithms. *Applied Energy, 221*, 386–405. doi:10.1016/j.apenergy.2018.02.069

Lago, J., Ridder, F., Vrancx, P., & Schutter, B. (2018b). Forecasting day-ahead electricity prices in Europe: The importance of considering market integration. *Applied Energy, 211*, 890–903. doi:10.1016/j.apenergy.2017.11.098

Lammgard, C. (2012). Intermodal train services: A business challenge and a measure for decarbonisation for logistics service providers. *Research in Transportation Business & Management, 5*, 48–56. doi:10.1016/j.rtbm.2012.11.001

Lam, P. T., Chan, A. P., Wong, F. K., & Wong, F. W. (2007). Constructability rankings of construction systems based on the analytical hierarchy process. *Journal of Architectural Engineering, 13*(1), 36–43. doi:10.1061/(ASCE)1076-0431(2007)13:1(36)

Langworthy, A., Howard, J., & Mawson, F. (2006). Building the relationship between agriculture and tourism-models and benefits of co-operation. Report Prepared for the Centre for Agriculture and Business, Swinburne University of Technology, Melbourne.

Laporte, G. (1992). The traveling salesman problem: An overview of exact and approximate algorithms. *European Journal of Operational Research, 59*(2), 231–247. doi:10.1016/0377-2217(92)90138-Y

Larranaga, P., Kuijpers, C. M. H., Murga, R. H., Inza, I., & Dizdarevic, S. (1999). Genetic algorithms for the travelling salesman problem: A review of representations and operators. *Artificial Intelligence Review, 13*(2), 129–170. doi:10.1023/A:1006529012972

Larsen, S., Wolff, K., Marnburg, E., & Øgaard, T. (2013). Belly full, purse closed: Cruise line passengers' expenditures. *Tourism Management Perspectives, 6*, 142–148. doi:10.1016/j.tmp.2013.02.002

Larson, L. N., Rovers, J. P., & MacKeigan, L. D. (2002). Patient satisfaction with pharmaceutical care: Update of a validated instrument. *Journal of the American Pharmaceutical Association, 42*(1), 44–50. doi:10.1331/108658002763538062 PMID:11833515

Lavelle, C., Schroth, R., & Wiltshire, W. A. (2004). Performance measures to improve the quality of orthodontic services and control expenditures. *American Journal of Orthodontics and Dentofacial Orthopedics, 126*(4), 445–449. doi:10.1016/j.ajodo.2003.10.030 PMID:15470347

Lawrence, S. R. (1994). Negotiating due-dates between customers and producers. *International Journal of Production Economics, 37*(1), 127–138. doi:10.1016/0925-5273(94)90013-2

Le Gouriérès, D. (1982). *Wind Power Plants Theory and Design*. Pergamon Press Ltd.

Lee, C. H., Zhao, X., & Lee, Y. C. (2019). Service Quality Driven Approach for Innovative Retail Service System Design and Evaluation: A Case Study. *Computers & Industrial Engineering, 135*, 275–285. doi:10.1016/j.cie.2019.06.001

Lee, D. W., & Shin, I. S. (2018). Critical quality evaluation of network meta-analyses in dental care. *Journal of Dentistry, 75*, 7–11. doi:10.1016/j.jdent.2018.05.010 PMID:29879456

Lee, M., Pham, H., & Zhang, X. (1999). A methodology for priority setting with application to software development process. *European Journal of Operational Research, 118*(2), 375–389. doi:10.1016/S0377-2217(98)00320-8

Lee, Y. S., & Tong, L. I. (2011). Forecasting energy consumption using a grey model ımproved by ıncorporating genetic programming. *Energy Conversion and Management, 52*(1), 147–152. doi:10.1016/j.enconman.2010.06.053

Lei, M., & Feng, Z. (2012). A proposed grey model for short-term electricity price forecasting in competitive power markets. *International Journal of Electrical Power & Energy Systems, 43*(1), 531–538. doi:10.1016/j.ijepes.2012.06.001

Lekakou, M. B., Pallis, A. A., & Vaggelas, G. K. (2009). *The potential of Piraeus as a major Mediterranean cruise home-port: A selection criteria analysis*. 4th International Conference Planning for the Future – Learning from the Past: Contemporary Developments in Tourism, Travel & Hospitality, Rhodes, Greece.

Lekakou, M. B., Pallis, A. A., & Vaggelas, G. K. (2010). Which homeport in Europe: The cruise industry's selection criteria. *Tourismos: An International Multidisciplinary Journal of Tourism, 4*(4), 215–240.

Lemmink, J., & Mattsson, J. (1998). Warmth during non-productive retail encounters: The hidden side of productivity. *International Journal of Research in Marketing, 15*(5), 505–517. doi:10.1016/S0167-8116(98)00016-0

Lenz, R. (2009). *Information Systems in Healthcare– State and Steps towards Sustainability*. IMIA and Schattauer GmbH. doi:10.1055-0038-1638640

Lewis, A., & Estefen, S. (2012). *Ocean Energy. In IPCC Special Report on Renewable Energy Sources and Climate Change Mitigation*. Cambridge University Press.

Lewis, H. F., & Slotnick, S. A. (2002). Multi-period job selection: Planning work loads to maximize profit. *Computers & Operations Research, 29*(8), 1081–1098. doi:10.1016/S0305-0548(00)00105-2

Liaudanskienė, R., Ustinovičius, L., & Bogdanovičius, A. (2015). Evaluation of construction process safety solutions using the TOPSIS method. *The Engineering Economist, 64*(4).

Li, G. D., Masuda, S., & Nagai, M. (2013). The prediction model for electrical power system using an improved hybrid optimization model. *International Journal of Electrical Power & Energy Systems, 44*(1), 981–987. doi:10.1016/j.ijepes.2012.08.047

Limbourg, S., Giang, H. T. Q., & Cools, M. (2016). Logistics service quality: The case of Da Nang city. *Procedia Engineering, 142*, 124–130. doi:10.1016/j.proeng.2016.02.022

Limpuangthip, N., Somkotra, T., & Arksornnukit, M. (2018). Modified retention and stability criteria for complete denture wearers: A risk assessment tool for impaired masticatory ability and oral health-related quality of life. *The Journal of Prosthetic Dentistry*, *120*(1), 43–49. doi:10.1016/j.prosdent.2017.09.010 PMID:29195820

Lin, H. T., & Chang, W. L. (2008). Order selection and pricing methods using flexible quantity and fuzzy approach for buyer evaluation. *European Journal of Operational Research*, *187*(2), 415–428. doi:10.1016/j.ejor.2007.03.003

Lin, W.-M., Gow, H.-J., & Tsai, M.-T. (2010). An enhanced radial basis function network for short-term electricity price forecasting. *Applied Energy*, *87*(10), 3226–3234. doi:10.1016/j.apenergy.2010.04.006

Lionis, C., Tsiraki, M., Bardis, V., & Philalithis, A. (2004). Seeking quality improvement in primary care in Crete, Greece: The first actions. *Croatian Medical Journal*, *45*(5), 599–603. PMID:15495288

Listl, S. (2019). Value-based oral health care: Moving forward with dental patient-reported outcomes. *The Journal of Evidence-Based Dental Practice*, *19*(3), 255–259. doi:10.1016/j.jebdp.2019.101344 PMID:31732101

Liu, S. & Lin, Y., (2006), *Grey Information: Theory And Practical Applications*. Springer.

Liu, H., Long, H., & Li, X. (2020). Identification of critical factors in construction and demolition waste recycling by the grey-DEMATEL approach: A Chinese perspective. *Environmental Science and Pollution Research International*, *27*(8), 1–19. doi:10.100711356-019-07498-5 PMID:31907809

Liu, S., & Lin, Y. (2010). *Grey Systems: Theory And Application*. Springer-Verlag. doi:10.1007/978-3-642-13938-3

Livas, P. C. (1994). *Input output analysis*. Stamoulis.

Li, X. B., & Reeves, G. R. (1999). A Multiple Criteria Approach to Data Envelopment Analysis. *European Journal of Operational Research*, *115*(3), 507–517. doi:10.1016/S0377-2217(98)00130-1

Li, X., Guo, S., & Yu, L. (2015). Skewness of fuzzy numbers and its applications in portfolio selection. *IEEE Transactions on Fuzzy Systems*, *23*(6), 2135–2143. doi:10.1109/TFUZZ.2015.2404340

Li, X., Qin, Z., & Kar, S. (2010). Mean-variance-skewness model for portfolio selection with fuzzy returns. *European Journal of Operational Research*, *202*(1), 239–247. doi:10.1016/j.ejor.2009.05.003

Lobo, R. (2012). *Agritourism definition, UC Small Farm Program, University of California Corporative Extension.* Retrieved from: http://sfp.ucdavis.edu/agritourism/Definitions/

Lohr, K. N., & Harris-Wehling, J. (1991). Medicare: A Strategy For Quality Assurance, I: A Recapitulation of the Study and a Definition of Quality of Care. *QRB. Quality Review Bulletin*, *17*(1), 6–9. doi:10.1016/S0097-5990(16)30413-4 PMID:1900934

Lundberg, D. E. (1990). *The tourist business* (6th ed.). Van Nostrand-Reinhold.

MacKay, D. (1992). Bayesian Interpolation. *Neural Computation*, *4*(3), 415–447. doi:10.1162/neco.1992.4.3.415

MacKeigan, L. D., & Larson, L. N. (1989). Development and validation of an instrument to measure patient satisfaction with pharmacy services. *Medical Care*, *27*(5), 522–536. doi:10.1097/00005650-198905000-00007 PMID:2725084

Mackenbach, J. P., Stirbu, I., Roskam, A. J. R., Schaap, M. M., Menvielle, G., Leinsalu, M., & Kunst, A. E. (2008). Socioeconomic inequalities in health in 22 European countries. *The New England Journal of Medicine*, *358*(23), 2468–2481. doi:10.1056/NEJMsa0707519 PMID:18525043

Maetzold, J. A. (2002). *Nature-based tourism and agritourism trends: Unlimited opportunities*. Retrieved from: https://www.agmrc.org/media/cms/agritourism_E6794269B3FF6.pdf

Maetzold, J. A. (2004). *Nature-based tourism and agritourism trends: Unlimited opportunities, National Alternative Enterprises and Agritourism Leader*. USDA/NRCS.

Maghsoodi, A. I., Saghaei, A., & Hafezalkotob, A. (2019). Service quality measurement model integrating an extended SERVQUAL model and a hybrid decision support system. *European Research on Management and Business Economics*, *25*(3), 151–164. doi:10.1016/j.iedeen.2019.04.004

Manavizadeh, N., Goodarzi, A. H., Rabbani, M., & Jolai, F. (2013). Order acceptance/rejection policies in determining the sequence in mixed model assembly lines. *Applied Mathematical Modelling*, *37*(4), 2531–2551. doi:10.1016/j.apm.2012.06.012

Manolitzas, P., Kostagiolas, P., Grigoroudis, E., Intas, G., & Stergiannis, P. (2018). Data on patient's satisfaction from an emergency department: Developing strategies with the Multicriteria Satisfaction Analysis. *Data in Brief*, *21*, 956–961. doi:10.1016/j.dib.2018.10.041 PMID:30426051

Marcus, M., & Spolsky, V. (1998). Concepts of quality and the provision of periodontal care: A survey. *Journal of Periodontology*, *69*(2), 228–240. doi:10.1902/jop.1998.69.2.228 PMID:9526923

Marino, R., Schofield, M., Wright, C., Calache, H., & Minichiello, V. (2008). Self-reported and clinically determined oral health status predictors for quality of life in dentate older migrant adults. *Community Dentistry and Oral Epidemiology*, *36*(1), 85–94. PMID:18205644

Markowitz, H. M. (1952). Portfolio Selection/Harry Markowitz. *The Journal of Finance*, *7*(1), 77–91.

Marques, B., & Cruse, R. (2015). Cruise tourism in Dominica: Benefits and beneficiaries. *ARA Journal of Tourism Research.*, *5*(2), 7–19.

Marshall, B. C., Cunny, K. A., & Lawson, K. A. (1997). HIV-positive males' satisfaction with pharmacy services. *Journal of the American Pharmaceutical Association*, *37*(1), 66–75. doi:10.1016/S1086-5802(16)30180-2 PMID:9040172

Marti, B. E. (2004). Trends in world and extended-length cruising (1985–2002). *Marine Policy*, *28*(3), 199–211. doi:10.1016/j.marpol.2003.09.004

Martínez, E., & Arredondo, F. (2010). Order acceptance for revenue management and capacity allocation in make-to-order batch plants. *Computer-Aided Chemical Engineering*, *28*, 1189–1194. doi:10.1016/S1570-7946(10)28199-3

Martinez, S. A., Beebe, L. A., Thompson, D. M., Wagener, T. L., Terrell, D. R., & Campbell, J. E. (2018). A structural equation modeling approach to understanding pathways that connect socioeconomic status and smoking. *PLoS One*, *13*(2), e0192451. doi:10.1371/journal.pone.0192451 PMID:29408939

Masiero, L. Transportation. In J. Jafari & H. Xiao (Eds.), *Encyclopedia of Tourism* (pp. 965–967). Springer International Publishing.

Masood, M., Newton, T., Bakri, N. N., Khalid, T., & Masood, Y. (2017). The relationship between oral health and oral health related quality of life among elderly people in United Kingdom. *Journal of Dentistry*, *56*, 78–83. doi:10.1016/j.jdent.2016.11.002 PMID:27825838

Mathe, N., Johnson, S. T., Wozniak, L. A., Majumdar, S. R., & Johnson, J. A. (2015). Alternation as a form of allocation for quality improvement studies in primary healthcare settings: The on-off study design. *Trials*, *16*(1), 1–10. doi:10.118613063-015-0904-x PMID:26303892

Matsui, M. (1982). Job-shop model: A M/(G, G)/1 (N) production system with order selection. *International Journal of Production Research*, *20*(2), 201–210. doi:10.1080/00207548208947761

Matsui, M. (1985). Optimal order-selection policies for a job shop production system. *International Journal of Production Research, 23*(1), 21–31. doi:10.1080/00207548508904688

Mavrovouniotis, M., & Yang, S. (2011). A memetic ant colony optimization algorithm for the dynamic travelling salesman problem. *Soft Computing, 15*(7), 1405–1425. doi:10.100700500-010-0680-1

Ma, W., Zhu, X., & Wang, M. (2013). Forecasting iron ore import and consumption of China using grey model optimized by particle swarm optimization algorithm. *Resources Policy, 38*(4), 613–620. doi:10.1016/j.resourpol.2013.09.007

Ma, X., Mei, X., Wu, W., Wu, X., & Zeng, B. (2019). A novel fractional time delayed grey model with grey wolf optimizer and ıts applications in forecasting the natural gas and coal consumption in Chongqing China. *Energy, 178*, 487–507. doi:10.1016/j.energy.2019.04.096

McKee, D. L., & Chase, G. L. (2003). The economic impact of cruise tourism on Jamaica. *Journal of Tourism Studies, 14*(2), 16–22.

MedCruise Yearbook 2013/14. (2014). *A Directory of Cruise Ports and Professionals in the Mediterranean and Adjoining Seas.* Academic Press.

Medicines sans Frontiers (MSF). (2017a). *A Dramatic Deterioration for Asylum Seekers on Lesbos.* Retrieved from https://msf.gr/sites/default/files/msfpublications//msf_report_vulnerable_lesvos_en.pdf

Medicines sans Frontiers (MSF). (2017b). *Confronting the mental health emergency on Samos and Lesvos: Why the containment of asylum seekers on the Greek islands must end.* Retrieved from https://msf.gr/sites/default/files/msfpublications//2017_10_mental_health_greece_report_lowres_spreads.pdf

Mehrotra, D. (2018). Quality Of Care In Dental Institutes. *Journal of Oral Biology and Craniofacial Research, 8*(2), 73. doi:10.1016/j.jobcr.2018.05.007 PMID:29892523

Menegaki, A. N. (2013). Growth and Renewable Energy in Europe: Benchmarking with Data Envelopment Analysis. *Renewable Energy, 60*, 363–369. doi:10.1016/j.renene.2013.05.042

Menger, K. (1932). Das botenproblem. *Ergebnisse eines mathematischen kolloquiums, 2*, 11-12.

Merks, P., Kaźmierczak, J., Olszewska, A. E., & Kołtowska-Häggström, M. (2014). Comparison of factors influencing patient choice of community pharmacy in Poland and in the UK, and identification of components of pharmaceutical care. *Patient Preference and Adherence, 8*, 715–726. doi:10.2147/PPA.S53829 PMID:24868150

Mestry, S., Damodaran, P., & Chen, C. S. (2011). A branch and price solution approach for order acceptance and capacity planning in make-to-order operations. *European Journal of Operational Research, 211*(3), 480–495. doi:10.1016/j.ejor.2011.01.002

Mielck, A., Vogelmann, M., & Leidl, R. (2014). Health-related quality of life and socioeconomic status: Inequalities among adults with a chronic disease. *Health and Quality of Life Outcomes, 12*(1), 58. doi:10.1186/1477-7525-12-58 PMID:24761773

Mikhailov, L. (2003). Deriving priorities from fuzzy pairwise comparison judgements. *Fuzzy Sets and Systems, 134*(3), 365–385. doi:10.1016/S0165-0114(02)00383-4

Mikhailov, L. (2004a). A fuzzy approach to deriving priorities from interval pairwise comparison judgements. *European Journal of Operational Research, 159*(3), 687–704. doi:10.1016/S0377-2217(03)00432-6

Mikhailov, L. (2004b). Group prioritization in the AHP by fuzzy preference programming method. *Computers & Operations Research, 31*(2), 293–301. doi:10.1016/S0305-0548(03)00012-1

Mikhailov, L., & Tsvetinov, P. (2004). Evaluation of services using a fuzzy analytic hierarchy process. *Applied Soft Computing*, *5*(1), 23–33. doi:10.1016/j.asoc.2004.04.001

Millá-Perseguer, M., Guadalajara-Olmeda, N., Vivas-Consuelo, D., & Usó-Talamantes, R. (2019). Measurement of health-related quality by multimorbidity groups in primary health care. *Health and Quality of Life Outcomes*, *17*(1), 1–10. doi:10.118612955-018-1063-z PMID:30634992

Mills, I., & Batchelor, P. (2011). Quality indicators: The rationale behind their use in NHS dentistry. *British Dental Journal*, *211*(1), 11–15. doi:10.1038j.bdj.2011.521 PMID:21738180

Mills, I., Frost, J., Cooper, C., Moles, D. R., & Kay, E. (2014). Patient-centred care in general dental practice-a systematic review of the literature. *BMC Oral Health*, *14*(1), 1–13. doi:10.1186/1472-6831-14-64 PMID:24902842

Mills, I., Frost, J., Moles, D. R., & Kay, E. (2013). Patient-centred care in general dental practice: Sound sense or sound-bite? *British Dental Journal*, *215*(2), 81–85. doi:10.1038j.bdj.2013.684 PMID:23887534

Minarikova, Malovecka, I. & Foltan, V. (2016). Patient choice of pharmacy and satisfaction with pharmaceutical care – Slovak regional comparison. *Farmacia*, *64*(3), 473–480.

Ministry of Migration and Asylum. (2020). *Monthly Statistics.* Retrieved from http://www.immigration.gov.gr/documents/20182/45177/StatsMetan_20200131.zip/adddf678-d929-4f44-a429-f0f9517eef0d

Misiorek, A., Trueck, S., & Weron, R. (2006). Point and interval forecasting of spot electricity prices: Linear vs. non-linear time series models. *Studies in Nonlinear Dynamics and Econometrics*, *10*, 1–34.

Mitarakis, N. (2020a). *Speech by the Minister of Migration and Asylum, Member of Parliament for Chios, Notis Mitarakis, during the discussion of Current Queries in the Hellenic Parliament.* Retrieved from https://www.mitarakis.gr/gov/migration/6115-dt-yma-vouli-epikairi-erwtisi

Mitarakis, N. (2020b). *JMD Restriction of Movement in the Accommodation centers of Asylum Seekers.* Retrieved from https://www.mitarakis.gr/gov/migration/1962-dt-yma-kya-periorismou

Mitarakis, N. (2020c). *New 6-measures bundle of the Ministry of Migration and Asylum for the protection of the coronavirus in the RICs on the Islands and other host structures.* Retrieved from https://www.mitarakis.gr/gov/migration/1970-dt-yma-nea-desmi-metrwn

Mitarakis, N. (2020d). *Visit of the Minister of Immigration and Asylum in Lesvos.* Retrieved from https://www.mitarakis.gr/gov/migration/6121-dt-yma-programma-lesvou

Mitarakis, N. (2020e). *Operation of Outpatient Clinics in KYT Chios.* Retrieved from https://www.mitarakis.gr/gov/migration/1958-dt-yma-kyt-vial-iatreia

Mitarakis, N. (2020f). *Measures to protect against coronavirus in the Reception and Identification Centers, in the accommodation centers and in the Asylum Service.* Retrieved from https://www.mitarakis.gr/gov/migration/1956-μέτρα-προστασίας-από-τον-κορωνοϊό-στα-κέντρα-υποδοχής-και-ταυτοποίησης,-στις-δομές-φιλοξενίας-και-στην-υπηρεσία-ασύλου

Mitarakis, N. (2020g). *We shielded the national borders against provocative actions (in Greek).* Retrieved from https://www.mitarakis.gr/gov/migration/1931-dt-yma-ant1-prwinoi-typoi

Mohajeri, N., & Amin, G. R. (2010). Railway station site selection using analytical hierarchy process and data envelopment analysis. *Computers & Industrial Engineering*, *59*(1), 107–114. doi:10.1016/j.cie.2010.03.006

Mohammed, K. I., Zaidan, A. A., Zaidan, B. B., Albahri, O. S., Albahri, A. S., Alsalem, M. A., & Mohsin, A. H. (2020). Novel technique for reorganisation of opinion order to interval levels for solving several instances representing prioritisation in patients with multiple chronic diseases. *Computer Methods and Programs in Biomedicine*, *185*, 105151. doi:10.1016/j.cmpb.2019.105151 PMID:31710981

Mohanty, B. K., & Singh, N. (1994). Fuzzy relational equations in analytical hierarchy process. *Fuzzy Sets and Systems*, *63*(1), 11–19. doi:10.1016/0165-0114(94)90141-4

Mohsin-Shaikh, S., Furniss, D., Blandford, A., McLeod, M., Ma, T., & Beykloo, M. Y. (2019). The impact of electronic prescribing systems on healthcare professionals' working practices in the hospital setting: A systematic review and narrative synthesis. *BMC Health Services Research*, *19*(742). PMID:31640689

Mon, D. L., Cheng, C. H., & Lin, J. C. (1994). Evaluating weapon system using fuzzy analytic hierarchy process based on entropy weight. *Fuzzy Sets and Systems*, *62*(2), 127–134. doi:10.1016/0165-0114(94)90052-3

Montgomery, J., & Randall, M. (2003). The accumulated experience ant colony for the traveling salesman problem. *International Journal of Computational Intelligence and Applications*, *3*(02), 189–198. doi:10.1142/S1469026803000938

Moser, C. A., & Kalton, G. (1971). *Survey methods in Social Investigation*. Gower.

Motaghifard, A., Omidvari, M., & Kaazemi, A. (2020). Introducing a conceptual model for evaluating health safety environmental performance of residential buildings using the fuzzy decision-making approach. *Environmental Monitoring and Assessment*, *192*(1), 19. doi:10.100710661-019-8004-0 PMID:31820115

Motorola. (1995). *Customer satisfaction assessment guide*. Motorola University Press.

Mubeen, S., Patel, K., Cunningham, Z., O'Rourke, N., Pandis, N., Cobourne, M. T., & Seehra, J. (2017). Assessing the quality of dental clinical practice guidelines. *Journal of Dentistry*, *67*, 102–106. doi:10.1016/j.jdent.2017.10.003 PMID:29031996

Muggli, M. E., Hurt, R. D., & Becker, L. B. (2004). *Turning free speech into corporate speech: Philip Morris' efforts to influence U.S. and European journalists regarding the U.S. EPA report on secondhand smoke*. Academic Press.

Murdoch, J., & Clark, J. (1994). Sustainable Knowledge. *Geoforum*, *25*(2), 115–132. doi:10.1016/0016-7185(94)90010-8

Myles, G. (1996). Imperfect competition and the optimal combination of ad valorem and specific taxation. *International Tax and Public Finance*, *3*(1), 29–44. doi:10.1007/BF00400145

Naik, P. P., Saini, B., & Armour, C. (n.d.). Measurement of patient satisfaction with community pharmacy services: A review. *International Journal of Clinical Pharmacology Research*, *31*(5), 92–100.

National Center for Social Solidarity. (2020). *Situation Update: Unaccompanied Children (UAC) in Greece 15 January 2020*. Retrieved from https://www.e-pronoia.gr/sites/default/files/ekka_dashboard_15-1-2020.pdf

National Centre for Social Solidarity/UNICEF. (2020). *Refugee and migrant children in Greece (As of 31 January 2020)*. Retrieved from https://www.unicef.org/eca/media/10861/file

National Public Health Organisation (NPHO/EODY). (2020a). *Update of accredited health journalists by the Crisis Management and Civil Protection Minister Nikos Hardalias and the representative of the Ministry of Health for the new coronavirus, Professor Sotiris Tsiodras April 1, 2020*. Retrieved from: https://eody.gov.gr/dienergeia-elegchoy-gia-sars-cov-2-sto-kfpm-ritsonas-kai-ichnilatisi-epafon-apo-to-prosopiko-toy-eody/ and https://eody.gov.gr/enimerosi-diapisteymenon-syntakton-ygeias-apo-ton-yfypoyrgo-politikis-prostasias-kai-diacheirisis-kriseon-niko-chardalia-kai-ton-ekprosopo-toy-ypoyrgeioy-ygeias-gia-to-neo-koronoio-kathigiti-sotir-5/

National Public Health Organisation (NPHO/EODY). (2020b). *Update of accredited health journalists by the Crisis Management and Civil Protection Minister Nikos Hardalias and the representative of the Ministry of Health for the new coronavirus, Professor Sotiris Tsiodras April 5, 2020.* Retrieved from: https://eody.gov.gr/klimakio-toy-eody-sto-kfpm-malakasas-gia-tin-ygeionomiki-prostasia-prosopikoy-kai-filoxenoymenon/

National Public Health Organisation (NPHO/EODY). (2020c). *Update of accredited health journalists by the Crisis Management and Civil Protection Minister Nikos Hardalias and the representative of the Ministry of Health for the new coronavirus, Professor Sotiris Tsiodras April 21, 2020.* Retrieved from: https://eody.gov.gr/0421_briefing_covid19/

National Public Health Organisation (NPHO/EODY). (2020d). *Update of accredited health journalists by the Crisis Management and Civil Protection Minister Nikos Hardalias and the representative of the Ministry of Health for the new coronavirus, Professor Sotiris Tsiodras May 14, 2020.* Retrieved from: https://eody.gov.gr/0514_briefing_covid19/

National Public Health Organization (NPHO/EODY). (2020e). *Current state of Covid-19 outbreak in Greece and timeline of key containment events. News, Press Releases. 04.03.2020.* Retrieved from https://eody.gov.gr/en/current-state-of-covid-19-outbreak-in-greece-and-timeline-of-key-containment-events/

National Public Health Organization (NPHO/EODY). (2020f). *Daily epidemiological surveillance report new coronavirus infection (COVID-19) Data until May 6, 2020, at 15:00.* Retrieved from https://eody.gov.gr/wp-content/uploads/2020/05/covid-gr-daily-report-20200506.pdf

National Public Health Organization (NPHO/EODY). (2020g). *Weekly Epidemiological Surveillance Report in points of healthcare of refugees/migrants Week 8/2020 (17 to 23 February 2020).* Retrieved from https://eody.gov.gr/wp-content/uploads/2020/03/Surveillance-refugees-weekly-GR-2020-08.pdf

Naumann, E., & Giel, K. (1995). *Customer satisfaction measurement and management: Using the voice of the customer.* Thomson Executive Press.

Neumann, A., Kalenderian, E., Ramoni, R., Yansane, A., Tokede, B., Etolue, J., Vaderhobli, R., Simmons, K., Even, J., Mullins, J., Kumar, S., Bangar, S., Kookal, K., White, J., & Walji, M. (2017). Evaluating quality of dental care among patients with diabetes: Adaptation and testing of a dental quality measure in electronic health records. *The Journal of the American Dental Association, 148*(9), 634–643. doi:10.1016/j.adaj.2017.04.017 PMID:28624074

Neves, M., do Amaral Giordani, J. M., Ferla, A. A., & Hugo, F. N. (2017). Primary care dentistry in Brazil: From prevention to comprehensive care. *The Journal of Ambulatory Care Management, 40*(2), 35–48. doi:10.1097/JAC.0000000000000186 PMID:28252501

ni Riain, A., Vahey, C., Kennedy, C., Campbell, S., & Collins, C. (2015). Roadmap for developing a national quality indicator set for general practice. *International Journal of Health Care Quality Assurance, 28*(4), 382-393.

Niaz, Q., Godman, B., Massele, A., Campbell, S., Kurdi, A., Kagoya, H. R., & Kibuule, D. (2019). Validity of World Health Organisation prescribing indicators in Namibia's primary healthcare: Findings and implications. *International Journal for Quality in Health Care, 31*(5), 338–345. doi:10.1093/intqhc/mzy172 PMID:30169688

Nitu, L., & Nitu, L. (2015). *A management system for sustainable development - A new challenge.* ResearchGate.

Niu, D. X., Zhao, L., Zhang, B., & Wang, H. F. (2007). The application of particle swarm optimization based grey model to power load forecasting. *Chinese Journal Of Management Science, 15*(1), 69–73.

Nobre, H., & Rodringuez, C. (2018). Loyalty programs in the pharmaceutical retail: The impact of a network loyalty program on the pharmacy loyalty. *International Journal of Business Excellence, 14*(4), 454–469. doi:10.1504/IJBEX.2018.090312

Nogales, F., Contreras, J., Conejo, A., & Espínola, R. (2002). Forecasting next-day electricity prices by time series models. *IEEE Transactions on Power Systems*, *17*(2), 342–348. doi:10.1109/TPWRS.2002.1007902

Nowotarski, J., & Weron, R. (2018). Recent advances in electricity price forecasting: A review of probabilistic forecasting. *Renewable & Sustainable Energy Reviews*, *81*, 1548–1568. doi:10.1016/j.rser.2017.05.234

Nunes, F. G., Anderson, J. E., & Martins, L. M. (2014). Patient reactions to community pharmacies' roles: Evidence from the Portuguese market. *Health Expectations*, *18*(6), 2853–2864. doi:10.1111/hex.12269 PMID:25229450

Obadan-Udoh, E. M., Calvo, J. M., Panwar, S., Simmons, K., White, J. M., Walji, M. F., & Kalenderian, E. (2019). Unintended consequences and challenges of quality measurements in dentistry. *BMC Oral Health*, *19*(1), 1–5. doi:10.118612903-019-0726-4 PMID:30823894

Ocampo, L., Alinsub, J., Casul, R. A., Enquig, G., Luar, M., Panuncillon, N., Bongo, M., & Ocampo, C. O. (2019). Public service quality evaluation with SERVQUAL and AHP-TOPSIS: A case of Philippine government agencies. *Socio-Economic Planning Sciences*, *68*, 100604. doi:10.1016/j.seps.2017.12.002

OECD. (1994). *Tourism strategies and rural development*. OECD Publication.

Office, I. O. M. (2020a). *IOM announcement on positive COVID 19 case in the temporary facility in Porto Heli*. Retrieved from https://greece.iom.int/en/news/iom-announcement-positive-covid-19-case-temporary-facility-porto-heli

Office, I. O. M. (2020b). *IOM Responding to New COVID-19 Outbreak in Greece*. Retrieved from https://www.iom.int/news/iom-responding-new-covid-19-outbreak-greece

Office, I. O. M. (2020c). *IOM Raises Concern Over Increasing COVID-19 Cases Recorded in Greece Mainland Refugee and Migrant Camp*. Retrieved from https://greece.iom.int/en/news/iom-raises-concern-over-increasing-covid-19-cases-recorded-greece-mainland-refugee-and-migrant

Oğuz, C., Salman, F. S., & Yalçın, Z. B. (2010). Order acceptance and scheduling decisions in make-to-order systems. *International Journal of Production Economics*, *125*(1), 200–211. doi:10.1016/j.ijpe.2010.02.002

Olave Quispe, S. Y., Traverso, M. T., Palchik, V., Garcıa Bermudez, E., La Casa Garcıa, C., Perez Guerrero, C., & Calero, M. (2011). Validation of a patient satisfaction questionnaire for services provided in Spanish community pharmacies. *International Journal of Clinical Pharmacy*, *33*(6), 949–957. doi:10.100711096-011-9566-2 PMID:22130784

Oliver, R. L. (1977). Effect of expectation and disconfirmation on postexposure product evaluations: An alternative interpretation. *The Journal of Applied Psychology*, *62*(4), 480–486. doi:10.1037/0021-9010.62.4.480

Oliver, R. L. (1980). A cognitive model of the antecedents and consequences of satisfaction decisions. *JMR, Journal of Marketing Research*, *17*(11), 460–469. doi:10.1177/002224378001700405

Oliver, R. L. (1997). *Satisfaction: A behavioral perspective on the customer*. McGraw-Hill.

Ong, C. S., Huang, J. J., & Tzeng, G. H. (2005). A novel hybrid model for portfolio selection. *Applied Mathematics and Computation*, *169*(2), 1195–1210. doi:10.1016/j.amc.2004.10.080

Oparah, A. C., & Kikanme, L. C. (2006). Consumer satisfaction with community pharmacies in Warri, Nigeria. *Research in Social & Administrative Pharmacy*, *2*(4), 499–511. doi:10.1016/j.sapharm.2006.02.004 PMID:17161808

Orueta, J. F., García-Alvarez, A., Grandes, G., & Nuño-Solinís, R. (2015). The origin of variation in primary care process and outcome indicators: Patients, professionals, centers, and health districts. *Medicine*, *94*(31), 1–10. doi:10.1097/MD.0000000000001314 PMID:26252315

Ouaarab, A., Ahiod, B., & Yang, X. S. (2014). Discrete cuckoo search algorithm for the travelling salesman problem. *Neural Computing & Applications*, *24*(7-8), 1659–1669. doi:10.100700521-013-1402-2

Ouhbi, S., Fernandez-Aleman, J. L., Toval, A., Pozo, J. R., & Idri, A. (2017). Sustainability requirements for connected health applications. *WILEY Software: Evolution and Process*.

Ouyang, X., Zhou, Y., Luo, Q., & Chen, H. (2013). A novel discrete cuckoo search algorithm for spherical traveling salesman problem. *Applied Mathematics & Information Sciences*, *7*(2), 777–784. doi:10.12785/amis/070248

Oxfam. (2019). *Vulnerable and abandoned.* Retrieved from https://oi-files-d8-prod.s3.eu-west-2.amazonaws.com/s3fs-public/file_attachments/2019-01_greece_media_briefing_final.pdf

Ozkok, B. A., & Pappalardo, E. (2013). Fuzzy Analytic Hierarchy Process and Its Application to E-Marketplace Selection. In Enterprise Business Modeling, Optimization Techniques, and Flexible Information Systems (pp. 26-44). IGI Global.

Ozkok, B. A., & Kocken, H. G. (2019). An Investment Decision Making Methodology Based on Type-2 Fuzzy AHP: A Case Study on BORSA ISTANBUL (BIST). *Proceedings of the International Conference on Applied Research in Management, Business and Economics*, 50-61.

Oztaysi, B. (2015). A Group Decision Making Approach Using Interval Type-2 Fuzzy AHP for Enterprise Information Systems Project Selection. *Journal of Multiple-Valued Logic & Soft Computing*, *24*(5).

Palcic, I., & Lalic, B. (2009). Analytical Hierarchy Process as a tool for selecting and evaluating projects. *International Journal of Simulation Modelling*, *8*(1), 16–26. doi:10.2507/IJSIMM08(1)2.112

Panapakidis, I., & Dagoumas, A. (2016). Day-ahead electricity price forecasting via the application of artificial neural network based models. *Applied Energy*, *172*, 132–151. doi:10.1016/j.apenergy.2016.03.089

Pan, G., Li, K., Ouyang, A., & Li, K. (2016). Hybrid immune algorithm based on greedy algorithm and delete-cross operator for solving TSP. *Soft Computing*, *20*(2), 555–566. doi:10.100700500-014-1522-3

Pansiri, J., & Mmereki, R. N. (2010). Using the SERVQUAL model to evaluate the impact of public service reforms in the provision of primary health care in Botswana. *Journal of African Business*, *11*(2), 219–234. doi:10.1080/15228916.2010.509005

Pao, H. T., Fu, H. C., & Tseng, C. L. (2012). Forecasting of co2 emissions, energy consumption and economic growth in china using an improved grey model. *Energy*, *40*(1), 400–409. doi:10.1016/j.energy.2012.01.037

Papadimitriou, C. H., & Steiglitz, K. (1982). *Combinatorial optimization* (Vol. 24). Prentice Hall.

Parasuraman, A., Zeithaml, V. A., & Berry, L. L. (1988). SERVQUAL: A multiple item scale for measuring consumer perceptions of service quality. *Journal of Retailing*, *64*(1), 14–40.

Parasuraman, A., Zeithaml, V. A., & Berry, L. L. (1991). Refinement and reassessment of the SERVQUAL scale. *Journal of Retailing*, *67*(4), 420–450.

Parra, M. A., Terol, A. B., & Urıa, M. R. (2001). A fuzzy goal programming approach to portfolio selection. *European Journal of Operational Research*, *133*(2), 287–297. doi:10.1016/S0377-2217(00)00298-8

Pastena, L. (2011). *Modernizing the Med.* World Cruise Industry Review.

Pavlič, D. R., Sever, M., Klemenc-Ketiš, Z., Švab, I., Vainieri, M., Seghieri, C., & Maksuti, A. (2018). Strength of primary care service delivery: A comparative study of European countries, Australia, New Zealand, and Canada. *Primary Health Care Research and Development*, *19*(3), 277–287. doi:10.1017/S1463423617000792 PMID:29307317

Peisley, T. (1992). *World Cruise Ship Industry in the 1990's*. Economist Intelligence Unit.

Pepermans, G. (2019). European energy market liberalization: Experiences and challenges. *International Journal of Economic Policy Studies*, *13*(1), 3–26. doi:10.100742495-018-0009-0

Peretti-Watel, P., Constance, J., Seror, V., & Beck, F. (2009). Cigarettes and social differentiation in France: Is tobacco use increasingly concentrated among the poor? *Addiction (Abingdon, England)*, *104*(10), 1718–1728. doi:10.1111/j.1360-0443.2009.02682.x PMID:19681803

Pérez-Rivas, F. J., Martín-García, Á., Sánz-Bayona, M. T., Fernández-Díaz, M. C., Barberá-Martín, A., Cárdenas-Valladolid, J., López-Palacios, S., & Rico-Blázquez, M., & Nurses at the Quality of Care Commission of the Community of Madrid, Spain. (2019). Establishing Technical Values for Nursing Diagnoses in Primary Healthcare. *International Journal of Nursing Knowledge*. PMID:31379131

Petrelis, M., Domeyer, P.R., (2016). 'HEALTH 2020' and Greece. The Contribution of Mass Pre-symptomatic Control. *To Vima tou ASKLIPIOU*, *15*(3).

Petrick, J. F., & Durko, A. (2016). Cruise tourism. In J. Jafari & H. Xiao (Eds.), *Encyclopedia of Tourism* (pp. 206–285). Springer International Publishing. doi:10.1007/978-3-319-01384-8_42

Philipoom, P. R., & Fry, T. D. (1992). Capacity-based order review/release strategies to improve manufacturing performance. *International Journal of Production Research*, *30*(11), 2559–2572. doi:10.1080/00207549208948176

Pi, D., Liu, J., & Qin, X. (2010). A grey prediction approach to forecasting energy demand in China. *Energy Sources. Part A, Recovery, Utilization, and Environmental Effects*, *32*(16), 1517–1528. doi:10.1080/15567030902780360

Pizam, A., Neumann, Y., & Reichel, A. (1978). Dimensions of tourist satisfaction with a destination area. *Annals of Tourism Research*, *5*(3), 314–322. doi:10.1016/0160-7383(78)90115-9

Popp, D., Hascic, I., & Medhi, N. (2011). Technology and The Diffusion of Renewable Energy. *Energy Economics*, *33*(4), 648–662. doi:10.1016/j.eneco.2010.08.007

Porterfield, A., Engelbert, K., & Coustasse, A. (2014). Electronic Prescribing: Improving the Efficiency and Accuracy of Prescribing in the Ambulatory Care Setting. *Perspectives in Health Information Management*, *11*(1g). PMID:24808808

Pourbabai, B. (1992). Optimal selection of orders in a just-in-time manufacturing environment: A loading model for a computer integrated manufacturing system. *International Journal of Computer Integrated Manufacturing*, *5*(1), 38–44. doi:10.1080/095119X9208547564

Prakash, S. V. (2010). Renewable Energy And Sustainable Development: An Overview. *Birla Institute OfTechnology&Science CURIE Journal*, *3*(2), 59–69.

Psimmenos, I. (2007). Work Culture and Migrant Women's Welfare Marginalization. *Epitheorese Koinonikon Ereunon*, *124*(3), 9–33. doi:10.12681/grsr.112

Psimmenos, I., & Kassimati, K. (2003). Immigration Control Pathways: Organizational Culture and Work Values of Greek Welfare Officers. *Journal of Ethnic and Migration Studies*, *29*(2), 337–371. doi:10.1080/1369183032000079639

Punnakitikashem, P., Buavaraporn, N., Maluesri, P., & Leelartapin, K. (2012, April). Health care service quality: case example of a hospital with lean implementation. In *POMS 23 rd Annual Conference* (pp. 20-23). Academic Press.

Puszka, S., Nagel, T., Matthews, V., Mosca, D., Piovesan, R., Nori, A., & Bailie, R. (2015). Monitoring and assessing the quality of care for youth: Developing an audit tool using an expert consensus approach. *International Journal of Mental Health Systems*, *9*(1), 1–12. doi:10.118613033-015-0019-5 PMID:26170899

Quercioli, C., Nisticò, F., Troiano, G., Maccari, M., Messina, G., Barducci, M., Carriero, G., Golinelli, D., & Nante, N. (2018). Developing a new predictor of health expenditure: Preliminary results from a primary healthcare setting. *Public Health, 163*, 121–127. doi:10.1016/j.puhe.2018.07.007 PMID:30142482

Raguž, I. V., Perucic, D., & Pavlic, I. (2012). Organization and Implementation of Integrated Management System Processes - Cruise Port Dubrovnik. *International Review of Management and Marketing*, *2*(4), 199–209.

Rahman, H. F., Sarker, R., & Essam, D. (2015). A real-time order acceptance and scheduling approach for permutation flow shop problems. *European Journal of Operational Research*, *247*(2), 488–503. doi:10.1016/j.ejor.2015.06.018

Ratiu, C., & Beverlee, A. B. (2015). The multiple identities of sustainability. *World Journal of Science, Technology and Sustainable Development, 2042*(5945).

Regina, A. K. (2007). *Energy Supply and Renewable Resources*. Infobase Publishing.

Reiners, F., Sturm, J., Bouw, L. J., & Wouters, E. J. (2019). Sociodemographic factors influencing the use of eHealth in people with chronic diseases. *International Journal of Environmental Research and Public Health*, *16*(4), 645. doi:10.3390/ijerph16040645 PMID:30795623

Reinert, C. D., & Murray, J. A. (2017). Successful Implementation of Pharmacy Retail Store Loyalty Reward Programs. *International Journal of Applied Management and Technology*, *16*(1), 152–165. doi:10.5590/IJAMT.2017.16.1.10

Reis, V., Meier, J., Pace, G., & Palacin, R. (2013). Rail and multi-modal transport. *Research in Transportation Economics*, *41*(1), 17–30. doi:10.1016/j.retrec.2012.10.005

REN21. (2019). *Renewables 2019 Global Status Report*. Paris: REN21 Renewables Now.

Renewable Energy World. (2020). *Ocean Energy Tech*. www.renewableenergyworld.com: https://www.renewableenergyworld.com/types-of-renewable-energy/ocean-energy-tech/#gref

Renggli, S., Mayumana, I., Mboya, D., Charles, C., Mshana, C., Kessy, F., Glass, T. R., Pfeiffer, C., Schulze, A., Aerts, A., & Lengeler, C. (2019). Towards improved health service quality in Tanzania: Appropriateness of an electronic tool to assess quality of primary healthcare. *BMC Health Services Research*, *19*(1), 1–16. PMID:30670011

Research Centre for Coastal Tourism. (2012). *Cruise Tourism: from a Broad Perspective to a Focus on Zeeland*. Retrieved from http://www.kenniscentrumtoerisme.nl/l/library/download/13920

Retail Independent Pharmacy. (2018). *Good medicine*. Cardinal Health. Retrieved from: https://www.readkong.com/page/good-medicine-5967197?p=1

Rezaei, J., Kothadiya, O., Tavasszy, L., & Kroesen, M. (2018). Quality assessment of airline baggage handling systems using SERVQUAL and BWM. *Tourism Management*, *66*, 85–93. doi:10.1016/j.tourman.2017.11.009

Righolt, A. J., Sidorenkov, G., Faggion, C. M. Jr, Listl, S., & Duijster, D. (2019). Quality measures for dental care: A systematic review. *Community Dentistry and Oral Epidemiology*, *47*(1), 12–23. doi:10.1111/cdoe.12429 PMID:30375669

Rizet, C., Cruz, C., & Mbacke, M. (2012). Reducing freight transport CO_2 emissions by increasing the load factor. *Procedia: Social and Behavioral Sciences*, *48*, 184–195. doi:10.1016/j.sbspro.2012.06.999

Robinson, J. (1949). On the Hamiltonian game (a traveling salesman problem). RAND Project (No. RAND/RM-303).

Rodrigue, J. P., & Notteboom, T. (2012). The geography of cruises: Itineraries, not destinations. *Applied Geograghy*, *38*, 31–42. doi:10.1016/j.apgeog.2012.11.011

Rodrigue, J.-P., Comtois, C., & Slack, B. (2013). *The geography of transport systems* (3rd ed.). Routledge. doi:10.4324/9780203371183

Roig-Tierno, N., Baviera-Puig, A., Buitrago-Vera, J., & Mas-Verdu, F. (2013). The retail site location decision process using GIS and the analytical hierarchy process. *Applied Geography (Sevenoaks, England)*, *40*, 191–198. doi:10.1016/j.apgeog.2013.03.005

Rom, W. O., & Slotnick, S. A. (2009). Order acceptance using genetic algorithms. *Computers & Operations Research*, *36*(6), 1758–1767. doi:10.1016/j.cor.2008.04.010

Roses, L. K., Hoppen, N., & Henrique, J. L. (2009). Management of perceptions of information technology service quality. *Journal of Business Research*, *62*(9), 876–882. doi:10.1016/j.jbusres.2008.10.005

Rosillo-Calle, F. (2016). A Review of Biomass Energy – Shortcomings and Concerns. *Journal of Chemical Technology and Biotechnology (Oxford, Oxfordshire)*, *91*(7), 1933–1945. doi:10.1002/jctb.4918

Roslan, N. A. A., Wahab, E., & Abdullah, N. H. (2015). Service Quality: A case study of logistics sector in Iskandar Malaysia using SERVQUAL Model. *Procedia: Social and Behavioral Sciences*, *172*, 457–462. doi:10.1016/j.sbspro.2015.01.380

Ruoning, X., & Xiaoyan, Z. (1992). Extensions of the analytic hierarchy process in fuzzy environment. *Fuzzy Sets and Systems*, *52*(3), 251–257. doi:10.1016/0165-0114(92)90236-W

Ruoning, X., & Xiaoyan, Z. (1996). Fuzzy logarithmic least squares ranking method in analytic hierarchy process. *Fuzzy Sets and Systems*, *77*(2), 175–190. doi:10.1016/0165-0114(95)00073-9

Rutkowski, L. (2008). *Computational intelligence: Methods and techniques*. Springer-Verlag. doi:10.1007/978-3-540-76288-1

Saab, S., Badr, E., & Nasr, G. (2001). Univariate modeling and forecasting of energy consumption: The case of electricity ın Lebanon. *Energy*, *26*(1), 1–14. doi:10.1016/S0360-5442(00)00049-9

Saaty, R. W. (1987). The analytic hierarchy process—What it is and how it is used. *Mathematical Modelling*, *9*(3-5), 161–176. doi:10.1016/0270-0255(87)90473-8

Saaty, T. L. (1980). The Analytic Hierarchy Process McGraw-Hill New York. *Agricultural Economics Review*, 70.

Sadorsky, P. (2009). Renewable Energy Consumption and Income in Emerging Economies. *Energy Policy*, *37*(10), 4021–4028. doi:10.1016/j.enpol.2009.05.003

Sağlam, Ü. (2018). *The Efficiency Assessment of Renewable Energy Sources with Data Envelopment Analysis. Assessment of Renewable Energy Sources.*

Sagnak, M. (2020). Selecting the Most Appropriate Supplier in the Green Environment: A Hybrid Multi-Criteria Decision-Making Application. In Handbook of Research on Industrial Applications for Improved Supply Chain Performance (pp. 101-124). IGI Global. doi:10.4018/978-1-7998-0202-0.ch005

Sahin, Y., & Karagul, K. (2019). Solving travelling salesman problem using hybrid fluid genetic algorithm (HFGA). *Pamukkale University Journal of Engineering Sciences*, *25*(1), 106–114. doi:10.5505/pajes.2018.81084

Saini, L. (2008). Peak load forecasting using Bayesian regularization: Resilient and adaptive backpropagation learning based artificial neural networks. *Electric Power Systems Research*, *78*(7), 1302–1310. doi:10.1016/j.epsr.2007.11.003

Saleh, S., Alameddine, M., Mourad, Y., & Natafgi, N. (2015). Quality of care in primary health care settings in the Eastern Mediterranean region: A systematic review of the literature. *International Journal for Quality in Health Care*, *27*(2), 79–88. PMID:25574040

Samartzis, L., & Talias, M. A. (2020). Assessing and improving the quality in mental health services. *International Journal of Environmental Research and Public Health*, *17*(1), 1–31. PMID:31905840

Sam, E. F., Hamidu, O., & Daniels, S. (2018). SERVQUAL analysis of public bus transport services in Kumasi metropolis, Ghana: Core user perspectives. *Case Studies on Transport Policy*, *6*(1), 25–31. doi:10.1016/j.cstp.2017.12.004

San Cristobal, J. R. (2011). A Multi Criteria Data Envelopment Analysis Model to Evaluate the Efficiency of The Renewable Energy Technologies. *Renewable Energy*, *36*(10), 2742–2746. doi:10.1016/j.renene.2011.03.008

Sanchez, J., Callarisa, L., Rodriguez, R., & Moliner, M. (2006). Perceived value of the purchase of a tourism product. *Tourism Management*, *27*(3), 394–409. doi:10.1016/j.tourman.2004.11.007

Sandberg, B., & Aarikka-Stenroos, L. (2014). What makes it so difficult? A systematic review on barriers to radical innovation. *Industrial Marketing Management*, *43*(8), 1293–1305. doi:10.1016/j.indmarman.2014.08.003

Sayareh, J., Iranshahi, S., & Golfakhrabadi, N. (2016). Service quality evaluation and ranking of container terminal operators. *The Asian Journal of Shipping and Logistics*, *32*(4), 203–212. doi:10.1016/j.ajsl.2016.12.003

Scambler, S., Delgado, M., & Asimakopoulou, K. (2016). Defining patient-centred care in dentistry? A systematic review of the dental literature. *British Dental Journal*, *221*(8), 477–484. doi:10.1038j.bdj.2016.777 PMID:27767160

Schalkoff, R. J. (1997). *Artificial neural networks*. McGraw-Hill.

Schiff, G., Mirica, M. M., Dhavle, A. A., Galanter, W. L., Lambert, B., & Wright, A. (2018). A Prescription For Enhancing Electronic Prescribing Safety. *Quality of Care*, *11*, 1877–1883. PMID:30395495

Schröder, P. J. H. (2004). The Comparison between Ad Valorem and Unit Taxes under Monopolistic Competition. *Journal of Economics*, *83*(3), 281–292. doi:10.100700712-004-0096-y

Schumacker, R. E., & Lomax, R. G. (2004). *A beginner's guide to structural equation modeling* (2nd ed.). Lawrence Erlbaum Associates Publishers. doi:10.4324/9781410610904

Seidl, A., Guiliano, F., & Pratt, L. (2006). Cruise Tourism and Economic Developmet in Central America and the Caribbean: The case of Costa Rica. *Revista de Turismo y Patrimonio Cultural*, *4*(2), 213–224. doi:10.25145/j.pasos.2006.04.015

Şenel, M. C., & Koç, E. (2015). Dünyada ve Türkiye'de Rüzgar Enerjisi Durumu- Genel Değerlendirme. *Mühendis ve Makina*, *56*(663), 46–56.

Shafie-khah, M., Parsa Moghaddam, M., & Sheikh-El-Eslami, M. K. (2010). Price forecasting of day-ahead electricity markets using a hybrid forecast method. *Energy Conversion and Management*, *52*(5), 2165–2169. doi:10.1016/j.enconman.2010.10.047

Shaikh, B. T., Mobeen, N., Azam, S. I., & Rabbani, F. (2008). *Using SERVQUAL for assessing and improving patient satisfaction at a rural health facility in Pakistan*. Academic Press.

Shaw, K., Shankar, R., Yadav, S. S., & Thakur, L. S. (2012). Supplier selection using fuzzy AHP and fuzzy multi-objective linear programming for developing low carbon supply chain. *Expert Systems with Applications*, *39*(9), 8182–8192. doi:10.1016/j.eswa.2012.01.149

Shimpi, N., Schwei, K., Cooper, S., Chyou, P. H., & Acharya, A. (2018). Understanding patients' oral health information needs: Findings of a survey on use of patient portals in dentistry. *The Journal of the American Dental Association*, *149*(3), 184–190. doi:10.1016/j.adaj.2017.09.051 PMID:29395012

Shin, K., & Cheong, J. (2011). Estimating transportation-related greenhouse gas emissions in the port of Busan, S. Korea. *Asian Journal of Atmospheric Environment*, *5*(1), 41–46. doi:10.5572/ajae.2011.5.1.041

Shuval, K., Leonard, T., Drope, J., Katz, D. L., Patel, A. V., Maitin-Shepard, M., ... Grinstein, A. (2017). Physical activity counseling in primary care: Insights from public health and behavioral economics. *CA: a Cancer Journal for Clinicians, 67*(3), 233–244.

Simou, E., Pliatsika, P., Koutsogeorgou, E., & Roumeliotou, A. (2015). Quality indicators for primary health care: A systematic literature review. *Journal of Public Health Management and Practice, 21*(5), 8–16. doi:10.1097/PHH.0000000000000037 PMID:24374361

Sinclair, T. M., & Stabler, M. (1997). *The economics of tourism*. Routledge. doi:10.4324/9780203195437

Sindakis, S., & Kitsios, F. (2016). Entrepreneurial Dynamics and Patient Involvement in Service Innovation: Developing a Model to Promote Growth and Sustainability in Mental Health Care. *Journal of the Knowledge Economy, 7*(2), 545–564. doi:10.100713132-014-0228-1

Singh, A., & Prasher, A. (2019). Measuring healthcare service quality from patients' perspective: Using Fuzzy AHP application. *Total Quality Management & Business Excellence, 30*(3-4), 284–300. doi:10.1080/14783363.2017.1302794

Siskos, Y., Grigoroudis, E., Zopounidis, C., & Saurais, O. (1998). Measuring customer satisfaction using a collective preference disaggregation model. *Journal of Global Optimization, 12*(2), 175–195. doi:10.1023/A:1008262411587

Skeath, S. E., & Trandel, G. E. (1994). A Pareto comparison of ad valorem and specific taxation in noncompetitive environments. *Journal of Public Economics, 53*(1), 53–71. doi:10.1016/0047-2727(94)90013-2

Slotnick, S. A. (2011). Order acceptance and scheduling: A taxonomy and review. *European Journal of Operational Research, 212*(1), 1–11. doi:10.1016/j.ejor.2010.09.042

Slotnick, S. A., & Morton, T. E. (1996). Selecting jobs for a heavily loaded shop with lateness penalties. *Computers & Operations Research, 23*(2), 131–140. doi:10.1016/0305-0548(95)00015-E

Slotnick, S. A., & Morton, T. E. (2007). Order acceptance with weighted tardiness. *Computers & Operations Research, 34*(10), 3029–3042. doi:10.1016/j.cor.2005.11.012

Soita, P. W. (2012). Measuring perceived service quality using SERVQUAL: A case study of the Uganda health and fitness sector. *International Journal of Business and Social Science, 3*(5).

Solanki, R., Darbari, J. D., Agarwal, V., & Jha, P. C. (2020). A Fuzzy Multi-criteria Decision Model for Analysis of Socio-ecological Performance Key Factors of Supply Chain. In *Soft Computing for Problem Solving* (pp. 671–685). Springer. doi:10.1007/978-981-15-0035-0_55

Song, W., Zhu, Y., & Zhao, Q. (2020). Analyzing barriers for adopting sustainable online consumption: A rough hierarchical DEMATEL method. *Computers & Industrial Engineering, 140*, 106279. doi:10.1016/j.cie.2020.106279

Souri, M. E., Sajjadian, F., Sheikh, R., & Sana, S. S. (2018). Grey SERVQUAL method to measure consumers' attitudes towards green products-A case study of Iranian consumers of LED bulbs. *Journal of Cleaner Production, 177*, 187–196. doi:10.1016/j.jclepro.2017.12.105

Sridharan, S. V. (1998). Managing capacity in tightly constrained systems. *International Journal of Production Economics, 56*, 601–610. doi:10.1016/S0925-5273(98)00045-0

Srikatanyoo, N., & Campiranon, K. (2010). Agritourist needs and motivations: The Chiang Mai case. *Journal of Travel & Tourism Marketing, 27*(2), 166–178. doi:10.1080/10548400903579795

Stam, A., Sun, M., & Haines, M. (1996). Artificial neural network representations for hierarchical preference structures. *Computers & Operations Research, 23*(12), 1191–1201. doi:10.1016/S0305-0548(96)00021-4

Statista. (2020). *Surface Area of Oceans Worldwide*. https://www.statista.com/: https://www.statista.com/statistics/558499/total-area-of-global-oceans/

Stefanakakis, S., Trevlakis, S., & Kitsios, F. (2017). Analyzing quality characteristics: The case of an e – health application. In *Proceedings of the 6th International Symposium and 28th National Conference on Operational Research* (pp. 246-251). Thessaloniki, Greece: Academic Press.

Stefanidaki, E., & Lekakou, M. (2012). Generated Economic Impact on Cruise Destinations: The Piraeus Case. In *Cruise Tourism and Society* (pp. 69–84). Springer-Verlag.

Stiitzle, T., & Hoos, H. (1997). The MAX-MIN ant system and local search for the traveling salesman problem. In *Proceedings of IEEE international conference on evolutionary computation* (pp. 309-314). Indianapolis, IN: IEEE. 10.1109/ICEC.1997.592327

Strąk, Ł., Skinderowicz, R., & Boryczka, U. (2018). Adjustability of a discrete particle swarm optimization for the dynamic TSP. *Soft Computing, 22*(22), 7633–7648. doi:10.100700500-017-2738-9

Stuurman-Bieze, A.G., Kokenberg, M.E., Tobi, H., de Boer W.O., van Doormaal, J.E., & Jong-van den de Berg, L.T. (2005). Complex pharmaceutical care intervention in pulmonary care: part B patient opinion and process survey. *Pharmacy World and Science, 27*(5), 385-392.

Styliadis, T. (2012). *Greek Sea is not in Crisis. An institutional approach of the economic impact cruise ships have on local societies: the case of two Ionian Islands* (MSc Thesis). Erasmus University, Rotterdam.

Suganthi, L., & Samuel, A. A. (2012). Energy Models for Demand Forecasting-A Review. *Renewable & Sustainable Energy Reviews, 16*(2), 1223–1240. doi:10.1016/j.rser.2011.08.014

Sun, C. C. (2010). A performance evaluation model by integrating fuzzy AHP and fuzzy TOPSIS methods. *Expert Systems with Applications, 37*(12), 7745–7754. doi:10.1016/j.eswa.2010.04.066

Sun, W., & Yan, Y. (2008). The model based on grey theory and pso for electricity consumption forecasting. *International Conference On Intelligent Computation Technology And Automation (Icıcta)*, 152-156. 10.1109/ICICTA.2008.40

Sureja, N. (2012). New inspirations in nature: A survey. *International Journal of Computer Applications & Information Technology, 1*(3), 21–24.

Swider, D., & Weber, C. (2007). Extended ARMA models for estimating price developments on day-ahead electricity markets. *Electric Power Systems Research, 77*(5-6), 583–593. doi:10.1016/j.epsr.2006.05.013

Syngelakis, A. I., & Kitsios, F. (2017). Evaluating occupational stress and job satisfaction in health sector during the austerity: the case of Greek NHS dentists. In *Proceedings of the 6th International Symposium and 28th National Conference on Operational Research* (pp. 252-256). Thessaloniki, Greece: Academic Press.

Syngelakis, A. I., Koletsi-Kounari, H., Polychronopoulou, A., Panagiotakos, D. B., & Lionis, C. (2013). Dentistry in Primary Health Care Centres: Attendance and access of users, professional satisfaction and security of dentists. *Stomatologia, 70*(3), 113–123.

Syngelakis, A., Kitsios, F., & Kamariotou, M. (2019). Work-related Stress in Public Healthcare: A Case Study in Greece during economic crisis. In D. P. Sakas & D. K. Nasiopoulos (Eds.), *Strategic Innovative Marketing* (pp. 211–216). Springer. doi:10.1007/978-3-030-16099-9_26

Syngelakis, A., Koletsi-Kounari, H., Polychronopoulou, A., Panagiotakos, D. B., & Lionis, C. (2012). Evaluation of the staffing and the organization of the Greek NHS Primary Health Care Centres Dental Departments. *Hellenic Hospital Dentistry, 5*, 11–20.

Szilágyi, T., & Chapman, S. (2004). Tobacco industry efforts to keep cigarettes affordable: A case study from Hungary. *Central European Journal of Public Health, 11*(4), 223–228. PMID:14768787

Tanaka, H., & Guo, P. (1999). Portfolio selection based on upper and lower exponential possibility distributions. *European Journal of Operational Research, 114*(1), 115–126. doi:10.1016/S0377-2217(98)00033-2

Tanaka, H., Guo, P., & Türksen, I. B. (2000). Portfolio selection based on fuzzy probabilities and possibility distributions. *Fuzzy Sets and Systems, 111*(3), 387–397. doi:10.1016/S0165-0114(98)00041-4

Taner, T., & Antony, J. (2006). Comparing public and private hospital care service quality in Turkey. *Leadership in Health Services, 19*(2), 1–10. doi:10.1108/13660750610664991 PMID:16875104

Taylor, J. W. (2003). Short-term electricity demand forecasting using double seasonal exponential smoothing. *The Journal of the Operational Research Society, 54*(8), 799–805. doi:10.1057/palgrave.jors.2601589

ten Kate, H. A. (1994). Towards a better understanding of order acceptance. *International Journal of Production Economics, 37*(1), 139–152. doi:10.1016/0925-5273(94)90014-0

TETC. (2019). *Türkiye Elektrik Üretim-İletim İstatistikleri.* www.teias.gov.tr: https://www.teias.gov.tr/tr-TR/turkiye-elektrik-uretim-iletim-istatistikleri

Tian-Cole, S., & Crompton, J. (2002). A conceptualization of the relationships between service quality and visitor satisfaction, and their links to destination selection. *Leisure Studies, 22*(1), 65–80. doi:10.1080/02614360306572

Tickle, M., & Campbell, S. (2013). How do we measure quality in primary dental care? *British Dental Journal, 215*(4), 183–187. doi:10.1038j.bdj.2013.789 PMID:23969661

Tickle, M., O'malley, L., Brocklehurst, P., Glenny, A. M., Walsh, T., & Campbell, S. (2015). A national survey of the public's views on quality in dental care. *British Dental Journal, 219*(3), 1–7. doi:10.1038j.bdj.2015.595 PMID:26271885

Tinanoff, N. (2012). Potential to improve oral health care through evidence, protocols, and payment models. *Journal of Public Health Dentistry, 72*, S48–S51. doi:10.1111/j.1752-7325.2012.00325.x PMID:22433101

Tinelli, M., Bond, C., Blenkinsopp, A., Jaffray, M., Watson, M., & Hannaford, P. (2007). Patient evaluation of a community pharmacy medications management service. *The Annals of Pharmacotherapy, 41*(12), 1962–1970. doi:10.1345/aph.1K242 PMID:17971403

Tiryaki, F., & Ahlatcioglu, B. (2009). Fuzzy portfolio selection using fuzzy analytic hierarchy process. *Information Sciences, 179*(1-2), 53–69. doi:10.1016/j.ins.2008.07.023

Tiryaki, F., & Ahlatcioglu, M. (2005). Fuzzy stock selection using a new fuzzy ranking and weighting algorithm. *Applied Mathematics and Computation, 170*(1), 144–157. doi:10.1016/j.amc.2004.10.092

Tiwari, G. N., & Mishra, R. K. (2012). *Advanced Renewable Energy Sources.* The Royal Society of Chemistry.

To Vima. (2019). *Vroutsis blocks simplified AMKA issuance for non-EU nationals.* Retrieved from https://www.tovima.gr/2019/07/12/international/vroutsis-blocks-simplified-amka-issuance-for-non-eu-nationals/

To Vima. (2020). *Civil protection authority: Kranidi quarantined after 150 migrants test positive for COVID-19.* Retrieved from https://www.tovima.gr/2020/04/21/english-edition/civil-protection-underminister-kranidi-quarantined-after-150-migrants-test-positive-for-covid-19/

Top, M., Sönmez, S., & Aslan, H. (2019). An Evidence-Based Approach to Outcome Measurement in Oral and Dental Health Services: Oral Health-Related Quality of Life and Oral Health Impact. *Worldviews on Evidence-Based Nursing, 16*(5), 408–415. doi:10.1111/wvn.12398 PMID:31544360

Torbianelli, J. (2010). *Mediterranean Cruises as Drivers of Investment and Relations between Ports and Cities.* University of Trieste.

Torppa-Saarinen, E., Suominen, A. L., Lahti, S., & Tolvanen, M. (2019). Longitudinal pathways between perceived oral health and regular service use of adult Finns. *Community Dentistry and Oral Epidemiology, 47*(5), 374–380. doi:10.1111/cdoe.12478 PMID:31179572

Torppa-Saarinen, E., Tolvanen, M., Suominen, A. L., & Lahti, S. (2018). Changes in perceived oral health in a longitudinal population-based study. *Community Dentistry and Oral Epidemiology, 46*(6), 569–575. doi:10.1111/cdoe.12393 PMID:29956841

Traverso, M. L., Salamano, M., Botta, C., Colautti, M., Palchik, V., & Perez, B. (2011). Questionnaire to assess patient satisfaction with pharmaceutical care in Spanish language. *International Journal for Quality in Health Care, 19*(4), 217–224. doi:10.1093/intqhc/mzm014 PMID:17545673

Tribe, J., & Snaith, T. (1998). From SERVQUAL to HOLSAT: Holiday satisfaction in Varadero, Cuba. *Tourism Management, 19*(1), 25–34. doi:10.1016/S0261-5177(97)00094-0

Truong, T. H., & Foster, D. (2006). Using HOLSAT to evaluate tourist satisfaction at destinations: The case of Australian holidaymakers in Vietnam. *Tourism Management, 27*(5), 842–855. doi:10.1016/j.tourman.2005.05.008

Tsai, H. K., Yang, J. M., Tsai, Y. F., & Kao, C. Y. (2004). Some issues of designing genetic algorithms for traveling salesman problems. *Soft Computing, 8*(10), 689–697. doi:10.100700500-003-0317-8

Tseng, F. M., Yu, H. C., & Tzeng, G. H. (2001). Applied hybrid grey model to forecast seasonal time series. *Technological Forecasting and Social Change, 67*(2), 291–302. doi:10.1016/S0040-1625(99)00098-0

Tsitsiloni, M., Grigoroudis, E., & Zopounidis, C. (2012). A service quality evaluation method in the tourism industry: A multicriteria analysis. *Journal of Computational Optimization in Economics and Finance, 3*(3), 185–202.

Uğur, A., Korukoğlu, S., Çalıskan, A., Cinsdikici, M., & Alp, A. (2009). Genetic algorithm based solution for TSP on a sphere. *Mathematical and Computational Applications, 14*(3), 219-228.

UNHCR. (2020). *Most common nationalities of sea arrivals (since 1 January 2020).* Retrieved from https://data2.unhcr.org/en/situations/mediterranean/location/5179

United Nations(1987*Our Common Future - Report of the World Commission on Environment and Development.* United Nations.

UNWTO. (2016). *Yearbook of tourism Statistics.* United Nations World Tourism Organization.

UNWTO. (2017). *UNWTO panel on indigenous Tourism: Promoting equitable partnerships.* Berlin: United Nations World Tourism Organization. Retrieved from: http://ethics.unwto.org/event/unwto-panel-indigenous-tourism-promotingequitable-partnerships

Vaggelas, G. (2011). Cruise Tourism: Economic Benefits, Sustainable Development and Port City. Plan Bleu Regional Seminar – Tourism and sustainable development in the Mediterranean, Genoa, Italy.

Vaidya, O. S., & Kumar, S. (2006). Analytic hierarchy process: An overview of applications. *European Journal of Operational Research, 169*(1), 1–29. doi:10.1016/j.ejor.2004.04.028

Van Bets, L. K., Lamers, M. A., & Van Tatenhove, J. P. (2017). Collective self-governance in a marine community: Expedition cruise tourism at Svalbard. *Journal of Sustainable Tourism, 25*(11), 1583–1599. doi:10.1080/09669582.2017.1291653

Van Laarhoven, P. J., & Pedrycz, W. (1983). A fuzzy extension of Saaty's priority theory. *Fuzzy Sets and Systems*, *11*(1-3), 229–241. doi:10.1016/S0165-0114(83)80082-7

Vardavas, C. L., & Kafatos, A. G. (2006). Smoking policy and prevalence in Greece: An overview. *European Journal of Public Health*, *17*(2), 211–213. doi:10.1093/eurpub/ckl094 PMID:17065175

Vavra, T. G. (1997). *Improving your measurement of customer satisfaction: A guide to creating, conducting, analyzing, and reporting customer satisfaction measurement programs*. ASQC Quality Press.

Vayá, E., Garcia-Sanchís, J. R., Murillo, J., Romaní, J., & Suriñach, J. (2016). *Economic Impact of Cruise Activity: the Port of Barcelona*. Retrieved from https://poseidon01.ssrn.com/delivery.php?ID=859095117066064115106126113118101 112023069025 0840900 220290 80122123075072073 02506508504504 300701702504302 607010511306807101900 30460020270350000960041130060801221160 97068050008064 10700700208700 210200212 5102087 0950 96 11906410701608107111 2111082008082125120&EXT=pdf

Veizis, A. (2020). "Leave No One Behind" and Access to Protection in the Greek Islands in the COVID-19 Era. *International Migration (Geneva, Switzerland)*, *58*(3), 264–266. Advance online publication. doi:10.1111/imig.12721 PMID:32518422

Villako, P., & Raal, A. (2007). A survey of Estonian consumer expectations from the the pharmacy service and a comparison with the opinions of pharmacists. *Pharmacy World & Science*, *29*(5), 546–550. doi:10.100711096-007-9102-6 PMID:17497224

Vina, L., & Ford, J. (1998). Economic impact of proposed cruise ship business. *Annals of Tourism Research*, *25*, 205–221.

Vogel, M. (2009). Onboard revenue: The secret of the cruise industry's success? In A. Papathanassis (Ed.), *Cruise sector growth: Managing emerging markets, human resources, processes and systems* (pp. 3–15). Gabler. doi:10.1007/978-3-8349-8346-6_1

Voinea-Griffin, A., Fellows, J. L., Rindal, D. B., Barasch, A., Gilbert, G. H., & Safford, M. M. (2010a). Pay for performance: Will dentistry follow? *BMC Oral Health*, *10*(1), 1–8. doi:10.1186/1472-6831-10-9 PMID:20423526

Voinea-Griffin, A., Rindal, D. B., Fellows, J. L., Barasch, A., Gilbert, G. H., & Safford, M. M.DPBRN Collaborative Group. (2010b). Pay-for-performance in dentistry: What we know. *Journal for Healthcare Quality*, *32*(1), 51–58. doi:10.1111/j.1945-1474.2009.00064.x PMID:20151592

Volume, C. I., Farris, K. B., Kassam, R., Cox, C. E., & Cave, A. (2001). Pharmaceutical care research and education project: Patient outcomes. *Journal of the American Pharmaceutical Association*, *4*(3), 411–420. doi:10.1016/S1086-5802(16)31255-4 PMID:11372906

Vozikis, A., Stavropoulou, L., & Patrinos, G. (2015). Community Pharmacists' Strategies in Greece: An Assessment of the Policy Environment and the Mapping of Key Players. *Health*, *7*(11), 1560–1577. doi:10.4236/health.2015.711170

Wang, J., Du, P., Lu, H., Yang, W., & Niu, T. (2018). An improved grey model optimized by multi-objective ant lion optimization algorithm for annual electricity consumption forecasting. *Applied Soft Computing*, *72*, 321–337. doi:10.1016/j.asoc.2018.07.022

Wang, J., & Lin, Y. I. (2003). A fuzzy multicriteria group decision making approach to select configuration items for software development. *Fuzzy Sets and Systems*, *134*(3), 343–363. doi:10.1016/S0165-0114(02)00283-X

Wang, J., Yang, J. Q., & Lee, H. (1994). Multicriteria order acceptance decision support in over-demanded job shops: A neural network approach. *Mathematical and Computer Modelling*, *19*(5), 1–19. doi:10.1016/0895-7177(94)90086-8

Wang, K. P., Huang, L., Zhou, C. G., & Pang, W. (2003). Particle swarm optimization for traveling salesman problem. In *Proceedings of the 2003 international conference on machine learning and cybernetics* (Vol. 3, pp. 1583-1585). Xi'an, China: IEEE. 10.1109/ICMLC.2003.1259748

Wang, X., Kerre, E. E., & Ruan, D. (1995). Consistency of judgement matrix and fuzzy weights in fuzzy analytic hierarchy process. *International Journal of Uncertainty, Fuzziness and Knowledge-based Systems*, *3*(01), 35–46. doi:10.1142/S0218488595000050

Wang, Y., Tian, D., & Li, Y. (2013). An improved simulated annealing algorithm for traveling salesman problem. In *Proceedings of the 2012 International Conference on Information Technology and Software Engineering* (pp. 525-532). Springer. 10.1007/978-3-642-34522-7_56

Wang, Z. X., Li, Q., & Pei, L. L. (2018). A seasonal gm (1, 1) model for forecasting the electricity consumption of the primary economic sectors. *Energy*, *154*, 522–534. doi:10.1016/j.energy.2018.04.155

Wang, Z., Qi, Y., Cui, H., & Zhang, J. (2019). A hybrid algorithm for order acceptance and scheduling problem in make-to-stock/make-to-order industries. *Computers & Industrial Engineering*, *127*, 841–852. doi:10.1016/j.cie.2018.11.021

Wanyonyi, K. L., Radford, D. R., & Gallagher, J. E. (2019). Electronic primary dental care records in research: A case study of validation and quality assurance strategies. *International Journal of Medical Informatics*, *127*, 88–94. doi:10.1016/j.ijmedinf.2019.04.007 PMID:31128836

Waraich, P., Saklikar, R. S., Aubé, D., Jones, W., Haslam, D., & Hamill, K. (2010). Quality measures for primary mental healthcare: A multistakeholder, multijurisdictional Canadian consensus. *BMJ Quality & Safety*, *19*(6), 519–525. doi:10.1136/qshc.2008.027839 PMID:20702436

Weaver, A. (2005). The McDonaldization thesis and cruise shipping. *Annals of Tourism Research, 32*(2), 346-366.

Webster, J., & Watson, R. T. (2002). Analyzing the past to prepare for the future: Writing a literature review. *Management Information Systems Quarterly*, *26*, 13–23.

Weck, M., Klocke, F., Schell, H., & Rüenauver, E. (1997). Evaluating alternative production cycles using the extended fuzzy AHP method. *European Journal of Operational Research*, *100*(2), 351–366. doi:10.1016/S0377-2217(96)00295-0

Weeden, C., Lester, J. A., & Thyne, M. (2011). Cruise tourism: Emerging issues and implications for a maturing industry. *Journal of Hospitality and Tourism Management*, *18*(1), 26–29. doi:10.1375/jhtm.18.1.26

Weron, R. (2014). Electricity price forecasting: A review of the state-of-the-art with a look into the future. *International Journal of Forecasting*, *30*(4), 1030–1081. doi:10.1016/j.ijforecast.2014.08.008

Wester, F. A. W., Wijngaard, J., & Zijm, W. H. M. (1992). Order acceptance strategies in a production-to-order environment with setup times and due-dates. *International Journal of Production Research*, *30*(6), 1313–1326. doi:10.1080/00207549208942959

Wetzstein, A., Hartmanna, E., Benton, W. C. Jr, & Hohenstein, N.-O. (2016). A systematic assessment of supplier selection literature – State-of-art and future scope. *International Journal of Production Economics*, *182*(C), 304–323. doi:10.1016/j.ijpe.2016.06.022

Whitehead, P., Atkin, P., Krass, I., & Benrimoj, S. (1999). Patient drug information and consumer choice of pharmacy. *International Journal of Pharmacy Practice*, *7*(2), 71–79. doi:10.1111/j.2042-7174.1999.tb00952.x

Whybark, D. C. (1994). Marketing's influence on manufacturing practices. *International Journal of Production Economics*, *37*(1), 41–50. doi:10.1016/0925-5273(94)90006-X

Wickramasingle, N., Bali, R. K., Kirn, S., & Suomi, R. (2012). *Critical Issues For The Development of Sustainable E-health Solutions*. Springer. doi:10.1007/978-1-4614-1536-7

Wicks, B. E., & Merrett, C. D. (2003). Agritourism: An economic opportunity for Illinois. *Rural Research Report*, *14*(9), 1–8.

Wiedenmayer, K., Summers, R. S., Mackie, C., Gous, A., Everard, M., & Tromp, D. (2006). *Developing pharmacy practice: A focus on patient care*. World Health Organization Department of Medicines Policy and Standards.

Wiederholt, J. B. (1987). Development of an Instrument to Measure Evaluative Criteria that Patients Use in Selecting a Pharmacy for Obtaining Prescription Drugs. *Journal of Pharmaceutical Marketing & Management*, *1*(4), 35–59. doi:10.3109/J058v01n04_04

Wiegmans, B., & Konings, R. (2007). Strategies and innovations to improve the performance of barge transport. *European Journal of Transport and Infrastructure Research*, *7*(2), 145–162.

Wild, G. P. (International) Ltd., B. (2012). The Cruise Industry: Contribution of Cruise Tourism to the Economies of Europe. European Cruise Council.

Wild, P., & Dearing, J. (2000). Development of and prospects for cruising in Europe. *Maritime Policy & Management*, *27*(4), 315–337. doi:10.1080/030888300416522

Wilkinson, P. (1999). Caribbean cruise tourism: Delusion? Illusion? *Tourism Geographies*, *1*(3), 261–282. doi:10.1080/14616689908721321

Wilkinson, R. G., & Marmot, M. (Eds.). (2003). *Social determinants of health: the solid facts*. World Health Organization.

Winfried, S. (2008). *Securing the product in the container, Container Handbook*. GDV.

Wirth, F., Tabone, F., Azzopardi, L. M., Gauci, M., Zarb-Adami, M., & Serracino-Inglott, A. (2011). Consumer perception of the community pharmacist and community pharmacy services in Malta. *Journal of Pharmaceutical Health Services Research*, *1*(4), 189–194. doi:10.1111/j.1759-8893.2010.00034.x

Wong, S. T., Browne, A. J., Varcoe, C., Lavoie, J., Smye, V., Godwin, O., Littlejohn, D., & Tu, D. (2011). Enhancing measurement of primary health care indicators using an equity lens: An ethnographic study. *International Journal for Equity in Health*, *10*(1), 1–12. doi:10.1186/1475-9276-10-38 PMID:21892956

Wood, R. E. (2000). Caribbean cruise tourism: Globalization at sea. *Annals of Tourism Research*, *27*(2), 345–370. doi:10.1016/S0160-7383(99)00073-0

Woodruff, R. B., & Gardial, S. F. (1996). *Know your customer: New approaches to understanding customer value and satisfaction*. Blackwell Publishers.

World Health Organization (WHO). (2010). *WHO Technical Manual On Tobacco Tax Administration*. WHO.

World Health Organization- Europe. (2016). *Greece-Profile of health and well-being*. Author.

World Health Organization. (2008). *WHO Report on the global tobacco epidemic. The MPOWER package*. WHO.

World Health Organization. (2009). *Global health risks: Mortality and burden of disease attributable to selected major risks*. World Health Organization.

World Health Organization. (2012). *Global Progress Report on the Implementation of the WHO Framework Convention on Tobacco Control*. WHO.

World Health Organization. (2014). Global status report on noncommunicable diseases 2014 (No. WHO/NMH/NVI/15.1). World Health Organization.

World Health Organization. (2016). *World health statistics 2016: monitoring health for SDGs sustainable development goals*. World Health Organization.

World Health Organization. (2018). *Global status report on road safety*. Retrieved from:https://www.who.int/publications-detail/global-status-report-on-road-safety-2018

World Health Organization. Management of Substance Abuse Unit. (2014). Global status report on alcohol and health, 2014. World Health Organization.

Wu, S. M., You, X. Y., Liu, H. C., & Wang, L. E. (2020). Improving quality function deployment analysis with the cloud MULTIMOORA method. *International Transactions in Operational Research*, *27*(3), 1600–1621. doi:10.1111/itor.12484

Xiao, L., Shao, W., Yu, M., Ma, J., & Jin, C. (2017). Research and application of a hybrid wavelet neural network model with the improved cuckoo search algorithm for electrical power system forecasting. *Applied Energy*, *198*, 203–222. doi:10.1016/j.apenergy.2017.04.039

Xiao, Y. Y., Zhang, R. Q., Zhao, Q. H., & Kaku, I. (2012). Permutation flow shop scheduling with order acceptance and weighted tardiness. *Applied Mathematics and Computation*, *218*(15), 7911–7926. doi:10.1016/j.amc.2012.01.073

Xia, Y., Liu, B., Wang, S., & Lai, K. K. (2000). A model for portfolio selection with order of expected returns. *Computers & Operations Research*, *27*(5), 409–422. doi:10.1016/S0305-0548(99)00059-3

Xie, N. M., Yuan, C. Q., & Yang, Y. J. (2015). Forecasting china's energy demand and self-sufficiency rate by grey forecasting model and Markov model. *International Journal of Electrical Power & Energy Systems*, *66*, 1–8. doi:10.1016/j.ijepes.2014.10.028

Xu, R. (2000). Fuzzy least-squares priority method in the analytic hierarchy process. *Fuzzy Sets and Systems*, *112*(3), 395–404. doi:10.1016/S0165-0114(97)00376-X

Yamalik, N. (2007). Quality systems in dentistry* Part 2. Quality Assurance and Improvement (QA/I) tools that have implications for dentistry. *International Dental Journal*, *57*(6), 459–467. doi:10.1111/j.1875-595X.2007.tb00150.x PMID:18265780

Yamalik, N. (2007). Quality systems in dentistry. Part 1. The increasing pressure for quality and implementation of quality assurance and improvement (QA/I) models in health care. *International Dental Journal*, *57*(5), 338–346. doi:10.1111/j.1875-595X.2007.tb00143.x PMID:17992920

Yang, B., & Geunes, J. (2007). A single resource scheduling problem with job-selection flexibility, tardiness costs and controllable processing times. *Computers & Industrial Engineering*, *53*(3), 420–432. doi:10.1016/j.cie.2007.02.005

Yang, J., & Lee, H. (1997). An AHP decision model for facility location selection. *Facilities*, *15*(9/10), 241–254. doi:10.1108/02632779710178785

Yang, J., Shi, X., Marchese, M., & Liang, Y. (2008). An ant colony optimization method for generalized TSP problem. *Progress in Natural Science*, *18*(11), 1417–1422. doi:10.1016/j.pnsc.2008.03.028

Yang, Z., Ce, L., & Lian, L. (2017). Electricity price forecasting by a hybrid model, combining wavelet transform, ARMA and kernel-based extreme learning machine methods. *Applied Energy*, *190*, 291–305. doi:10.1016/j.apenergy.2016.12.130

Yao, A. W., & Chi, S. C. (2004). Analysis and design of a Taguchi–grey-based electricity demand predictor for energy management systems. *Energy Conversion and Management*, *45*(7-8), 1205–1217. doi:10.1016/j.enconman.2003.08.008

Yao, A. W., Chi, S. C., & Chen, J. H. (2003). An improved grey-based approach for electricity demand forecasting. *Electric Power Systems Research*, *67*(3), 217–224. doi:10.1016/S0378-7796(03)00112-3

Yazdani, M., & Payam, A. F. (2015). A comparative study on material selection of microelectromechanical systems electrostatic actuators using Ashby, VIKOR and TOPSIS. *Materials & Design*, *65*, 328–334.

Yazdi, M., Nedjati, A., Zarei, E., & Abbassi, R. (2020). A novel extension of DEMATEL approach for probabilistic safety analysis in process systems. *Safety Science*, *121*, 119–136. doi:10.1016/j.ssci.2019.09.006

YEGM. (2020). *Biyokütle Enerjisi Nedir?* Yenilenebilir Enerji Genel Müdürlüğü: http://www.yegm.gov.tr/yenilenebilir/biyokutle_enerjisi.aspx

Yeon, A. A. A., Kamaruddin, A. I., Harun, N. A., & Kusumawardani, A. (2019). Parents' satisfaction and the quality of dental care provided by dental students at International Islamic University Malaysia. *Materials Today: Proceedings*, *16*, 2341–2350.

Yesilada, F., & Direktör, E. (2010). Health care service quality: A comparison of public and private hospitals. *African Journal of Business Management*, *4*(6), 962.

Yfantopoulos, J. (2001c). Health-Related Quality of Life. *Archives of Hellenic Medicine*, *19*, 131–146.

Yfantopoulos, J., & Sarris, M. (2001). *Health-Related Quality of Life* – Methodology of measurement. *Archives of Hellenic Medicine*, *18*(3), 218–229.

Yın, X., Zhou, Y., He, Y. G., & Chen, J. (2010). Power load forecasting based on adaptive particle swarm optimization grey model. *Proceedings Of The Chinese Society Of Universities For Electric Power System And İts Automation*, 4.

Yu, C. S. (2002). A GP-AHP method for solving group decision-making fuzzy AHP problems. *Computers & Operations Research*, *29*(14), 1969–2001. doi:10.1016/S0305-0548(01)00068-5

Yuksel, A. (2001). Managing customer satisfaction and retention: A case of tourist destinations, Turkey. *Journal of Vacation Marketing*, *7*(2), 153–168. doi:10.1177/135676670100700205

Yu, L., & Goulden, M. (2006). A comparative analysis of international tourists' satisfaction in Mongolia. *Tourism Management*, *27*(6), 1331–1342. doi:10.1016/j.tourman.2005.06.003

Yurekli, A., & Beyer, J. (2011). *Design and Administer Tobacco Taxes. World Bank Economics of Tobacco Toolkit. Tool 4*. World Health Organization.

Yu, Y., Chen, Y., & Li, T. (2011, April). A new design of genetic algorithm for solving TSP. In *2011 Fourth International Joint Conference on Computational Sciences and Optimization* (pp. 309-313). IEEE. 10.1109/CSO.2011.46

Zadeh, L. A. (1965). Fuzzy sets. *Information and Control*, *8*(3), 338–353. doi:10.1016/S0019-9958(65)90241-X

Zadeh, P. E., & Tremblay, M. C. (2016). A review of the literature and proposed classification on e-prescribing: Functions, assimilation stages, benefits, concerns, and risks. *Research in Social & Administrative Pharmacy*, *12*(1), 1–19. doi:10.1016/j.sapharm.2015.03.001 PMID:25847858

Zarei, A., Arab, M., Froushani, A. R., Rashidian, A., & Tabatabaei, S. M. G. (2012). Service quality of private hospitals: The Iranian Patients' perspective. *BMC Health Services Research*, *12*(1), 31. doi:10.1186/1472-6963-12-31 PMID:22299830

Zavras, D., Tsiantou, V., Pavi, E., Mylona, K., & Kyriopoulos, J. (2012). Impact of economic crisis and other demographic and socio-economic factors on self-rated health in Greece. *European Journal of Public Health*, *23*(2), 206–210. doi:10.1093/eurpub/cks143 PMID:23093716

Zbuchea, A., Pinzaru, F., Busu, M., Stan, S.-O., & Bargaoanu, A. (2018). Sustainable Knowledge Management and Its Impact on the Performances of Biotechnology Organizations. *Sustainability*.

Zhang, J.-L., Zhang, Y.-J., Li, D.-Z., Tan, Z.-F., & Ji, J.-F. (2019). Forecasting day-ahead electricity prices using a new integrated model. *International Journal of Electrical Power & Energy Systems*, *105*, 541–548. doi:10.1016/j.ijepes.2018.08.025

Zhang, L., Liu, R., Jiang, S., Luo, G., & Liu, H. C. (2020, March). Identification of Key Performance Indicators for Hospital Management Using an Extended Hesitant Linguistic DEMATEL Approach. *Health Care*, *8*(1), 7. PMID:31881773

Zhang, L., Lu, L., & Yuan, J. (2010). Single-machine scheduling under the job rejection constraint. *Theoretical Computer Science*, *411*(16-18), 1877–1882. doi:10.1016/j.tcs.2010.02.006

Zhang, W. G., Wang, Y. L., Chen, Z. P., & Nie, Z. K. (2007). Possibilistic mean–variance models and efficient frontiers for portfolio selection problem. *Information Sciences*, *177*(13), 2787–2801. doi:10.1016/j.ins.2007.01.030

Zhao, H. X., & Magoulès, F. (2012). A review of the prediction of building energy consumption. *Renewable & Sustainable Energy Reviews*, *16*(6), 3586–3592. doi:10.1016/j.rser.2012.02.049

Zhao, L., Lin, H. C., Lo, E. C. M., & Wong, M. C. M. (2011). Clinical and socio-demographic factors influencing the oral health-related quality of life of Chinese elders. *Community Dental Health*, *28*(3), 206–210. PMID:21916355

Zhong, X., Ou, J., & Wang, G. (2014). Order acceptance and scheduling with machine availability constraints. *European Journal of Operational Research*, *232*(3), 435–441. doi:10.1016/j.ejor.2013.07.032

Zhou, D., Yanagida, F., Chakravorty, U., & Leung, P. (1997). Estimating the economic impacts from tourism. *Annals of Tourism Research*, *24*(1), 76–89. doi:10.1016/S0160-7383(96)00035-7

Zhou, P. A. B. W., Ang, B. W., & Poh, K. L. (2006). A trigonometric grey prediction approach to forecasting electricity demand. *Energy*, *31*(14), 2839–2847. doi:10.1016/j.energy.2005.12.002

Zhou, X., Wang, J., Yang, X., Lev, B., Tu, Y., & Wang, S. (2018). Portfolio selection under different attitudes in fuzzy environment. *Information Sciences*, *462*, 278–289. doi:10.1016/j.ins.2018.06.013

Zhu, K. J., Jing, Y., & Chang, D. Y. (1999). A discussion on extent analysis method and applications of fuzzy AHP. *European Journal of Operational Research*, *116*(2), 450–456. doi:10.1016/S0377-2217(98)00331-2

Ziel, F., Steinert, R., & Husmann, S. (2015). Forecasting day ahead electricity spot prices: The impact of the EXAA to other European electricity markets. *Energy Economics*, *51*, 430–444. doi:10.1016/j.eneco.2015.08.005

Zimmermann, H. J. (2011). *Fuzzy set theory—and its applications*. Springer Science & Business Media.

About the Contributors

Evangelos Grigoroudis is Associate Professor on management of quality processes in the School of Production Engineering and Management of the Technical University of Crete, Greece (2002-). He followed postgraduate studies in Technical University of Crete, Greece from where he received his Ph.D degree in 1999. He has received distinctions from the Hellenic Operational Research Society, the Academy of Business and Administrative Sciences, the Foundation of Ioannis and Vasileia Karayianni, the Technical University of Crete, and the State Scholarships Foundation of Greece. He acts as reviewer for more than 60 scientific journals, and he is Associate Editor and member of the Editorial Board of several scientific journals. He is author of 10 books on the measurement of service quality and the business strategy and management, as well as of a significant number of research reports and papers in scientific journals and conference proceedings. His research interests include service quality measurement processes, customer and employee satisfaction, performance evaluation, business excellence, operational research, multicriteria decision analysis, data analysis, and marketing.

Nikolaos Matsatsinis is a full Professor of Information and Decision Support Systems in the School of Production Engineering and Management of the Technical University of Crete, Greece. He is Director of DSS Lab and former President of the Hellenic Operational Research Society (HELORS). He has contribute as scientific or project coordinator on over of fifty national and international projects. He is chief editor of the Operational Research: An International Journal and International Journal of Decision Support Systems and member of the international advisory board of three scientific journals. He is the author or co-author of 21 books and over of 105 articles in international scientific journals and books. He has organized and participated in the organization of over of 80 scientific conferences and he has over of 180 presentations in international and national scientific conferences. His research interests fall into the areas of Intelligent Decision Support Systems, Recommendation Systems, Multicriteria Decision Analysis, Group Decision Making, Operational Research, Electronic Business, Marketing, Consumer Behavior Analysis, Analytics, Business Intelligence, Workflow Systems.

* * *

Christina Arampantzi is an Operational Research Scientist in Amazon and also a Visiting Lecturer at the Department of Financial & Management Engineering at the University of the Aegean. Christina has a first-class honors Bachelor degree in the Environmental Engineering field and a first-class honors Master degree in the Financial and Management Engineering field receiving fellowship and institutional awards. Her Doctoral research focused on the Design of Efficient and Sustainable Supply Chain

Networks for which she developed combinatorial optimization models considering all-important costs, environmental and social aspects to address real-life applications. Dr. Arampantzi has also published many papers and studies in academic journals and conferences.

Archontoula Armoutaki is a Ph.D. candidate at the Department of Economics at University of Piraeus in Greece. She has been sponsored by the Karelia Foundation for her doctoral dissertation. She received a B.S. in Marketing and Communication from Athens University of Economics and Business in 2007 and a M.S. in Financial and Business Strategy from University of Piraeus in 2012. Since 2012 she has been working as an auditor in the Attica Customs Audit Office in Greece.

Talip Arsu still works as an assistant professor in Aksaray University, Vocational School of Social Sciences, Tourism and Hotel Management program. He started working as a research assistant in Aksaray University, Department of Business Administration in 2012. In 2013, he received his MS in Quantitative Methods in Dokuz Eylül University, Turkey, and in 2018 he received his Ph.D. in Business Administration from Aksaray University, Turkey. He has academic studies on multi-criteria decision-making techniques and efficiency measurement methods.

Emre Aslan was born in 1981. Graduated Department of Business Administration in Ege University in 2004. Completed his MBA in 2007 and PhD in 2014 from Erciyes University Department of Business Administration. Worked as a research assistant between 2004-2014 and as an instructor till 2014 in Tokat Gaziosmanpaşa University.

Erdal Aydemir is an assistant professor in the Department of Industrial Engineering at Suleyman Demirel University (SDU), Isparta, Turkey since 2013. He was awarded a BS in Industrial Engineering (IE) from Selcuk University in 2005 and an MS in Industrial Engineering (IE) from Suleyman Demirel University in 2009. In 2013, he was awarded a Ph.D. in Mechanical Engineering (ME) with the degree of greyness and its effects on economic production quantity levels from SDU. His research interest includes the modeling of production-service systems and its adaptation with grey system theory, artificial intelligence, decision making, etc. on the various fields. Dr. Aydemir who is a volunteer for the science of grey systems and applications also works as one of the deputy general secretaries of International Association of Grey Systems and Uncertainty Analysis (IAGSUA), a member of Grey Technical Committee on IEE Systems, Man and Cybernetics Group, a member of Chinese Grey System Association, and founder of Turkish Association of Grey Systems Theory.

Chrystala Charalambous was born in Cyprus. In 1995, she graduated from the Dental School of the National and Kapodistrian University in Athens. In 2004 she received a Master degree in Dental Public Health from the Kings' College, University of London. She is also holding a Master degree in Healthcare Management and a PhD in Health Policy and Planning from the Cyprus's Open University. Chrystala Charalambous is working for the Public Dental Services in Cyprus (Ministry of Health) and the last years is responsible for the organization and implementation of various preventive programs for the promotion of oral health in Cyprus. Since 2017 Chrystala Charalambous is working as Scientific Collaborator at the Dental School of the European University and she is teaching the course "Community and Preventive Dentistry". Oral health policy and prevention as well as epidemiology of oral diseases are among her research interests.

Konstantinia Daskalou holds a Diploma in Electrical and Computer Engineering (2013) and Master in Business Administration (2019). During her undergraduate studies, she penetrated in the energy sector and completed her dissertation in the Greek Wholesale Electricity Market. During 2014, she worked as Junior Electrical Engineer designing electromechanical installations, while at the end of the same year she moved at WATT + VOLT as Energy Analyst Engineer with main tasks the participation in the Greek Wholesale Electricity Market and cost analysis. In 2016, she was upgraded to Senior Energy Analyst taking over, in addition to her other tasks, the project of company's entrance to the Greek Gas Market. Since 2019, Ms. Daskalou holds the position of Head of Supply Portfolio Management, undertaking, under her responsibility, the company's electrical and gas supply portfolio.

Christina Diakaki holds a Diploma in Production and Management Engineering (1991), M.Sc. in Operations Management (1993) and PhD in Decision Systems and Operations Research (2000). Since 1991, she has participated in numerous projects sponsored by the European Commission, and public and private organizations, while since 1999 she has been also teaching at several Universities. Currently, she is Adjunct Professor at the Hellenic Open University, the Technical University of Crete, Greece, and the European University Cyprus. Dr. Diakaki serves as reviewer for several scientific journals and conferences, she is Deputy Editor-in-Chief of the IET (Institute of Engineering and Technology) Intelligent Transport Systems journal and author and co-author of numerous research reports and papers in scientific and technical journals, and conference proceedings. Her research interests include decision analysis and operations research, and systems' analysis, modelling, simulation and optimization with applications in the fields of transportation, energy and environment, and management.

Hasan Dinçer is an Associate Professor of finance at Istanbul Medipol University, Faculty of Economics and Administrative Sciences, Istanbul-Turkey. Dr. Dinçer has BAs in Financial Markets and Investment Management at Marmara University. He received PhD in Finance and Banking with his thesis entitled "The Effect of Changes on the Competitive Strategies of New Service Development in the Banking Sector". He has work experience in finance sector as portfolio specialist and his major academic studies focusing on financial instruments, performance evaluation, and economics. He is the executive editor of the International Journal of Finance and Banking Studies (IJFBS) and the founder member of the Society for the Study of Business and Finance (SSBF).

Iris Efthymiou-Egleton is an author, speaker, and entrepreneur whose wisdom and eclectic knowledge have earned her a reputation as a visionary. With publications spanning a wide range of topics (parenting, the environment, health, amongst many others), she believes that combining different disciplines produces a more well-rounded view of life.She studied Economics at the University of Athens and earned her 'Masters' in Health Economics and Management, at the University of Piraeus, where she is now a PhDc. She has worked as a college director, created her own Public Affairs business, worked with politicians, ambassadors, entrepreneurs, and spoken at the UN Headquarters in Geneva. In recent years, as her family priorities changed, she has published a number of papers and written 11 books. Iris' other interests include being President of the Interdisciplinary Committee of HAPSc (has Special Consultative Status with ECOSOC and is a Member of UNAI). She is, also, a Scientific Associate of the Laboratory of Health Economics and Management (LabHEM) of the University of Piraeus, and a Member of the Board of Womanitee.

Theodoros Fouskas, Sociologist, is currently teaching at the School of Public Health of the University of West Attica (Greece). He has taught Sociology of Migration, Migration Policy and Integration Policies and Health of Populations on the Move at many institutions; published extensively on migration, precarious employment, access to healthcare, social integration and exclusion; since 2005, participated in many research programs for migrants, refugees and other vulnerable groups.

Niki Glaveli is an Associate Professor in the Department of Business at the University of the Aegean, Greece. She holds two PhDs in Management/HRM Management-OB awarded by the University of Macedonia (Scholar of A. Onassis Foundation) and the Aristotle University of Thessaloniki, respectively. For her second PhD she received the 2014 Emerald/EFMD Outstanding Doctoral Research Award in Human Resource Management from Personnel Review. Her research focuses primarily on issues related to the Strategic Behavior and Business Development, in the private, public and social sector, Strategic HRM/Organizational Behavior and Corporate Social Responsibility.

Yaşar Gökalp is research assistant of health management at Istanbul Medipol University, Faculty of Health Sciences, Istanbul-Turkey. He has BAs in Health Management at Sivas Cumhuriyet University. He is PhD candidate in İstanbul Medipol University, the department of health management. His research interests lie in health management, service quality in the hospitals and financial analysis in health industry.

Hale Gonce Kocken is an Associate Professor in the Department of Mathematical Engineering at the Yildiz Technical University (YTU), Istanbul, Turkey. She has been a faculty member of YTU since 2004. She completed her Ph.D. entitled "Fuzzy approaches to network analysis" in Applied Mathematics (2011) from the same department. Her current area of research is mathematical Programming, supply chain management and some related Operational Research subjects in multi criteria and fuzzy environments. Additionally, she has doing research about multi criteria decision making methods with type-2 fuzzy numbers and also fractional programming.

Emmanouil-Nektarios Kalligeris is a 2nd year Ph.D. candidate at the Department of Statistics and Actuarial - Financial Mathematics, University of the Aegean, Samos, Greece. His research interests lie among others in Multivariate Analysis, Information Criteria, Dynamical Systems, Neural Networks, Time-Series, Statistical Programming, Model Selection, Hidden (semi-) Markov processes, and Generalized Linear Mixed Models. He is currently working on his Ph.D. thesis entitled "New Developments on Modelling Techniques for Dynamical Systems with Applications" (supervisor: Prof. Alex Karagrigoriou).

Maria Kamariotou is researcher associate at the Department of Applied Informatics, School of Information Sciences, University of Macedonia, Greece. She graduated from the University of Macedonia with a degree in Applied Informatics Department. She is author and also acts as reviewer for scientific journals and conference proceedings. Her scientific research interests mainly focus on information systems planning and innovation management.

Kenan Karagül studied industrial engineering for his Bachelor degree and business administration for M.S. and PhD degrees. His field of study includes operations research, logistics, vehicle routing problems, metaheuristics, and quantitative models. He worked at various firms between 1996 and 2001. He worked at Pamukkale University as an instructor until 2013. He has been working at the same university as an assistant professor since 2013. He was awarded the best Ph.D. thesis on Graduate Tourism Students Congress in Kuşadası (2014).

Fotis Kitsios is an Associate Professor in Business Strategy and Innovation Management at the Department of Applied Informatics, School of Information Sciences, University of Macedonia, Greece. He holds a PhD in Strategic Innovation Management with a specialization in Services from Technical University of Crete. He has consulted several SMEs and has also been involved in the implementation and coordination of European research projects. He also lectured at several Universities in Greece and Cyprus. He is author and also acts as reviewer for scientific journals, books, research reports and conference proceedings.

Maria Lachnidaki is a graduate of the Department of Applied Mathematics of the University of Crete, Greece, with a Master in Business Administration (MBA) of the School of Production Engineering and Management of the Technical University of Crete, Greece. Her research interest includes satisfaction analysis and her thesis title is "Agro-tourism customer satisfaction analysis based on the Kano model".

Panagiotis Manolitzas, is an Assistant Professor at the Ionian University. He holds a Bachelor in Public Administration (specialization in Management Science) from the University of Athens, MSc in Public Management (specialization in Management Science) from the University of Athens and a PhD in Management Science from the Technical University of Crete. He has over 70 publications in international, national scientific conferences and scientific journals. His research interests lie in the areas of Strategic Management, Strategic Decision Making, simulation, decision support systems for the health care organizations, health services research, process mining, multiple criteria analysis and tourism management.

Burhan Oran was born in Istanbul in 1994, and graduated from Computer Engineering of Pamukkale University, in 2018. After graduation, He is working in healtcare industry as a computer engineer.

Panagiotis Otapassidis is a Research Associate at the Department of Financial & Management Engineering at the University of the Aegean. His research interests focus on Supply Chain Management with emphasis on transport operations. Panagiotis is active in publishing articles and participating in research projects. He is a member of the Technical Chamber of Greece.

Beyza Ahlatcioglu Ozkok is a Professor Dr. of Department of Business Administration at Yildiz Technical University, Turkey, where she has been since 2019. During 2010-2011 she was a Visiting Researcher at Center for Applied Optimization of Industrial and Systems Engineering at the University of Florida. She received her B.S. in Mathematics from Istanbul University in 2002, and an M.S. from Istanbul University and Marmara University in 2004. She received her Ph.D. in Quantitative Methods from Istanbul University in 2009. Her research interests are in fuzzy mathematical programming, decision making, multi criteria decision making, game theory and supply chain modeling.

Georgia Papadopoulou is an Assistant Professor of Economics and Project Management at the Al Ghurair University in Dubai. She received a BSc degree in Maritime Studies from the University of Piraeus (2006), MBA in Operations Management from the University of the Aegean (2009), an MSc in Tourism Business Administration from the Hellenic Open University (2012) and PhD degree in Tourism

Economics from the University of Piraeus (2015). She has published papers and participated in international conferences in the area of tourism economics and management. Georgia Papadopoulou collected the data, prepared the literature review and contributed to the concluding remarks.

Yannis A. Pollalis is a professor in the Department of Economics at the University of Piraeus, Greece since 1999. He holds a PhD in Business Strategy & MIS from the University of Pittsburgh, an MBA from the University of Scranton and a BSc in Economics from the University of Piraeus. During 1993-1999 he was a professor of Strategic Information Systems at Syracuse University, while collaborating in research projects with the University of Southern California's Annenberg School for Communication. For the last 25 years he has served as a Keynote & Motivational Speaker and a consultant in Strategic Management, Marketing & Branding, Leadership, Digital Transformation & Disruption, Growth Mindset, Motivation, Emotional Intelligence and Train-the-Trainer workshops. He is also the founder of the M.Sc. Program in Economic & Business Strategy at the University of Piraeus in Greece. He is also the founder and director of the Research & Training Center in Digital Transformation & Strategic Leadership (iLEADS Lab). Professor Pollalis teaches in undergraduate, graduate and doctoral programs, as well as, in executive development & training courses and has received numerous Teaching Excellence Awards. He has published 17 books in the areas of management, strategic planning, information management, e-business, marketing, organizational behavior and corporate strategy. His research has been funded by the Onassis Public Benefit Foundation, NATO's Science & Research Program and the European Network for SMEs Research.

Argy Polychronopoulou is a Professor and Chair at the Department of Preventive and Community Dentistry, National and Kapodistrian University of Athens, School of Dentistry. She holds a Doctoral Degree in Dental Surgery from the National and Kapodistrian University of Athens, School of Dentistry, a Master of Science in Pediatric Dentistry from The University of Michigan, School of Dentistry, a Post-Doctoral Certificate in Pediatric Dentistry from Columbia University, School of Dental and Oral Surgery, and, a Master of Science in Epidemiology and a Doctor of Science in Epidemiology, both from Harvard University, School of Public Health. Dr. Polychronopoulou has been active in major research projects in both the medical and the dental fields and has published over one hundred papers in refereed journals, covering topics in public health, chronic disease epidemiology, research methodology and oral epidemiology, including publications in The Lancet, International Journal of Cancer, Community Dentistry and Oral Epidemiology, Journal of Dental Education, Journal of Periodontology, and, American Journal of Orthodontics and Dentofacial Orthopedics. Her research papers have received a total of over four thousand citations. Her primary research interests include methodological issues and novel methodological applications in oral epidemiology. Dr. Polychronopoulou is involved in pre-doctoral and post-doctoral teaching of preventive dentistry, community dentistry, oral epidemiology, research methodology, biostatistics, and in supervision of Master's theses and Doctoral dissertations in oral epidemiology.

Yusuf Şahin received the M.S. degree in industrial engineering from Pamukkale University in 2009 and a PhD degree in business administration from Suleyman Demirel University in 2014. He has been an assistant professor of business administration at Burdur Mehmet Akif Ersoy University since 2014. His field of study includes operations research, logistics, warehouse management, vehicle routing, metaheuristics, and quantitative models.

Aristomenis I. Syngelakis holds a Degree in Dentistry (DDS) as well as a Degree in Political Sciences & Public Administration (both from the University of Athens). He also holds an MSc in Community Dentistry (University of Athens) as well as an MSc in Health Management (Hellenic Open University). He received his PhD in Community Dentistry (University of Athens). His thesis was on the evaluation of the Primary Dental Care provided by the Greek NHS during the period of economic crisis and austerity. At the moment he is teaching "Community and Preventive Dentistry" as well as "Dental Ethics" (Department of Dentistry, School of Medicine, European University Cyprus) and he participates in teaching of "Introduction in Dental Science and History of Dentistry" as well as "Community Dentistry" (Faculty of Dentistry, School of Health Sciences, University of Athens). He is also a scientific collaborator of the University of Western Attica (MSc Programme "Management of Health and Welfare Services"), as well as of the University of Thessaly (Faculty of Medicine, Department of Hygiene and Epidemiology). Since 2003 he works in the Greek Ministry of Health (Strategic & Operational Planning of the co-financed by the EU and the Greek State health projects).

Michael A. Talias is an Assistant Professor in the Healthcare Management Program at the Open University of Cyprus. His research interests fall into the broad area of Healthcare Management, Health Economics, Decisions under uncertainty and Operational Research. He is committed to advancing the newly statistical developed methodology to large and complex databases to address critical questions in Health Management and to impact policy. He Published more than 50 articles in scientific journals.

Sezai Tokat received the BSc and PhD degrees in control and computer engineering from Istanbul Technical University, Istanbul, Turkey, in 1994 and 2003, respectively. He received his MSc degree in systems and control engineering from Bogazici University, Istanbul, in 1997. Since 2011 he has been with Pamukkale University, Turkey as associate professor. His research interests include intelligent control techniques, robust control, nonlinear control, optimization, vehicle routing problems.

Adem Tüzemen was born in Erzurum, Turkey in 1976. After completing his education there, he started to work at Tokat Gaziosmanpaşa University in 2002 as a Research Assistant. In 2014, he received his doctorate from Izmir Dokuz Eylül University Institute of Social Sciences. In the same year, he started to work at Tokat Gaziosmanpaşa University Faculty of Economics and Administrative Sciences as Assistant Professor Doctor. He still works at the same university. Research areas; production/operations management, operations research, management information systems, forecasting, enterprise resource planning, multi-criteria decision making, project management.

Athanassios Vozikis is Associate Professor of Health Economics and Information Systems at the Department of Economics of the University of Piraeus. He holds a degree in Public Administration from the Panteion University, a M.Sc. in Health Informatics from National Kapodistriako University of Athens and a Ph.D. in Health Economics and Information Systems from the University of Piraeus. He is Director of The Laboratory of Health Economics and Management of University of Piraeus, and has served as member of the National Council of Public Health and Chair of the Negotiation Committee of National Organization for the Healthcare Services Provision (EOPYY). He is member of various scientific associations and organizations and member of the Editorial Board of several academic journals. He has participated in several research projects, Committees and Advisory Boards related to Health Technology Assessment, Business and Economic Re-engineering Processes in the Health Care.

Serhat Yüksel is Assistant Professor of finance in İstanbul Medipol University. Before this position, he worked as a senior internal auditor for seven years in Finansbank, Istanbul-Turkey and 1 year in Konya Food and Agriculture University as an assistant professor. Dr. Yüksel has a BS in Business Administration (in English) from Yeditepe University (2006) with full scholarship. He got his master degree from the economics in Boğaziçi University (2008). He also has a PhD in Banking from Marmara University (2015). His research interests lie in banking, finance and financial crisis. He has more than 70 publications (books, book chapters, scientific articles, etc.).

Vasileios Zeimpekis is an Assistant Professor at the Department of Financial & Management Engineering at the University of the Aegean and Vice President of Administration at the BoD of the Hellenic Logistics Association. His research interests focus on Supply Chain Management with emphasis on logistics operations & Information Systems. He has published 7 edited volumes in the management of Supply Chains and more than 90 papers and studies in academic journals and conferences. During the last 20 years, Vasileios has participated in more than 50 research and 150 industrial projects. Dr. Zeimpekis is a member of the Technical Chamber of Greece.

Stelios Zimeras holds a BSc. (Hons) on Statistics and Insurance Sciences from University of Piraeus (5fth in the rank) and Ph.D. on Statistics from the University of Leeds, U.K. He had received a full scholarship (fees and maintenance) during his research studies (1993-1997) from the University of Leeds. Since 2012, he is a full time staff member (Assistant Prof) on statistics and probabilities at the Department of Mathematics, Division of Statistics and Financial-Actuarial Mathematics, University of the Aegean, Samos, Greece. He had worked as post-doc at University of Leeds, Department of Electrical Engineering (funded by EPSRC), at IRISA/INRIA-VISTA PROJECT institute, France (funded by EUMETSAT-ESA) and at MedCom GmbH, Germany (funded by Marie-Curie grants). Finally 2011-2012 he had worked during his sabbatical in University of Piraeus, Department of Digital Systems, in computer virus modelling.

Constantin Zopounidis is the Director of the Financial Engineering Laboratory. Prof. Zopounidis received his Doctorat D'État (1986) in management science, a D.E.A. (1982) in financial management from the University of Paris-IX Dauphine and a B.A. in Business Administration from the University of Macedonia. Prof. Zopounidis has been with the School of Production Engineering and Management of the Technical University of Crete, since 1987 and served as Chairman of the Department during the period 2001-2005. He has published over 300 papers in premier international journals, edited volumes, and conference proceedings. He has also published several books and edited volumes on financial engineering and operations research. Prof. Zopounidis is Editor-in-Chief and member of the editorial board of several international journals.

Index

A

Absolute Error 16
Access to Healthcare 283, 288
Accommodation Center 280, 285
Accommodation Centre 288
Ad Valorem Tax 313, 315, 318, 323-324
Advice Provider 178-179, 182-183, 190
Agency for the Cooperation of Energy Regulators 3
Agro-Tourism 290-292, 300-308, 312
AHP 30-32, 34-35, 37, 39-40, 43-45, 48, 50-55, 57-60, 64-70, 216-218, 226-227, 231
Analytical Hierarchy Process 30, 51, 53, 57, 64, 69-73, 75, 231
Analytical Hierarcy Process 56
ANP 226, 231
Ant Colony Optimization 94, 96, 102, 110-113
Artificial Intelligence 5-6, 29, 111-112, 148
Artificial Neural Network 1-2, 25, 27, 29, 33, 54, 153
Asylum Seeker 272, 282, 289
Asylum Seekers 270-274, 276-278, 280-283, 286
Attractive Quality 290, 292, 297-298, 303, 307, 309, 312

B

Behavioral Risk Factors 252, 254
Better-Worse Diagram 292, 299, 301-304, 307
Big data 76-77, 79
Bi-Objective Multiple Criteria Data Envelopment Analysis (BiO-MCDEA) 155-156, 169
Biomass Energy 155, 160, 164, 171, 174, 176

C

Categorical data analysis 76-77, 79
Chronic Diseases 229, 242, 252-253, 255-256, 258-259, 261-268
Cigarette Taxation 313-315
Combined freight transportation 114
Community Health Promoter 178-179, 182, 190
Community pharmacies 177-179, 185-189
Community Pharmacy 177-180, 184-190
Community pharmacy customers 177
Community pharmacy selection criteria 177
Correspondence analysis 76-77, 79, 81-82, 84, 92-93
COVID-19 270-271, 273, 277-285, 287-288
Cruise Lines International Association 330-331, 337, 339
cruise passengers 329-330, 332-334, 336, 339
cruise ships 329-335, 337-338, 340-341
Customer Prioritization 58, 64, 71, 75
Customer Satisfaction 59, 178-180, 183-186, 202, 241, 290-293, 295-302, 304-305, 308-312

D

Day Ahead Electricity Price 1-2, 5-7, 13, 25, 29
Decision Maker 37, 45, 212, 220, 232-233
Demand Management 58, 62, 71, 74-75
DEMATEL 40, 50, 213, 215, 217, 220, 223, 226-228, 230-231
Dentistry 192-196, 200, 202-212
Desired Quality 297, 299, 304, 307, 312
Dijkstra's algorithm 114-115, 118, 121-122, 127, 131

E

Economic Evaluation 234
Electricity Consumption 2, 136-139, 147-148, 152-154
Electricity Market Coupling 1, 29
EU Electricity Target Model 29
European Cruise Council 331, 333, 336-337, 339-340, 342
European cruise sector 329, 331-332, 338
European Network for Electricity 1
Expected Quality 297, 303, 305, 307, 312

F

Fluid Genetic Algorithms 94, 101
Forecasting 1-2, 4-11, 13, 15-16, 18-20, 22-29, 59, 65, 71, 136-139, 142-143, 145-146, 148, 152-154
Fossil Fuels 156-160, 171, 176
Fuzzy Analytical Hierarchy Process 30, 72
Fuzzy Decision Making 32, 56
Fuzzy DEMATEL 40, 50, 213, 215, 217, 220, 226-228

G

Genetic Algorithms 31, 73, 94, 96, 99, 101, 110-113
Geothermal Energy 155, 158, 161, 163-164, 170-173, 176
Greece 1, 4, 11, 19, 22-23, 25-26, 76, 114, 116-117, 122, 124-125, 128, 175, 178, 180-181, 185, 187, 189-190, 192, 206, 209, 252, 254-256, 264-265, 268-288, 290, 300, 307, 313-318, 320, 324-327, 330, 336, 340
Grey Modeling 136, 141-142

H

Health Industry 213, 217-218, 223, 226-227
Health Information Systems 234, 236, 250-251
Health Policy 200, 205, 212, 250
Health service 192, 196, 208
Hydroelectric Energy 161, 176

I

Investment Decision 30, 40, 54

K

Kano Model 227, 290, 292, 296-300, 307-308, 312

M

Make to Order 75
Make to Stock 75
market growth 329
Measurement 33, 60, 64, 141, 157, 165, 168, 179-180, 182, 185, 187-188, 192, 199, 202, 204, 206-207, 210-212, 215, 228-229, 256, 269, 290-291, 296, 309-310
Medicines Supplier 178-179, 182-184, 190
Mediterranean 194, 209, 288, 329-331, 336-337, 339-341
Migrants 270-283, 285, 287-288
Multicriteria Decision Analysis 290, 309, 312
Multi-Criteria Decision Making 31, 33, 57, 75
MUSA Method 290, 292, 294, 296, 299-300, 302, 305-308, 312
My Regular Pharmacy 177, 191

N

Neural Network 1-2, 13, 25-29, 33, 54, 73, 139-140, 153

O

Order Management 62-63, 75
Order Prioritization 57-58, 61-62, 65, 67-69, 75

P

Performance 6-7, 13-17, 19, 22-24, 32-35, 37, 39, 44, 48, 52, 54, 58, 60, 64-65, 67, 69-70, 73, 94, 101, 105, 108-109, 128, 132, 134, 138, 140, 146-148, 152, 188, 192-193, 203-206, 210-211, 228-231, 240-243, 245, 248, 295-297, 299, 304-307, 312, 337-338
Planning 4-5, 60-63, 65, 69-70, 72, 99, 137, 152, 162, 172, 175, 193, 202, 205-206, 229, 245, 247, 308, 312, 340
Policy Maker 212
Portfolio Selection Problem 30-32, 45, 48, 52-53, 55-56
Primary health care 190, 192, 194, 202, 206-210, 229
Primary Healthcare 192-194, 196, 199-200, 202, 205, 207-208, 212
Prioritization 53, 57-58, 61-65, 67-69, 71, 75, 292, 307
PUN Price 2, 9-11, 16, 18-20, 22-24, 29

Q

Quality 15, 23, 34-35, 54, 61, 98, 156, 178, 184, 189, 192-197, 199-219, 223, 226-231, 235-239, 242, 244-246, 248-257, 263-264, 266-269, 277, 290-292, 296-299, 301, 304-305, 307-312, 315, 328
Quality of Life 199, 203, 205-207, 210-211, 214, 252-257, 263-264, 266-268

R

Reception and Identification Centers 270-272, 274-275, 284, 286
Refugees 270-277, 279, 281-283, 287-288
Renewable Energy 3, 155-158, 162-163, 165, 172-176
Renewable Energy Sources (RES) 3, 155-156
Reprofessionalize 178, 191

Revenu Management 75
route selection 114

S

Self-perceived Health 252
Service Delivery 208, 212, 236
SERVQUAL 213, 215-219, 227-231, 291, 296, 310, 312
SHARE 50+ 252, 255-256, 266
shortest path trans-shipment problem (SPTP) 114-115, 118, 130
Socioeconomic Factors 252, 255-256
Solar Energy 155, 158-159, 162, 164, 167, 171-172, 176
Specific Tax 313, 315, 318, 323-324
Spherical Surface 94, 97-98, 110
Spot Electricity Price 1
Sustainability 58, 64, 69, 155, 161, 172, 175, 206, 209, 234, 239, 242-243, 250-251, 332, 340
SWOT Analysis 312

T

Target Model 1, 3-4, 27, 29
Technique for Order Preference by Similarity to Ideal Solution 57, 64, 69
Theory of Attractive Quality 290, 292, 297-298, 307, 309
Third-Country National 289
Tobacco Companies 313-314, 318, 323
Tobacco Retailers Revenue Shares 313
Tobacco Supply Chain 313-315, 324
TOPSIS 35, 39-40, 51-52, 54, 57-58, 60, 64-65, 67-72, 74-75, 217-218, 227
Triangulation 185-186, 191
Trigonometric Grey Modeling 136
Type-1 Fuzzy Set 56
Type-2 Fuzzy Numbers 30, 39
Type-2 Fuzzy Set 30, 40-42, 44, 50, 56

W

Wind Energy 140, 152, 155, 159-160, 164, 170, 173, 176

www.ingramcontent.com/pod-product-compliance
Lightning Source LLC
LaVergne TN
LVHW060404300926
849162LV00006B/1339